# WCDMA Design Handbook

Developed out of a successful professional engineering course, this practical handbook provides a comprehensive explanation of the Wideband CDMA (Code Division Multiple Access) air interface of third-generation UMTS cellular systems. The book addresses all aspects of the design of the WCDMA radio interface from the lower layers to the upper layers of the protocol architecture. The book considers each of the layers in turn, to build a complete understanding of the design and operation of the WCDMA radio interface including the physical layer, RF and baseband processing, MAC, RLC, PDCP/BMP, Non-Access Stratum and RRC. An ideal course book and reference for professional engineers, undergraduate and graduate students.

**Andrew Richardson** has many years of experience in digital communication systems, having worked for Philips, Nokia and Simoco on both second- and third-generation mobile phone systems. Since 1999 he has run his own consultancy, Imagicom Ltd, offering design and training services in telecommunication systems technology.

# WCDMA Design Handbook

Developed out of a successful professional engineering course, this practical handbook provides a comprehensive explanation of the Wideband CDMA (Code Division Multiple Access) air interface of third-generation UMTS cellular systems. The book addresses all aspects of the design of the WCDMA radio interface from the lower layers to the upper layers of the protocol architecture. The book considers each of the layers in turn to build a complete understanding of the design and operation of the WCDMA radio interface, including the physical layer, RF and baseband processing, MAC, RLC, PDCP, BMP, Non-Access Stratum and RRC. An ideal course book and reference for professional engineers, undergraduate and graduate students.

Andrew Richardson has many years of experience in digital communication systems, having worked for Philips, Nokia and Simoco on both second- and third-generation mobile phone systems. Since 1999 he has run his own consultancy, Imagicom Ltd, offering design and training services in telecommunication systems technology.

# WCDMA Design Handbook

**Andrew Richardson**
Imagicom Ltd

PUBLISHED BY THE PRESS SYNDICATE OF THE UNIVERSITY OF CAMBRIDGE
The Pitt Building, Trumpington Street, Cambridge, United Kingdom

CAMBRIDGE UNIVERSITY PRESS
The Edinburgh Building, Cambridge CB2 2RU, UK
40 West 20th Street, New York, NY 10011-4211, USA
477 Williamstown Road, Port Melbourne, VIC 3207, Australia
Ruiz de Alarcòn 13, 28014 Madrid, Spain
Dock House, The Waterfront, Cape Town 8001, South Africa

http://www.cambridge.org

© Andrew Richardson 2005

This book is in copyright. Subject to statutory exception
and to the provisions of relevant collective licensing agreements,
no reproduction of any part may take place without
the written permission of Cambridge University Press.

First published 2005

First South Asian Edition 2005

Reprinted 2007

Printed at Brijbasi Art Press Ltd., I-72, Sector-9, Noida, U.P. India.

*Typefaces* Times 10.5/14 pt and HelveticaNue   *System* $L^AT^EX2_E$   [TB]

*A catalogue record for this book is available from the British Library*

*Library of Congress Cataloguing in Publication Data*

Richardson, Andrew, 1961
WCDMA Design Handbook / Andrew Richardson.
    p. cm.
Includes bibliographical references and index.
ISBN: 0-521-67037-3
1. Code division multiple access—Handbooks, manuals, etc. 2. Wireless communication systems-Handbooks, manuals, etc. 3. Mobile communication systems-Handbooks, manuals, etc. I. Title.

TK5103.452.R53    2004
621.3845 — dc22   2003058670

ISBN: 0-521-67037-3 paperback
ISBN: 978-0-521-67037-1 paperback

This edition is for sale in South Asia only, not for export elsewhere.

To my wife and family, Alex, Beth, Emma and Evie,
and also to my parents Peter and Marea.

# Contents

| | | |
|---|---|---|
| Preface | page | xiii |
| Acknowledgements | | xv |
| List of abbreviations | | xvi |

## 1  Introduction — 1

1.1 Concepts and terminology — 1
1.2 Major concepts behind UMTS — 4
1.3 Release 99 (R99) network architecture — 8
1.4 R4 and R5 network architecture — 16
1.5 Services provided by UMTS and their evolution from GSM/GPRS services — 19
1.6 Summary — 23

## 2  WCDMA in a nutshell — 24

2.1 Protocol architecture — 24
2.2 SAPs — 29
2.3 Principles of the physical layer — 33
2.4 Principles of the upper layers — 42
2.5 Radio and data connections — 47
2.6 Security issues — 51
2.7 Summary of the operation of the radio interface — 59

## 3  Spreading codes and modulation — 64

3.1 Introduction — 64
3.2 Introducing WCDMA spreading functions — 66

| | | |
|---|---|---|
| 3.3 | Channelisation codes | 71 |
| 3.4 | Scrambling codes | 87 |
| 3.5 | Modulation | 97 |
| 3.6 | Downlink spreading and modulation | 102 |
| 3.7 | Uplink spreading and modulation | 108 |

## 4 Physical layer 115

| | | |
|---|---|---|
| 4.1 | Introduction | 115 |
| 4.2 | Physical channel mapping | 115 |
| 4.3 | Uplink channels | 115 |
| 4.4 | Downlink channels | 122 |
| 4.5 | Spreading and scrambling codes | 128 |
| 4.6 | Cell timing | 134 |
| 4.7 | PRACH timing and CPCH timing | 136 |
| 4.8 | Summary | 136 |

## 5 RF aspects 137

| | | |
|---|---|---|
| 5.1 | Frequency issues | 137 |
| 5.2 | UE transmitter specifications | 140 |
| 5.3 | Node B transmitter specifications | 143 |
| 5.4 | Received signals | 146 |
| 5.5 | Node B receiver characteristics | 154 |
| 5.6 | Node B receiver performance | 165 |
| 5.7 | UE receiver characteristics | 169 |
| 5.8 | UE receiver performance tests | 174 |
| 5.9 | UMTS transceiver architecture study | 176 |

## 6 Chip rate processing functions 184

| | | |
|---|---|---|
| 6.1 | Introduction | 184 |
| 6.2 | Analogue to digital converter (ADC) | 184 |
| 6.3 | Receive filtering | 187 |
| 6.4 | Rake receiver overview | 189 |
| 6.5 | Channel estimation | 204 |

|  |  |  |
|---|---|---|
| 6.6 | Searcher | 206 |
| 6.7 | Initial system acquisition | 208 |

# 7 Symbol rate processing functions — 217

|  |  |  |
|---|---|---|
| 7.1 | WCDMA symbol rate transmission path | 217 |
| 7.2 | Convolutional error correction codes | 229 |
| 7.3 | Turbo codes as used in WCDMA | 235 |
| 7.4 | The performance of the WCDMA turbo code via examples | 247 |

# 8 Layer 2 – medium access control (MAC) — 248

|  |  |  |
|---|---|---|
| 8.1 | MAC introduction | 248 |
| 8.2 | MAC architecture | 251 |
| 8.3 | MAC functions and services | 257 |
| 8.4 | MAC PDUs and primitives | 261 |
| 8.5 | MAC operation | 264 |
| 8.6 | Random access procedure | 264 |
| 8.7 | Control of CPCH | 277 |
| 8.8 | TFC selection in uplink in UE | 282 |

# 9 Layer 2 – RLC — 300

|  |  |  |
|---|---|---|
| 9.1 | Introduction | 300 |
| 9.2 | TM | 300 |
| 9.3 | UM | 306 |
| 9.4 | AM | 314 |
| 9.5 | Summary | 335 |

# 10 PDCP and BMC protocols — 337

|  |  |  |
|---|---|---|
| 10.1 | PDCP architecture and operation | 337 |
| 10.2 | Broadcast/multicast control | 344 |
| 10.3 | CBS PDU summary | 347 |
| 10.4 | Summary | 348 |

# 11 Layer 3 – RRC                                                         349

11.1  Introduction                                                        349
11.2  System information broadcasting                                     352
11.3  Paging and DRX                                                      358
11.4  RRC connection establishment                                        362
11.5  Direct transfer procedure                                           374
11.6  RB setup                                                            377
11.7  Handover                                                            379
11.8  Miscellaneous RRC procedures                                        391
11.9  Summary                                                             394

# 12 Measurements                                                          395

12.1  Introduction                                                        395
12.2  Measurement control                                                 400
12.3  Measurement variables                                               404
12.4  Cell signal measurement procedures                                  406
12.5  Reporting the measurement results                                   414
12.6  Measurements for interoperation with GSM                            425
12.7  Location services measurements                                      433
12.8  Summary                                                             436

# 13 NAS                                                                   437

13.1  Introduction                                                        437
13.2  NAS architecture                                                    437
13.3  MS classes and network modes                                        441
13.4  MM protocol entity                                                  442
13.5  Call control protocol                                               456
13.6  GMM protocol states                                                 467
13.7  GMM procedures                                                      476
13.8  SM protocol and PDP contexts                                        483
13.9  SMS protocol                                                        498

# 14 Idle mode functions                                                   508

14.1  USIM architecture and operation                                     509
14.2  Idle mode overview                                                  514

| 14.3 | Idle mode substate machine | 515 |
| 14.4 | NAS idle mode functions and interrelationship | 519 |
| 14.5 | AS idle mode functions and interrelationship | 527 |
| 14.6 | Example of idle mode procedures | 537 |
| 14.7 | Summary | 541 |

*Appendix*   542
*References*   551
*Index*   553

# Preface

The *WCDMA Design Handbook* addresses the subject of wideband code division multiple access (WCDMA) as defined by the Third Generation Partnership Project (3GPP) and provides a detailed review of the architecture and the operation of the system. In particular, the focus of the book is the radio interface, from the physical layer through to the upper layers of the non-access stratum. This text either offers a complete 'end-to-end' explanation of the system operation, or alternatively allows the reader to focus on any aspects of the system which are of specific interest and relevance. For this reason, the material is presented in a modular fashion, with the overlap and interlinking of the chapters kept to a minimum to allow the chapters to be as self-standing as possible in order to facilitate a 'pick and mix' approach to the book where required.

The structure of the book is intended to provide a solid introduction to the basic principles for the operation of the complete system and then to focus on the specific details in each of the relevant chapters. The major principles for the operation of the WCDMA system are considered throughout the different chapters, including the use of codes and multiplexing in the physical layer, the procedures for transport format combination control in layer 2 and the radio interface control procedures either within the radio resource control (RRC) protocol in the access stratum, or within the mobility management and service management protocols in the non-access stratum. One of the key methods of examining the system is the use of examples to demonstrate the operation of specific procedures or processes.

At the lower layers, the book focuses on the FDD mode of the WCDMA system. The use of the TDD mode is considered to a greater degree as the higher layer protocols are considered. The emphasis is on the first release of the WCDMA specifications (Release 99).

Written for a professional audience, the book is relevant to practising engineers and managers, and graduate and undergraduate students. Like most texts at this level, it is beneficial for the reader to have had some previous exposure to cellular radio systems such as GSM. It is assumed that the reader is comfortable with the technical nature of the information in this technical book.

The book can be considered as being in four parts. Part 1 comprises Chapters 1–3 and is a general introduction; Part 2, Chapters 4–7, covers mainly the physical layer; Part 3, Chapters 8–12, covers layers 2 and 3 in the access stratum; and Part 4, Chapters 13 and 14, covers the non-access stratum protocols. The reading of these four parts will depend upon the specific interests of the reader. For RF, DSP, ASIC and hardware engineers Parts 1 and 2 are recommended. For protocol designers/software designers and protocol test engineers who are focussing on the operation of the access stratum of the WCDMA system, Parts 1 and 3 are the most appropriate. Both protocol designers/software designers and protocol test engineers concentrating on the operation of the non-access stratum of the WCDMA system should read Parts 1 and 4. Finally, for an interested reader, or for a graduate or undergraduate course, the chapters can be taken in order. The book closely follows the 3GPP specifications; for completeness the relevant specifications are outlined in the Appendix.

The *WCDMA Design Handbook* is based on the experience and knowledge gained over a 20-year period by the author. The detail has been honed during the process of presenting the material in the form of a number of training courses on WCDMA, from layer 1 through to the non-access stratum. It is the first in what is planned to be a series of books by the author following the development of the UMTS and wireless cellular market place, with an emphasis on a detailed understanding of the design and operation of the technology. Dr Richardson is a director of the established training and consultancy company Imagicom Ltd (www.imagicom.co.uk), which specialises in delivering regular advanced level technical training courses, the material for which is constantly updated and presented both via a range of scheduled public courses usually held in Cambridge UK and to major players in the telecommunications industry on an in-house basis.

# Acknowledgements

I would like to give special thanks to my wife Alex for her enduring support over the many hours that it has taken to bring this volume from conception into existence; without her this book could not exist. I love her deeply.

# Abbreviations

| | |
|---|---|
| 2G | second generation |
| 3G | third generation |
| 3G-MSC/VLR | third generation mobile switching centre/visitor location register |
| 3GPP | 3rd Generation Partnership Project |
| 3G-SGSN | third generation serving GPRS support node |
| AC | access class |
| ACK | acknowledgement |
| ACLR | adjacent channel leakage ratio |
| ACS | adjacent channel selectivity |
| ADC | analogue to digital converter |
| ADF | application dedicated files |
| AGC | automatic gain control |
| AI | acquisition indicator |
| AICH | acquisition indication channel |
| AID | application identifier |
| AK | anonymity key |
| AM | acknowledged mode |
| AMD | acknowledged mode data |
| AMF | authentication and key management field |
| AMR | adaptive multirate |
| AP | access preamble |
| APN | access point name |
| ARQ | automatic repeat request |
| AS | access stratum |
| ASC | access service class |
| ASIC | application specific integrated circuit |
| ATM | asynchronous transfer mode |
| ATT | AICH transmission timing |
| ATT | attach flag |
| AUTN | authentication token |
| AV | authentication vector |

## Abbreviations

| | |
|---|---|
| AWGN | additive white Gaussian noise |
| BBF | baseband filter |
| BC | broadcast control |
| BCCH | broadcast control channel |
| BCD | binary coded decimal |
| BCFE | broadcast channel functional entity |
| BCH | broadcast channel |
| BER | bit error rate |
| BGCF | breakout gateway control function |
| BLER | block error rate |
| BMC | broadcast and multicast control protocol |
| BO | buffer occupancy |
| BPF | band pass filter |
| BPSK | binary phase shift keyed |
| BS | base station |
| BSC | base station controller |
| BSS | base station system |
| BTS | base transceiver station |
| C/I | carrier to interference ratio |
| C/T | control/traffic |
| CA | channel assignment |
| CAI | channel assignment indicator |
| CAMEL | customised application for mobile network enhanced logic |
| CBC | cell broadcast centre |
| CBS | cell broadcast service |
| CC | call control |
| CCC | CPCH control channel |
| CCCH | common control channel |
| CCDF | complementary cumulative distribution function |
| CCTrCH | coded composite transport channel |
| CD | collision detection |
| CD/CA-ICH | collision detection/channel assignment indicator channel |
| CDMA | code division multiple access |
| CFN | connection frame number |
| CID | context identifier |
| CK | cipher key |
| CKSN | cipher key sequence number |
| CLI | calling line identification |
| CLIR | calling line identification restriction |
| CM | connection management |
| CN | core network |

| | | |
|---|---|---|
| CP | control protocol | |
| CPBCCH | compact packet BCCH | |
| CPCH | common packet channel | |
| CPICH | common pilot channel | |
| CRC | cyclic redundancy check | |
| CRNC | controlling radio network controller | |
| c-RNTI | cell radio network temporary identifier | |
| CS | circuit switched | |
| CSCF | call session control function | |
| CSICH | CPCH status indication channel | |
| CTCH | common traffic channel | |
| CTFC | calculated transport format combination | |
| CTS | cordless telephony system | |
| CW | continuous wave | |
| D/C | data/control | |
| DAC | digital to analogue converter | |
| DC | dedicated control | |
| DCCH | dedicated control channel | |
| DCF | digital channel filter | |
| DCFE | dedicated control functional entity | |
| DCH | dedicated transport channel | |
| DCS1800 | digital cellular network at 1800MHz | |
| DC-SAP | dedicated control SAP | |
| DECT | digital enhanced cordless telecommunications | |
| DF | dedicated files | |
| DPCCH | dedicated physical control channel | |
| DPCH | dedicated physical channel | |
| DPDCH | dedicated physical data channel | |
| DRAC | dynamic resource allocation control | |
| DRNC | drift radio network controller | |
| DRNS | drift radio network subsystem | |
| DRX | discontinuous reception | |
| DSCH | downlink shared transport channel | |
| DSP | digital signal processor | |
| DTCH | dedicated traffic channel | |
| DTX | discontinuous transmission | |
| EDGE | enhanced data rates for GSM evolution | |
| EF | elementary file | |
| EGC | efficient Golay correlator | |
| EIR | equipment identity register | |
| e-MLPP | enhanced multilevel precedence and preemption | |

| | |
|---|---|
| EMS | extended message service |
| EOT | end of transmission |
| EPC | estimated PDU counter |
| ETSI | European Telecommunications Standards Institute |
| EVM | error vector magnitude |
| FACH | forward access channel |
| FBI | feedback mode indicator |
| FCT | frame count transmitted |
| FDD | frequency division duplex |
| FDMA | frequency division multiple access |
| FER | frame error rate |
| FFT | fast Fourier transform |
| FHT | fast Hadamard transform |
| FIR | finite impulse response |
| G3 | Group 3 |
| GC | general control |
| GERAN | GSM/EDGE radio access network |
| GGSN | gateway GPRS support node |
| GMM | GPRS mobility management |
| GMMAS-SAP | GPRS mobility management SAP |
| GMSC | gateway mobile switching centre |
| GPRS | general packet radio service |
| GSM | global system for mobile communications |
| GSMS | GPRS short message service |
| GTP | GPRS tunnelling protocol |
| HC | header compression |
| HCS | hierarchical cell structures |
| HE/AuC | home environment/authentication centre |
| HFN | hyper frame number |
| HLR | home location register |
| HPLMN | home PLMN |
| HPSK | hybrid PSK |
| HSDPA | high speed downlink packet access |
| HSS | home subscriber server |
| HTTP | hypertext transfer protocol |
| I-CSCF | interrogating call session control function |
| IE | information element |
| IK | integrity key |
| IMEI | international mobile equipment identity |
| IMS | internet protocol multimedia subsystem |
| IMSI | international mobile subscriber identity |

| | | |
|---|---|---|
| IMT2000 | | International Mobile Telecommunications 2000 |
| IP | | internet protocol |
| IPDL | | idle period on the downlink |
| ISDN | | integrated services digital network |
| ITU | | International Telecommunication Union |
| KSI | | key set identifier |
| LA | | location area |
| LAC | | location area code |
| LAI | | location area identifier |
| LAPP | | log *a-posteriori* probability |
| LAU | | location area update |
| LI | | length indicator |
| LLC | | logical link control |
| LLR | | log likelihood ratio |
| LNA | | low noise amplifier |
| LO | | local oscillator |
| LR | | location registration |
| LSB | | least significant bit |
| MAC | | message authentication code |
| MAC | | medium access control |
| MAC-b | | MAC – broadcast |
| MAC-c/sh | | MAC – common or shared |
| MAC-d | | MAC – dedicated |
| MAC-hs | | MAC – high speed |
| MAP | | maximum a-posteriori probability |
| MASF | | minimum available spreading factor |
| MCC | | mobile country code |
| ME | | mobile equipment |
| MF | | master file |
| MGCF | | media gateway control function |
| MGW | | media gateway |
| MIB | | master information block |
| MLSE | | maximum likelihood sequence estimation |
| MM | | mobility management |
| MN | | mobile network |
| MNC | | mobile network code |
| MO | | mobile originated |
| MRC | | maximum ratio combining |
| MRF | | media resource function |
| MRFC | | media resource function controller |
| MRFP | | media resource function processor |

| | |
|---|---|
| MRW | move receive window |
| MS | mobile station |
| MSB | most significant bit |
| MSC | mobile switching centre |
| MSE | mean square error |
| MSIN | mobile subscriber identifier number |
| MT | mobile terminated |
| MUX | multiplex |
| NACK | negative acknowledgement |
| NAS | non-access stratum |
| NSAPI | network service access point identifier |
| NW | network |
| OCQPSK | orthogonal complex QPSK |
| OSI | open systems interconnection |
| OTDOA | observed time difference of arrival |
| OVSF | orthogonal variable spreading factor |
| PCCC | parallel concatenated convolutional code |
| PCCH | paging control channel |
| PCCPCH | primary common control physical channel |
| PCDE | peak code domain error |
| PCF | policy control function |
| PCH | paging channel |
| PCPCH | physical common packet channel |
| P-CPICH | primary common pilot channel |
| PCs | personal communication system |
| P-CSCF | proxy call session control function |
| PD | protocol discriminator |
| PDC | personal digital cellular |
| PDCP | packet data convergence protocol |
| PDN | packet data network |
| PDP | packet data protocol |
| PDSCH | physical downlink shared channel |
| PDU | protocol data unit |
| PI | paging indicator |
| PICH | paging indication channel |
| PID | packet identifier |
| PIN | personal identification number |
| PLMN | public land mobile network |
| PMM | PS mobility management |
| PN | pseudo-noise |
| PNFE | paging and notification functional entity |

| | |
|---|---|
| PRA | PCPCH resource availability |
| PRACH | physical random access channel |
| PS | packet switched |
| PSC | primary synchronisation code |
| P-SCH | primary synchronisation channel |
| PSK | phase shift keying |
| PSTN | public switched telephone network |
| PTM | point to multipoint |
| P-TMSI | packet temporary mobile subscriber identity |
| PTP | point to point |
| QoS | quality of service |
| QPSK | quadrature phase shift keying |
| R4 | Release 4 |
| R5 | Release 5 |
| R6 | Release 6 |
| R99 | Release 99 |
| RA | routing area |
| RAB | radio access bearer |
| RABM | radio access bearer manager |
| RAC | radio access capability |
| RACH | random access channel |
| RAI | routing area identifier |
| RAT | radio access technology |
| RAU | routing area update |
| RB | radio bearer |
| RES | response |
| RL | radio link |
| RLC | radio link control |
| RLS | radio link set |
| RLS | recursive least squares |
| RM | rate match |
| RNC | radio network controller |
| RNS | radio network subsystem |
| RNTI | radio network temporary identifier |
| ROHC | robust header compression |
| RPLMN | registered PLMN |
| RRC | radio resource control |
| RRC | root raised cosine |
| RR-SAP | radio resource SAP |
| RSCP | receive signal code power |

| | |
|---|---|
| RTT | round trip time |
| S/P | serial to parallel |
| SAP | service access point |
| SCCPCH | secondary common control physical channel |
| SCH | synchronisation channel |
| S-CPICH | secondary common pilot channel |
| SCR | source controlled rate |
| S-CSCF | serving call session control function |
| SDP | session description protocol |
| SDU | service data unit |
| SF | spreading factor |
| SFN | system frame number |
| SGSN | serving GPRS support node |
| SHCCH | shared channel control channel |
| SI | status indicator |
| SI | stream identifier |
| SIB | system information block |
| SIBn | system information broadcast type $n$ ($n = 1, \ldots, 18$) |
| SID | silence descriptor |
| SIP | session initiation protocol |
| SIR | signal to interference ratio |
| SISO | soft in soft out |
| SLF | subscription location function |
| SM | session management |
| SMC-CS | short message control – circuit switched |
| SMC-GP | short message control – GPRS protocol |
| SM-RL | short message relay layer |
| SMS | short message service |
| SMSMM | SMS mobility management |
| SM-TL | short message transfer layer |
| SNR | signal to noise ratio |
| SOVA | soft output Viterbi algorithm |
| SQN | sequence number |
| SRB | signalling radio bearer |
| SRNS | serving radio network subsystem |
| s-RNTI | serving radio network temporary identifier |
| SS | supplementary service |
| S-SCH | secondary synchronisation channel |
| SSDT | site selection diversity transmission |
| STTD | space time transmit diversity |

| | |
|---|---|
| SUFI | super fields |
| TACS | total access communications system |
| TAF | terminal adaptation function |
| TCP | transmission control protocol |
| TCTF | target channel type field |
| TCTV | traffic channel transport volume |
| TDD | time division duplex |
| TDMA | time division multiple access |
| TE | terminal equipment |
| TF | transport format |
| TFC | transport format combination |
| TFCI | transport format combination indicator |
| TFCS | transport format combination set |
| TFS | transport format selection |
| TFT | traffic flow template |
| TG8/1 | Task Group 8/1 |
| TGMP | transmission gap sequence measurement purpose |
| TI | transaction identifier |
| TIA | Telecommunications Industry Association |
| TM | transparent mode |
| TMD | transport mode data |
| TMSI | temporary mobile subscriber identity |
| ToS | type of service |
| TPC | transmit power control |
| TTI | transmission time interval |
| TVM | traffic volume measurement |
| Tx | transmit |
| UARFCN | UTRA absolute radio frequency channel number |
| UDP | user datagram protocol |
| UE | user equipment |
| UICC | universal integrated circuit card |
| UM | unacknowledged mode |
| UMTS | Universal Mobile Telecommunications System C304 |
| URA | UTRAN registration area |
| URL | uniform resource locator |
| u-RNTI | UTRAN radio network temporary identifier |
| US | update status |
| USAT | USIM application toolkit |
| USCH | uplink shared channel |
| USIM | universal subscriber identity module |
| UTRAN | UMTS terrestrial radio access network |

| | |
|---|---|
| VAD | voice activity detection |
| VCAM | versatile channel assignment mode |
| VGCS | voice group call service |
| VLR | visitor location register |
| WCDMA | wideband code division multiple access |
| XMAC | expected message suthentication code |
| XRES | expected response |

# 1 Introduction

## 1.1 Concepts and terminology

This book is concerned with an exploration of the design and operation of a technology referred to as wideband code division multiple access (usually abbreviated to WCDMA or W-CDMA).

Before starting the journey exploring the design and operation of the WCDMA technology, we first need to define what we mean by WCDMA. To do this, it is useful to examine first the terminology that surrounds the WCDMA system and which is put into context in Figure 1.1. To understand the diagram, we need to start at the outer ring.

The outer ring encompasses all of the technology that is viewed as being part of the third generation (often shortened to the term 3G) of mobile phone technology. This third generation follows (obviously) from second-generation (2G) technologies such as: the Global System for Mobile (GSM) communications, which is deployed in Europe and many other countries throughout the world; IS54/IS136, which is a 2G standard developed in the USA and also used in a number of countries throughout the world; Personal Digital Cellular (PDC) developed and deployed in Japan; and IS95, the USA developed CDMA standard, which is deployed in the USA and many other countries.

3G is a term used to reflect a number of different technologies being defined as successors to 2G technology. In many cases, the 3G technologies can be seen as an evolution of the 2G networks. Another general term for 3G is International Mobile Telecommunications 2000 (IMT2000).

IMT2000 is a name that was defined by Task Group 8/1 (TG8/1), a standardisation group set up in 1985 by the International Telecommunication Union (ITU) to define world standards for 3G mobile technologies. The role of TG8/1 was essentially to define the requirements for 3G, and to facilitate the process of defining a radio technology that meets them. The term IMT2000 represents the family of technologies that are contained within the outer ring, which have been defined to meet these requirements. The outer ring in Figure 1.1, therefore, is intended to illustrate the extent of 3G technologies and the different terms that are commonly applied to define 3G.

# Introduction

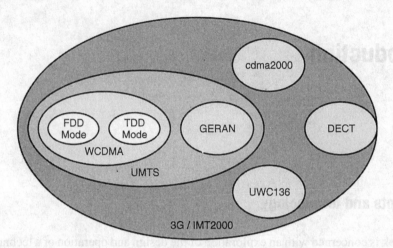

**Figure 1.1** 3G technology relationship.

Within the outer ring we can see a number of smaller rings. These smaller rings represent the technologies that have been defined and which form 3G. The original objective for TG8/1 was to define a single standard for the 3G radio access technology (RAT). The resulting situation of having multiple technologies was a consequence of the very difficult environment that existed prior to the definition of 3G. As mentioned previously, there are a number of 2G technologies throughout the world, each of which should have some evolutionary path to 3G. Due to the differences in the technologies as well as some political differences, the best compromise was the creation of a number of different 'technology modes' for 3G leading to a family of technologies.

### 1.1.1 Universal Mobile Telecommunications System (UMTS)

UMTS is the name of the telecommunication system that is being defined by the 3rd Generation Partnership Project (3GPP). The functionality included within UMTS encompasses all that is required to support the service requirements that were outlined for 3G by TG8/1 and also includes some that are new and evolved.

UMTS represents the complete system, including elements such as the user's mobile equipment, the radio infrastructure necessary to support a call or data session, and the core network equipment that is necessary to transport that user call or data from end to end, as well as billing systems, security systems etc.

Within Figure 1.1, as in this book, we are focussing predominantly on RAT (we will see later that this is not strictly correct, as we will also be considering upper layer protocols that extend beyond the radio access network into the UMTS core network). As a consequence, the core network equipment, which is equally a part of UMTS, is not shown in the diagram.

## 1.1 Concepts and terminology

In UMTS there are two different types of RAT. The first is the subject of this book, namely WCDMA. The second is an evolution of the 2G GSM system that is generically referred to as the GSM/EDGE radio access network (GERAN). Different rings illustrate these two technologies. The focus of this book is WCDMA and as a consequence we won't consider GERAN in much more detail, except where interworking between the two technologies requires us to do so.

Within the WCDMA ring there are two additional rings called frequency division duplex (FDD) mode and time division duplex (TDD) mode. These two modes represent two quite different (different from the perspective of the physical layer of the protocol) technologies. The FDD mode of WCDMA separates users employing codes and frequencies with one frequency for the uplink and a second frequency for the downlink. The TDD mode on the other hand separates users employing codes, frequencies and time and uses the same frequency for both the uplink and the downlink. The emphasis of our considerations is on the FDD mode, although we will consider aspects of the TDD mode.

### 1.1.2 Other IMT2000 technologies

To complete the picture illustrated by Figure 1.1, we should also consider the other technologies that form a part of our 3G family of technologies referred to as IMT2000. They are cdma2000, UWC136 and DECT.

Cdma2000 has evolved from the 2G technology commonly referred to as IS95. Cdma2000 is a CDMA technology that supports the service requirements defined by TG8/1 for 3G mobile technology. Although cdma2000 and WCDMA have elements within them that are similar, the two systems are sufficiently different to require separate consideration. As a consequence, we will not consider cdma2000 in any greater depth except when we need to address issues that relate to intersystem operation.

UWC136 is a technology derived from the IS54 and IS136 2G digital cellular standards defined in the USA by the Telecommunications Industry Association (TIA). UWC136 is a time division multiple access (TDMA) technology, similar in concept to that used by GSM. In fact UWC136 in one of its modes of operation includes elements of the evolved GSM technology to provide for high data rates. Again, however, we will not consider UWC136 in any greater depth as it is a subject that lies beyond the scope of this book.

Digitally enhanced cordless telephone (DECT) is the final member of the 3G-technology family. DECT was defined by the European Telecommunication Standards Institute (ETSI) as a low power wireless communications system. DECT uses a combination of FDMA, TDMA and TDD to allow users to access the services provided by DECT. Again, DECT lies outside the scope of this book and as a consequence will not be considered in greater detail.

## 1.2 Major concepts behind UMTS

A question that is often asked regarding UMTS is why there needs to be a new radio interface technology and why an evolution of 2G technology such as enhanced data rates for GSM evolution (EDGE) cannot be used in its place.

The answer, as could be anticipated, is not a simple one. There are many issues that we could consider, such as spectrum efficiency, that could be used (arguably) to justify a new radio technology solution for 3G. There is, however, another line of reasoning, which is less controversial as to the reasons for the existence of WCDMA within the UMTS technology family. When TG8/1 set out the objectives and requirements for 3G, there were a number of elements that made the reuse of 2G technology in a 3G context very difficult without significant evolution of the 2G technology itself (this is the case for cdma2000 and UWC136 described earlier). Three of the most significant elements are outlined below and defined in greater detail in the subsections that follow:

- Support of general quality of service (QoS).
- Support of multimedia services.
- Support of 2 Mb/s.

### 1.2.1 Support of general QoS

We will see in Section 13.9 that a vast majority of services provided by UMTS use a technique known as packet switched communications. The mechanism behind the success of packet switched communications is dynamic resource sharing, whereby a number of users can share the same transmission resource (this could be anything from a radio signal to an optical fibre). One consequence of dynamic resource sharing in a packet switched network is the dynamic variation in the resources that are available to a user. If nothing is done to overcome the effects of dynamic resource allocation, then this will lead to queuing delays and inefficient use of resources in the network. To help circumvent this problem, packet switched communication systems define a concept known as quality of service (QoS). QoS cannot, on its own, overcome the problem of queuing delays, but it can quantify the requirements that need to be achieved using other procedures.

From the outset, UMTS was designed to support services with an arbitrary QoS. Figure 1.2 illustrates an example of such a service as well as the basic quantities that are used to define the service characteristics. Before considering the specifics of the example service illustrated in Figure 1.2, we must first consider how we define QoS. QoS in general is defined in terms of three quantities: data rate, delay and error characteristics. We will see in Section 13.8 that UMTS defines QoS using a number of parameters, but

## 1.2 Major concepts behind UMTS

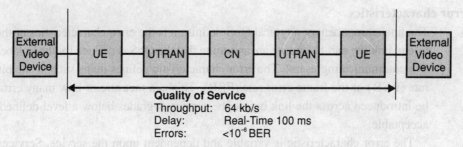

**Figure 1.2** Example service and QoS relationship.

that broadly speaking these parameters reduce to these three quantities. Let us begin by considering what is meant by these three quantities, and then we can consider the specific example shown in the figure.

### Data rate

In a packet switched network, data rate is usually defined in terms of average data rate and peak data rate (both measured over some defined time period). It is the objective of the packet network to offer a service to a user that lies within the requested data rate. If the service were a constant data rate service such as a 57.6 kb/s modem access, then the peak and the average data rates would tend to be the same. If, on the other hand, the service is an Internet access data service, then the peak and average data rates could be quite different.

### Delay

The second element of QoS is some measure of delay. In general, delay comprises two parts. First, there is the type of delay, and second there is some measure of the magnitude of the delay.

The type of delay defines the time requirements of the service such as whether it is a real-time service (such as voice communications) or a non-real-time service (such as e-mail delivery). We will see in Chapter 9 that in UMTS the delay characteristic is referred to as the traffic class, and that there are four traffic classes currently defined.

The magnitude of the delay defines how much delay can be tolerated by the service. Bi-direction services such as voice communications in general require low delay, typically in the region of tens of milliseconds. Uni-directional services such as e-mail delivery can accept higher delays, measured in terms of seconds.

The objective for the delay component of QoS, therefore, is to match the user's service requirements for the delay to the delay that can be delivered by the network. In a system operating correctly, the delay of the data should correspond to the delay specified within the QoS.

# 6 Introduction

**Error characteristics**

The final component of a typical QoS definition is the error characteristics of the end-to-end link (by end-to-end we are assuming that the QoS is defined for the link between two communicating users). The error characteristic defines items such as the bit error rate (BER) or the frame error rate (FER). This is a measure of how many errors can be introduced across the link before the service degrades below a level defined to be acceptable.

The error characteristic is variable and dependent upon the service. Services such as speech, for example, have been found to be quite tolerant to errors. Other services, such as packet data for Web access, are very sensitive to errors.

**Example service**

Let us return to the example service shown in Figure 1.2. This example is not applicable to the UMTS circuit switched services that carry voice and video; rather it applies to the evolved services based on packet switched connections. In the example we are considering some type of video service (for instance as part of a video conference). For this specific service, we require a peak data rate of 64 kb/s. Although a low data rate, this is typical of what could be required for video conferencing using the very small display screens likely on pocket UMTS video phones. For video communications, the peak and average data rates are not necessarily the same. Most video coding algorithms operate by sending a complete image of data infrequently, and frequently sending an update to this complete image. This procedure results in a data rate whose peak and average values are different.

The delay characteristic for the video service reflects the fact that the service is a conversational real-time service with a need for a low end-to-end delay. The delay that is acceptable will depend on user perceptions of the delay, but we could imagine that end-to-end delays, typically, must be less than a few hundred milliseconds.

The error characteristics of the service are defined by the sensitivity of the video codecs to errors. Typically, a video codec operates with an error rate in the region of $1 \times 10^{-6}$. This terminology means that the video codec can operate acceptably as long as there is (on average) not more than one error for every million bits received.

## 1.2.2 Support of multimedia services

The second significant element introduced by UMTS that pushes the requirement for a new radio interface technology further is the support of multimedia services.

Multimedia is one of those terms that is never clearly and consistently defined. Here, and throughout this book, we define multimedia as a collection of data streams between the user and some other end user(s) or application(s). The data streams that comprise this multimedia connection will, in general, have differing QoS characteristics.

## 1.2 Major concepts behind UMTS

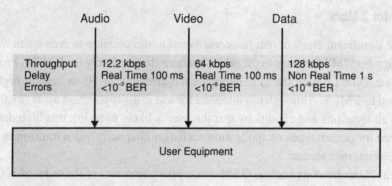

**Figure 1.3** Example multimedia service.

It is the objective of the UMTS standard to allow the user to have such a multimedia connection, which comprises a number of such data streams with different QoS characteristics, but all of them multiplexed onto the same physical radio interface connection.

> **Data stream**
>
> A data stream in the context of this book refers to a flow of information from a user to a destination user or possibly to some destination application. For a specific user there could be a number of these data streams, but not necessarily going to the same destination end-point.

Figure 1.3 illustrates an example of a multimedia connection comprising a number of data streams. In this example, we see that there are three components to this multimedia service – voice, video and packet data. For each of these services, the QoS is defined and different.

- For the voice service, we need a low data rate (typically 12.2 kb/s), low and constant delay and a moderate error characteristic (a BER in the region of $10^{-3}$ should be acceptable for most voice communication systems).
- For the video service, the data rate needs to be greater (maybe in the region of 64 kb/s as defined earlier), the delay perhaps similar to the voice connection and the error characteristic probably in the region of $10^{-6}$, reflecting the greater sensitivity that the video service has to errors.
- For the packet data communications, we could envisage data rates in the region of 128 kb/s for high speed Internet access, a delay characteristic that reflects the type of Internet access and an error characteristic with a BER better than $10^{-9}$.

Obviously, this is a somewhat artificial example, but what it serves to illustrate is that the UMTS system needs to be able to combine services with totally different QoS requirements onto the same physical radio connection, but do so in a very efficient manner.

## 1.2.3 Support for 2 Mb/s

The final significant element that has contributed to the decision to create a new radio technology for UMTS relates to the peak data rates that are likely for UMTS. From the very outset TG8/1 within the ITU defined a requirement for 3G to support high data rates of up to 2 Mb/s. Although not intended for use in all application areas (by this we mean in all locations and conditions that the user is likely to be in), this high data rate is required for certain types of application such as high quality video transmission and high speed internet access.

Current 2G technology such as GSM can support data rates of the order of 100 kb/s. With enhancements, such as the change in modulation schemes proposed by EDGE, this provides an upper data rate in the region of 384 kb/s.

> **EDGE**
>
> EDGE is part of the ongoing evolution of the GSM standard, in this case to support higher data rates. EDGE achieves the higher data rates through the use of a higher order modulation scheme based on a technique called 8-PSK (phase shift keying).

EDGE, therefore, can achieve data rates approaching those required for UMTS. To achieve the 2 Mb/s data rate, however, we need to reconsider the design of the radio interface.

The limits of the GSM radio interface technology are perhaps some of the main reasons why the decision was made to reconsider the definition of a new radio technology. Ultimately, within UMTS, this decision led to the definition of the radio interface technology that we now know as WCDMA.

## 1.3 Release 99 (R99) network architecture

The UMTS specifications are being issued in different releases. The purpose of this is to stagger the introduction of new services and technologies, which in turn will reduce the problems with installation and commissioning of the system as a whole. The first release of the UMTS specifications is referred to as the R99 specifications. The subsequent releases are referred to as Release 4 (R4), Release 5 (R5) and Release 6 (R6).

Perhaps the main objective of R99 was the introduction of the WCDMA radio technology within the radio interface. WCDMA introduces a significant degree of complexity in the design and operation of the radio interface, and is the main focus of this book. R4, R5 and R6 are subsequent releases that add additional functionality and services to the R99 standard.

In this section we want to examine the basic network architecture for the R99 series of specifications. A good understanding of the architecture of the UMTS network is very

## 1.3 Release 99 (R99) network architecture

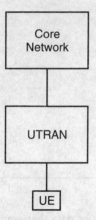

**Figure 1.4** High level representation of UMTS network elements.

important when it comes to understanding the design and operation of the remainder of the network.

### 1.3.1 Fundamental architecture concepts

**Basic network structure**

Figure 1.4 shows, from a very high level, the basic structure of the UMTS system. The structure is split into three main components: the core network (CN); the UMTS terrestrial radio access network (UTRAN); and the user equipment (UE).

The CN is responsible for the higher layer functions such as user mobility, call control, session management and other network centric functions such as billing, security control. The UTRAN is responsible for functions that relate to access, such as the radio access, radio mobility and radio resource utilisation. The decision to divide the fixed network into two distinct networks was a very deliberate and considered step in the overall design of the UMTS system.

In separating radio access from the other network functions, it becomes feasible to evolve the radio access network and the CN independently from each other. In doing this, as part of R99, we can introduce a new radio access technology (WCDMA) and reuse existing CN technology from the GSM and general packet radio service (GPRS) networks. Subsequent releases of the CN (e.g. R4 and R5) can then modify the CN, without necessarily introducing significant changes to the UTRAN.

**Access stratum (AS) and non-access stratum (NAS)**

A consequence of this decision to separate the UTRAN and the CN is the division in the layers of the protocol stacks, which is illustrated figuratively in Figure 1.5. The diagram illustrates a division in the protocol architecture between the UE, the UTRAN and the CN. The protocols are separated into what are called the AS and NAS. This division applies to both the signalling messages and the user data messages.

# 10  Introduction

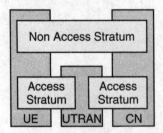

**Figure 1.5**  AS and NAS protocol split.

> **Signalling and user data**
>
> Throughout the book we will be referring to different types of data that are passed through the system. The term signalling is often used to define the type of data that is used to send special control messages that are used to control the system in some manner. The other type of information, the user data, refers to the actual user information that is sent from some source user to a destination user or destination application.
>
> In general, the signalling data passes through what is called the control plane. The user data passes through the user plane.

The AS carries all of the signalling and user data messages that relate to the access technology used across a specific interface in that part of the system. Across the radio interface, the AS protocols are the lower level protocols between the UE and the UTRAN, and between the UTRAN and the CN.

The NAS carries the signalling messages and user data messages that are independent of the underlying access mechanism. These signalling and user data are passed between the UE and the CN and, conceptually, pass transparently through the UTRAN.

Examples of the types of signalling messages that are carried via the AS are messages that control the power control loops in the system, that control the handover procedures or that allocate channels to a user for use, for instance in a speech call. An example of an NAS signalling message would be one associated with a call setup request, where the call setup messages are independent of the underlying access mechanism. In this example, the call setup message would come from the CN, and be routed transparently through the AS.

## 1.3.2  Service domains

Figure 1.6 illustrates the functional decomposition of the UMTS fixed network architecture into the UTRAN and the CN service domains. For the R99 architecture there are three of these service domains, referred to as the circuit switched domain (CS domain), the packet switched domain (PS domain) and the broadcast control domain (BC domain).

## 1.3 Release 99 (R99) network architecture

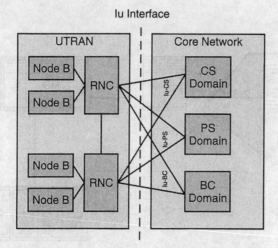

**Figure 1.6** UTRAN and CN service domains.

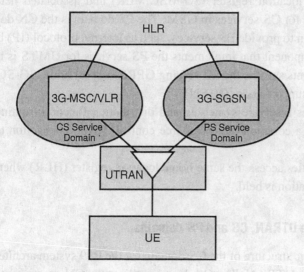

**Figure 1.7** CS and PS service domains.

In Figure 1.6, the radio network controllers (RNCs) in the UTRAN are connected to each of the three domains in the CN. This connection is via an interface that is called the Iu interface. There are three variants of the Iu interface and the corresponding protocol stacks associated with the interface. The Iu–CS interface is between the UTRAN and the CS domain, the Iu–PS interface is between the UTRAN and the PS domain, and the Iu–BC interface is between the UTRAN and the BC domain.

Figure 1.7 illustrates the main service control entities within the CS and PS domains, and their link to the UTRAN and the UE. The CS domain is the CN domain that is traditionally known to provide CS services such as speech calls. For R99, the CN component that implements the CS services for UMTS is the third generation mobile

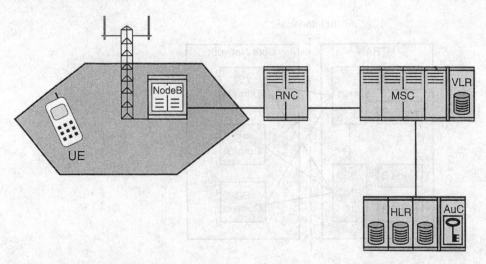

**Figure 1.8** CS domain architecture.

switching centre/visitor location register (3G-MSC/VLR) and associated network elements that are defined for CS services in GSM. The PS domain is the CN domain that is traditionally known to provide PS services, such as internet protocol (IP) based user traffic. The CN component that implements the PS services for UMTS is based on the GPRS CN elements such as the 3G serving GPRS support node (3G-SGSN) and the gateway GPRS support node (GGSN).

The UTRAN provides a service switching and distribution function, routing the traffic that is carried on a common radio resource control (RRC) connection to the correct service domain.

Both CN domain entities access the same home location register (HLR) where the user subscription information is held.

### 1.3.3 Detailed structure of the UTRAN, CS and PS domains

Figure 1.8 shows the basic structure of the CS domain for the R99 system architecture and its connections to the UTRAN. The CS domain utilises the CN components that are a derivative of the GSM system and which we consider briefly below. For the UTRAN, there are two new components: the RNC, and the Node B.

**CS domain**

The mobile switching centre (MSC) and the home location register/authentication centre (HLR/AuC) are the main network elements in the CS domain and are responsible for establishing and controlling circuit switched connections to and from a UE. The MSC is based on the GSM switch, which in turn is based on an integrated services digital network (ISDN) switch with 64 kb/s circuits arriving at and leaving from the switch. This 64 kb/s I/O characteristic introduces an upper limit on CS data services

## 1.3 Release 99 (R99) network architecture

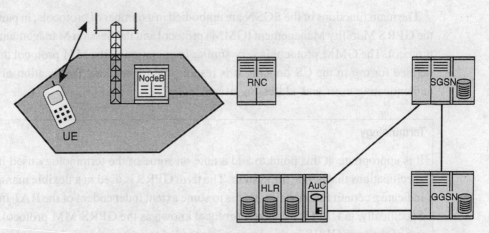

**Figure 1.9** PS domain architecture.

via the CS domain for the R99 standard. To support the 3G mode of operation, the MSC is modified to support the Iu–CS interface, which is based on an asynchronous transfer mode (ATM transport technology [1], [2]).

In an identical way to the GSM, the MSC provides a number of different protocol functions. Across the radio interface, these protocol functions can be grouped together into categories often referred to as the mobility management (MM) functions, connection management (CM) functions, call control (CC) functions and short message service (SMS) functions. If we recall the protocol layering into AS and NAS that we considered previously, all of these functions (MM, CM, CC and SMS) reside in the NAS and are based on equivalent GSM protocols. In the next chapter, we will consider the basic purpose for each of these protocols and in later chapters examine the detail in the design and operation of these protocols. The HLR/AuC provides a central permanent store for subscriber specific information and security information that is used to ensure the security of the system.

## PS domain

Figure 1.9 illustrates the network architecture for the PS domain and its interconnection to the UTRAN. The CN PS domain comprises the SGSN, the GGSN and the HLR/AuC. The HLR/AuC is the same entity as in the CS domain; it has the capability to store PS domain subscriber and security information in addition to the CS domain information.

The SGSN functions are similar to those of the MSC in the CS domain, the main difference being due to the packet nature of the connections. The SGSN is responsible for controlling packet sessions rather than for management of calls. Managing the mobility of the UE is the main area of commonality between the SGSN and the MSC. This MM involves location tracking, authentication and addressing of the UE.

The main functions of the SGSN are embodied in a number of protocols, in particular the GPRS Mobility Management (GMM) protocol and the Session Management (SM) protocol. The GMM protocol is very similar in operation to the MM protocol that was defined for use in the CS domain. It is responsible for tracking the location of users, authenticating users and addressing users.

> **Terminology**
>
> It is appropriate at this point to add a note on some of the terminology used in the specifications that define the system. The term GPRS is used in a flexible manner to indicate a certain functionality that is to some extent independent of the RAT in use. Specifically, in UMTS we have a protocol known as the GPRS MM protocol. The use of the term GPRS in this context is intended to convey that the protocol applies to the PS domain; in this instance it is not intended to indicate that the GPRS radio interface is involved.

The SM protocol is similar in concept to the CC protocol in the CS domain. Where the CC protocol is responsible for establishing and controlling CS connections, the SM protocol is responsible for establishing and controlling PS data sessions, referred to in GPRS and UMTS as packet data protocol (PDP) contexts.

The SGSN interacts with the UE and GGSN to provide an end-to-end packet transfer service. The SGSN is based on the GPRS SGSN, but supports the Iu–PS interface, which for R99 uses an ATM transport technology.

The SGSN is responsible for detecting new UEs arriving within the radio coverage area covered by the SGSN. The SGSN notifies the GGSN of the presence of a new mobile, and all of the data packets are tunnelled by the GGSN to the new SGSN for delivery to the mobile station (MS).

> **Tunnelling**
>
> Within GPRS and the PS domain of UMTS there is a concept known as tunnelling. At the simplest level, tunnelling involves the receipt of a packet of data and encapsulating it within another packet of data. There are a number of reasons for doing this. The first is to manage the problems associated with the mobility of a user and the fact that IP addresses are usually associated with a fixed location. The second main reason is that the tunnel could use a secure connection potentially across a public network and still be able to guarantee the security of the connection.

The GGSN is the gateway GPRS support node. It is responsible for providing interconnection between the PS domain and external data networks such as X.25 or the Internet (in reality there seems to be a move away from X.25, and so the Internet and

## 1.3 Release 99 (R99) network architecture

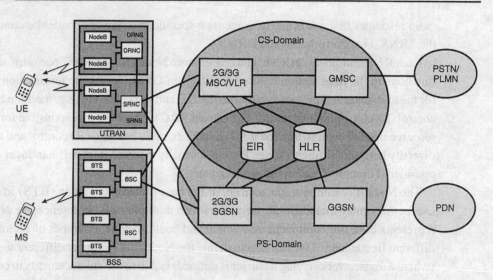

**Figure 1.10** R99 network architecture.

networks based on the Internet family of protocols are the most likely networks for the GGSN to connect to).

The functions provided by the GGSN include tunnelling of the protocol data units (PDUs) to the UE via the SGSN, as well as managing address mapping between the external data networks and the addresses used between the UE and the PS domain.

## UTRAN

The structure of the UTRAN is illustrated in Figure 1.10 and is common to both the CS domain and the PS domain. The figure includes the existing 2/2.5G radio network, which forms part of the UMTS system concept outlined in Section 1.1.1. The UTRAN comprises all of the fixed network entities that are required to operate and control the AS, such as one or more radio network subsystems (RNSs), which are conceptually similar to the base station system (BSS) in the GSM system. The RNS comprises an RNC and one or more Nodes B.

A key UTRAN architecture design requirement was that multiple RNSs within the UTRAN should be interconnected as shown in Figure 1.10. This type of interconnection is not present in GSM and is a consequence of two things: first the desire to decouple the radio access from the CN, and second the need to support soft-handover in which a UE may be connected to more than one RNS at the same time.

The RNS is regarded as being either a drift RNS (DRNS) or a serving RNS (SRNS). The distinction comes from where a UE makes a connection to the UTRAN when making or receiving a call. If the UE moves towards a new cell served by a different RNS, this will become the DRNS that the UE is connected to. Physically there is no difference between the DRNS and the SRNS. The SRNS is responsible for the RRC connection that a UE makes to the UTRAN, whilst the DRNS is responsible for the radio resources that a UE needs to access a specific cell that is outside the control of the SRNS, but controlled by the DRNS.

Each RNS comprises an RNC and one or more Nodes B. The RNC performs similar functions to the base station controller (BSC) in GSM, and is the termination point for the high layer AS protocols between the UE and the UTRAN (e.g. medium access control (MAC), radio link control (RLC) and RRC). The RNC is responsible for radio resource management within the RNS, handover, radio bearer (RB) control and macro diversity between different Nodes B (this is usually referred to as soft-handover and is considered in more detail in the next chapter).

The Node B is similar to the concept of the base transceiver station (BTS) in GSM except that it may include multiple carriers with multiple cells. In essence, the Node B represents base site equipment as a whole and could include a number of sectors and different frequencies. The decision to define the Node B in a manner different to GSM (where a transceiver carrying a carrier is defined) is due to the enhancements in cellular radio technology.

## 1.4  R4 and R5 network architecture

In this section we examine some of the main changes to the architecture of the UMTS network. The main focus of these changes is the removal of the MSC from the R99 CS domain, replacing it with a media gateway (MGW) and an MSC server that acts in part as the media gateway control function (MGCF). The R4 network architecture removes the data rate bottleneck caused by the MSC for the CS domain data connections.

The R5 network architecture takes a further step by introducing a new subsystem component referred to as the All IP multimedia subsystem (IMS). This new network component is present to allow very flexible control over a range of multimedia services.

### 1.4.1  R4 network architecture

Figure 1.11 illustrates the changes to the UMTS architecture for R4. The main change that is common to both R4 and R5 is the MSC, which is replaced with an MSC server, and a MGW.

**MGW**

What is not visible from Figure 1.11 is the transmission mechanism that is used to carry the signalling and user data (called traffic in this specific case). For the R4 and subsequently the R5 architecture, the signalling and traffic may be carried via an IP network connection within the R4/R5 architecture.

If the destination user for a call is external to the R4 network, then the traffic and the signalling streams will need to be converted from the PS transport to the normal

## 1.4 R4 and R5 network architecture

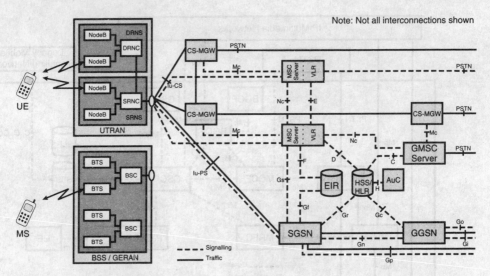

**Figure 1.11** R4 network architecture.

CS transport. To achieve this, the MGWs are used. These are devices that can connect a PS media flow to a CS equivalent and vice-versa depending on the direction of the media stream.

### MSC server

The MSC server is present in the architecture to absorb the role of the media gateway control function (MGCF). It is conventional to use an MGCF to control an MGW. In R4 this role is taken by the MSC server, which is also responsible for CS domain signalling messages from the UTRAN, and for signalling messages from external CS networks such as the PSTN. As far as the UE is concerned, the MSC server appears as a normal MSC. The MSC server terminates all of the CC, MM and SMS protocols. If the user is attempting to establish a CS connection (for instance for a speech call), then the MSC server will interact with the MGW to create an appropriate media stream between the MGW and the destination mode. The connection between the MGW and the UTRAN is via the Iu–CS interface.

### 1.4.2 R5 IMS architecture

Figure 1.12 illustrates the functional architecture of the IMS. The IMS represents quite a departure from traditional CS telecommunications. In the first instance, all of the multimedia streams are transported via a PS connection. Second, the control of the multimedia streams uses a completely different control protocol philosophy to the one we might expect in an equivalent CS network.

In a traditional telecommunications network, calls are established using a CS CC protocol, usually a derivative of an ITU protocol referred to as Q.931. GSM, for

# 18 Introduction

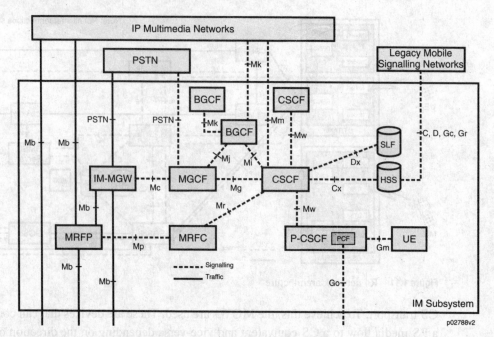

**Figure 1.12** R5 network architecture.

example, utilises most of the call setup procedures defined in this protocol for the call control protocol across the radio interface. In the R5 IMS, however, this set of CC procedures has been replaced with a set of CC functions that are based on a protocol that comes from the Internet family of protocols. The alternative to the GSM CC protocol is called the session initiation protocol (SIP). The SIP is a client server protocol whose roots lie in the Web access protocol HTTP (hyper text transfer protocol).

The SIP provides the session control messages, but it also requires an associated protocol called the session description protocol (SDP) to define the media types that are being controlled by SIP.

The SDP, as the name suggests, defines the media types that are used for a specific multimedia session. One of the great benefits of the SDP is the flexibility it supports in terms of the establishment of multimedia sessions, allowing a rich selection of media types to be combined together to form some elaborate multimedia service.

To support the IMS including its use of SIP and SDP some new architectural elements are required for the R5 IMS. Figure 1.12 illustrates some of these new architectural elements.

## Call session control function (CSCF)

At the centre of the session control protocol there is a new architectural entity referred to as the CSCF. There are a number of different types of CSCF in the network, for instance there is a proxy-CSCF (P-CSCF), a serving CSCF (S-CSCF) and an interrogating-CSCF (I-CSCF). The decision on which of these should be used depends on what part of the multimedia session we are considering. A P-CSCF is the first point of contact for the UE in the IMS, and is in the same network as the GGSN. The S-CSCF provides service control and is located in the home network. The I-CSCF is the first point of contact in the target network and is used to identify to which S-CSCF the UE should be assigned.

All of these new network entities use the SIP/SDP family of protocols to perform functions such as registration and multimedia session control.

**MGW and MGCF**

The MGW and the MGCF provide the same basic functions as their counterparts in the R4 network architecture for the CS domain, except that we are now considering multimedia streams rather than a single media stream. The role of the MGW is to provide media conversion between the IMS and the external transport networks. The MGCF is responsible for configuring and activating these MGW streams under the direction of the CSCF.

**Home subscriber server (HSS)**

For the IMS, the HLR has been replaced with a device called an HSS. The HSS takes on the role of the HLR including service authorisation and user authentication. The HSS is a superset of the HLR and includes all HLR functions and interfaces. To access subscriber data from the HSS by the IMS the internet DIAMETER [3] protocol is used.

**Media resource function (MRF)**

The MRF provides support for multiparty and multimedia conferencing functions. It is responsible for multimedia conference calls. In addition, it will communicate with the CSCF for service validation for multiparty and multimedia sessions.

## 1.5 Services provided by UMTS and their evolution from GSM/GPRS services

### 1.5.1 Teleservices

A teleservice is a complete telecommunication service. This means that the functions in the mobile necessary to provide the service are defined in addition to the transmission functions between the end users. Table 1.1 illustrates the main GSM teleservices that are available.

**Table 1.1.** *UMTS teleservices*

| Bearer number | Description | GSM R98 | UMTS R99/R4 | UMTS R5 |
|---|---|---|---|---|
| 11 | Telephony | Y | Range of codecs | Range of codecs |
| 11 | + Wideband adaptive multirate codec | | | Y |
| 12 | Emergency calls | Y | Y | Y |
| 21 | Short message MT/PTP | Y | Y | Y |
| 22 | Short message MO/PTP | Y | Y | Y |
| 23 | Short message cell broadcast | Y | Y | Y |
| 61 | Alternate speech and fax group 3 (G3) | Y | Y | Y |
| 62 | Automatic fax G3 | Y | Y | Y |
| 91 | Voice group call service | Y | GSM only | GSM only |
| 92 | Voice broadcast service | Y | GSM only | GSM only |

An example of the GSM teleservices is the speech service. Speech is a teleservice because the functions necessary in the mobile such as the speech codec are defined as well as the means to transport the speech across the network between the users. The speech teleservice uses an evolution of the GSM speech codec and is called the AMR speech codec.

## SMS

The GSM SMS is a teleservice because the functions necessary to send and receive an SMS are defined in the mobile as well as the transmission functions. The GSM SMS service has two basic modes of operation – mobile originated (MO) SMS and mobile terminated (MT) SMS.

The characteristics of the SMS are as follows:
- SMS messages can be sent or received whilst a speech or data call is in progress.
- SMS messages are guaranteed for delivery by the network.
- Messages up to 160 characters in length can be sent.

SMS has been extended to include the extended message service (EMS), which also incorporates the ability to send and receive:
- text formatted messages,
- small pictures,
- melodies.

## Cell broadcast SMS

The GSM cell broadcast SMS is a teleservice. With the GSM cell broadcast SMS, a group of users within a cell are able to receive a cell broadcast. Cell broadcast can be used for a range of services such as news and sports updates, stock market updates,

## 1.5 Services provided by UMTS

**Table 1.2.** *UMTS CS bearer services*

| Bearer number | Description | GSM R98 | UMTS R99/R4 | UMTS R5 |
|---|---|---|---|---|
| 20 | Asynchronous general bearer service | Y | Y | Y |
| 21 | Asynchronous 300 b/s | Y | | |
| 22 | Asynchronous 1.2 kb/s | Y | | |
| 23 | Asynchronous 1200/75 b/s | Y | | |
| 24 | Asynchronous 2.4 kb/s | Y | | |
| 25 | Asynchronous 4.8 kb/s | Y | | |
| 26 | Asynchronous 9.6 kb/s | Y | | |
| 30 | Synchronous general bearer service | Y | Y | Y |
| 31 | Synchronous 1.2 kb/s | Y | | |
| 32 | Synchronous 2.4 kb/s | Y | | |
| 33 | Synchronous 4.8 kb/s | Y | | |
| 34 | Synchronous 9.6 kb/s | Y | | |
| 61 | Alternate speech/data | Y | | |
| 81 | Speech followed by data | Y | | |
| | Multimedia service | | Y | Y |

**Table 1.3.** *UMTS PS bearer services*

| | | UMTS | |
|---|---|---|---|
| Description | GSM R98 | R99/R4 | UMTS R5 |
| Point-to-point (PTP) connectionless network service | Y | Y | Y |
| Point-to-multipoint (PTM) group call | – | Y | Y |
| IP multicast | – | Y | Y |
| Multimedia service | – | – | Y |

weather and advertising. The cell broadcast service can transmit a message up to 93 characters in length. In addition multiple message 'pages' can be sent to create longer messages. In total up to 15 pages can be sent, leading to a total maximum length of 1395 characters.

### 1.5.2 Bearer services

A bearer service is a telecommunication service that provides the user with the ability to transfer data between two points in the network (e.g. mobile to mobile or mobile to PC). The difference between a bearer service and a teleservice is that a bearer service does not define all of the characteristics of the mobile terminal. In general, therefore, the bearer service only defines the data transfer characteristics; all other elements of a service using the bearer service will be defined by other protocols and procedures.

**Table 1.4.** *UMTS CS supplementary services*

| SS | Description | GSM R98 | UMTS R99/R4 | UMTS R5 |
|---|---|---|---|---|
| eMLPP | Enhanced multilevel precedence and pre-emption | Y | Y | Y |
| CD | Call deflection | Y | Y | Y |
| CLIP | Calling line identification presentation | Y | Y | Y |
| CLIR | Calling line identification restriction | Y | Y | Y |
| COLP | Connected line identification presentation | Y | Y | Y |
| COLR | Connected line identification restriction | Y | Y | Y |
| CFU | Call forward unconditional | Y | Y | Y |
| CFB | Call forward on mobile subscriber busy | Y | Y | Y |
| CFNRy | Call forwarding on no reply | Y | Y | Y |
| CFNRc | Call forwarding on MS not reachable | Y | Y | Y |
| CW | Call waiting | Y | Y | Y |
| HOLD | Call hold | Y | Y | Y |
| MPTY | Multiparty service | Y | Y | Y |
| CUG | Closed user group | Y | Y | Y |
| AoCI | Advice of charge (information) | Y | Y | Y |
| AoCC | Advice of charge (charging) | Y | Y | Y |
| UUS | User-to-user signalling | Y | Y | Y |
| BAOC | Barring of all outgoing calls | Y | Y | Y |
| BOIC | Barring of outgoing international calls | Y | Y | Y |
| BOIC-exHC | Barring of outgoing interneational calls except home PLMN country | Y | Y | Y |
| BIC | Barring of incoming calls | Y | Y | Y |
| BIC-Roam | Barring of incoming calls when roaming outside the HPLMN country | Y | Y | Y |
| ECT | Explicit call transfer | Y | Y | Y |
| CCBS | Completion of calls to busy subscribers | Y | Y | Y |
| SPNP | Support of private numbering plan | Y | Y | Y |
| CNAP | Calling name presentation | Y | Y | Y |
| MSP | Multiple subscriber profile | Y | Y | Y |
| MC | Multicall | | Y | Y |

Table 1.2 illustrates the bearer services defined for GSM from R98 and for UMTS from R99.

Table 1.3 illustrates the PS bearer services that are available for GSM and UMTS. There is a range of support for the different types of PS bearer service. Additionally, PS multimedia services are available for the R5 series of specifications.

### 1.5.3 Supplementary services

A supplementary service is a service that modifies or supplements a basic telecommunication service (such as a GSM teleservice). Table 1.4 illustrates the types of supplementary services that are supported by the GSM system. Supplementary services are implemented in software within the GSM network. They are based on supplementary services that were originally defined for the fixed telephone network.

A common example of a supplementary service is the calling line identification (CLI) service. This service is commonly available in GSM mobiles and allows the called party to see who is calling prior to answering. The CLI service also includes a number of other similar services that can be used to define how the CLI service is operated (such as the ability of a user to prevent their identity being presented).

The other supplementary services are defined in Table 1.4.

## 1.6 Summary

In this introductory chapter, we have considered the origins of the WCDMA system, particularly in the context of the previously existing technologies, and we have looked briefly at the system architecture, and services. Let us now move on to look in more detail at the system as a whole, beginning with a quick review of the structure and operation of the radio interface protocol architecture and WCDMA in general.

# 2 WCDMA in a nutshell

In this chapter, we address some of the fundamental concepts behind the design and operation of the WCDMA system. Many of the issues addressed here will be considered in greater detail in later chapters.

We start by reviewing the basic protocol architecture and principles behind its operation. Next, we consider aspects of the design and operation of the physical layer such as the concepts behind the physical channels, the ideas behind soft-handover, and consideration of other physical layer techniques such as power control. Following this, we address the operation of the radio interface from a higher layer perspective by considering the modes and states that are defined for the WCDMA system, and subsequently the methods used to identify and address the UE. We then consider some of the operational aspects of the radio interface, including the definitions of radio links, RBs, radio access bearers and signalling radio bearers (SRBs). Next, we consider the principles of the security architecture. This extends over a range of layers and so it is worth considering it in one section in detail. We conclude with a summary of the operation that a UE will follow in the establishment of some typical services. This section provides some context in which we can consider the remainder of the book.

## 2.1 Protocol architecture

Figure 2.1 illustrates the protocol architecture that exists across the Uu interface between the UE and the network. Following on from Chapter 1, we can see the separation between AS functions and NAS functions.

The figure illustrates the different layers in the protocol architecture and the service access points (SAPs) that define the interfaces between the layers. The architecture is viewed from the perspective of the UE. In the network, the different layers terminate in different entities. The protocol architecture is separated into a control plane on the left used by the control signalling messages and the user plane on the right that is used by the user data messages such as speech and packet data. Shown in Figure 2.1 are the control SAPs between the RRC protocol and the lower layers of the AS. The RRC is

## 2.1 Protocol architecture

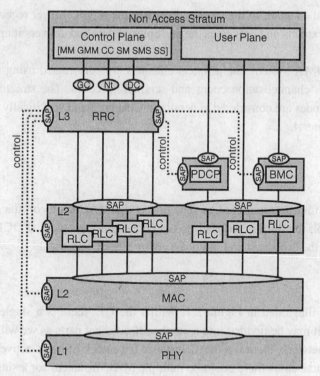

**Figure 2.1** WCDMA radio interface protocol architecture. (PHY denotes the physical layer.)

responsible for configuring and controlling the protocol architecture via these control SAPs.

In this section we consider the different layers in the protocol architecture before moving on to look at the different SAPs. In the lower layers, the SAPs are referred to as channels, and so we will consider the many different types of channels defined within the architecture. We start, however, by examining the protocol layers, commencing at the lowest level with layer 1, better known as the physical layer.

### 2.1.1 Physical layer

The physical layer is responsible for the entire physical layer processing functions across the Uu interface. The physical layer is located in the UE (as shown in Figure 2.1) and on the network side it is mainly located in the Node B (there are a few scenarios, where technically the physical layer can extend to the RNC). The functions of the physical layer are extensive, and a large proportion of this book (Chapters 3–7) is devoted to the physical layer and its operation. These functions include RF processing aspects, chip rate processing, symbol rate processing and transport channel combination.

In the transmit direction, the physical layer receives blocks of data from the higher layers (transport blocks via transport channels from the MAC layer) and multiplexes them onto a physical channel. In the receive direction, the physical layer receives the physical channels, extracts and processes the multiplexed data and delivers it up to the MAC.

Within the WCDMA system, the physical channels are constructed using special codes referred to as channelisation codes and scrambling codes. The structure and operation of these codes are considered generally in Chapter 3 and specifically for the FDD mode in Chapter 4.

### 2.1.2 Layer 2 protocols

Layer 2 comprises four protocol entities. The first is the MAC protocol; the second is the RLC protocol; the third is the packet data convergence protocol (PDCP); and finally the fourth is the broadcast and multicast control (BMC) protocol.

**MAC protocol**

The MAC layer is illustrated in Figure 2.1. Within the UE, there is a single MAC instance (although it may be divided into a number of different parts as we will see in Chapter 8). In the network, there is a MAC instance for each UE that is active within the RNC, and in some cases there could be multiple MAC instances for a single UE (one in a CRNC and one in an SRNC for instance).

The MAC provides some important functions within the overall radio interface architecture. The MAC is responsible for the dynamic resource allocation under the control of the RRC layer. Part of the resource allocation requires the MAC to use relative priorities between services to control the access to the radio interface transmission resources. These functions comprise the mapping between the logical and the transport channels, the transport format selection and priority handling of data flows. The MAC is responsible for UE identification management in order to facilitate transactions such as random access attempts and the use of downlink common channels.

When the RLC is operating in transparent mode, i.e. when the data are passing through the RLC layer without any header information and when ciphering is enabled, it is the function of the MAC actually to perform the ciphering task. The MAC is also responsible for traffic volume measurements across the radio interface. To achieve this, the MAC monitors the buffer levels for the different RLC instances that are delivering data to the MAC.

In general, there are multiple channels entering the MAC (referred to as logical channels) and there are multiple channels leaving the MAC (referred to as transport channels). The number of logical channels coming in and the number of transport channels leaving are not necessarily the same. The MAC can provide a multiplexing function that results in different logical channels being mapped onto the same transport channels.

## RLC protocol

The RLC illustrated in Figure 2.1 is above the MAC. There will be a number of RLC instances; in general, there will be one per service (there are a few exceptions such as speech and signalling where we might expect a number of RLC entities per service). Each RLC entity can be configured differently from other RLC entities. The configuration of the RLC will depend upon the specific QoS that is desired for the service.

The RLC provides a number of different types of transport service: the transparent, the unacknowledged, or the acknowledged mode of data transfer, and we will see in more detail in Chapter 9 what these different modes are, and how they work. Each mode has a different set of services that define the use of that mode to the higher layers. Example services provided by the RLC include segmentation and reassembly, which allows the RLC to segment large PDUs into smaller PDUs. A concatenation service is provided to allow a number of PDUs to be concatenated.

The acknowledged mode data transfer service provides a very reliable mechanism for transferring data between two peer RLC entities. In addition, the acknowledged mode also provides flow control and in-sequence delivery of PDUs.

Error correction is provided by an automatic repeat request (ARQ) system, where PDUs identified as being in error can be requested to be retransmitted. Flow control is the procedure by which the transfer of PDUs across the radio interface can be governed to prevent buffer overload, for instance at the receiving end. Because the ARQ system that is used could result in PDUs arriving out of sequence, the sequence number in the PDU can be used by the RLC to ensure that all PDUs arrive in the correct order.

## Packet data convergence protocol (PDCP)

Figure 2.1 illustrates the location of the PDCP layer, which is defined for use with the PS domain only. At the inputs to the PDCP layer are the PDCP SAPs. The PDCP layer for R99 provides header compression (HC) functions and support for lossless SRNS relocation. HC is defined using standard algorithms of which RFC 2507 is the one defined for R99.

Lossless SRNS relocation is used when the SRNC is being changed, and it is required that no data are lost. Data that are not acknowledged as being correctly received are retransmitted once the new SRNC is active.

## Broadcast and multicast control (BMC) protocol

Figure 2.1 illustrates the protocol layer for the BMC. The BMC provides support for the cell broadcast SMS. The BMC messages are received on a common physical channel. The messages are periodic with a periodicity defined by parameters that are broadcast to the UE in SIB5 and SIB6 messages. The UE is able to select and filter the broadcast messages according to settings defined by the user.

### 2.1.3 Layer 3 – RRC protocol

The RRC layer is the main AS control protocol. The RRC protocol is responsible for the establishment, modification and release of radio connections between the UE and the UTRAN. The radio connections are commonly referred to as the RRC connections. The RRC connection is used to transfer RRC signalling messages. The RRC protocol also provides transportation services for the higher layer NAS protocols that use the RRC connections created by the RRC protocol. In the UE there is a single instance of the RRC protocol. In the UTRAN there are multiple instances of the RRC protocol, one per UE. The RRC entity in the UE receives its configuration information from the RRC entity in the UTRAN.

In addition to establishing an RRC connection that is used by the various sources of signalling, the RRC protocol is also responsible for the creation of user plane connections, referred to as radio access bearers (RABs). The RABs are created to transport user plane information such as speech or packet data across the radio interface from the UE to the core network. More of the details of RABs are considered later in this chapter and in Chapter 11.

The RRC protocol provides radio mobility functions including elements such as the control of soft-handover to same-frequency UMTS cells, hard-handover to other UMTS cells and hard-handover to other RAT cells such as GSM. Cell updates and UTRAN registration area (URA) updates are procedures that are used to allow the UTRAN to track the location of the UE within the UTRAN. Both of these concepts are considered in more detail later in this chapter.

### 2.1.4 NAS protocols

The NAS protocols, which we met briefly in Chapter 1, are the higher layer protocols that exist between the UE and the core network. The NAS protocols are initially based on the GSM and GPRS protocols that existed prior to the definition of UMTS. Over time, however, new NAS protocols are defined to provide new services to new core networks.

The NAS protocols are used to provide features such as user service control, registration, identification, authentication and MM functions.

Figure 2.1 illustrates some examples of the NAS protocols that are defined for the initial release of the UMTS standard. The NAS protocols are based on the GSM upper layer protocols (MM, CC, supplementary services (SS) and SMS protocols) and the GPRS upper layer protocols (GMM and SM protocols).

The NAS protocols that are based on GSM are broadly the same as those defined for GSM. The NAS protocols based on the GPRS upper layer protocols differ in some respects. The details for the NAS protocols are considered in Chapters 13 and 14.

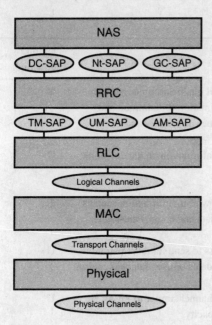

**Figure 2.2** SAPs for UE control plane.

## 2.2 SAPs

In this section, we will consider the interlayer connections. These are the SAPs, which at the lower layers are referred to as channels.

Figure 2.2 illustrates the basic SAPs that exist in the control plane for the radio interface. The user plane has the same SAPs from the input to the RLC down and also SAPs for two additional entities (PDCP and BMC) that are not shown. Coming from the physical layer there are the physical channels, between the physical layer and the MAC there are the transport channels, and between the MAC and the RLC there are the logical channels.

In this section, we will examine some of these channels, highlighting some of their key characteristics, which we will consider when we look at the structure and operation of the radio interface in Chapters 4 and 8. We will start by considering the physical channels.

### 2.2.1 Physical channels

Table 2.1 illustrates the physical channels that are available from the physical layer in both the uplink and the downlink. Most of these channels have a very specific purpose and characteristic. From the table, it can be seen that they are not all present in

**Table 2.1.** *Summary of physical channels*

| Name | Purpose | Uplink/downlink | Mode |
|---|---|---|---|
| Common pilot channel (CPICH) | Used in FDD mode for cell phase and time reference and also for channel estimation for common and dedicated channels. Comes as primary and secondary. | D | FDD |
| Primary common control physical channel (PCCPCH) | Carries system broadcast information on the downlink. | D | FDD TDD |
| Secondary common control physical channel (SCCPCH) | Carries general purpose downlink information such as paging messages, user data and control messages. | D | FDD TDD |
| Physical random access channel (PRACH) | Used on the uplink by UEs to send initial registration messages, cell/URA updates or user data and control information. | U | FDD TDD |
| Physical common packet channel (PCPCH) | Uplink contention based channel similar to PRACH, but with greater data capacity. | U | FDD |
| Dedicated physical data channel (DPDCH) | Carries dedicated traffic and control information between the network and UE. Also requires the dedicated physical control channel (DPCCH) to transport physical control information. | U D | FDD TDD |
| Synchronisation channel (P-SCH and S-SCH) | Primary and secondary synchronisation channels used for initial system acquisition by the UE. | D | FDD TDD |
| Physical downlink shared channel (PDSCH) | Additional downlink channel associated with a DCH. Used on a shared basis by UEs within a specific cell. | D | FDD TDD |
| Paging indication channel (PICH) | Channel used to carry the paging indicators (PIs). The UE needs only to monitor this channel to identify if a paging message is present for that UE. It then needs to read the SCCPCH for the paging message. | D | FDD TDD |
| Acquisition indication channel (AICH) | Downlink physical channel used to carry the acquisition indicators (AIs) that are used as part of the PRACH assignment procedure. | D | FDD |
| Physical uplink shared channel (PUSCH) | Carries the USCH transport channel. | U | TDD |
| Access preamble acquisition indicator channel (AP-AICH) | Carries access preamble indicators (APIs) for the PCPCH. Similar to the AICH but applied to the PCPCH instead of the PRACH. | D | FDD |
| CPCH status indicator channel (CSICH) | Carries CPCH status information. This is used by the UE to identify available capacity on the PCPCH. | D | FDD |
| Collision detection/channel assignment indicator channel (CD/CA-ICH) | Carries CD and CA indicators. Is used as part of the PCPCH resource allocation procedures. | D | FDD |

both uplink and downlink; and equally, the TDD system and the FDD system do not necessarily support all of the specific channels.

In general the physical channels are constructed through the use of the channelisation codes and scrambling codes that are defined later in this section and in greater detail in Chapter 3. The channels tend to be active simultaneously and rely on the codes to separate the transmissions. In Chapter 4 we will explore the structure and use of most of these physical channels.

### 2.2.2 Transport channels

A transport channel is the name given to an SAP that is at the output of the MAC and at the input of the physical layer. The transport channels are a mechanism by which the MAC can send messages via the physical layer to the MAC in the receiving system. The choice of which of the transport channels to use depends on the specific requirements of the message that needs to be transmitted.

In general, the transport channels map onto specific physical channels, and each one of these transport channels will have specific characteristics in terms of the direction, i.e. whether it is an uplink or downlink only or both an uplink and a downlink channel, and also in terms of data capacity i.e. what is the variation in data rate that the channel can support.

Table 2.2 presents the transport channels that are available in the FDD and TDD modes of operation. The top seven transport channels are referred to as common transport channels in that they are intended for use by a number of users in a cell. In a similar way to the physical layer, the common transport channels are not available in both uplink and downlink; and equally, not all the channels are available for TDD mode and FDD mode of operation.

Table 2.2 also illustrates the transport channels that are available for dedicated use. 'Dedicated' means that there is essentially a point-to-point link between the UE and the UTRAN, unlike the common channels, where there was a point-to-multipoint link between the UE and the UTRAN.

We will see later that the UE has different transport channels assigned to it at different points in time. In addition, the UE may also have multiple transport channels assigned, depending upon the nature of the connection for the UE.

### 2.2.3 Logical channels

Table 2.3 indicates the logical channels that are available across the radio interface between the UE and the UTRAN. The logical channels define the type of information that flows across the radio interface, i.e. is it user traffic, or control, and is it dedicated or common. The logical channels are between the RLC and the MAC. The use of the different logical channels is defined by the radio interface protocols.

**Table 2.2.** *Common and dedicated transport channels*

| Name | Purpose | Uplink/downlink | Mode |
|---|---|---|---|
| Random access channel (RACH) | Initial access or transfer of small amounts of data | U | FDD TDD |
| Common packet channel (CPCH) | Contention channel for bursty data | U | FDD |
| Forward access channel (FACH) | Transfer of user data and control signalling messages | D | FDD TDD |
| Downlink shared channel (DSCH) | Shared channel carrying dedicated control or user data | D | FDD TDD |
| Uplink shared channel (USCH) | Shared channel carrying dedicated control or user data | U | TDD |
| Broadcast channel (BCH) | Broadcast channel to all UEs in a cell | D | FDD TDD |
| Paging channel (PCH) | Broadcast of paging and notification messages while allowing UE to use sleep mode | D | FDD TDD |
| Dedicated channel (DCH) | Dedicated channel for transfer of bi-directional traffic and control | U D | FDD TDD |

**Table 2.3.** *Logical channels*

| Name | Purpose | Uplink/downlink | Mode |
|---|---|---|---|
| Dedicated traffic channel (DTCH) | Used to carry user plane data such as speech or packet data | U D | FDD TDD |
| Dedicated control channel (DCCH) | Used to carry control signalling messages | U D | FDD TDD |
| Shared control channel (SHCCH) | Used to carry control messages | U D | TDD |
| Common control channel (CCCH) | Used to carry control messages, but on common channels | U D | FDD TDD |
| Common traffic channel (CTCH) | Used to carry user traffic such as the cell broadcast SMS | D | FDD TDD |
| Paging control channel (PCCH) | Used to carry paging messages to the UE | D | FDD TDD |
| Broadcast control channel (BCCH) | Used to carry the broadcast messages to the UE | D | FDD TDD |

Figure 2.3 summarises the channel mapping for the logical to transport to physical channels. The diagram illustrates that there is a wide range of mapping options.

The multimapping that is present with channels such as the common packet channel (CPCH) shows that the transport and physical channels can be used to carry a number of different logical channels. It should be noted, however, that in some cases one transport

## 2.3 Principles of the physical layer

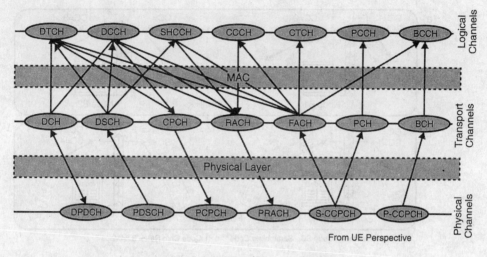

**Figure 2.3** Physical channel–transport channel–logical channel mapping.

channel can only carry one logical channel, whereas other transport channels such as the dedicated transport channel (DCH) can be used in parallel to carry a number of simultaneous logical channels.

One thing to note is that not all of the physical channels that are presented in Table 2.1 are present in Figure 2.3. The reason for this is that a number of physical channels do not carry higher layer data and, as a consequence, there is no transport channel or logical channel associated with them.

What is not obvious from the diagram is what mapping options might exist at a specific point in time. When we examine the RRC protocol (Chapter 11) we will review some of the issues that define what channel mappings exist at a specific point in time.

## 2.3 Principles of the physical layer

The physical layer provides the WCDMA FDD mode with some of its key differentiating features. Before considering the details of some of the higher layer functions of the WCDMA radio interface, it is worth reviewing some of these features, starting with the use of the channelisation and scrambling codes to provide the spreading and scrambling functions. Soft-handover is another differentiating feature of the WCDMA radio interface and so we will consider the basic principles associated with soft-handover. Compressed mode is required to allow a UE the ability to make measurements on different frequencies whilst using a dedicated physical channel. We consider some of the basic features of compressed mode. Finally in this section, we consider the subject

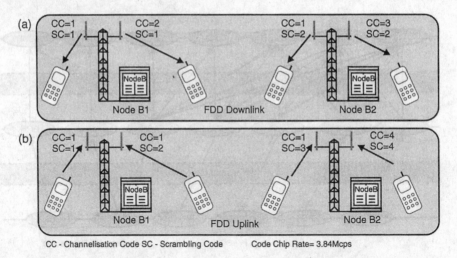

**Figure 2.4** Use of codes in FDD mode: (a) uplink and (b) downlink.

of power control, which is another important characteristic of the WCDMA radio interface.

### 2.3.1 Spreading and scrambling

Figure 2.4 illustrates the basic use of the different types of CDMA codes in the uplink and downlink of the UMTS FDD mode. In this example we are considering the specific case of the use of the dedicated physical data channel (DPDCH). In reality there are other channels present in both the uplink and the downlink. This example is intended to indicate the basic principles used to separate users and cells in the FDD mode in both the uplink and the downlink.

The upper diagram (Figure 2.4(a)) illustrates the situation for the FDD mode downlink. In this example, within a cell the users are allocated a specific channelisation code to provide the data rate they require. The channelisation code on the downlink is used to control the data rate for the user and also to separate the users on the downlink, with each user within the cell being allocated a different channelisation code.

The same channelisation code could be allocated between the cells on the downlink, and so a scrambling code is assigned to a cell to separate its transmissions from transmissions from adjacent cells. Each adjacent cell is defined to use a different scrambling code to help reduce the effects of interference between the cells.

As shown in Figure 2.4(b), in the uplink of the FDD mode the UE selects the channelisation code based on the data rate requirement for that UE at that point in time (this can change across the duration of a specific connection). The UE is assigned a unique scrambling code by the RNC.

## 2.3 Principles of the physical layer

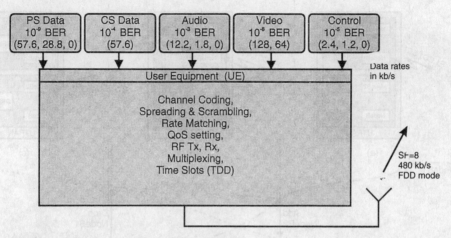

**Figure 2.5** Transport channel combining functions.

As we can see in Figure 2.4(b), users in the same cell can use the same channelisation code, but will use different scrambling codes within a cell and also between cells.

### 2.3.2 Transport channel combining

Figure 2.5 illustrates some of the other functions of the physical layer of the UMTS system. The physical layer receives a number of data streams from a range of sources. Each one of these data streams has different characteristics in terms of the data rate (and variation in data rate) as well as requirements for reliability (expressed in terms of BER).

In the example shown, the physical layer is responsible for multiplexing all of these data streams onto a single physical channel. In this case these are the functions occurring in the UE and so a spreading factor of 8 corresponding to a channel data rate of 480 kb/s is used (assuming FDD mode of operation).

In addition to the multiplexing and spreading functions, the physical layer provides other functions including channel coding, RF transmission/reception and rate matching.

The objective of the transport channel combiner is to bring together all the different data streams, changing them where appropriate, and to multiplex them onto the physical channel (in this example at 480 kb/s).

Changing the data rate of the individual channels is the main function of the rate matcher. The rate matcher is instructed as to which data rates to use by the RNC. The change in data rate is achieved either through puncturing (too many bits) or repetition (too few bits). If puncturing is used, the channel decoder in the receiver knows which bits were punctured (or deleted) and attempts to correct for them (treating them as errors).

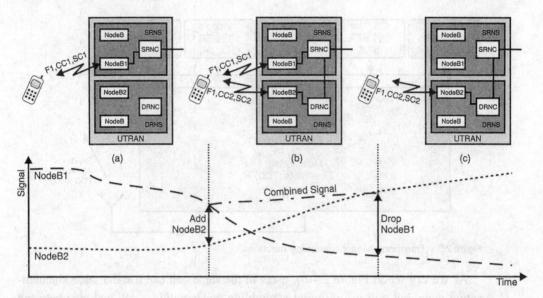

**Figure 2.6** Soft-handover example scenario: (a) before soft-handover; (b) during soft-handover; (c) after soft-handover.

### 2.3.3 Soft-handover

One of the slightly unusual aspects of the WCDMA FDD mode is referred to as soft-handover. Soft-handover is where the UE transmits to/receives from multiple cells.

**Soft-handover on downlink**

Soft-handover is illustrated in Figure 2.6. In Figure 2.6(a), the UE is connected to Node B1 and is receiving data from the SRNC via Node B1 using frequency F1, channelisation code CC1 and scrambling code SC1. The UE is measuring the adjacent cells such as Node B2 as shown, and reporting the measurements back to the SRNC. In this instance the signal level/quality from Node B2 is improving (as seen in Figure 2.6(b)), until it reaches a certain threshold below that from Node B1, at which point the SRNC decides to put the UE into soft-handover (the official term is to add the radio link from Node B2 to the active set).

To do this, the SRNC will request the drift radio network controller (DRNC) to activate a radio link from Node B2 towards the UE. Next, the SRNC notifies the UE, via the control signalling channel, that it is being put into soft-handover, and indicates that channelisation code CC2 and scrambling code SC2 are being used for the transmission from Node B2. The UE, using the rake receiver, is now able to receive the data from both Node B1 and Node B2. The rake receiver combines the energy from these two signals to produce a third signal whose composite energy is greater than either of the original signals.

## 2.3 Principles of the physical layer

The UE remains in soft-handover until the SRNC decides to drop one of the radio links. In the case shown in Figure 2.6(c), the radio link from Node B1 is dropped and so the connection to the UE is via the SRNC, the DRNC and Node B2. This situation remains until it is decided to perform a procedure known as SRNS relocation, at which point the DRNC becomes the SRNC for that specific UE, and the SRNC is removed from the signalling path.

### Soft-handover on uplink

On the uplink in the soft-handover example considered previously, the Node B detects the data from the UE and transports it to the SRNC, but attaches a quality estimate that is based on the estimated BER for the uplink data. At the SRNC, when there are multiple uplink paths from all of the Node Bs present in the soft-handover process, the SRNC selects the radio link from the Node B that has the lowest BER. In this case, the SRNC for the uplink is only selecting the best signal.

### Softer-handover on uplink

The alternative to soft-handover on the uplink is what is known as softer-handover. In this scenario, the UE is connected to more than one cell at the Node B.

The UE is transmitting to these cells. Because it is at the same Node B, the Node B can combine the data at the chip rate like the UE does on the downlink. In this case the combination is a soft combination, rather than the selection that occurs in the previous example.

The radio links (this is the name given to the radio connections from each cell to the UE) that the UE is receiving on the downlink comprise what is known as a radio link set (RLS). One of the features of such an RLS is that the links in the set coming from the same Node B have the same power control commands, which means that the UE can combine these power control bits to improve upon the performance of the power control loop (this is considered a little later when we consider power control).

### 2.3.4 Compressed mode

The WCDMA FDD mode is a full duplex system. This means that in normal operation it is transmitting and receiving continuously. Unlike GSM, there is no idle frame in which it can make measurements on adjacent cells and on other systems.

If a UE needs to perform a handover to a UMTS cell on a different frequency, or to a different system on a different frequency (such as GSM), it will need to make measurements on the target cells. To facilitate this, a technique referred to as compressed mode has been developed. Compressed mode inserts an artificial gap in transmission in either the uplink or the downlink, or both. During this gap the UE is able to switch to the alternative frequency, make a measurement and switch back in time to continue with the transmission and reception of the signals from the serving cell.

**Need for compressed mode**

It is a UE capability as to whether the UE needs to support a compressed mode for the different measurements. If the UE is designed with a second measuring receiver it may not need to support the compressed mode in the downlink. If the noise generated by the UE transmitter is low enough and the transceiver architecture is such as to allow the UE to make measurements on the downlink whilst transmitting on the uplink, it may not need to support the uplink compressed mode.

**Compressed mode methods**

There are three basic mechanisms that are used to achieve the compressed mode pattern. The methods are referred to as puncturing, spreading factor reduction and higher layer signalling. All of the methods apply to the Node B, but only spreading factor reduction and higher layer signalling apply to the UE.

With compressed mode via puncturing, a technique known as puncturing, which is defined in greater detail in Chapter 6, is used. With puncturing, some of the encoded data to be transmitted is deleted. By deleting the data, a transmission gap is created that will allow the UE to make measurements during the gap, only needing to return prior to the start of the transmission of the data at the end of the gap.

With compressed mode via spreading factor reduction, the transmitted data rate on the uplink and downlink is doubled by changing the length of the channelisation code (see Chapter 3 for details of the effects of code lengths on data rates). In doing this, the UE or the Node B only needs to transmit for half the time, creating a transmission gap that can be used to make measurements.

The final technique uses what is known as higher order signalling. With higher order signalling, a transmission gap can be created by temporarily restricting the transmission rate that is allowed. This reduction in the data to be transmitted can be used to create a transmission gap in which to make the measurements.

**Slot structures**

Figure 2.7 illustrates the basic concept of compressed mode. In order to keep the overall QoS the same during the compressed mode period, the transmission power may need to be increased to compensate for the increase in code rate or the lowering in spreading factor.

**Compressed mode structure**

The size and location of the compressed mode gaps are well defined and illustrated in Figure 2.7, with the parameters that define the gaps illustrated by Table 2.4.

## 2.3 Principles of the physical layer

**Table 2.4.** *Parameters that define compressed mode gaps*

TGSN – Transmission gap starting slot number
TGL1 – Transmission gap length 1
TGL2 – Transmission gap length 2
TGD – Transmission gap start distance
TGPL1 – Transmission gap pattern length 1
TGPL2 – Transmission gap pattern length 2
TGPRC – Transmission gap repetition count
TGCFN – Transmission gap connection frame number

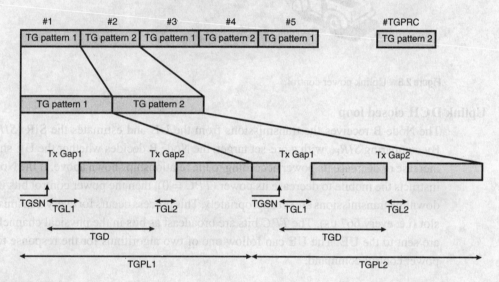

**Figure 2.7** Compressed mode pattern structure.

### 2.3.5 Power control

**Uplink power control**

Figure 2.8 illustrates the basic principles associated with uplink power control. The closed loop power control illustrated here has two parts: the inner power control loop between the UE and the Node B, and the outer power control loop between the RNC and the Node B. For the inner loop, a UE transmits to the Node B. The Node B estimates the received signal to interference ratio (SIR) and compares this with a target level. Based on the target level, the Node B will instruct the UE to either increase or decrease its transmit power level. For the outer loop, the RNC receives measurements from the Node B for a specific UE. Based on these measurements, the RNC can decide how to set the target value used by the Node B.

**Figure 2.8** Uplink power control.

### Uplink DCH closed loop

The Node B receives the transmissions from the UE and estimates the SIR ($SIR_{est}$). By comparing $SIR_{est}$ with a pre-set target, the Node B decides whether the UE should increase or decrease its power according to the relationship shown above. If the Node B instructs the mobile to decrease its power ($TPC = 0$), then the power control bits in the downlink transmissions are set appropriately. This process occurs for every transmission slot (i.e. every 667 µs). The $TPC$ bits are broadcast as bits in the physical channel that are sent to the UE. The UE can follow one of two algorithms for the response to the power control command.

### Uplink algorithm 1

When it is not in soft-handover, the UE should only receive a single power control command per slot, in which case the power control commands should be interpreted as shown in Figure 2.8.

If the UE is in soft-handover, there are two additional possibilities. The first is that the UE is in softer-handover, in which case the power control commands are coming from the same Node B (this is referred to as the same RLS) in which case the power control commands should be the same and can be combined prior to being acted upon.

If the UE is in soft-handover and needs to combine power control commands that come from different RLSs, then the following procedure is used. The UE processes the individual power control commands from the different cells to produce a composite power control command. The function that produces the composite power control command should bias the resultant combined power control towards a reduction in power level, i.e. there should be a greater probability that the power control level will decrease given a random set of power control commands from all the elements of the RLSs.

In addition, the composite power control command is such that if all the elements are thought to reliably be 1 (increase power) then the output composite will be 1, if any of the power control commands is thought to reliably be 0 (decrease power) then the composite power control command should also be 0.

## Uplink algorithm 2

With algorithm 2, the power control rate is reduced and in addition, there is a possibility to have no change in power level ($TPC\_cmd = 0$). Like in algorithm 1, there are three possible scenarios, which we will consider in turn. The first is no soft-handover, the second is soft-handover with radio links in the same RLS and the third is soft-handover with radio links in different RLSs.

If the UE is not in soft-handover the UE will receive the power control commands that are received across five consecutive slots and process them according to the following procedure. For the first four slots, the $TPC\_cmd$ is set to 0 (no change in power control). For the fifth slot the $TPC\_cmd$ is set to 1 (increase power) if all five received power control commands indicate an increase in power, the $TPC\_cmd$ is set to $-1$ if all five received power control commands indicate a decrease in power, else it is set to zero (no change in power) for any other combinations. This algorithm will result in an increase or decrease in power control only if all five slots indicate an increase or decrease, else the power level will remain the same.

If the UE is in a soft-handover with $N$ cells and the $N$ cells are in different RLSs, then the UE will receive $N$ sets of power control commands for the five slots in question. As before for the no soft-handover case, for the first four slots the power control command is set to zero indicating no change in power level. For the fifth slot the power control command is a composite power control command derived from the $N$ power control commands obtained from the $N$ cells in the active set. The composite is derived first by creating a temporary power control command for each of the $N$ radio links as follows. If all five power control commands are 1 $TPC\_temp = 1$ (increase power), if all five power control commands are 0 $TPC\_temp = -1$ (decrease power), and for all other combinations $TPC\_temp = 0$ (the power level remains the same). Next, the temporary power control commands are combined by taking the average of the $N$ values. If the average is greater than 0.5 the $TPC\_cmd$ is set to 1 (increase power), if the average is less than $-0.5$ $TPC\_cmd$ is set to $-1$ (decrease power) and for all other values $TPC\_cmd = 0$ (no power change).

If two or more of the radio links are from the same RLS, then the UE should perform a soft combination of the power control commands for each slot for all the members of that RLS before combining these with the power control commands from the other RLSs as outlined above. The result of this combination is that a majority of the elements in the RLSs have to indicate either an increase in power or a decrease in power before the change is acted upon, else the power will remain the same.

**Uplink outer loop power control**

The outer loop power control is between the Node B and the RNC. The objective of the outer loop power control is to ensure that the QoS for a specific UE is maintained within defined limits.

To achieve this, the Node B periodically sends quality information to the RNC. This quality information is based on the received BER and block error rate (BLER) for that specific UE. Using this information, the RNC can decide either to increase the link quality by raising the target value, or alternatively to reduce the link quality by lowering the target value.

**Downlink power control**

The downlink power control is similar to the uplink power control except the roles of the UE and the Node B are switched. In the downlink power control, the UE is responsible for making measurements and telling the Node B to increase or decrease power levels. This is the inner loop. The outer loop consists of the measurements being sent from the UE to the RNC, and the RNC setting the target based on these measurements.

## 2.4 Principles of the upper layers

In this section we consider some of the higher layer features and design concepts for the WCDMA radio interface. As in the previous section, we want to present a broad-brush overview to the radio interface, so that we can then proceed in the following chapters to consider some of the detailed design and operational aspects.

We start by considering the UTRAN RRC modes. These RRC modes are used to define the type of connection that a specific UE will have with the UTRAN. An investigation into these modes is very useful to give an insight into what a UE is doing at a specific point in time, and why it is doing what it is doing.

Next, we consider the definitions of the different geographical areas that are defined in the WCDMA network to introduce some of the MM concepts that are defined in the system. These concepts include when and where a UE will be paged and when and where a UE will need to perform some kind of location update. One of the key differences we will see between GSM and UMTS is that the UTRAN takes on more of the MM role than that defined for GSM.

In the final part of this section, we examine some of the addressing mechanisms that are defined to address a specific UE within the network. Once again, we will see some slight differences between the approaches employed in GSM and those defined for use in UMTS.

## 2.4 Principles of the upper layers

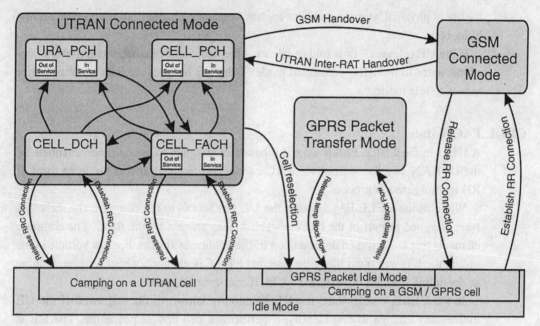

**Figure 2.9** UTRAN RRC modes and states.

### 2.4.1 UTRAN RRC modes and states

Figure 2.9 illustrates the RRC modes and states that the UE can be in, including those that are relevant to dual mode UEs (UMTS and GSM/GPRS). The diagram illustrates all the states as well as the transitions that are possible between the states. In the UTRAN idle state, the only way that a UE can move to the connected state is to request an RRC connection. Once in the connected state, there are a number of possible states the UE can be in.

In this section, we consider some of the key attributes of the UTRAN RRC modes and the transitions between these different states within the RRC connected mode.

**CELL_DCH state**

A UE is in the CELL_DCH state when it is assigned a dedicated physical channel (DPCH) either in response to the RRC connection request, or through some UTRAN controlled bearer reconfiguration procedure.

In the CELL_DCH state, the UE location is known by the UTRAN to the cell level. In the CN, the UE location is known to the serving RNC.

Whilst in the CELL_DCH state the UE has access not only to the allocated dedicated resource but also to any shared resources that were allocated. The UE can dynamically utilise these resources within the confines of the transport format combination (TFC) that was assigned by the RNC. The UE is identified in the CELL_DCH state through the use of physical layer mechanisms such as frequency, channelisation code and scrambling code.

If the UE releases – or is told to release – the dedicated channel, it can move to any of the states in the RRC connected mode. If the UE releases the RRC connection, it moves to idle mode.

### CELL_FACH state

A UE is in the CELL_FACH state if it has been assigned to use common channels by the UTRAN as either part of the RRC connection request process, or some form of RB reconfiguration process.

While in the CELL_FACH state, the UE has access to the common channels that were assigned as part of the bearer establishment process by the RNC. The common channels that are assigned depend upon traffic volume estimates. For low volume in the uplink the UE will most likely be assigned a RACH channel, whilst for high volume in the uplink it might be assigned a CPCH transport channel.

In the CELL_FACH state, the UE location is known to the cell level. If the UE notices any change in cell identity, it performs a cell update procedure. The UE is addressed by the cell radio network temporary identifier (c-RNTI) (see Section 2.4.3) from the RNC that is controlling the cell that the UE is present in.

The UE can move to any of the other connected states as directed by the UTRAN. Such moves are a consequence of changes in traffic loading in either the UE or the cell.

### CELL_PCH state

A UE is in the CELL_PCH state if it was previously in the CELL_FACH or Cell_DCH state and the flow of traffic stopped. In the CELL_PCH state, the UE has no uplink resources that it can use. The UE monitors the paging channel using the assigned discontinuous reception (DRX) parameters. If the UE notices that the cell has changed, it will initiate a cell update procedure after moving back to the CELL_FACH state. If the UE is paged by the UTRAN, the UE must move back to the CELL_FACH state to respond with an appropriate uplink access.

The purpose of the CELL_PCH state is to allow the UE to use DRX to conserve battery power, but still allow the network rapid access to the UE by knowing that only one cell needs to be paged.

### URA_PCH state

The URA_PCH state is very similar to the CELL_PCH state except that the location of the UE is now only known to the URA level. The URA is a collection of cells similar to a GSM location area. The UE monitors the URA and performs URA updates whenever it notices that the URA has changed. The UE has no uplink access in the URA_PCH state, and so it has to move to the CELL_FACH state to perform uplink accesses.

## 2.4 Principles of the upper layers

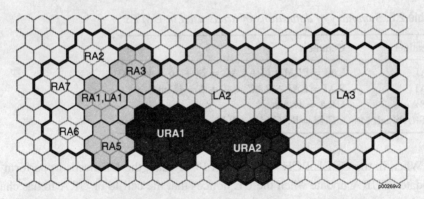

**Figure 2.10** Relationship between LAs, RAs and URAs.

The URA_PCH state is used by UEs that are moving too quickly (too many cell updates if it was in the CELL_PCH state), but which need to use DRX to conserve battery power.

### 2.4.2 Areas

Figure 2.10 illustrates the relationships that can exist between location areas (LAs), routing areas (RAs) and URAs. The GSM and GPRS systems define groups of cells as, respectively, an LA or an RA. The purpose of these areas is to allow the core network to track the movement of the UE through the coverage area to facilitate the paging of the mobile when an active radio connection is not available. In UMTS, the same concepts of LA and RA are used.

The RA and LA are used in particular when the UE is in the idle mode and the UE does not have an active RRC connection. As the UE moves through the cells, it will identify the LA identifier (LAI) and the RA identifier (RAI). Whenever the LAI or the RAI changes, the UE needs to perform either an LA update (LAU) or an RA update (RAU).

Within the UTRAN, similar MM procedures have been defined. In this case the UTRAN tracks the movement of the UE to either a cell or a URA. The URA is similar in concept to the LA or RA. It is a collection of cells and is hierarchical; a particular cell can be in up to eight different URAs, each of a different size (i.e. a different number of cells).

The objective of having URAs of different sizes is to match the URA size to the velocity of the UE. Fast moving UEs make frequent URA updates and should be in a large URA. Slow moving UEs do not make as many updates and can therefore be in a smaller URA. There is a trade-off in the URA size between the amount of paging traffic and the amount of URA update traffic. The optimum URA size depends upon the velocity of the UE.

**Table 2.5.** *Summary of definitions of areas used in UMTS*

| Identifier | Comment |
|---|---|
| LAI | LAI = PLMN Id + LAC = 24 + 16 = 40 bits |
| RAI | RAI = LAI + RAC = 40 + 8 = 48 bits |
| URA | URA = 16 bits |
| UTRAN cell Id | UC Id = RNC Id + cell Id = 12 + 16 = 28 bits |

When in the CELL_PCH or the URA_PCH state, the UE needs to perform a cell update or URA update when the UE notices that the cell Id or URA Id has changed. In either case, the UE moves to the CELL_FACH state and performs the update.

Table 2.5 summarises the definitions for the LA, RA, cell Id and URA Id.

### 2.4.3 Addressing

In this section, we consider the mechanisms defined to address a specific UE within the UMTS network. These addressing mechanisms utilise addressing mechanisms defined for GSM and GPRS, and also introduce some new addresses that are used within the UMTS network.

We start with the CN addresses before moving to the UTRAN addresses. The CN addresses that we consider are the international mobile subscriber identity (IMSI), the temporary mobile subscriber identity (TMSI) and the packet TMSI (P-TMSI). The UTRAN addresses are the UTRAN radio network temporary identifier (u-RNTI), the serving radio network temporary identifier (s-RNTI) and the c-RNTI.

**IMSI**

The IMSI identifies the subscriber (the universal subscriber identity mode (USIM) in this case) within the network. It is the permanent address that is used to reference information stored in the HLR. The IMSI is defined as a 15-digit (or less) number and comprises an MNC + MCC + mobile subscriber identification number (MSIN). The mobile network code (MNC) and the mobile country code (MCC) define the public land mobile network (PLMN) as mentioned previously, and the MSIN defines the identity of the subscriber within the network. The mobile networks use the IMSI to identify a subscriber and check their location.

**TMSI**

The TMSI uses the same concept as defined in GSM. It defines the identity of the subscriber and, together with the LAI, uniquely identifies the subscriber. The TMSI is used to mask the identity of the subscriber to reduce the possibility of fraudulent activity by hiding the subscriber's true identity. The TMSI is defined as a 32-bit number and is allocated by the serving network.

### The P-TMSI

The P-TMSI uses the same concept as the TMSI except towards the PS domain. It is the same as the P-TMSI in GPRS. The P-TMSI is a 32-bit number and is used as a temporary identity within the PS domain.

### The s-RNTI

The s-RNTI is the RNTI assigned by the serving RNC. It uniquely identifies the UE within the SRNS. The s-RNTI identifies the UE to the SRNC and is used by the UE, SRNC or DRNC. It is 20 bits in length. The s-RNTI is normally used with the RNC Id of the SRNC that allocated it. In this case, it becomes the u-RNTI.

### The u-RNTI

The u-RNTI is assigned by the SRNC and uniquely identifies the UE within the UTRAN. It is intended to be used when a UE cannot be uniquely identified in the UTRAN through other mechanisms. Examples of these occasions are the first access after a UE has gone through a cell change (cell update or URA update) or for UTRAN originated paging. The u-RNTI is a 32-bit number and comprises the RNC Id and the s-RNTI.

### The c-RNTI

The c-RNTI is the identifier allocated by the CRNC when a UE accesses a new cell. It is only valid within the cell to which it was allocated. The c-RNTI is 16 bits in length, and is used by the UE as the main identity whilst the UE is in a cell and while it is not using DPCHs. It is only if the UE leaves the cell that it needs to use a UTRAN unique identifier such as the u-RNTI.

### UE ID dedicated channels

When the UE is operating using DPCHs (CELL_DCH state) it is addressed explicitly using the physical layer parameters such as the frequency channelisation code, the scrambling code and the timeslot (TDD only). In these situations (such as in a speech call), the UE is addressed purely through the physical layer. The u-RNTI and the c-RNTI are not used.

## 2.5 Radio and data connections

In this section, we review the various types of connections that the UE can have with the network. We see that the connections that the UE has can be of different types and also at different layers within the protocol architecture. We will introduce the concepts of an RRC connection, RLs, RBs, RABs, SRBs and PDP. These concepts are used throughout the book to define the type of connection that the UE has to the network.

## 2.5.1 RRC connection

We have already described the basic concepts behind the RRC connection. An RRC connection exists when a UE has gone through the connection establishment procedure and has been allocated resources in the UTRAN and a temporary UTRAN identity (u-RNTI).

As part of the RRC connection, the UE is also allocated some SRBs that are used by the UE to pass the signalling control messages between the UE and the UTRAN and also between the UE and the CN.

If the UE proceeds to establish a service such as a speech service, an RAB is created for that specific service. The RAB comprises one or more RBs and is used to carry the user plane information between the UE and the core network.

If the UE requests a packet data connection via the PS domain, then the RAB that is assigned for the packet data connection will be extended from the SGSN to the GGSN, at which point it becomes a PDP context.

In all of the above radio and data connections, there exists a lower layer connection that links the UE to the UTRAN access points in the Node B. This lower layer link is defined as a physical layer connection and is referred to as an RL.

We can proceed by considering the RLs and some of the characteristics of the RLs.

## 2.5.2 RLs

The RL represents the physical connection between the UE and some access point in the UTRAN. Figure 2.11 illustrates an example of the use of RLs. In this particular example (FDD mode), there are two radio links to two UTRAN access points. Each radio link is defined by its RF frequency channelisation code and scrambling code. The UTRAN access points are the Nodes B and the RLs are terminated in separate Nodes B which in this example are connected to different RNCs (SRNC and DRNC).

In this example, because the RLs are from different Nodes B, they are not in the same RLS. If the UE were in softer-handover as defined earlier, then the RLs could be in the same RLS.

Figure 2.12 illustrates two RLs that are going to two users at the same time. The RLs are part of the same RF carrier, but are separated by the different codes. In this example we can see that the two users have multiple services active within the RLs.

## 2.5.3 RBs

Above the RLs we have a number of higher layer connections between the UE and the network. The next major connection point is the RB. The RB is a layer 2 connection between the UE and the RNC. It is used for both control signalling and user data.

## 2.5 Radio and data connections

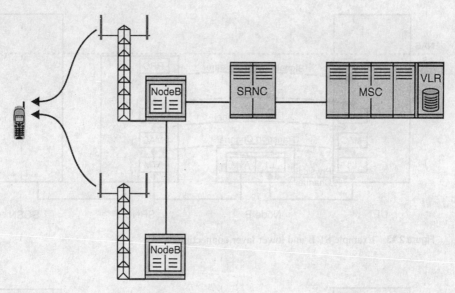

**Figure 2.11** RLs to a UE in soft-handover.

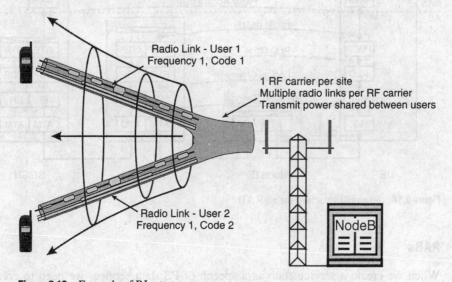

**Figure 2.12** Example of RLs to two users.

In the example shown in Figure 2.13 we see an RB used for signalling, called an SRB. The SRB is located at the top of layer 2 and is used by the RRC protocol for the transfer of control messages. As we can see in the figure, the SRB comprises an RLC connection, a logical channel connection, a transport channel connection and a physical channel connection.

We will see later that when an RRC connection is created four SRBs may be created at the same time.

## 50 WCDMA in a nutshell

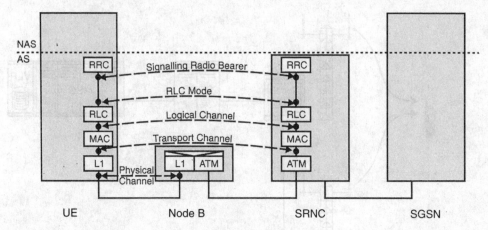

**Figure 2.13** Example SRB and lower layer connections.

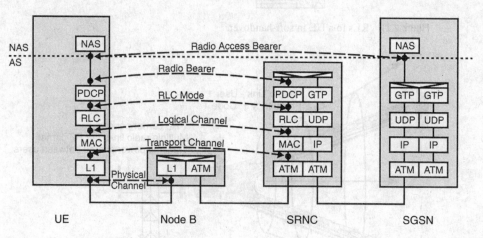

**Figure 2.14** Example structure for an RAB.

### 2.5.4 RABs

When we create a service such as a speech or PS data service, we need to create a connection in the user plane to transport the data across the radio interface to the CN. To provide this user plane data transfer, we have a higher layer entity referred to as an RAB.

An example RAB is illustrated in Figure 2.14. In this example, the RAB is connected to the PS domain. The RAB comprises an RB and a connection from the SRNC to the SGSN that is referred to as an Iu bearer.

The RAB is created on request from the SGSN with a specific QoS objective. For a given UE there can be multiple RABs, one per NAS service, assuming that multiplexing in the AS is not performed.

### 2.5.5 PDP contexts

For the PS domain, the next hierarchical step up the data connectivity ladder is a PDP context, which is a user plane data connection to the PS domain. The PDP context comprises an RAB and a CN bearer and extends between the UE and the GGSN. The PDP context is created with a defined QoS that is requested by the UE and granted by the network.

## 2.6 Security issues

### 2.6.1 Components of security architecture

The security architecture within UMTS is broken into four main components: user identity confidentiality; authentication; integrity protection; and ciphering. In this section, we start by reviewing the basic principles behind each of these components.

**User identity confidentiality**

GSM and GPRS use the concept of a temporary identity to mask the true identity of a subscriber. The temporary identity is referred to as a TMSI or a P-TMSI. UMTS uses exactly the same concepts for user identity confidentiality. The TMSI applies to the CS domain and the P-TMSI to the PS domain.

The principle behind the TMSI and P-TMSI is that it is an identifier assigned by the network (usually via a ciphered connection) and which does not disclose the identity of a subscriber. A TMSI or P-TMSI only has significance within the MSC or SGSN serving that particular subscriber. The network changes the TMSI and P-TMSI frequently. In Chapter 13 we review the basic structure and mechanisms that are used to assign and change the temporary identities.

**Authentication**

Authentication is a procedure that is performed between the UE and the network to validate a specific entity. Within UMTS, this entity validation is bi-directional. The network authenticates the UE and the UE authenticates the network.

The authentication procedure is based on the knowledge of a secret key. Only the network and the USIM within the UE know the secret key. In Section 2.6.2 we review the basic principles of authentication, and the algorithms and the parameters that are input into the authentication procedure.

**Integrity protection**

Integrity protection is a new service to UMTS provided by the AS. Integrity protection is applied to a number of message exchanges between the UE and the UTRAN. The intention of integrity protection is to provide a mechanism that can be used to detect whether messages have been corrupted.

The integrity protection scheme is based on the knowledge of a secret key that is only known in the UE and the home network. In Section 2.6.4 we consider the details of the operation of the integrity protection scheme.

**Ciphering**

Ciphering, or encryption as it is sometimes called, is the procedure that is used to protect the data across the RL. With ciphering, the data are manipulated at the transmitter using special procedures that utilise the knowledge of a secret key that is only available at the transmitter and the intended recipient (the UE and the RNC in this case). Without knowledge of the secret key, it is not computationally feasible to perform the reverse procedure. The details of the ciphering procedure are considered in more detail in Section 2.6.5.

### 2.6.2 Authentication

In this section, we consider the basic principles of the authentication procedure. In Chapter 13 we consider the specific message exchanges and the context of authentication. Here, we focus on the methodology, parameters and functions.

**Basic authentication principles**

Figure 2.15 illustrates the basic authentication mechanism applied in UMTS. The scheme is similar to that employed in GSM. In UMTS a collection of security information is created in the home network and sent on request to the serving network. The security information is referred to as an authentication vector (AV). The AV is an authentication quintuplet, instead of an authentication triplet as in GSM. In the next section we review the contents of the AV in more detail.

The home network, on request, supplies a set of AVs to be used by the serving network. Each of these AVs is used for a single authentication and key agreement step before being discarded, i.e. they are only used once. Next, the VLR/SGSN selects one of the AVs and sends two of the elements of the selected AV (a random number (RAND) and authentication token (AUTN)) to the UE.

The UE checks that AUTN is valid and a counter within the AUTN is in the correct range (authenticating the CN). The UE then computes the response (RES) using an algorithm within the USIM. RES is sent to the CN where it is compared with the expected response (XRES) that was received from the home network. If the two agree, the UE has been successfully authenticated.

**AV generation**

Figure 2.16 illustrates the generation of an AV within the home environment/ authentication centre (HE/AuC). The algorithms f1, f2, f3, f4 and f5 are home network

## 2.6 Security issues

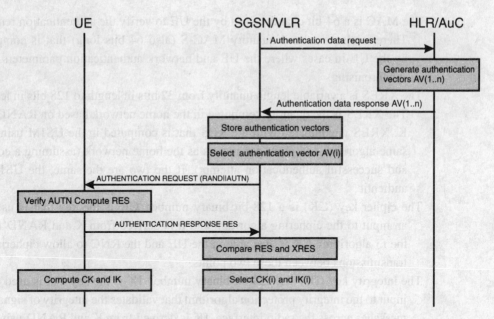

**Figure 2.15** Basic authentication mechanism in UMTS.

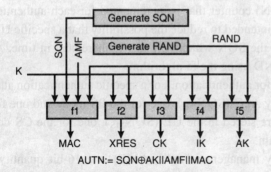

**Figure 2.16** Authentication vector generation procedure.

specific proprietary algorithms, although Ref. [4] does make some suggestions for algorithms f1 and f2.

The elements of the AV have the following definitions:

K is the secret key, also called the authentication key. K is a 128-bit number that is stored in the USIM and in the AuC in the home network. The secrecy of K and the computational robustness of the cryptographic algorithms are the basis of the security procedures utilised in UMTS.

RAND is a random number, also referred to as a random challenge, that is generated in the AuC in the home network from a random number generation procedure. RAND is a 128-bit binary sequence that is used with K in a number of algorithms to generate the various cryptographic quantities required in UMTS.

The MAC is a 64-bit quantity used by the UE to verify the authentication request. There is also a related quantity MAC-S (also 64 bits long) that is computed by the UE in cases where the UE and network authentication parameters need synchronising.

The XRES is a variable length quantity from 32 bits in length to 128 bits in length. The XRES is the quantity computed in the home network based on RAND and K. XRES is compared with the RES that is computed in the USIM using the same algorithm and input parameters as the home network (assuming a correct and successful authentication attempt). If the two are the same, the USIM is authentic.

The cipher key (CK) is a 128-bit binary number. CK is the key that is used as an input to the ciphering algorithm and it is derived from K and RAND using the f3 algorithm. CK is available in the UE and the RNC to allow ciphering of transmissions between these two points.

The integrity key (IK) is a 128-bit binary number. IK is the key that is used as an input to the integrity protection algorithm that validates the integrity of signalling messages across the radio interface. IK is derived from K and RAND using the f4 algorithm.

The anonymity key (AK) is a 48-bit number. AK is used to mask the contents of a sequence number (SQN) counter that is incremented for each authentication attempt. The masking is intended to reduce the possibility that a specific UE can be tracked by observing the SQN which will slowly increment in time. AK is derived from K and RAND using the f5 algorithm.

The SQN is a counter used for authentication. For a specific authentication attempt, two SQNs are created by AuC ($SQN_{HE}$), one for the PS mode and one for the CS mode. Two SQNs are stored in the UE ($SQN_{MS}$), one for the CS domain and one for the PS domain.

The authentication and key management field (AMF) is a 16-bit quantity. The AMF is used for the management of keys. It is the intention that the AMF field can be used to indicate which algorithm and key were used to generate a specific AV.

The AUTN is 128 bits long and is a composite sequence derived from SQN, AK, AMF and MAC using the following representation:

$$AUTN = SQN \oplus AK \parallel AMF \parallel MAC$$

The characters $\parallel$ mean that the quantities are serially concatenated to form a binary string of 128 bits.

The AV is created as the serial concatenation of the quantities described above. The AV is a sequence given by:

$$AV = RAND \parallel XRES \parallel CK \parallel IK \parallel AUTN$$

## 2.6 Security issues

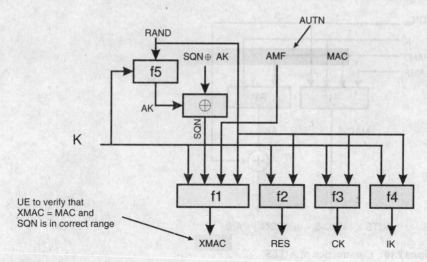

**Figure 2.17** UE actions in authentication.

### 2.6.3 UE actions for authentication

Figure 2.17 illustrates the actions within the UE regarding a specific authentication attempt (in Chapter 13 we will consider the specific protocol events surrounding authentication; here we are simply reviewing the basic procedures of authentication that apply equally to the CS domain and the PS domain).

The UE receives the RAND and the AUTN. First, the USIM checks the validity of the message using the MAC. Using the f1 algorithm, the UE derives XMAC (expected MAC). The XMAC should be the same as the MAC received in the AUTN. If the XMAC and MAC are different it is due to either transmission errors in the message or the deliberate modification of the message. In either case, the UE will indicate to the network that the authentication procedure failed due to a MAC failure.

From the AUTN, RAND, K and the f5 algorithm, the USIM can derive the $SQN_{HE}$ (the sequence number from the home environment). The UE can compare the $SQN_{HE}$ with sequence number $SQN_{MS}$ that is stored in the USIM. The sequence number check is used to allow the UE to estimate whether the network is valid or invalid.

An authentication failure message can be sent by the UE when the SQN is out of range. An SQN out of range occurs if the $SQN_{HE}$ differs from the $SQN_{MS}$ by more than 32.

As part of the rejection, a quantity AUTS is returned to the network. AUTS is similar to AUTN, but includes the sequence counter $SQN_{MS}$, concealed using a different AK derived from RAND and K using a different algorithm f5*. The objective in doing this is to send the $SQN_{MS}$ quantity to the AuC where it can be recovered. The AuC can then decide whether the existing network AVs can be used or whether they will

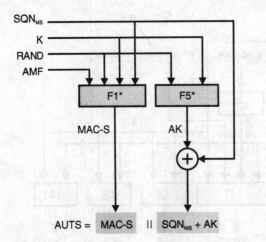

**Figure 2.18** Construction of AUTS.

be regarded by the UE as out of sequence. If the AVs can be used, a new set can be sent to the serving network for a new attempt at authentication. If the AVs cannot be used (they are regarded by the UE as being out of sequence), then the AuC resets the $SQN_{HE}$ to equal the $SQN_{MS}$, recomputes a set of AVs and proceeds as before. The construction of the AUTS is illustrated in Figure 2.18.

Finally, the UE can compute the RES, CK, and IK, returning the RES to the network to allow the network to authenticate the UE.

### 2.6.4 Message integrity protection

Message integrity protection is applied to all RRC messages except those included in Table 2.6. The objective for the integrity protection procedure is to allow the receiver of an RRC message (this includes NAS as well as other L3 AS messages) to decide whether the message has been corrupted either deliberately or accidentally.

The integrity protection procedure operates by passing the message to be protected through the protection algorithm to generate a quantity referred to as the MAC. The MAC is appended to the message and sent to the receiver. The receiver can then perform the same procedure. If the received MAC and the locally computed MAC are the same, the assumption is that the message integrity has not been violated. If the received MAC and the computed MAC are different, then the assumption is that the message integrity has been compromised either deliberately or accidentally.

Figure 2.19 illustrates the basic procedures associated with the message integrity protection that occurs in both the UE and the UTRAN. Message integrity protection is achieved using the f9 algorithm into which the message and a number of parameters are input. These parameters are:

## 2.6 Security issues

**Table 2.6.** *List of RRC messages not covered by integrity protection*

HANDOVER TO UTRAN COMPLETE
PAGING TYPE1
PUSCH CAPACITY REQUEST
PHYSICAL SHARED CHANNEL ALLOCATION
RRC CONNECTION REQUEST
RRC CONNECTION SETUP
RRC CONNECTION SETUP COMPLETE
RRC CONNECTION REJECT
RRC CONNECTION RELEASE (via CCCH)
SYSTEM INFORMATION (BROADCAST INFO)
SYSTEM INFORMATION CHANGE INDICATION

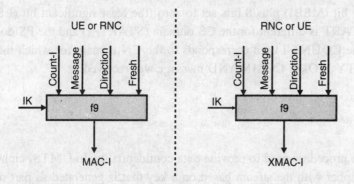

**Figure 2.19** Use of integrity protection.

- The IK, which is computed as part of the authentication procedure in the UE, and received in the SRNC from the CN domain. There is an IK for the CS domain $IK_{CS}$ and one for the PS domain $IK_{PS}$.
- COUNT-I, which comprises the RRC hyper frame number (HFN) (28 bits) plus the RRC sequence number (4 bits).
- FRESH, a network generated random value (32 bits) that is passed to the UE via the RRC SECURITY MODE COMMAND sent to the UE when a CN signalling connection is being established. At SRNS relocation a new FRESH is allocated.
- DIRECTION, which defines the direction of the message being either uplink or downlink.
- MESSAGE, which is the message to be integrity protected.

The message is transmitted across the radio interface with the MAC and at the receive end of the link the reverse procedure occurs to validate that the message has not been corrupted.

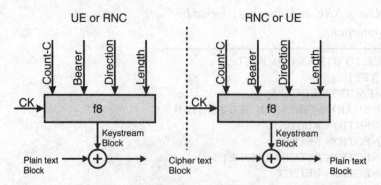

**Figure 2.20** Principles of ciphering across radio interface.

The counter COUNT-I is created from a 28-bit RRC HFN plus a 4-bit RRC sequence number. The 28-bit HFN is created from a 20-bit number called START (the most significant bit (MSB)) plus 8 bits set to zero (the least significant bit (LSB)). The quantity START is different for the CS domain ($START_{CS}$) and the PS domain ($START_{PS}$). The COUNT-I used corresponds to the CN domain for which the last RRC SECURITY MODE COMMAND message was received.

### 2.6.5 Ciphering

Ciphering is the procedure used to provide data confidentiality. In UMTS, ciphering uses a stream cipher with the stream based on a key that is generated as part of the authentication procedure (CK) and also involves K. The cipher functions in the UMTS system are located in either the RLC or the MAC in the UE and UTRAN. The exact location depends upon the type of RLC connection. A transparent RLC connection puts the ciphering in the MAC, in all other situations it is in the RLC.

Figure 2.20 illustrates the basic structure of the ciphering function. There are four inputs to the cipher algorithm (apart from the data to be ciphered). These inputs include a counter COUNT-C, the cipher key CK, the bearer channel ID (to allow multiple parallel ciphered links) and the direction (uplink or downlink).

The meaning of these parameters is defined as follows:

COUNT-C is a 32-bit counter derived from the RLC HFN or MAC-d HFN depending on where ciphering is being performed plus a frame or PDU counter as shown below:
- RLC TM: COUNT-C = MAC-d HFN (25 bits) + CFN (7 bits)
- RLC UM: COUNT-C = RLC HFN (25 bits) + RLC SN (7 bits)
- RLC AM: COUNT-C = RLC HFN (20 bits) + RLC SN (12 bits)

The HFN is either a 25- or a 20-bit number incremented each RLC SN cycle or transmitted frame. HFN is initialised to the value START at RRC connection establishment.

CK is 128 bits long, and there may be one CK for the CS domain ($CK_{CS}$) and one for the PS ($CK_{PS}$), with $CK_{CS}$ and $CK_{PS}$ stored in the USIM.
BEARER is four bits long and is a RB identifier.
LENGTH is 16 bits long and defines the length of the required key stream block.
DIRECTION is 1 bit and defines whether it is the uplink or the downlink that is being ciphered (0 is uplink, 1 is downlink).

### 2.6.6 Key creation and operation

In the previous section, we saw that the ciphering and the integrity protection functions were based on counters COUNT-C for ciphering and COUNT-I for integrity protection. The counters are both 32 bits in length. As we saw earlier, both counters are made from an HFN and a counter that is specific to the layer at which the function is being applied (this would be an RRC SN for integrity protection, and RLC SN or MAC HFN for ciphering).

The 20 MSB of both of these counters is initialised to a value called the START value. In addition, the START value is defined individually for the CS and PS domains, $START_{CS}$ and $START_{PS}$. The START values are stored in the USIM and read into the ME on activation. The ME transfers $START_{CS}$ and $START_{PS}$ to the RNC as part of the establishment of a radio connection (this is the RRC CONNECTION SETUP COMPLETE message, which is defined in detail in Chapter 11).

When a specific RB (either part of a RAB or part of a SRB) is initialised and includes integrity protection or ciphering, the START value for that specific connection is based on the maximum value of COUNT-I and COUNT-C for that specific CN domain. In this way, the START value increments over time.

To limit the use of CK and IKs beyond a certain time period, there is a check to enforce the change of the keys. This check is linked to the START values. A quantity THRESHOLD is defined by the network operator and stored in the USIM. As soon as the radio connection is released (RRC CONNECTION RELEASED), the values for $START_{CS}$ and $START_{PS}$ are compared with THRESHOLD. If $START_{CS}$ and $START_{PS}$ have reached THRESHOLD the CK and IKs are deleted and the UE requires a new set of keys at the establishment of the next connection to the CN. If $START_{CS}$ and $START_{PS}$ are below THRESHOLD, then the values are stored in the USIM for use the next time the radio connection is established.

## 2.7 Summary of the operation of the radio interface

In this section, we summarise some of the key functions that a UE passes through as it is activated and starts to request services from the network. This section summarises many of the details that will be considered throughout the remainder of the book, and provides a reference point as to what the different stages are in the activities of the UE.

We start with a UE that is activated and consider some of the basic stages that the UE needs to go through.

### 2.7.1 Locate cell and system camp on cell

When a UE is switched on it needs to go through a number of stages to identify the cell and the system before making a decision as to whether it is the correct cell in the correct system. This procedure starts with the higher layers in NAS deciding upon PLMN preferences and requesting the AS to perform cell searches and camping on cells; this is considered in Chapter 14, and encompasses many other topics that are covered throughout the book.

The AS needs to locate the cells, decode the transmissions from the cells and read the contents of the broadcast information from the cell in order to identify that the cell is the correct cell with suitable signal power and quality and with the correct PLMN identity.

The initial cell acquisition procedure is considered in Chapter 6, the data decoding and transport channel recovery are considered in Chapter 7, and the contents and meaning of the broadcast messages are considered in Chapter 11.

Once the cell is defined as suitable, the UE can camp on the cell and proceed to the next main stage of the activity which is the establishment of an RRC connection to the UTRAN.

### 2.7.2 Request RRC connection

The NAS needs to register with the appropriate CN domains. Before the UE can do this, however, an RRC connection to the UTRAN is required. The establishment of an RRC connection registers the UE with the UTRAN, creates signalling connections (SRBs) between the UE and the UTRAN and allocates radio network resources for the use of the UE. The process of establishing an RRC connection is considered in detail in Chapter 11. The RRC connection establishment will result in the creation of a number of SRBs that in turn will require the creation of RLC connections, MAC connections and physical channel connections.

The characteristic of the physical channels and the transport channels depends upon whether the UE is put into the CELL_DCH state or the CELL_FACH state as described earlier.

Chapter 4 considers the many issues that need consideration when we are establishing the RRC connection between the UE and the UTRAN. These concepts also apply to the transfer of other signalling messages, as well as user data messages that occur in subsequent exchanges between the UE and the network.

Once the SRBs are available, the UE is able to proceed with the process of registration with the CN.

### 2.7.3 Registration and authentication with CS domain (IMSI attach)

This procedure can include registration with either, or both, of the CS domain and the PS domain. Registration with the CS domain is often referred to as IMSI attach, registration with the PS domain as GPRS attach. Both of these procedures are NAS functions that are considered in Chapter 13. Part of this functionality includes the authentication and integrity protection functions that are considered in this chapter.

The NAS messages that perform the registration and authentication are carried employing a procedure known as direct transfer, which is the mechanism used to carry an NAS message across the AS to the CN. Direct transfer is considered in Chapter 11.

### 2.7.4 Request MO speech call

If the user, after successfully registering with the CS domain, wishes to establish a CS connection such as a speech call, first the NAS must request the establishment of a MM connection. Next, using this high layer signalling connection, the CC protocol is used to request the establishment of a speech call. As part of the call establishment procedure, the CS domain requests the establishment of an RAB to carry the speech traffic.

The establishment of the MM connection and the CC call setup signalling are considered in Chapter 13. The establishment of the RAB for the speech service is considered in Chapter 11.

### 2.7.5 Request PDP context

The user may wish to establish a packet data connection, or PDP context to the PS domain. This will require a GMM context for the signalling and the creation of a PDP context. The establishment of the GMM signalling connection and the PDP context are considered in Chapter 13. The creation of the PDP context requires the establishment of an RAB, which is considered in Chapter 11.

### 2.7.6 On-going mobility UE procedures

With the different services active, there are many on-going functions and services that the UE needs to consider. Some of these depend upon the state that the UE is in and the location of the UE.

Chapters 11 and 12 review the on-going operation and maintenance of the radio interface, considering issues such as the measurements that the UE needs to make and the signalling that may be required if a handover to a new UMTS cell or new radio technology is required. Equally, if the calls are cleared and the UE moves into a low power state, then the measurements are still on-going, in which case Chapter 14

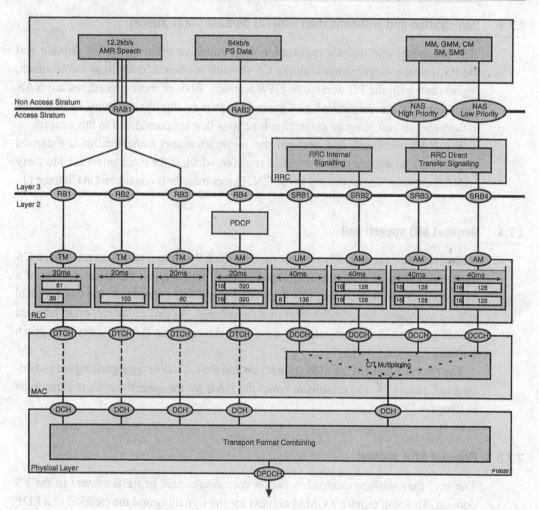

**Figure 2.21** Summary of radio interface structure for SRBs, CS and PS services.

considers the decisions that the UE needs to make whilst operating in a idle mode, such as the cell reselection and PLMN selection and reselection issues.

### 2.7.7 Summary of UE configuration

Let us assume that the UE has a number of SRBs active, an RAB for speech and an RAB for the PS data. Each of these entities requires configuration within layers 1, 2 and 3 of the radio interface. Figure 2.21 summarises the likely structure of the radio interface after the creation of these entities. The remainder of the book is focussed on examining in greater detail the structure and operation of the WCDMA radio interface protocol as a whole. The book addresses the functions in layers, but as we have seen

## 2.7 Summary of the operation of the radio interface

many of the procedures are interlinked between the layers and consequently we bend the definitions of the layers in places in order to consider a specific topic as a whole.

We start the remainder of the book at the physical layer, considering the definition and use of the codes that are a fundamental component in the design and operation of the WCDMA radio interface. From there we will move up the protocol stack, considering the roles that the higher layer entities perform in the overall operation of the radio interface.

Although the book is written to be read end-to-end, it is expected that the reader may concentrate only on specific sections and layers within the protocol architecture. It is hoped that the structure of the book will facilitate either approach.

# 3 Spreading codes and modulation

## 3.1 Introduction

In this chapter we consider the basic principles of operation of the different codes used in the WCDMA system and the modulation scheme that is closely associated with the codes. To start we consider both of these from a general perspective before applying the concepts to the specific situation of the WCDMA system.

We introduce the concepts of spreading and modulation. Spreading is the process fundamental to the operation of the WCDMA radio interface, so we shall spend some time considering what it is and how it works. Modulation is the mechanism for superimposing the spread information onto an RF carrier. Whilst this part of the procedure is very similar to that in similar digital communication systems, spreading introduces some additional complications that we need to consider.

First, we will consider the basic principles of spreading. To do this we need to review what we are trying to achieve in using these spreading functions and subsequently how we can achieve this.

In studying these concepts, we need to introduce something generically referred to as a spreading code (we will see later that there are a number of different types of codes under this general umbrella). We should be careful not to become confused between these codes and the other codes that exist in the WCDMA system. The other codes, often referred to as error correcting or error detecting codes, are there to protect the data being transmitted across the radio interface. We will consider these codes in more detail in Chapter 7.

The spreading codes are grouped into two basic types: channelisation codes and scrambling codes. We review what these basic types of codes are, as well as how the codes are used within the transmission path both for the uplink and the downlink. We will see that the use of the codes differs quite appreciably in the uplink and the downlink of the system.

Channelisation codes are relatively short in length and are made from something referred to as an orthogonal function or waveform. It is the properties of orthogonality that are particularly important for the channelisation codes and so we will review what we mean by orthogonality and how orthogonality is achieved from the codes being considered.

Scrambling codes, in contrast to channelisation codes, are quite long (with the exception in the uplink of the short scrambling code). Scrambling codes are created from streams that are generally referred to as pseudo-noise sequences. We will see later that it is the noise-like behaviour of the scrambling codes that gives them the properties that are desirable in the role that they fulfil in the transmission path.

The use of channelisation codes and scrambling codes is different on the uplink and the downlink. This means that the radio interface is asymmetrical, with different functions provided by the codes in the different directions of the links.

The second major function that we consider in this chapter is that of modulation. The subject of modulation is well understood for modern digital communication systems and there are many excellent texts [5,6], so we will only consider the specific type of modulation used within WCDMA and focus on those aspects that are significant in either the implementation or the performance.

A key element of the modulation scheme is the spectrum shaping functions that are required to achieve the necessary adjacent channel performance that is essential for reasons of co-habitation of the radio system with other radio systems in the vicinity. Within this chapter we consider the spectrum shaping issues as well as a number of other issues that relate to the implementation and performance of the modulation scheme.

A key element to the performance of the modulation scheme is the different performance metrics that are used to assess how well a specific modulator is performing. These performance metrics are referred to as the error vector magnitude (EVM), the complementary cumulative distribution function (CCDF) and the adjacent channel leakage ratio (ACLR). These performance metrics extend across the RF domain as well as the baseband modulation domain. In this chapter we explore how the baseband processing functions impact these performance criteria, and in Chapter 5 we will examine how the RF stages can also affect these performance criteria.

In the final part of this chapter we bring together the channelisation codes, the scrambling codes and the modulation. In this case we will be focussing very specifically on the WCDMA uplink and downlink. In this final section we see that the modulation and channelisation and scrambling codes interact. First we examine how this interaction is detrimental to the overall transmission path and then we examine a specific technique in the uplink that helps to overcome some of these detrimental aspects of the combination of the WCDMA codes and the modulation.

**Table 3.1.** *Summary of multiple access techniques*

| Acronym | Name | Typical use |
|---|---|---|
| FDMA | Frequency division multiple access | 1G cellular |
| TDMA | Time division multiple access | 2G cellular |
| CDMA | Code division multiple access | 3G cellular |
| CSMA | Carrier sense multiple access | Ethernet |
| OFDMA | Orthogonal frequency division multiple access | DVB-T |
| ODMA | Opportunity driven multiple access | 3G cellular |
| SDMA | Space division multiple access | Beam forming |

## 3.2 Introducing WCDMA spreading functions

### 3.2.1 Multiple access techniques

Before considering the details of the different types of codes used in the WCDMA system it is worth just stepping back and reviewing exactly what the objective is for the use of these codes.

One of the main criteria in a multiuser communication system such as WCDMA is to provide what is generically referred to as a multiple access technique. Over the years, many different multiple access techniques have been defined, some of which are summarised in Table 3.1. The fundamental objective for the multiple access techniques is to separate the access to some transmission resource such as the radio carrier that is used in the wireless communication systems.

For radio systems, the first three techniques presented in Table 3.1 are historically the most important. Let us consider each of these techniques in turn, and with reference to Figure 3.1, which illustrates the main differences between the various multiple access schemes used to separate users employing frequency, time and codes. These are the three main multiple access techniques that have been used in the three generations of cellular radio systems. The following descriptions define each of the three different techniques in more detail.

**FDMA**

Figure 3.1(a) illustrates the radio access technique that is referred to as frequency division multiple access (FDMA). With FDMA, assigning a different frequency to each user at a specific cell site separates the users. The frequency allocation is traditionally paired, with one frequency used for the mobile transmissions and a second frequency used for the mobile reception.

## 3.2 Introducing WCDMA spreading functions

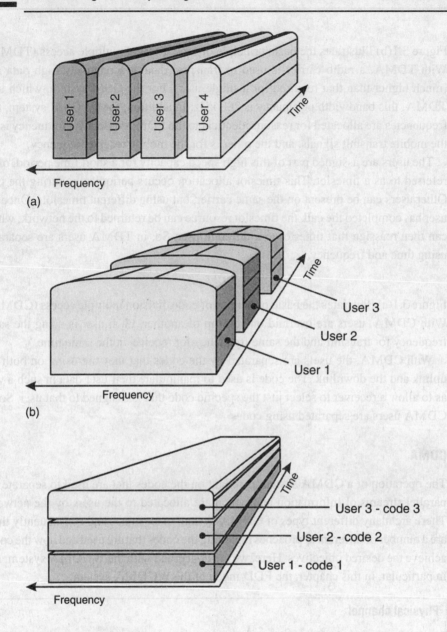

**Figure 3.1** Basic principles of: (a) FDMA (e.g. analogue TACS); (b) TDMA (used in GSM); and (c) CDMA (e.g. WCDMA).

FDMA was used for first generation cellular radio technologies, and typically the users were assigned a radio carrier which occupied a 25–30 kHz slice of the frequency spectrum. The users were allocated the frequency pair for the duration of a call, and once the call was completed, the frequencies were returned to the pool of frequencies ready to be allocated to other users. So, FDMA achieves multiple access by separating users in frequency.

## TDMA

Figure 3.1(b) illustrates the basic principle of time division multiple access (TDMA). With TDMA, a radio carrier is used to transport data at a relatively high data rate (much higher than that required for a single user). For the GSM system, which uses TDMA, this bandwidth occupancy is 200 kHz. In addition, for the GSM system, two frequencies are allocated for reasons identical to the FDMA case. One frequency is for the mobile transmit signals, and the other is for the mobile receive frequency.

The users are assigned part of this high speed capacity for a short time period, often referred to as a timeslot. This timeslot allocation occurs periodically during the call. Other users can be present on the same carrier, but using different timeslots. Once the user has completed the call, the timeslot resource can be returned to the network, which can then reassign that timeslot to a different user. So, in TDMA users are separated using time and frequency.

## CDMA

Figure 3.1(c) illustrates the basic principles of code division multiple access (CDMA). With CDMA, users are overlaid one on top of another. Each user is using the same frequency for transmit and the same frequency for receive, at the same time.

With CDMA, the users are separated by the codes that they are using on both the uplink and the downlink. The code is used to manipulate their user data in such a way as to allow a receiver to select just the specific code that is assigned to that user. So, in CDMA users are separated using codes.

### 3.2.2 CDMA

The operation of a CDMA system is reliant on the codes that are used to separate the parallel streams of information and which are allocated to the users by the network. There are many different types of CDMA system in existence and consequently there are a number of different approaches regarding the codes that are used and how the codes achieve the desired objectives. Here, we are concerned with the WCDMA system, and in particular, in this chapter, the FDD mode of the WCDMA system.

> **Physical channel**
>
> A physical channel is a stream of data that is being transmitted from either the UE or the Node B. The physical channels are separated through the use of special waveforms, generically referred to as spreading codes. Within WCDMA, there are numerous different physical channels and consequently different types of spreading codes are employed.
>
> Some of the physical channels are defined in terms of a pair of codes referred to as a channelisation code and a scrambling code. Other physical channels are defined just in terms of special waveforms. (Chapter 4 considers the details of the different physical channel types.)

## 3.2 Introducing WCDMA spreading functions

**Table 3.2.** *FDD mode code types and their use in uplink and downlink*

| Code type | Code use in uplink | Code use in downlink |
|---|---|---|
| Channelisation code | Data rate control | User separation |
| | | Data rate control |
| Scrambling code | User separation | Interference mitigation |
| | Interference mitigation | |

Within WCDMA we have the concept of a physical channel (see inset for description). In this section we concentrate on the physical channel that is used for user traffic and control signalling. An example of this type of physical channel is referred to as a dedicated physical data channel (DPDCH). There are two types of codes used for this physical channel: the first is called a channelisation code, and the second a scrambling code. These different codes are used for different reasons, and the reasons for using the codes are different for the uplink and the downlink.

### 3.2.3 Types of codes

We are considering the DPDCH. The DPDCH is used on both the uplink and the downlink, and is created from channelisation codes and scrambling codes. In this section we examine why these two types of codes are used on the uplink and the downlink and in doing so we will see that the codes are used for different reasons in the uplink and the downlink.

Table 3.2 presents the two different types of codes used in the WCDMA system for a number of different types of physical channel such as the dedicated physical channel (DPCH). In the uplink, the channelisation code is used to control the data rate that a user transmits to the cell. We will see in Section 3.3.2 how this is achieved. The scrambling code is used for two main reasons (when we look at the detail later we will see that these two reasons are actually related). The first reason is to separate the users on the uplink with each user being assigned a scrambling code that is unique within the UTRAN. The second reason is to provide a mechanism to control the effects of interference on the uplink both from within the cell (intracell interference) as well as from other adjacent cells (intercell interference).

In the downlink, the channelisation code is used to separate users within a cell and also to control the data rate for that user. The scrambling code on the downlink is used to control the effects of interference. As with the uplink, this interference can be either intracell interference or intercell interference.

Figure 3.2 summarises the code allocations for a few example UEs spread across two cells. In the diagram, CC indicates that it is a channelisation code and SC a scrambling

# Spreading codes and modulation

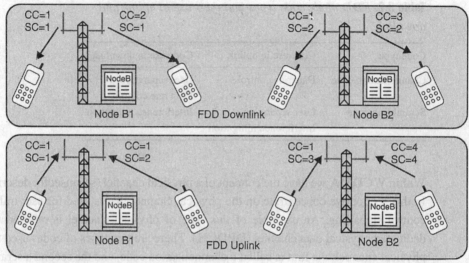

**Figure 3.2** Example DPCH code allocations.

code. The choice of code numbering is arbitrary. In Chapter 4 we will go through how the different codes are numbered in greater detail.

### 3.2.4 Code allocation

Staying with Figure 3.2 for a moment, it is worth considering where the code allocations are made. For the uplink, the channelisation code is selected by the UE based on the amount of data required for transmission. The specifications define a relationship between required data rate and code selection. The scrambling code on the uplink is selected and assigned by the UTRAN when the physical channel is established or possibly when it is reconfigured.

For the downlink, the channelisation code is allocated by the UTRAN from the ones that are available in that cell at that point when an allocation is required. The scrambling code on the downlink is used within that cell and by more than one UE in the cell. The scrambling code is likely to be assigned as part of the radio planning function, but a cell receives the allocation from the operations and maintenance entity. Table 3.3 summarises the different code allocations.

### 3.2.5 Chip rate

Before moving on to consider the specific details behind the channelisation codes and the scrambling codes, it is useful to consider a quantity that is common to both and a fundamental characteristic of the system – the chip rate.

## 3.3 Channelisation codes

**Table 3.3.** *DPCH code allocations*

| Code type | Uplink | Downlink |
|---|---|---|
| Channelisation code | UE based on data rate | UTRAN |
| Scrambling code | UTRAN | Radio planning function |

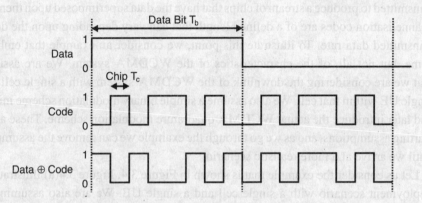

**Figure 3.3** Relationship between chips and bits.

In a CDMA system of the type we are considering here, user data (we are assuming that the data may include overheads from other layer 1 functions such as channel coding) are not transmitted directly as would be the case in a system such as GSM. Instead, in the CDMA system, user data are transmitted in pieces, with each piece transmitted using what is called a chip. The concept of a chip as a fraction of a bit is illustrated in Figure 3.3 along with some numerical values. In WCDMA, a chip is the fundamental unit of transmission across the radio interface. It has a well-defined chip rate, which is the reciprocal of the chip duration and equal to 3.84 Mc/s (million chips per second).

The chip rate for the FDD mode of WCDMA is fixed (at least for R99 and immediate follow-on releases at 3.84 Mc/s). For the TDD mode, for R99 the same chip rate is used, but for the R4 specification a lower chip rate of 1.28 Mc/s is also defined. We will see in the following section that the chip rate and the transmitted bit rate are related via a quantity that is commonly referred to as the spreading factor (SF).

## 3.3 Channelisation codes

In this section we begin exploring the physical structure and the use of the channelisation codes. To simplify things we start by considering only the codes as applied to the WCDMA downlink. Once we have considered the downlink scenario, we can then move on to the slightly different case of the uplink. We start with a brief qualitative description of what the channelisation codes are, before going on to examine them in more detail.

### 3.3.1 Single cell single user example – binary modulation

The channelisation codes are sequences of chips that are applied to the data to be transmitted to produce a stream of chips that have the data superimposed upon them. The channelisation codes are of a defined length that can vary depending upon the desired transmitted data rate. To illustrate this point, we consider an example that embodies some, but not all, of the characteristics of the WCDMA system. We are assuming that we are considering the downlink of the WCDMA system with a single cell and a single UE within that cell. We also assume a simple binary modulation scheme initially, and later introduce the actual WCDMA quadrature modulation scheme. These are our starting assumptions, and as we go through the example we can remove the assumptions until we arrive at a more realistic scenario.

Let us consider the example that is shown in Figure 3.4. Figure 3.4(a) illustrates the deployment scenario with a single cell and a single UE. We are also assuming that the two are directly linked, so that we can (temporarily) ignore the effects of radio path fading and the details of the modulation scheme. Figure 3.4(b) illustrates what is happening within the cell site transmitter of our simplistic example.

> **Spreading**
>
> Here and throughout the remainder of this section the term 'spreading' is used to define the procedure of taking a data sequence and a channelisation code and producing a sequence of chips. This definition, however, is not completely valid. The term 'spreading' comes from the physical manipulation of the spectrum of the transmitted data signal. The problem is that some of the channelisation codes that are allowable have little if no effect on the data spectrum, whilst other codes can expand the spectrum considerably, all of this independent of the commodity spreading factor.
>
> It is only after the scrambling code is applied that the spectrum reliably fills the expected spread bandwidth. All this said, however, we will still refer to the first stage as the spreading stage – there is some justification for this in terms of the wording applied in the specifications which define a spreading factor based on the channelisation code length.

We are assuming that the transmitter has obtained an input data sequence from somewhere. The input data are sequentially repeated to form a sequence of chips such that there is exactly the same number of chips of data as there are chips in the channelisation

## 3.3 Channelisation codes

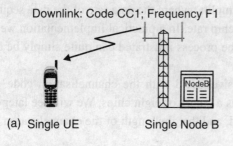
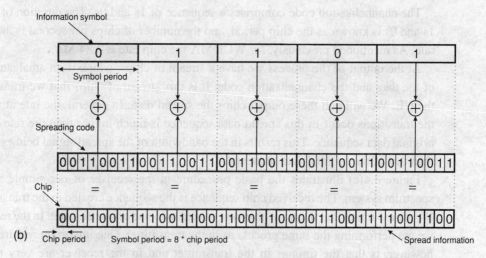
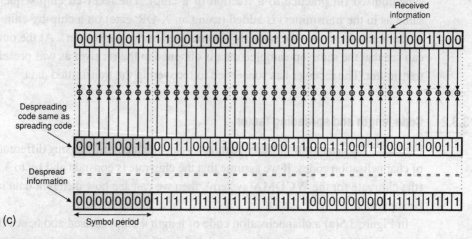

**Figure 3.4** Application of channelisation codes to a data sequence: (a) general scenario; (b) spreading at the transmitter; (c) despreading at the receiver.

code. This process is a type of up-sampling where the data rate of the data sequence is increased until it is the same as the chip rate. In a practical implementation we do not need to perform this repetition, as the process illustrated can quite simply be done in one stage.

Next, the repeated data is exclusive OR'd with the channelisation code. In this example, the channelisation code has a length of eight chips. We will see later exactly where these eight chips are obtained, and that the length of the channelisation code is not restricted to eight.

The channelisation code comprises a sequence of 1s and 0s. The duration of these 1s and 0s is known as the chip period, and the number of chips per second is the chip rate. As mentioned previously, for WCDMA the chip rate is 3.84 Mc/s.

At the output of the process we have a stream of chips, which is an amalgamation of the data and the channelisation code. It is this stream of chips that we transfer to the UE. We will call these output chips the spread data. In general, the rate at which the transitions occur in this spread data sequence is much higher than the rate in the original data sequence. This results in the bandwidth of the spread signal being greater than that of the original data sequence.

Figure 3.4(c) illustrates the basic procedure in the receiver of our simple spread spectrum system. The received chip sequence is the sequence created in the transmitter. We are assuming that there are no errors caused by noise or interference. In the receiver we are performing the same process as in the transmitter. One important requirement, however, is that the timings in the transmitter and in the receiver are very tightly synchronised (in practice to a fraction of a chip). The received chip sequence (the same as in the transmitter) is added (using an X-OR gate) on a chip-by-chip basis to a channelisation code (the same as the one used in the transmitter). At the output, we can see that the same up-sampled data is obtained in the receiver as was present in the transmitter. The receiver has succeeded in recovering the transmitted data.

### 3.3.2 Code length and spreading factor

Figure 3.5 illustrates how different data rates can be supported by using different lengths of channelisation codes. If we assume that the chip rate is constant and set to 3.84 Mc/s (the chip rate for the WCDMA system), then we can see how different data rates can be supported.

In Figure 3.5(a) a channelisation code of length 4 chips is used and hence the input data rate is 960 kb/s. It should be noted at this point that this example does not reflect the WCDMA system correctly. In WCDMA the modulation scheme also impacts the effective transmitted data rate resulting in a factor of 2 increase in the rate above that shown in Figure 3.5.

Now it is useful to define a quantity known as the spreading factor. This specifies the number of chips that are used to transmit a single bit of information. In the example

## 3.3 Channelisation codes

**Table 3.4.** *Downlink data rates – simple example and WCDMA downlink*

| Code length | Example downlink data rate | WCDMA downlink data rate |
|---|---|---|
| 4 | 960 kb/s | 1.92 Mb/s |
| 8 | 480 kb/s | 960 kb/s |
| 16 | 240 kb/s | 480 kb/s |
| 32 | 120 kb/s | 240 kb/s |
| 64 | 60 kb/s | 120 kb/s |
| 128 | 30 kb/s | 60 kb/s |
| 256 | 15 kb/s | 30 kb/s |
| 512 | 7.5 kb/s | 15 kb/s |

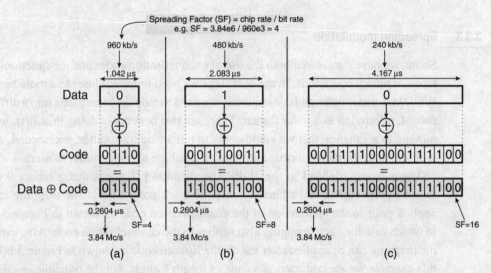

**Figure 3.5** Effects of code length on data rate for: (a) a code of length 4 chips; (b) a code of length 8 chips; (c) a code of length 16 chips.

shown in Figure 3.5(a), the spreading factor is 4. In the example shown in Figure 3.5(b), the data rate is 480 kb/s and the channelisation code has a length of 8 chips. In this case the spreading factor is 8. In the case shown in Figure 3.5(c), the channelisation code has a length of 16 chips (spreading factor of 16) and hence the transmitted data rate is reduced by a factor of 2 to 240 kb/s. The significance of this result is that it is relatively simple to change the transmitted data rate. The data rate is changed by altering the length of the channelisation code (and consequently the spreading factor). In doing this, none of the RF front end bandwidths need to change.

Table 3.4 summarises the effects of a change of the channelisation code length on the downlink data rate. The code lengths that are defined on the downlink range from 4 chips to 512 chips. In addition, the code lengths increase by a factor of 2 for each increment of code length. The data rate in the simple example system is shown in the middle column, and that for the WCDMA system in the column on the right hand side. That there are differences between the two is due to the modulation scheme. The simple example case represents a binary type modulation. In the real system, a quadrature modulation scheme is used, leading to a factor of 2 increase in the data rate. The use of the quadrature modulation scheme for the downlink is considered in Section 3.6 and that for the uplink in Section 3.7.

One final point on the downlink data rates – the WCDMA downlink data rates shown in the table are the total downlink data capacity for a given code length. Some of this capacity is required for physical channel overheads such as pilot bits and power control bits. As a consequence, the actual usable downlink data rates are correspondingly less than the values shown in the right hand column of Table 3.4.

### 3.3.3 Spreading modulation

So far, we have considered both the use of channelisation codes and the functioning in terms of binary operations. In the next stage we need to move closer to a more realistic WCDMA transmitter, and consequently we need to map the binary data into a different format, referred to as a polar format. There are two benefits in doing this: first, we are moving to a situation that we could apply to a realistic transmitter; and second, as we will see, this alternative format allows us to combine together multiple users.

The mapping, defined in the 3GPP specification [7], states that a binary 0 maps to a $+1$ polar signal, and a binary 1 maps to a $-1$ polar signal. The mapping can be applied prior to the application of the channelisation code as shown in Figure 3.6(a), in which case the same mapping also applies to the channelisation code. Alternatively, the mapping can be applied after the channelisation code as shown in Figure 3.6(b). In this example we are considering a code of length 8 chips, but the principle is valid for any other code of any other length.

### 3.3.4 Multiuser spreading

Figure 3.7 illustrates one of the first basic principles of spreading: that of transporting data for multiple users using a single composite spread waveform. In the figure, each user is assigned a specific channelisation code. We will see later how these codes are defined, and what specific properties they require. The data for user 1 are spread and mapped in the same manner as seen previously; similarly for user 2, but in this case user 2 uses a different channelisation code. Notice that the time alignment of these channelisation codes is coincident – this is an important requirement. The resulting composite signal is no longer a binary or polar signal, and starts to exhibit an amplitude variation. The important point about this result is that we have combined the transmissions from

## 3.3 Channelisation codes

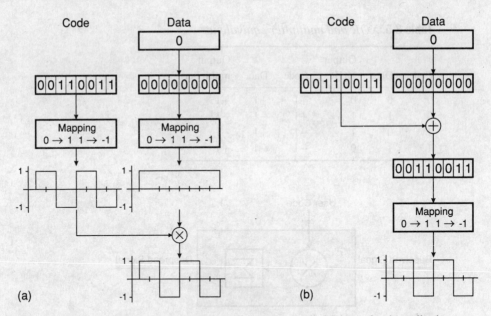

**Figure 3.6** Spreading modulation conversion: (a) mapping applied prior to the channelisation code; (b) mapping applied after the channelisation code.

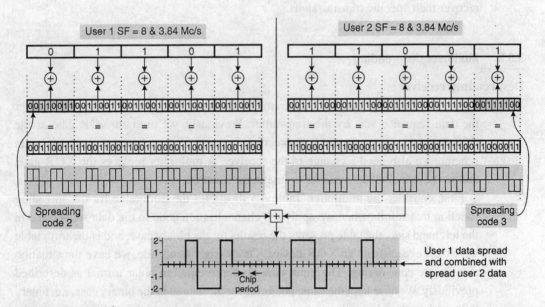

**Figure 3.7** Multiuser spreading modulation conversion.

**Table 3.5.** *XOR and multiplier equivalence*

| Code | Data | Output XOR | Code | Data | Output multiplier |
|------|------|------------|------|------|-------------------|
| 0 | 0 | 0 | +1 | +1 | +1 |
| 0 | 1 | 1 | +1 | −1 | −1 |
| 1 | 0 | 1 | −1 | +1 | −1 |
| 1 | 1 | 0 | −1 | −1 | +1 |

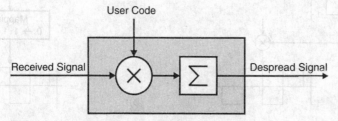

**Figure 3.8** Multiuser despreading receiver.

the two users onto a single carrier signal, and the users use this single carrier signal to recover their specific transmissions.

### 3.3.5 Multiuser despreading

**Despreading receiver**

Figure 3.8 illustrates the block diagram for the despreading receiver that can recover the data sequence sent by the transmitter. Previously, we used an XOR gate as the receiving device to perform the despreading function. With a change to the modulation scheme, we also need a change to the receiver. A multiplier replaces the XOR gate, and we have included a summer or integrator function.

First, consider the multiplier. Table 3.5 illustrates the truth table for the spreading function that results when we apply the channelisation code to the data sequence. On the left hand side, the table presents the results for the binary case, and is the truth table we would expect from an XOR device. On the right hand side, we have the situation that we are considering. The input data are represented in polar format as described previously. We have kept the same positions for the data as in the binary case, i.e. binary 0,0 translates to +1,+1. The output of the multiplier is simply the multiplication of the two inputs and in the first case is equal to +1. If we follow this process through for all elements of the table, we end up with the result in the far right column of the table.

If we compare the result from the multiplier using a polar input and the result from the XOR gate with a binary input we see that they are the same once we take the

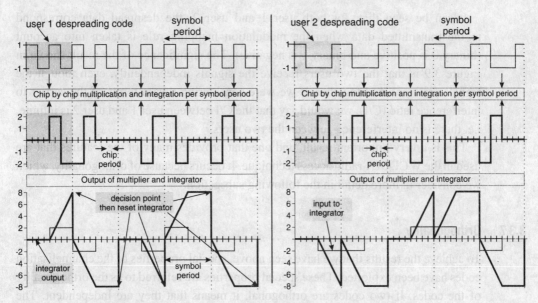

**Figure 3.9** Multiuser despreading.

mapping into account. So a multiplier using polar signals is equivalent to the XOR gate and binary signals.

Next, let us consider the summer or integrator. The object of our receiver is to collect the energy that was transmitted to the receiver in the most efficient way possible. In our simple model, we do not need to consider any phase effects for the received signal and so the only function we need to perform is the summation of the received energy from each received chip. To do this we use an integrator, but any device that collects the received energy would be suitable.

### 3.3.6 Two users receiving signals

Figure 3.9 illustrates the basic procedures for despreading, which are applied for both the spread signals received for user 1 and the spread signals received for user 2. In each case, the composite received spread signal is multiplied (on a chip-by-chip basis) by the despreading code. For user 1 the despreading code is the same as the channelisation code used for user 1 in the transmitter, i.e. code 2, and for user 2 the despreading code is the same as the channelisation code for user 2, i.e. code 3. The output from the multiplication can be seen in the lowest graph along with the integrator output. The integrator works by integrating the received signal across the symbol period and then resetting the integrator. The despread signal is detected at the time just prior to the resetting of the integrator. For both user 1 and user 2, the output is of amplitude $+8$ or $-8$. This is due to the spreading factor being 8, and each chip having an amplitude of $+1$ or $-1$.

It can be seen that, for both user 1 and user 2, the despread data correspond to the transmitted data when the modulation mapping rule is taken into account (binary 0 is positive and binary 1 is negative). The significance of the result shown in Figure 3.9 is that the two users receive the signals independently, even though the signals are combined together. If we were considering this in terms of the carrier to interference ratio (C/I), we would say that the C/I between user 1 and user 2 is infinite, i.e. there is no interference between the two users.

This is a very important result, and one that provides many of the features that are used in the WCDMA radio access technique. It occurs because of orthogonality, which we consider in more detail in the following subsection.

### 3.3.7 Orthogonality

To achieve the results that we have seen above, special properties of the channelisation codes have been exploited. These special properties are referred to as the orthogonality of the codes. If two codes are orthogonal, it means that they are independent. The simplest example of an orthogonal system is three-dimensional space, where each dimension is completely independent of the others. This means that a change in one dimension is not noticed in any of the other dimensions.

For WCDMA, the orthogonality can be expressed as follows. Consider two length $N$ polar channelisation codes $C_i$ and $C_j$ defined by (3.1):

$$C_i = (C_{i0}, C_{i1}, C_{i2}, C_{i3}, \ldots C_{i(N-1)})$$
$$C_j = (C_{j0}, C_{j1}, C_{j2}, C_{j3}, \ldots C_{j(N-1)}) \qquad (3.1)$$

where

$$C_{in}, 0 < n \leq N-1 \quad \text{and} \quad C_{in} \in \{+1, -1\}$$

$$\sum_{n=0}^{n=N-1} C_{in} C_{jn} = \begin{cases} 0 & i \neq j \\ N & i = j \end{cases} \qquad (3.2)$$

The two codes are of length $N$ chips. The orthogonality of the codes is expressed in (3.2). If we take the two codes $C_i$ and $C_j$, multiply them together chip by chip and sum them over the $N$ chips of their lengths, then the result is zero. This property of the codes is referred to as the orthogonality property. The physical interpretation of this result is that we can detect one code from one user and reject the information from a second user.

Consider the situation illustrated on the left hand side of Figure 3.10 where we have two users at the transmitter. The data from user 1 ($D_1$) is converted into chip rate samples ($d_{n1}$) (this simply consists of the repetition of the data bit $N$ times, where $N$ is the spreading factor). Each of the data chips is equal to $d_{n1}$ for user 1 ($n$ is the number of the chip from the data bit) and for user 2 the data chip is $d_{n2}$. At the transmitter

## 3.3 Channelisation codes

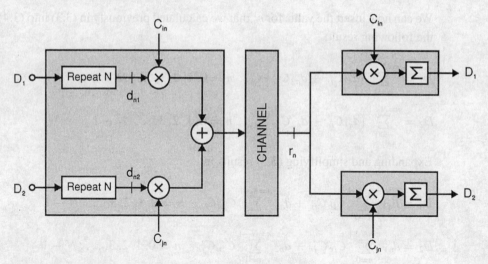

**Figure 3.10** Example multiuser spreading and despreading.

we multiply each of the data chip sequences by the channelisation code for that user. We then add together the chips from user one and the chips from user two to form the transmitted data.

The receivers for user 1 and user 2 are shown in the upper right and lower right of Figure 3.10, respectively. Assuming for the moment that the channel is transparent, the input to the receiver is the same as the output from the transmitter and is given by the received chip sequence $r_n$ that is defined by:

$$r_n = d_{n1}C_{in} + d_{n2}C_{jn}, \quad n = 0, 1, 2, 3, \ldots, N-1 \tag{3.3}$$

Equation (3.3) states that the received signal is the addition of the data chip from user 1 multiplied by the code for user 1 and the data chip from user 2 multiplied by the code for user 2.

At the output of the receiver for user 1, we want to recover the data bit ($D_1$) for user 1. To achieve this we multiply the received signal by the channelisation code used for user 1 ($C_i$) and then sum the result across the $N$ transmitted chips that were used to transmit the data bit. We do the same thing for user 2 to recover the data bit ($D_2$), but in this case we use the channelisation code for user two ($C_j$). The output for the two receivers is summarised by:

$$D_1 = \sum_{n=0}^{n=N-1} r_n C_{in}, \quad n = 0, 1, 2, 3, \ldots, N-1$$
$$D_2 = \sum_{n=0}^{n=N-1} r_n C_{jn}, \quad n = 0, 1, 2, 3, \ldots, N-1 \tag{3.4}$$

We can now insert the value for $r_n$ that we calculated previously in (3.3) into (3.4), with the following result:

$$D_1 = \sum_{n=0}^{n=N-1} (d_{n1}C_{in} + d_{n2}C_{jn}) C_{in}, \quad n = 0, 1, 2, 3, \ldots, N-1$$
$$D_2 = \sum_{n=0}^{n=N-1} (d_{n1}C_{in} + d_{n2}C_{jn}) C_{jn}, \quad n = 0, 1, 2, 3, \ldots, N-1$$
(3.5)

Expanding and simplifying (3.5) results in

$$D_1 = d_{n1} \sum_{n=0}^{n=N-1} C_{in}C_{in} + d_{n2} \sum_{n=0}^{n=N-1} C_{jn}C_{in}, \quad n = 0, 1, 2, 3, \ldots, N-1$$
$$D_2 = d_{n1} \sum_{n=0}^{n=N-1} C_{in}C_{jn} + d_{n2} \sum_{n=0}^{n=N-1} C_{jn}C_{jn}, \quad n = 0, 1, 2, 3, \ldots, N-1$$
(3.6)

Referring back to (3.2), we see that we have similar terms to those that define the orthogonality of the channelisation codes. We can apply those relationships in (3.2) to the case shown in (3.6) to see what happens for users 1 and 2. The result is

$$D_1 = d_{n1}N$$
$$D_2 = d_{n2}N$$
(3.7)

Equation (3.7) states that the output data bit for user 1 is simply $N$ times the magnitude of the individual data chip $d_{n1}$. The other terms that were present for $d_{n2}$ from user 2 have disappeared due to the relationship in (3.2) that stated that dissimilar codes have a product of zero. The same is also true for user 2, except that the information from user 1 is what has now been removed from the expression for user 2. The output, therefore, is the data that were transmitted to each of the users with no interference present from the other user.

The other interesting point to note is the magnitude of the output signals. The magnitude at the output is $N$ times the individual data chip amplitudes. $N$ is the spreading factor, or length of the codes, and consequently, as we increase the length of the codes, we are also increasing the output amplitude of the despread data signal.

### 3.3.8 Orthogonal variable spreading factor (OVSF) tree

**Creation of WCDMA channelisation codes**

Having considered a simple example of the use of the channelisation codes, we can now look at where these codes come from. We have already seen that they are from a class of codes referred to as orthogonal codes.

There are many different types of orthogonal code (Walsh and Hadamard are the two most prominent), but the one that is chosen for WCDMA is the OVSF code.

## 3.3 Channelisation codes

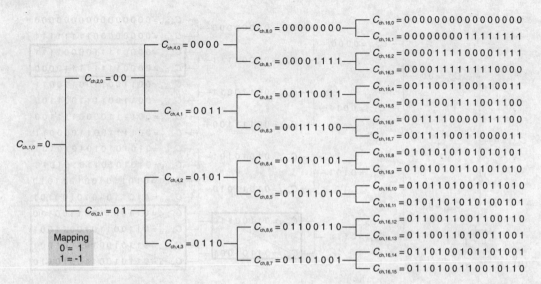

**Figure 3.11** OVSF code tree.

If we compare the OVSF code with the Walsh and Hadamard codes, we see that they all contain the same code sequences. The differences between the codes come from how we index them.

Figure 3.11 illustrates a family of channelisation codes ranging from a spreading factor of 1 (on the left hand side) to a spreading factor of 16 (on the right hand side). This family of channelisation codes is referred to as the OVSF code tree. The creation of the OVSF code tree can be defined with the following simple recursive algorithm. Using Figure 3.11, we start at the bottom of the tree with a value of binary 0. This corresponds to a channelisation code with a spreading factor of 1 (i.e. not spread). To move to the code of length 2 chips, we create two new branches on the tree. The algorithm for deciding the codes for these new branches can be summarised as follows:
- Moving up a branch repeat the parent node sequence twice.
- Moving down a branch repeat the parent node sequence twice, but inverting the second sequence.

From this we end up with two codes of length 2. By recursively applying this algorithm we can build four codes of length 4 and then eight codes of length 8.

### Variable data rate properties of OVSF codes

OVSF codes have some interesting properties in addition to the orthogonality property that was defined previously. The OVSF codes are defined to allow the support of simultaneous variable data rate channels. If the spreading factors are assigned according to a simple rule, then a multirate service can be supported. The rule for assigning a specific spreading code is that a code cannot be assigned with a parent or grand-parent code on the same branch of the tree.

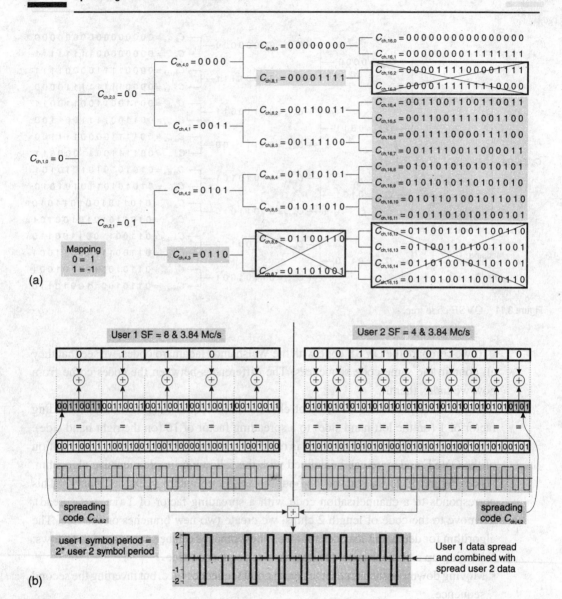

**Figure 3.12** OVSF code tree illustrating code restriction.

Consider Figure 3.12(a), in which the codes shown can all coexist at the same time, as they do not come from the same branch of the tree. The OVSF diagram illustrates which codes cannot be assigned simultaneously and which would result in 'code clashing'. By examining the typical spread waveforms that would result in using $C_{ch,4,3}$ for instance, it is clear why $C_{ch,8,6}$ and $C_{ch,8,7}$ cannot be used as that would result in the same codewords being used. Similar comments apply to the spreading factor 16 codes coming from the same branch.

## 3.3 Channelisation codes

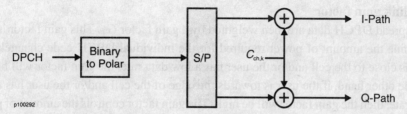

**Figure 3.13** Example downlink transmission path.

Figure 3.12(b) illustrates an example of how variable data rates can be implemented using the OVSF codes. User 1 is using a spreading factor 8 code. User 2 is using a spreading factor 4 code with the same chip rate but twice the data rate. The different codes selected come from different branches of the tree and hence there is no problem with the codes clashing; they maintain their orthogonality.

### 3.3.9 Spreading codes with WCDMA downlink quadrature modulation

In Section 3.3.1 and subsequent sections, we used a simple example to demonstrate how the channelisation codes are used in a system that is similar to WCDMA. As mentioned previously, however, we were only considering a simple binary modulation scheme.

In this section, we now want to extend the analysis presented there to also include a more realistic representation of the downlink modulation. The modulation that is used in the downlink is a type of quadrature modulation (this is a four-level modulation as opposed to the simple binary modulation that we considered initially). The quadrature modulation scheme consists of two binary modulation transmission paths referred to as the in-phase path (I) and the quadrature path (Q). In Sections 3.6 and 3.7 we will see the details of how the I and Q paths are modulated onto an RF carrier.

**DPCH data conditioning and spreading**

Figure 3.13 presents an example of a transmission path. In this example, we are assuming that the data to be transmitted are a dedicated channel to a user within the coverage area and which is called a DPCH. The DPCH to be transmitted on the downlink is first converted from binary format to polar format. Next the data pass through a serial to parallel converter with the first bit going through the I path and the second bit through the Q path.

The data bits are now converted into chips by repeating the data bit $n$ times, where $n$ is the length of the channelisation code used. In the example shown, the code is of length 4 chips, and consequently the data are repeated four times. In the next step, the data chips are multiplied by the channelisation code, and the same code is used for both the I path and the Q path. The output of the multiplication will be the spread DPCH data.

## Use of downlink gain factor

The spread DPCH data are then weighted by a gain factor $G_1$. This gain factor is used to define the amount of power required for the individual DPCH code channels. If a user is close to the cell and/or the user has a low data rate, the gain factor will be low. On the other hand, if the user is towards the edge of the cell and/or the user has a high data rate, then the gain factor will be high. The gain factor controls the amount of power that is allocated to a specific code and hence to a specific user.

We will see in Section 5.4 that the amount of received power required at the receiver depends upon the data rate of the DPCH that is being transmitted and not on the chip rate of the transmitted signal. The transmitted power, therefore, needs to be adjusted to account for the data rate of the transmission prior to the combination of the coded channels with any other coded channels to be transmitted on the downlink.

### 3.3.10 Uplink channelisation codes

Having considered the use of the OVSF codes in the downlink, we will now look at their use in the uplink. In the FDD mode of the WCDMA system the uplink and the downlink are asymmetrical. The use of the OVSF codes in the downlink is different from their use in the uplink. We saw that, for the downlink, the OVSF codes are used to separate the users, and also to define the data rate for that user. In the uplink, however, the OVSF codes are only used to define the transmission data rate for a specific user. The separation of the users on the uplink uses a different mechanism, which is based on the scrambling code that is also applied in the uplink.

## Uplink OVSF code use rationale

To understand why the uplink OVSF codes are not used to separate users, consider the situation presented in Figure 3.14. The diagram shows two users accessing the cell. The first user is close to the cell centre (distance $d_1$ away) and the second user is towards the cell edge (distance $d_2$ away). In the FDD mode, there is no uplink timing control and consequently the timing of the transmissions from user 2 and the timing of the transmission from user 2 are based on the received transmission on the downlink. If the user is close to the cell centre, the time of flight from the cell to the UE and back to the cell is quite small and, consequently, the uplink transmissions arrive close to some defined cell start time ($t_1$ seconds offset from ideal). For the UE at the edge of cell, however, the arrival time from the cell to the UE and then from the UE back to the cell will be much greater due to the propagation delay. This means that the transmissions from UE2 will arrive much later, at time $t_2$.

The two signals from UE1 and UE2 arrive at the Node B at different times. The time difference destroys the orthogonality property of the codes and consequently we cannot use the channelisation codes to separate the users on the uplink and we therefore

## 3.4 Scrambling codes

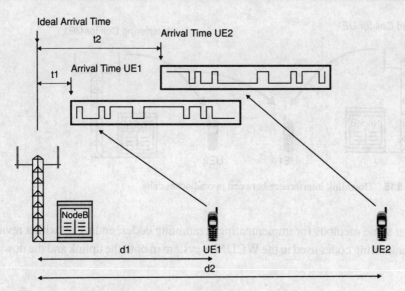

**Figure 3.14** Uplink transmission arrival times.

have to consider an alternative technique. In Section 3.4, we will explore the use of the scrambling codes as a means of isolating the users in the uplink.

### 3.3.11 Channelisation code peak-to-mean issues

**Peak-to-mean effects**

There is one important issue that concerns the use of the OVSF codes in both the uplink and the downlink. This issue relates to the peak-to-mean power ratio of the envelope of the signal when multiple OVSF codes are combined together.

Code combination in the downlink is a consequence of the multiple users being active within the same cell. Code combination in the uplink is a consequence of the user requiring multiple code channels to achieve high data rates. In either case, the consequence of the combination of the codes is an increase in the peak power of the RF signal as a ratio to the average power. This increase is particularly important for the RF power amplifier, which needs to be able to linearly amplify the transmitted signal, and also for the receiver, which needs to have sufficient receive dynamic range for correct signal reception.

## 3.4 Scrambling codes

In this section we start by explaining the need for scrambling codes, first in the downlink and then in the uplink. Once the need for scrambling codes is established, we then address some of the properties that are desirable for a scrambling code. Finally, we

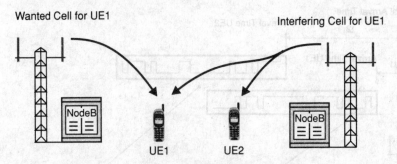

**Figure 3.15** Downlink interference between two adjacent cells.

look at some methods for implementing scrambling codes, ending up with a review of the scrambling codes used in the WCDMA system in both the uplink and the downlink.

### 3.4.1 Introduction to scrambling codes and requirements

**Downlink intercell interference**

Until now, we have considered the downlink of a WCDMA cell that is isolated from any other cells in the system. In reality, there are many cells in the system, and these cells are all using the same frequency. All the intercell transmissions are asynchronous, and so the interference between them needs to be considered. In addition, there are a finite number of orthogonal codes and we need to have the facility to use all of these codes within each cell.

The basic interference model is illustrated in Figure 3.15. Two signals arrive at a mobile with nominally equivalent power (equal power overlap between the two cells). The delays between the two signals are an arbitrary offset due to the asynchronous nature of the transmissions. Without modifications, the transmission scheme illustrated in Figure 3.15 will result in unacceptable interference between the two users. The solution is to introduce a second code in the transmission path on the downlink.

Figure 3.16 illustrates a modification to the basic transmission and reception paths from the cell to the user equipment. The addition to the transmission path is the scrambling code. The scrambling code is essentially a pseudo-random sequence of chips with amplitude of $+1/-1$. The code is applied at the chip rate to the spread data. The scrambling code is selected to have carefully controlled statistical properties.

In the receiver, the same scrambling code is applied. This has the effect of descrambling the data, returning it to its normal order. Subsequently, the despreading code can be applied and the data recovered.

Before considering the rationale for the success of a scrambling code, we need to consider what a scrambling code does in terms of its manipulation of the various signals that are being considered.

## 3.4 Scrambling codes

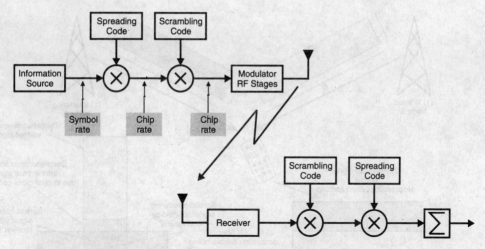

**Figure 3.16** Downlink transmission path including scrambling code.

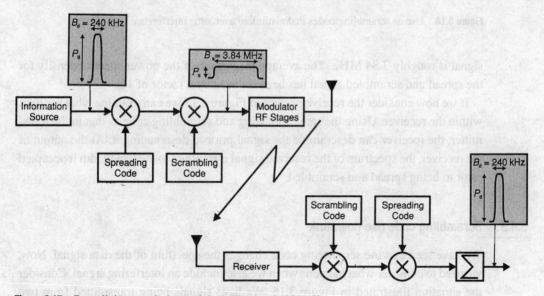

**Figure 3.17** Downlink transmission path including bandwidth effects.

### 3.4.2 Energy and spectrum considerations for codes

Let us consider Figure 3.17, which includes the spectrum of a filtered 240 kb/s data signal prior to spreading, scrambling and modulation in the transmitter. Figure 3.17 also shows the effects that the scrambling code has on the signal after spreading but prior to modulation. The scrambling code is applied with a spreading factor of 16, i.e. there are 16 chips from the scrambling code for every bit of data. Comparing the two stages, we can see that the spectrum of the data signal without the use of a spreading and scrambling code is roughly 240 kHz and the spectrum of the spread and scrambled

## 90  Spreading codes and modulation

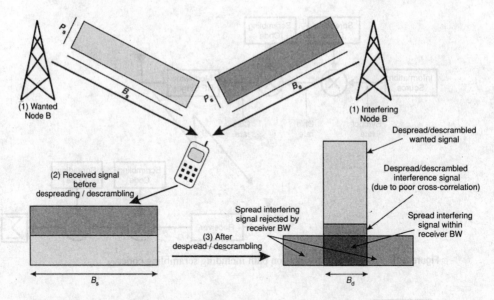

**Figure 3.18** Use of scrambling codes in downlink to overcome interference.

signal is roughly 3.84 MHz. The average magnitude of the power spectral density for the spread and scrambled signal has been reduced by a factor of 16.

If we now consider the receive portion of Figure 3.17 we can examine what happens within the receiver. Using the same spreading and scrambling code as that in the transmitter, the receiver can descramble the signal prior to despreading it. At the output of the receiver, the spectrum of the received signal is returned to the bandwidth it occupied prior to being spread and scrambled.

### 3.4.3  Scrambling code use downlink

We have seen that the scrambling code changes the spectrum of the data signal. Now, we need to consider what happens when we also include an interfering signal. Consider the situation illustrated in Figure 3.18. We have signals being transmitted from two cells to two different mobiles. The signals interfere with each other within the mobiles. In this section we want to consider the effects of this interference on the performance of the system, first for the downlink as illustrated and then on the uplink.

We now examine the processing steps as they occur in the receiver. In Figure 3.18, we see that the wanted signal and the interfering signal add together within the antenna of the UE. We are assuming that the spectrum density of the two signals is roughly (equal distance from the two cells) the same and that they occupy the same approximate bandwidth as seen at the input to the receiver.

After the despreading and descrambling, the spectrum of the wanted signal is returned to its original despread bandwidth of around 240 kHz. The interfering signal from cell 2,

## 3.4 Scrambling codes

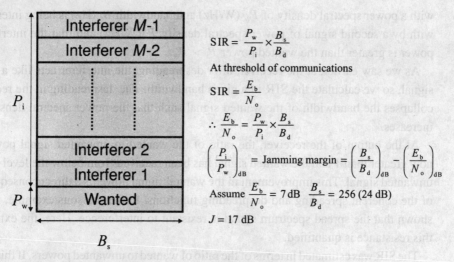

**Figure 3.19** Uplink use of scrambling codes for interference control.

on the other hand, is using a different scrambling code (code 2) to the code used by cell 1. This difference in the scrambling code means that the signal after descrambling in the receiver using code 1 will still remain scrambled. This is illustrated in the spectrum plots after the despreading stages in Figure 3.18.

Next, in the receiver there is an integrator and possibly a low pass filter (the integrator itself acts as a low pass filter, but we may wish to tighten up the cut-off using an additional filter in cascade). At the output of the filters we are left with the wanted signal plus the interference from cell 2 that lies within the wanted signal bandwidth.

From this result, we can see that the receiver, using the spreading and scrambling codes, is able to reject a significant portion of the interfering signal. The amount of the interfering signal that remains is related to the spreading factor and also the statistical properties of the scrambling code, some of which we consider in Section 3.4.7.

### 3.4.4 Scrambling code use uplink

We mentioned earlier that the uplink and downlink of the FDD mode WCDMA system are quite different, and that OVSF channelisation code in WCDMA is used to control data rates rather than separate the users (see Section 3.7.1). The scrambling code is used to separate the users on the uplink. Each user in the UTRAN can be assigned a unique scrambling code (assuming that there are sufficient codes available – which there should be). The unique scrambling code ensures that the transmissions on the uplink using the scrambling codes are identifiable to a specific UE and that interference from other UEs active at the same time can be rejected.

Figure 3.19 illustrates the basic principles of the spread spectrum from the perspective of the frequency domain. It is assumed that the wanted spread and scrambled signal with a power spectral density of $P_w$ (W/Hz) and bandwidth $B_s$ (Hz) is being interfered with by a second signal of power spectral density $P_i$ (W/Hz) and that the interfering power is greater than the wanted power $P_i > P_w$.

As we saw earlier, at the receiver, after despreading, the interferer acts like a noise signal, so we calculate the SIR in a given bandwidth. The despreading in the receiver collapses the bandwidth of the wanted signal such that the power spectral density $P_r$ increases.

At the output of the receiver, the ratio of the wanted to unwanted signal power is now greater than 1 and the wanted signal has been received from below the level of the unwanted signal. This improvement in the wanted signal power is a direct consequence of the coherent spreading and despreading functions. In the previous example, it was shown that the spread spectrum signal is resistant to interference. Here, the extent of this resistance is quantified.

The SIR was estimated in terms of the ratio of wanted to unwanted powers. If this ratio is set to equal the $E_b/N_o$ required for adequate communication, then the maximum ratio of the wanted to unwanted signals can be estimated, and hence from this the jamming margin. Assuming a spreading factor of 256, we see from Figure 3.19 that the jamming margin is 17 dB. This means that the interfering signal can be 17 dB greater than the wanted signal before the communications link is degraded below the threshold set by $E_b/N_o$.

**Capacity equation**

To assess how the jamming margin translates into system capacity for the CDMA system, we need to make some simplifying assumptions: namely that all interferers arrive at the base station with the same power level, and that all interference is from within the cell and there is no other cell interference.

The jamming margin estimated earlier can be directly translated into the carrier to interference margin at the base station. The total interference is given as $(M - 1)C$ (where $C$ is the carrier power). Therefore the ratio $I/C$ is equivalent to the jamming margin and so can be directly related to the spreading factor and SNR per bit as shown above.

From this simple analysis, we can see that, in the uplink, the scrambling codes are used to isolate the users, and that the number of users supported is finite and depends on factors such as the spreading factor and the sensitivity of the receiver ($E_b/N_o$). In this simplistic example, for the values illustrated in Figure 3.19, the capacity $M$ is in the region of 49 users.

### 3.4.5 Uplink power control

An additional important requirement for the uplink relates to power control. In the preceding analysis, we assumed that the power level for the different users was the same. If the power levels are not the same, the uplink capacity is significantly degraded.

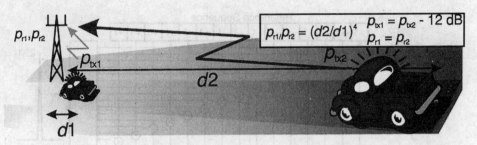

**Figure 3.20** Requirements and implementation of power control in uplink.

Consider the situation shown in Figure 3.20, where a wanted mobile is twice the distance from the base station than an unwanted mobile. Assuming an inverse fourth power law for the propagation loss, and equal transmit power for the mobiles, then the received signal power for the unwanted mobile is 12 dB higher than that for the wanted mobile. The signals are asynchronous, therefore only the spreading factor described earlier can reduce the effects of the unwanted mobile. By using the spreading factor in this manner, the capacity of the system is significantly reduced. This situation is called the near–far problem.

To overcome the near–far problem on the uplink, a technique called power control is used. Power control sets the power of both mobiles 1 and 2 so that the received power at the cell site is nominally the same for each.

The power can be set using either of two basic methods: by the mobile using the power received from the base station, or by the base station measuring the mobile power and explicitly telling the mobile what level to set it to. Later, we will see that there are still some problems associated with power control, and it is one of the factors that defines the uplink system capacity.

### 3.4.6 Scrambling code properties

Having reviewed the need for scrambling codes in the uplink and the downlink, in this section we examine the design and some of the properties of scrambling codes and in particular the scrambling codes that are used in the WCDMA system for both the uplink and the downlink.

An ideal scrambling code would be a truly random binary sequence, but the problem with such a scheme is making the sequence available in both the transmitter and the receiver. The solution is to use pseudo-random sequences based on some algorithm that will allow the creation of the pseudo-random sequence in both the transmitter and the receiver. This type of sequence is also commonly referred to as a pseudo-noise sequence (or PN sequence). To achieve its objectives, the scrambling code needs to have certain properties. The two main properties are called autocorrelation and cross correlation. Autocorrelation is a statistical property of the scrambling code that relates to the code on its own. Cross correlation is a property of two interacting scrambling codes.

# 94  Spreading codes and modulation

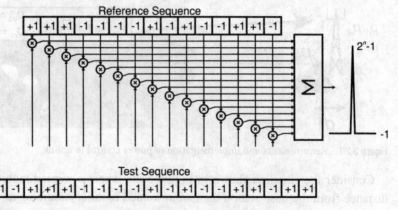

**Figure 3.21** Scrambling code autocorrelation and cross correlation.

In the section that follows we will examine the definitions for autocorrelation and cross correlation and then we go on to explore different candidates for the scrambling code.

## 3.4.7 Autocorrelation and cross correlation

Consider Figure 3.21. In the figure we see two scrambling codes being multiplied together on a chip-by-chip basis. The upper sequence is a reference sequence obtained from the scrambling code generator that pseudo-randomly takes the values of $\{+1$ or $-1\}$, and the lower sequence is the free-running output from the same scrambling code generator, which also pseudo-randomly takes the values of $\{+1$ or $-1\}$. After multiplication, the product of the two sets of sequences is summed with all other chip-by-chip multiplications. The output of the summer is referred to as the autocorrelation function of the scrambling code.

When the reference sequence and the free-running sequence perfectly align there is a peak signal at the output of the summer that is equal to the number of chips in the sequences being multiplied and summed. When the two sequences are not aligned the output is less than this peak level. The waveform shown at the output of the autocorrelation detection device in Figure 3.21 is typical for an autocorrelation function.

Cross correlation is similar to autocorrelation except that the reference sequence and the test sequence are different. Cross correlation is a measure of the differences between the two sequences. For systems such as WCDMA, the ideal cross correlation is one that has a low value.

**Desirable autocorrelation properties**

For a system such as WCDMA, the ideal scrambling code has a large autocorrelation peak when sequences are aligned and a small autocorrelation when sequences are not

## 3.4 Scrambling codes

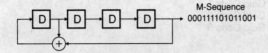

**Figure 3.22** Simple M-sequence generator and resultant sequence.

aligned. The rationale for such a requirement is driven mainly by the operation of the rake receiver that is considered in greater detail in Section 6.4. One of the functions of the rake is to reject interference from multipath reflections that occur with more than one chip delay. The autocorrelation function can be thought of as some type of transfer function for the rake receiver.

When the received signal is within one chip of some defined arrival time it is received by the rake with a high magnitude; when the multipath is more than one chip delayed, the rake receives a low magnitude. The autocorrelation function, therefore, needs to have a high peak and a low off-peak magnitude.

### Desirable cross correlation properties

For cross correlation we are measuring the amount of interfering signal that is coming from one transmission from one code into a second transmission using a different code. An ideal cross correlation performance, therefore, would be to have very low cross correlation for all time offsets of the code. In practice, cross correlation has some non-zero value and so the objective is to keep this to a minimum.

### 3.4.8 Scrambling code implementation

#### Types of PN sequences

There are a number of different types of PN sequences, each with different properties. Maximal length (M-sequences) sequences are the basic type. Gold codes and Kasami codes are derived from M-sequences and have slightly better properties in some regards.

#### Example PN sequence

A simple PN sequence based on M-sequences is shown in Figure 3.22. The length of the sequence is $2^4 - 1$ bits.

M-sequences are often used in CDMA systems for spreading and scrambling.

- M-sequences have very good, well-defined autocorrelation properties. These are described in greater detail later.
- M-sequences do not have very good cross correlation properties. These are described later.
- M-sequences are derived from an equation called a primitive polynomial. The structure and properties of primitive polynomials are considered next.

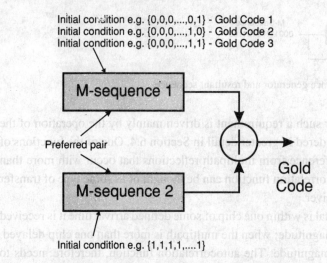

**Figure 3.23** Creation of Gold code from two M-sequences.

## Properties of M-sequences

M-sequences, based on primitive polynomials, have the following characteristics.

The length of an M-sequence is $2^n - 1$, where $n$ is the order of the primitive polynomial and the number of bits in the shift register used to create it.

In a complete sequence of $2^n - 1$ bits, the number of 1s and the number of 0s differs by at most 1.

Except for all 0 code words, every possible combination of $n$-bit words exists somewhere in the sequence.

The autocorrelation of an M-sequence has very good properties, as defined above and shown in Figure 3.22.

The cross correlation between different sequences of the same length, however, is not as good.

For use in a CDMA system, particularly when multiple M-sequences are used, both the autocorrelation and the cross correlation properties are very important, and both should be considered.

## Gold codes

Figure 3.23 illustrates the basic construction of a Gold code. Two separate and different M-sequences (referred to as preferred pairs) are used to create a Gold code. Different starting states for, say, the upper shift register (whilst retaining the same state for the lower register) result in a different Gold code being created.

Consider a pair of M-sequences each of length $2^n - 1$ and modulo-2 added as shown in the Figure 3.23. If this pair of M-sequences corresponds to what is referred to as a preferred pair, then the combined sequence is a Gold code.

## 3.5 Modulation

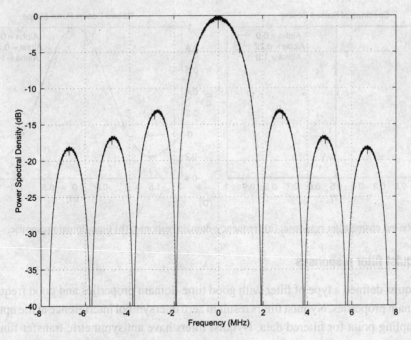

**Figure 3.24** WCDMA transmit signal spectrum.

$$t(n) = 2^{(n+1)/2} + 1 \quad n \text{ odd}$$
$$t(n) = 2^{(n+2)/2} + 1 \quad n \text{ even}$$

For a Gold code of length $2^n - 1$, the cross correlation takes on the three values $\{-1, -t(n), t(n) - 2\}$, where $t(n)$ is given above.

As an example, consider the case where $n = 12$. The worst-case cross correlation is $t(n) = 129$ and the autocorrelation is 4095.

## 3.5 Modulation

### 3.5.1 Introduction

Figure 3.24 shows the spectrum of a quadrature phase shift keying (QPSK) signal of the type used in WCDMA, excluding the filtering. The spectrum has adjacent channel sidelobes. The adjacent channel sidelobes introduce interference into users of the adjacent frequencies. The users of adjacent frequencies could be the same operator or could be other operators. We need to consider the use of filtering in the transmitter to overcome the effects of the adjacent channel sidelobes. When considering filtering we must also bear in mind its effects on the time-domain representation of the signal.

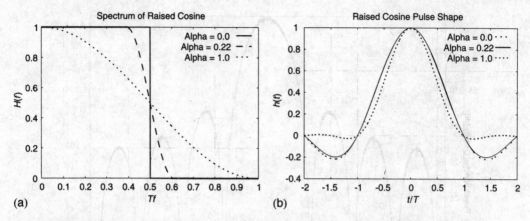

**Figure 3.25** Raised cosine filter response: (a) frequency-domain response; (b) time-domain response.

### 3.5.2 Nyquist filter responses

Nyquist defined a type of filter with good time-domain properties and good frequency-domain properties. Nyquist filters result in zero intersymbol interference at the optimum sampling point for filtered data. Nyquist filters have antisymmetric transfer functions about their cut-off point.

A practical example of a Nyquist filter is a raised cosine filter.

### 3.5.3 Raised cosine filter

The raised cosine filter is a type of Nyquist filter. Equation (3.8) presents the raised cosine in the frequency domain.

$$H(f) = \begin{cases} T & 0 \leq |f| \leq \dfrac{(1-\alpha)}{2T} \\ \dfrac{T}{2}\left(1 + \cos\left[\dfrac{\pi T}{\alpha}\left(|f| - \dfrac{1-\alpha}{2T}\right)\right]\right) & \dfrac{(1-\alpha)}{2T} \leq |f| \leq \dfrac{(1+\alpha)}{2T} \\ 0 & |f| > \dfrac{(1+\alpha)}{2T} \end{cases} \qquad (3.8)$$

The raised cosine filter is defined in terms of the quantity alpha. Alpha goes by many names: it is called the cut-off, the roll-off, the excess bandwidth. For WCDMA alpha takes the value of 0.22.

Figure 3.25(a) shows the raised cosine frequency response for alpha = 0, 0.22 and 1. Observe the antisymmetrical spectrum about its cut-off. The smaller alpha, the sharper the cut-off in the frequency domain. Figure 3.25(b) shows raised cosine time-domain response for alpha = 0, 0.22 and 1. Observe also that the impulse responses all pass through zero at the same point – zero intersymbol interference. Observe also that the magnitude of the ringing in the time domain becomes greater as the value of alpha becomes smaller.

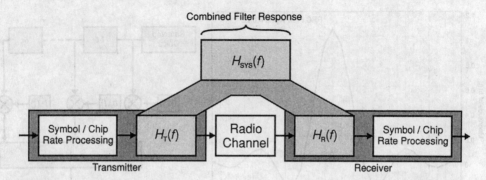

**Figure 3.26** Filter distribution between transmit and receive functions to meet Nyquist criteria.

## 3.5.4 Filter transfer function split between the transmitter (Tx) and the receiver (Rx)

In addition to the filtering in the transmitter, we also need filtering in the receiver. The filtering in the receiver is required to reject the noise and interference present in the receiver. The effects of the transmit filter and the receive filter are combined together.

It is the combined effects of the transmit filter and the receive filter that must be considered and which must have the properties of the Nyquist filter. Consider Figure 3.26, which shows the combination of the transmit and the receiver filters. $H_T(f)$ is the transfer function for the transmit filter in the frequency domain. $H_R(f)$ is the transfer function for the receive filter in the frequency domain. The combined effect of the two filters is given by the multiplication of the two transfer functions, as

$$H_{SYS}(f) = H_T(f)H_R(f) \tag{3.9}$$

We ignore any effects of the frequency-domain transfer function introduced by the radio channel. We now assume that the same filter is used in the receiver as in the transmitter $(H_R(f) = H_T(f))$; this is what is referred to as a matched filter and it has optimum performance. The combined transfer function of the transmitter and the receiver is then given by

$$H_{SYS}(f) = [H_T(f)]^2 \tag{3.10}$$

The transfer function for the transmitter is given by the square root of the combined transfer function. From the previous discussion we require the combined transfer function to be a raised cosine ($H_{SYS}(f) = RC$).

$$H_T(f) = \text{Sqrt}\,[H_{SYS}(f)] = \text{Sqrt}\,(RC) \tag{3.11}$$

From equation (3.11), therefore, the transmit filter should be the square root of a raised cosine, or a root raised cosine (RRC).

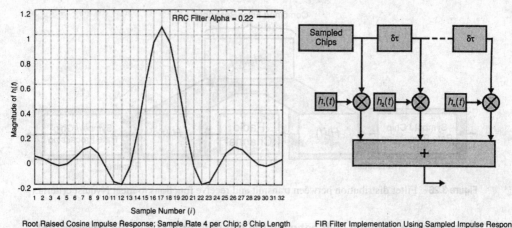

Figure 3.27 FIR filter implementation.

### 3.5.5 RRC filter

The RRC filter defines the transfer function for the filter that is to be used in the transmitter for the WCDMA system. The time-domain representation for the RRC is [8]

$$h(t) = \frac{\sin\left[(1-\alpha)\pi\frac{t}{T}\right] + 4\alpha\left(\frac{t}{T}\right)\cos\left[(1+\alpha)\pi\frac{t}{T}\right]}{\left(\pi\frac{t}{T}\right)\left[1 - \left(4\alpha\frac{t}{T}\right)^2\right]} \qquad (3.12)$$

where

$$T = \frac{1}{\text{chip rate}} = \frac{1}{3.84 \times 10^6} \approx 0.260\,\mu s$$

### 3.5.6 Implementation aspects: finite impulse response (FIR) filter, sampling rate, impulse response length

The simplest method of implementing a filter such as the RRC filter is to use an FIR filter. The FIR filter is shown in Figure 3.27. The FIR consists of a delay line that delays samples of the signal to be filtered. The output from each delay element is multiplied by a quantity $h(i)$. The output of the multipliers is then summed to produce the output of the filter, which is the filtered version of the input signal.

The quantity $h(i)$ is a sample of the impulse response of the filter being implemented. In this case it is the RRC filter. The input to the filter is the signal to be filtered, sampled at an appropriate rate. The sampling rate of the signal and the sampling rate used to create the filter impulse response must be the same.

## 3.5 Modulation

In designing the filter the sampling rate and the length of the impulse response both need to be considered. The higher the sampling rate the easier it is after a digital to analogue converter to remove high frequency components generated in the sampling process. The higher the sampling rate, the faster the converter must operate, which in general will lead to greater cost and greater power consumption.

A good design will trade-off sampling rate with cost and complexity to provide an optimum compromise. The next criterion to consider is the length of the FIR. The longer the FIR used, the better the filter performance in both the frequency and the time domain. The longer the FIR, the more elements in the FIR filter and hence a greater cost and greater power consumption will arise. Once again it is a compromise between cost/power consumption and performance.

### 3.5.7 I/Q modulation phase representation

In this section we consider the phase representation of signals. We use this technique to see how the modulation, spreading and scrambling affect the phase of the transmitted signal.

We start with a consideration of a simple RF carrier and then develop this into a more realistic example.

**RF carrier representation as a phasor**

Let us consider just an RF carrier on its own. The RF carrier can be represented as a sine wave. We can represent the characteristics of the RF carrier using a phasor diagram. A phasor diagram illustrates how the phase and amplitude of the RF carrier change. Figure 3.28(a) is a phasor diagram of the RF carrier. It is a circle, and the RF carrier traverses the circle at a rate that is equal to its frequency.

When considering modulation schemes, the continuous part of the phasor diagram is not of much interest, and we tend to focus on the slower variations in the RF carrier. To observe the slow variations, we stop the fast variations of the RF carrier by ignoring the rotations of the phasor diagram at the carrier frequency (we actually do this by subtracting from the RF carrier a phasor that is rotating at the same carrier frequency but in the opposite direction). The result is a single point on the phasor diagram as shown in Figure 3.28(b). The single point represents the phase of the RF carrier. In Figure 3.28(b) we define the phase of the RF carrier as being 0° for this cosine waveform.

**Simple binary phase modulation**

Consider the simple binary modulation illustrated in Figure 3.28(c). The phase of the RF carrier changes each time the polarity of the modulating signal changes. The RF carrier is represented as a cosine function and should have the phases 0° and 180° when

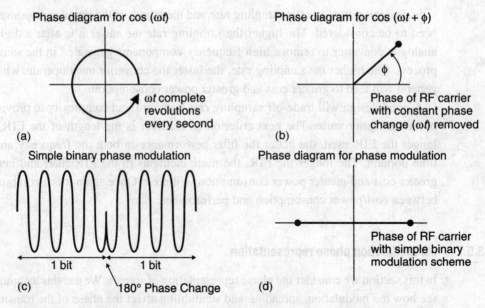

**Figure 3.28** Phasor diagram and its use presenting phase changes: (a) phasor diagram for an RF carrier; (b) phasor diagram showing only constant phase offsets; (c) sample binary phase modulation of an RF carrier in the time domain; (d) phasor diagram of simple binary phase modulation.

the modulating signal takes the values $+1$ and $-1$ respectively. Figure 3.28(d) shows the phase of the signal in the phase plane.

## 3.6 Downlink spreading and modulation

A WCDMA signal can be considered in terms of a simple example. Using Figure 3.29, which illustrates a simplified version of the WCDMA downlink, we will consider parts of the WCDMA downlink system. The WCDMA downlink consists of a DPCH.

The DPCH is a stream of binary information. The binary DPCH is converted into a polar format using the mapping rule defined in Section 3.3.3. Next the DPCH passes through a serial to parallel (S/P) converter, which passes the data alternately to two streams. The upper stream of polar symbols is referred to as the in-phase plane (I plane) and the lower stream of polar symbols is referred to as the quadrature plane (Q plane). The I plane is spread using a channelisation code that relates to the required data rate. The Q plane is spread using exactly the same channelisation code.

The I plane data are passed through an RRC filter to remove the high frequency variations in the transmitted stream of chips. The Q plane data are also passed through

## 3.6 Downlink spreading and modulation

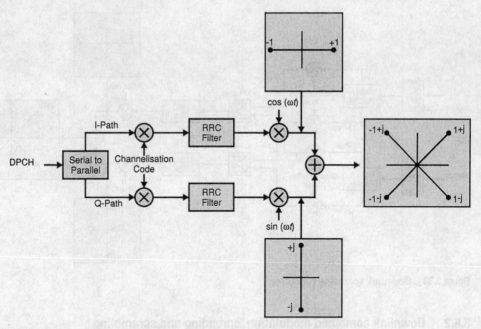

**Figure 3.29** WCDMA downlink transmission phase and constellations.

an RRC filter. An RF carrier then multiplies the I-plane; in this case it is a cosine function as seen for the binary modulation scheme.

The Q plane is multiplied by an RF carrier that is 90° rotated to the cosine, i.e. a sine function. Figure 3.29 illustrates the phases for the I plane signal on its own, and the Q plane signal on its own. For a real RF signal, the phase of the filtered signal will vary continuously.

I takes the values of 0° and 180°, Q takes the values of 90° and −90°. The sum of I and Q is a signal that has the values which correspond to the vector sum of I and Q, i.e. 45°, 135°, −45° and −135° as shown in Figure 3.29.

We can represent the two I and Q signals using complex arithmetic: I takes the values +1 or −1, Q takes the values of +j or −j. The four constellation points can also be represented in polar coordinates as $A \exp(j\psi)$, where $A$ is sqrt(2) and $\psi$ in degrees is: {45, 135, −45, −135}.

### 3.6.1 Treatment of I and Q planes

I and Q planes are transmitted together, but we can consider them independently. I and Q planes are orthogonal and therefore independent. The receiver should recover I and Q planes separately before recovering the transmitted data stream.

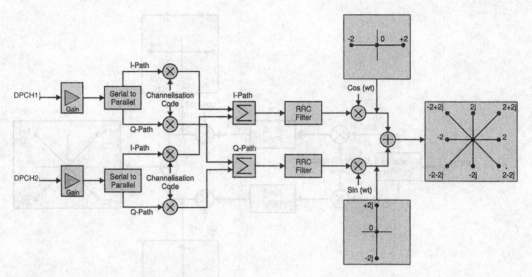

**Figure 3.30** Downlink spreading functions.

### 3.6.2 Downlink combined modulation, spreading and scrambling

This section considers the addition of the scrambling code to the downlink transmissions. In Figure 3.29, we saw that the phase of the downlink modulated RF carrier can be represented as four points on the phase constellation diagram. As the data and the channelisation code changed, the phase of the RF carrier moved between these four points. Up until now, we have only considered one user and hence one code in the downlink. With multiple users there are multiple codes in the downlink. Consider the case, where there are two users, shown in Figure 3.30. The data are separately split into I and Q planes and then spread using the channelisation code.

The I plane data for user 1 and user 2 is then summed. The Q plane data for user 1 and user 2 are then summed. In this example we are assuming that the gain block shown in Figure 3.30 is set to 1 for both users. At the output of either the I-plane summer or the Q plane summer, the waveform takes the values $+2$, $0$ or $-2$ assuming that the input data was $\pm 1$ for both users.

Figure 3.31(a) shows the possible constellation points that result if we allow all possible combinations of data and codes for users 1 and 2. If we had four users, at the output of the summer, we would expect to see waveforms with amplitudes $+4, +2, 0 -2 -4$. The constellation diagram for four users is shown in Figure 3.31(b). In general, as we increase the number of users, the constellation becomes more complex due to the combinations of the different codes for the different users.

**Amplitudes of different users**

Until now we have assumed that the gain block in the transmitter is unity for all users considered. In reality, the gain block is not unity for all users for the following reasons.

## 3.6 Downlink spreading and modulation

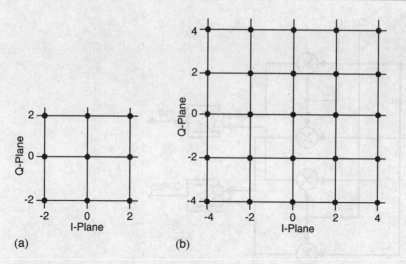

**Figure 3.31** Phase constellation diagram for downlink multiuser signals with: (a) two users; (b) four users.

If a user is close to a cell, that user will need less power than if they were at the extremes of the cell. Consequently, the gain is lower for users close to the cell centre and higher for users at the extremes of the cell. If a user has a low data rate, the effective bandwidth of the signal is low and therefore the required transmit power is low. If the user has a high data rate, the effective bandwidth is higher and consequently the transmit power needs to be higher.

The gain factor changes between users depending upon the data rate of the users and the location of the users in the cell. For the remainder of this section, however, we will assume that the gain factor is the same for all users.

### Complex scrambling and modulation

In the section on spreading and scrambling we saw that we need to use a scrambling code for various reasons not least of which is to provide interference rejection in the receiver. In WCDMA, the scrambling code is what is referred to as a complex scrambling code. With a complex scrambling code we have both an I-plane component and a Q-plane component.

Next we will review what effects the complex scrambling code has on the phase constellation that we saw for the modulated and spread case on its own.

### Introduce complex scrambling

Figure 3.32 shows the previous case, but with the addition of a complex scrambling code. The complex scrambling code is represented as two chip streams $I_{sc}$ and $Q_{sc}$. To explain the structure of Figure 3.32 consider expression (3.13) for complex scrambling of a complex spread signal. The complex spread signal is $(I_{sp} + jQ_{sp})$ and the complex

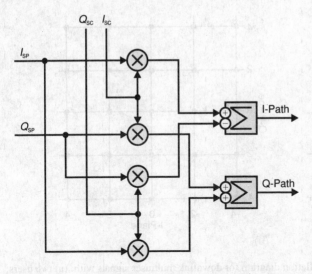

**Figure 3.32** Downlink transmission path including complex scrambling.

scrambling signal is $(I_{sc} + jQ_{sc})$. The product of these two codes is

$$(I_{sp}I_{sc} - Q_{sp}Q_{sc}) + j(I_{sp}Q_{sc} + Q_{sp}I_{sc}) \tag{3.13}$$

From (3.13), we can see where the cross multiplication between the I plane and the Q plane shown in Figure 3.32 comes from. Representing the spread data as $A_{sp}\exp(j\psi_{sp})$ and the complex scrambling code as $A_{sc}\exp(j\psi_{sc})$ the product of these is

$$A_{sp}A_{sc}\exp(j\psi_{sp})\exp(j\psi_{sc}) = A_{sp}A_{sc}I_{sc}\exp(j\psi_{sp} + j\psi_{sc}) \tag{3.14}$$

The magnitude of the resultant signal is the product of the magnitudes of the individual signals, and the phase of the signals is the sum of the phase of the individual signals.

## Phase diagram effects of complex scrambling for one user

Now we wish to explore the effect of the scrambling code on the phase constellation. First consider the constellation of the spread modulated signal for a single user shown in Figure 3.33(a). The constellation of the scrambling code is shown in Figure 3.33(b). The product of the spread modulation and the scrambling code is shown in Figure 3.33(c).

Two things are immediately obvious. First the amplitude of the combined constellation is greater (this is due to the multiplication of the amplitudes seen in (3.14)). If $A_{sp} = \text{sqrt}(2)$ and $A_{sc} = \text{sqrt}(2)$ then the product is 2. The second obvious difference is the phase changes of the constellation due to the phase of the scrambling code. Consider (3.14) again. The output phase after scrambling is the input phase plus the phase of the scrambling code. The phase of the scrambling code can take the values $\{+45°, +135°, -45°, -135°\}$. The scrambling code changes the output phase by these amounts. This results in the entire constellation appearing to have a $45°$ phase shift.

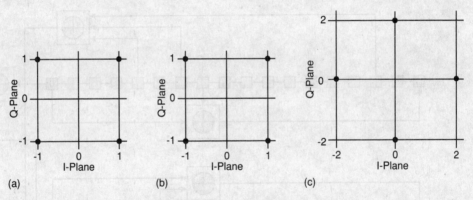

**Figure 3.33** Downlink phase constellation: (a) spread signal phase diagram; (b) scrambling code phase diagram; (c) spread scrambled signal phase diagram.

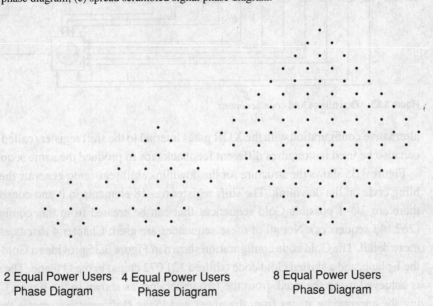

**Figure 3.34** Constellation diagrams for two, four and eight users.

## Constellation diagrams for two, four and eight users

With multiple users, the number of codes increases, the constellation changes and therefore the scrambled constellation is different. Figure 3.34 shows the constellations for two, four and eight users assuming that each spreading code has the same amplitude.

### 3.6.3 Structure and operation of downlink complex scramble code generator

In this section, we consider the structure and design for the downlink complex scrambling code. The specifications define the shift register structure in what is known as a Fibonacci configuration, with the XOR gates external to the shifting operation. The

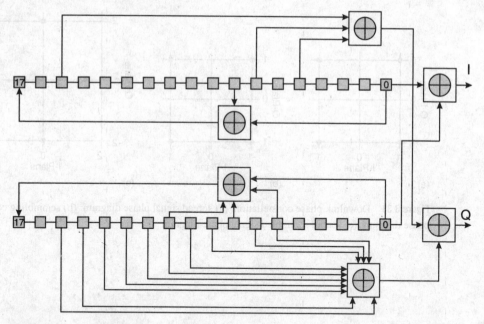

**Figure 3.35** Downlink Gold code generator.

alternative configuration with the XOR gates internal to the shift register (called Galois) can also be used but requires different feedback taps to produce the same sequences.

Figure 3.35 shows the structure for the downlink shift register to generate the scrambling code on the downlink. The shift register has 18 elements in it and consequently there are $2^{18}-1$ possible Gold sequences that can be created from this configuration (262 143 sequences). Not all of these sequences are used; Chapter 4 discusses this in more detail. The Gold code configuration shown in Figure 3.35 provides a Gold code in the I-plane, and a shifted Gold code (shifted 131 072 chips) in the Q plane. The shifting is achieved using the mask from the two generators as shown in the figure. Combining the appropriate stages from the upper and lower shift registers creates the mask. The locations of these stages are defined in the specifications. The two shift registers comprise 18 stages. The upper shift register is initialized to have 17 binary zeros and a single one, the lower shift register is initialized to have 18 ones.

The $n$th sequence from the shift register is defined by cycling the upper shift register through $n$ iterations to define the initial conditions for the upper shift register. The lower shift register starts with all elements equal to 1. When the upper and lower shift registers are combined we produce the I plane and the Q plane scrambling codes ($I_{sc}$, $Q_{sc}$).

## 3.7 Uplink spreading and modulation

In this section, we consider the uplink spreading and scrambling code generation. The uplink spreading and scrambling codes are implemented in a manner that is different

## 3.7 Uplink spreading and modulation

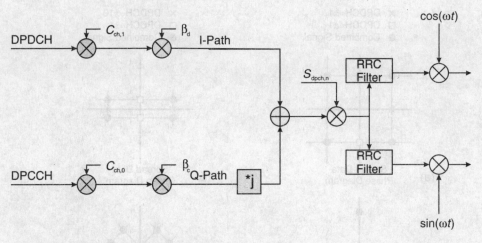

**Figure 3.36** Uplink transmission block diagram.

from the downlink. First we consider the use of spreading and how it applies to the different physical channels.

### 3.7.1 Uplink spreading and modulation

**DPDCH and DPCCH uplink codes**

Figure 3.36 shows the typical structure for the uplink spreading and modulation. In this section we consider a single code and in Chapter 4 we will extend the structure to include multiple physical channels.

The basic structure is similar to the downlink except that the I plane and the Q plane carry different types of information. The I plane carries a channel called the DPDCH. Typically, the DPDCH is a mixture of user traffic and control signalling. The spreading factor for the DPDCH varies between 4 and 256 and the physical channel data rate varies between 960 kb/s and 15 kb/s.

The physical channel data rate is not what the user will obtain: channel coding and other layer overheads need to be accounted for. For a single DPDCH the channelisation code number $k$ for the DPDCH channel is given in terms of the spreading factor ($SF$) in

$$k = SF/4 \qquad (3.15)$$

The Q-plane carries the dedicated physical control channel (DPCCH). The DPCCH carries pilot bits, power control (TPC) bits, feed back mode indicator (FBI) bits and transport format combination indicator (TFCI) bits. The spreading factor for the DPCCH is always 256, resulting in a physical channel data rate of 15 kb/s. The DPCCH uses channelisation code 0, spreading factor 256 from the OVSF tree.

# Spreading codes and modulation

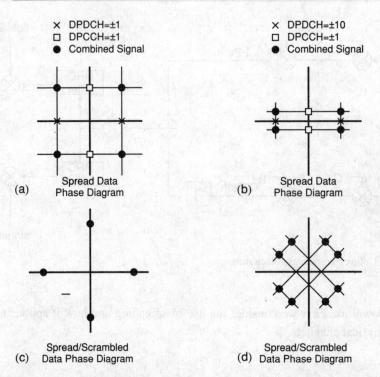

**Figure 3.37** Uplink signal constellations for varying gain factors: (a) spread signal equal in power in I and Q paths; (b) spread signal with unequal powers in I and Q paths; (c) spread and scramble signal with equal power in I and Q paths; (d) spread and scrambled signal with unequal powers in I and Q paths.

## Uplink modulation

After spreading, there is the scrambling stage, which we will consider in detail in the following section. After scrambling there is the modulation. The modulation for the uplink is similar to that for the downlink. The modulator uses an I and a Q plane with filtering of the I plane and Q plane signals using an RRC filter. The uplink uses a quadrature up-conversion procedure identical to that used in the downlink.

## DPDCH and DPCCH reasons for amplitude differences

When the DPDCH and the DPCCH have the same spreading factor of 256 the amplitude weights $\beta_d$ and $\beta_c$ are roughly the same. When the spreading factor for the DPDCH changes (decreases) due to higher uplink data rates for the DPDCH, the amplitude weight $\beta_d$ must increase to keep the DPDCH and DPCCH links balanced. DPDCH with a higher data rate experiences greater noise power in the receiver due to the increase in the receiver bandwidth. As a consequence of this, the constellation of the spread data does not look like the constellations we saw for the downlink. Figure 3.37(a) shows the constellation for the spread signal and modulation when $\beta_d$ and $\beta_c$

are nominally the same. Figure 3.37(b) shows the constellation for the spread signal with modulation when $\beta_d$ and $\beta_c$ are different by a factor of 10.

## Effects of zero crossings

We need to consider also the effects of zero crossings. A zero crossing is a transition that the modulation makes under certain circumstances and in which the phase constellation of the signal passes very close to the zero amplitude position. The zero crossings are particularly important for the design of the power amplifier in the handset for two reasons. The first reason is that the dynamic range over which the amplifier needs to be linear is much greater. The second reason is that there is a large overshoot of the modulation caused by the zero crossing which further increases the strain on the design of the linear amplifier.

In the next section we consider a technique defined for the uplink that helps overcome this problem. The technique is called hybrid PSK (HPSK) and it is also known as orthogonal complex QPSK (OCQPSK).

### 3.7.2 HPSK

HPSK helps in the design of linear power amplifiers in the uplink. It achieves this by controlling the phase variations of the spreading codes and the scrambling codes to only permit certain transitions. HPSK is implemented using a number of elements in the spreading code stage, a first scrambling code and a decimated second scrambling code.

HPSK works by limiting the phase transitions between pairs of adjacent chips, but between these pairs the transitions are not limited.

## Spreading code stage

The allowed spreading codes for HPSK are controlled to ensure that pairs of adjacent chips do not change value in either the I plane or the Q plane. There is no constraint over the phase of these pairs of chips; they can take any of the four allowed values ($+45°$, $-45°$, $+135°$, $-135°$). Between these pairs, the phase in the I plane and Q plane can change depending upon the codes. To achieve this constant phase across pairs, only codes that come from the upper part of the OVSF code tree should be used for HPSK. This is because they always have two consecutive chips the same. To select the codes, the DPCCH uses code 0, which is a code where the spreading chips are the same. The DPDCH uses code *SF*/4, which is always in the upper part of the code tree.

## Scrambling code considerations

Having considered the spreading code, we now need to consider the scrambling code. The HPSK benefits from the scrambling code can be considered in three stages.

## Spreading codes and modulation

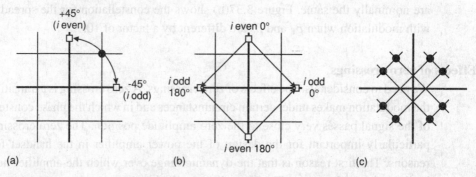

**Figure 3.38** Constellation changes introduced by elements of the uplink scrambling code: (a) phase change (±45°) from the first stage phase rotator; (b) phase transitions even to odd due to the second stage scrambling code; (c) complete constellation diagram.

### First scrambling code stage

The requirement for the first stage of the scrambling code is that, for the pair of chips of the spreading code that remain the same, the phase change from the scrambling code is +45° for the first chip and −45° for the second. Figure 3.38 shows a constellation point and the effects of the first element of the scrambling code. This is achieved by multiplying the spreading code by $(1 + j(-1)^i)$, where $i$ defines the chip number. This expression results in a waveform whose phase changes by +45° and −45° every alternate chip.

### Second scrambling code stage

The second stage of the scrambling code is a PN sequence based on a Gold code. This PN sequence is real-valued and so the phase of the combined signal changes by 0° or 180° depending upon the value of the code. To achieve this, a real scrambling code multiplies both the I-plane and the Q-plane by the same real quantity (either +1 or −1).

The second scrambling code is pseudo-random and so the first chip and the second chip are in general different. Between the pairs of chips we are considering the phase change is fixed at 90° from the first stage and is changed either by 0° or 180° from the second stage. The resultant phase change between the pair of adjacent chips is then either 90° or 270°. If the phase difference is 270°, this is the same as a −90° phase shift. With a phase shift of 90° or −90° the phase constellation does not pass through the zero crossing point. Figure 3.37(c) shows the four possible scrambling codes assuming the same spreading code.

### Third scrambling code stage

The third scrambling code stage consists of a real code decimated in time by 2. The decimation ensures that the code remains constant between adjacent pairs of chips. Outside of adjacent pairs of chips this final code is only applied to the Q plane component and will therefore cause a phase rotation of either +45° or −45°. This part

### 3.7 Uplink spreading and modulation

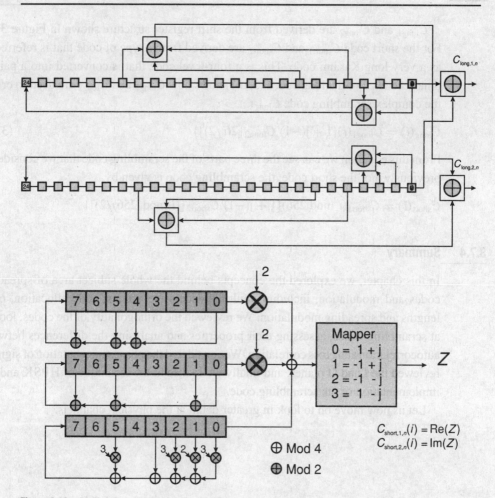

**Figure 3.39** Uplink long and short scrambling code generators.

of the spreading code is to ensure that the power in the I-plane and the Q-plane is equally distributed on average.

This power distribution is beneficial for equipment design and also reduces the impact of interference on the uplink by averaging the transmit power into both the I and Q planes. Figure 3.37(d) shows the effects of this complex scrambling phase rotation on the in-balance constellation that was considered previously.

#### 3.7.3 Scrambling code implementation

The complete scrambling code is implemented using either a short code or a long code. The long code is defined by two elements $C_{long1}$ and $C_{long2}$. $C_{long1}$ is a real-valued code of length $2^{25}-1$, $C_{long2}$ is the same code as $C_{long1}$ but with a 16 777 232 chip code phase offset between them. This means that $C_{long2}$ code is a delayed version of the $C_{long1}$ code.

$C_{\text{long1}}$ and $C_{\text{long2}}$ are derived from the shift register structure shown in Figure 3.39. For the short code $C_{\text{short1}}$ and $C_{\text{short2}}$ are derived from a type of code that is referred to as a very long Kasami code. This is a four-level code that is converted into a pair of binary codes. For the long codes (3.16) shows how $C_{\text{long1}}$ and $C_{\text{long2}}$ are used to create the complex scrambling code $C_{\text{long}}$:

$$C_{\text{long}}(i) = C_{\text{long1}}(i)\{1 + j(-1)^i C_{\text{long2}}\lfloor 2(i/2)\rfloor\} \tag{3.16}$$

From this equation we can see the three parts of the scrambling code that we considered previously. For the short code, the scrambling code is given by

$$C_{\text{short}}(i) = C_{\text{short1}}(i \bmod 256)\{1 + j(-1)^i C_{\text{short2}}\lfloor (i \bmod 256)/2\rfloor\}$$

## 3.7.4 Summary

In this chapter, we explored the concepts behind the whole subject area of spreading codes and modulation, including multiple access techniques, code allocation, code lengths and spreading modulation. We reviewed the orthogonality of the codes, looked at scrambling codes, assessing their properties and analysing the differences between autocorrelation and cross correlation. We considered the phase representation of signals, reviewed the I and Q planes, and finally we looked at the technique of HPSK and the implementation of the scrambling code.

Let us now move on to look in greater detail at the physical channels.

# 4 Physical layer

## 4.1 Introduction

In this chapter, we see how the specific spreading and scrambling codes are created for the WCDMA system, and how the physical channels are structured. We start by considering the physical channels defined for the uplink and then the physical channels defined for the downlink.

## 4.2 Physical channel mapping

Figure 4.1 illustrates the mapping between the transport channels and the physical channels. The figure shows which transport and physical channels exist in both the uplink and the downlink, and also whether or not these channels are bi-directional or single-directional.

Some of the physical channels are used only by the physical layer, and have no transport channel associated with them. In the sections that follow, we consider some of the physical channels in greater detail.

## 4.3 Uplink channels

### 4.3.1 Physical random access channel (PRACH)

The PRACH transports the RACH transport channel across the radio interface. As the name implies, the PRACH uses a random access process based on a modified form of the slotted aloha protocol [9].

The PRACH consists of the transmission of an access preamble in a specific part of an access frame, as shown in Figure 4.2. If a positive acknowledgement is received on the acquisition indication channel (AICH), then the UE can proceed to transmit what is known as the message part.

## Physical layer

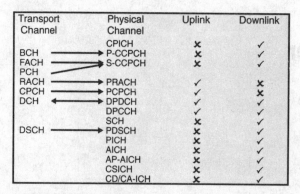

**Figure 4.1** Mapping between transport channels and physical channels.

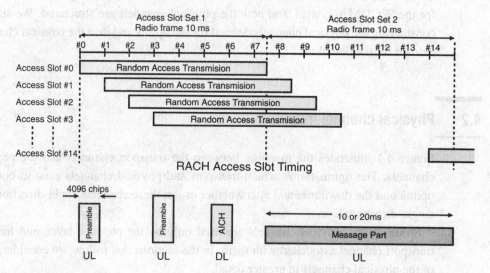

**Figure 4.2** Transmission of an access preamble in a specific part of an access frame.

For the PRACH to support a large number of simultaneous users, a number of different features are employed. First, there are 16 special waveforms called the access preambles. Each of these access preambles can be transmitted within segments of the access frame, called access slots. In total, there are 15 access slots per access frame, with each access frame having the duration of two radio frames (a radio frame is 10 ms in duration).

Each access slot is 5120 chips long and the access preambles are 4096 chips long. The random access process consists of the transmission of one or several access preambles and, following a successful acknowledgement by the UTRAN via the AICH, there is the transmission of the message part of the random access burst, which contains the layer 3 message.

## 4.3 Uplink channels

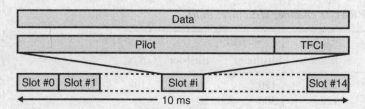

**Figure 4.3** Basic structure for the message part of the PRACH.

### PRACH slots

Figure 4.2 illustrates the basic random access process. In the upper part of the diagram, we see the random access slots distributed evenly within the radio frame structure. Access slot 0 starts at position 0 within the access frame, access slot 1 at position 1, and so on. With this structure, 15 access slots are provided on the PRACH access channel.

The lower part of the diagram illustrates the basic principle of the random access process. The UE transmits in selected access slots a selected PRACH access preamble until it receives an acknowledgement from the AICH. Chapter 8 describes the procedure used for selecting the access slot and access preamble. The AICH on the downlink can carry two specific types of indication, either a positive acknowledgement (ACK) or a negative acknowledgement (NACK). The ACK/NACK are matched to the access preamble; thus there are 16 AICH waveforms.

On successful receipt of an ACK from the UTRAN, the UE transmits the message part of the random access information. The message part for the PRACH channel is of length 10 ms or 20 ms (one or two radio frames). The message length is selected by higher layers, described in Chapter 11.

### PRACH message structure

Figure 4.3 illustrates the basic structure for the message part of the PRACH. The PRACH message can be constructed from either a single 10 ms frame or two 10 ms frames. If two 10 ms frames are used, the structure for each of these frames is the same as for the single 10 ms frame illustrated in Figure 4.3. There are two elements to the PRACH message: the data and the physical control information. Figure 4.3 illustrates the basic structure for the data at the top and the physical control information below it. The data and the physical control information are transmitted together, i.e. they are transmitted in parallel.

From Table 4.1(a) it can be seen that it is possible to have a variable data rate on the PRACH channel, ranging from 15 kb/s up to 120 kb/s.

For the physical control channel part of the PRACH, the data rate is fixed at 15 kb/s, corresponding to a spreading factor of 256. In total, there are eight pilot bits per slot and two TFCI bits per slot, which are spread by a factor of 256. Pilot bits are used

**Table 4.1(a).** *PRACH channel structure: data*

| Channel bit rate (kb/s) | Spread factor | Bits/frame | Bits/slot | $N_{data}$ |
|---|---|---|---|---|
| 15 | 256 | 150 | 10 | 10 |
| 30 | 128 | 300 | 20 | 20 |
| 60 | 64 | 600 | 40 | 40 |
| 120 | 32 | 1200 | 80 | 80 |

**Table 4.1(b).** *PRACH channel structure: physical control*

| Channel bit rate (kb/s) | Spread factor | Bits/frame | Bits/slot | $N_{pilot}$ | $N_{TFCI}$ |
|---|---|---|---|---|---|
| 15 | 256 | 150 | 10 | 8 | 2 |

for channel estimation by the Node B receiver; the TFCI bits are used for defining the message transport format.

## PRACH TFCI

Within the physical control part of the PRACH message, there are 2 bits per slot that are allocated for the coded TFCI bits. With 15 slots per 10 ms frame, there are 30 bits assigned for the coded TFCI per 10 ms message. These 30 bits are used to define the transport format that is used in the PRACH.

The selected TFCI is defined by 10 bits. These 10 TFCI bits are coded to 32 coded TFCI bits using a (32,10) Reed Muller code. Then two of the coded TFCI bits are punctured to reduce this to a 30 bit code word. The 30 bit code word is then divided into 15 segments with 2 bits per segment. Each of these segments is transmitted in the TFCI field of the physical control information part of the PRACH message. There are 15 TFCI fields per 10 ms frame and consequently the 30 bit code word is transmitted within a single 10 ms frame.

### 4.3.2  Physical common packet channel (PCPCH)

The PCPCH transports the CPCH transport channel. It is a common channel shared between a number of users. The channel uses access procedures similar to those employed by the PRACH, but also includes collision detection and optional channel assignment.

The PCPCH can be used in two different modes. In the first mode, the UE selects a specific PCPCH based on the transport formats (TFs) it supports, and attempts to access it on that basis. In the alternative mode, the UE selects a data rate requirement,

## 4.3 Uplink channels

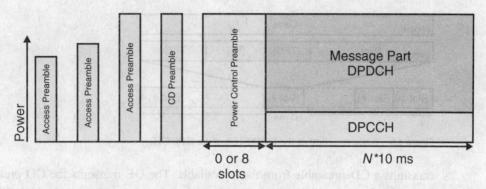

**Figure 4.4** Basic process through which the PCPCH operates.

and is assigned a specific channel. To achieve this, a CPCH status indication channel (CSICH) is used (discussed in detail next). The operation of the PCPCH function is discussed in Chapter 8.

In using the PCPCH, the UE potentially transmits a sequence of several access preambles, and also a collision detection preamble, followed by a 0 or 8 slot power control preamble, and then a variable length message.

The CPCH is a common transport channel that a UE in the RRC CELL_FACH state is assigned for instances where the requested user capacity is greater than that readily supported by a RACH. PCPCH message and power control preamble slot structures are the same as for the normal uplink DPDCH/DPCCH. Figure 4.4 illustrates the basic process through which the PCPCH operates.

We will outline the basic principle for the use of the PCPCH here, and study the detail of its operation when we consider the MAC (Chapter 8).

The PCPCH has two basic modes of operation: the first when channel assignment is not active, the second when channel assignment is active. When channel assignment is not active, there can be up to 16 PCPCHs in a PCPCH set. The UE can detect the TFs (and hence data rates) that have been assigned to each of the PCPCHs from broadcast channels. The availability of the 16 PCPCHs – and data rates – is defined through the CSICH. The UE detects the availability of the PCPCH from the CSICH. Each of the 16 PCPCHs corresponds to a specific access preamble (AP). Based on its data rate requirement, the UE selects its PCPCH and hence the AP.

In the second mode, when channel assignment is active, the UE is told through a broadcast channel the data rate that can be supported on the PCPCHs. In addition, the CSICH defines the availability of the data rates and the availability of each of up to 57 channels. The UE selects the AP that provides the required data rate, provided that the CSICH indicates that the data rate is available.

The next stage is common to both modes. Whichever mode the UE is using, it transmits the selected AP and waits for the AP-AICH. On receipt of an AP-ACK, the UE moves into the collision detector (CD) phase, in which the UE randomly selects and

# 120  Physical layer

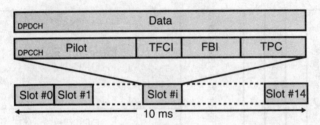

**Figure 4.5** Basic structure of the DPDCH and the DPCCH.

transmits a CD preamble from those available. The UE transmits the CD preamble, and waits for the collision detection/channel assignment indicator channel (CD/CA ICH).

In the first mode, with the CA not active, the CD/CA ICH indicates that the specific PCPCH selected has been allocated. In the second mode, with the CA active, the CD/CA ICH acknowledges the CD AP and at the same time includes one of 16 channel assignments ICH. The combination of the initial selected AP and CA-ICH defines to the UE which channel out of the possible 57 has been allocated.

After the successful reception of the preamble, there is a 0 or 8 slot power control preamble, again starting at nominally the same power and using a special downlink physical control channel (DL-DPCCH) for the power control. This special channel has the same basic structure as the spreading factor 512 downlink DPCH (considered in Section 4.4.9) but includes the CPCH control channel (CCC) bits in place of the data bits. The CCC bits (4 per slot) are used for layer 1 messages such as the start of message indicator and higher layer commands such as emergency stop.

It is possible for the UTRAN to adjust the power levels for the PCPCH message parts, such that the power levels are within acceptable limits when arriving at the UTRAN. Once the power control preamble has finished, and the power levels have been correctly set by the UTRAN, then the UE transmits the message part on the data channel, simultaneously with the physical control channel part. In total, this message can be a multiple of 10 ms in duration, providing anything from one 10 ms frame of data, to sixty-four 10 ms frames of data and thereby supporting the requirements for bursty traffic via the CPCH.

### 4.3.3 Uplink DPDCH and DPCCH

The uplink dedicated physical channel comprises a DPDCH and a DPCCH. The DPDCH transports the DCH. The DPCCH carries the physical control information necessary to provide successful layer 1 operation. This information is transmitted using a slotted radio frame structure, whose radio frame length is 10 ms. Figure 4.5 illustrates the basic structure of the DPDCH and the DPCCH. As can be seen, the DPDCH consists of a data burst transmitted in parallel with the DPCCH. The DPCCH includes a number of pilot bits, the TFCI bits, the FBI bits, as well as the TPC bits.

## 4.3 Uplink channels

**Table 4.2.** *Uplink DPDCH structure*

| Slot format | Channel bit rate (kb/s) | Channel symbol rate (ks/s) | SF | Bits/frame | Bits/slot | Data |
|---|---|---|---|---|---|---|
| 0 | 15 | 15 | 256 | 150 | 10 | 10 |
| 1 | 30 | 30 | 128 | 300 | 20 | 20 |
| 2 | 60 | 60 | 64 | 600 | 40 | 40 |
| 3 | 120 | 120 | 32 | 1200 | 80 | 80 |
| 4 | 240 | 240 | 16 | 2400 | 160 | 160 |
| 5 | 480 | 480 | 8 | 4800 | 320 | 320 |
| 6 | 960 | 960 | 4 | 9600 | 640 | 640 |

The pilot bits in the DPCCH are used in the Node B for channel estimation, frequency tracking, time synchronisation and tracking. The TFCI bits are used for transport format detection by the receiver.

The FBI bits are used as a means to convey information on site selection diversity (SSDT) or feedback mode transmit diversity if it is being used at that time. This FBI field is divided into two parts, the S field and the D field. The S field is used for the SSDT signalling and the D field for the feedback mode transmit diversity signalling. In general, there are, at most, two bits available for FBI signalling. Therefore, only under certain slot formats can both the S field and the D field be used. Under other formats, the S field or the D field is used individually. In some situations, neither is used, in which case the FBI bits are not present.

Table 4.2 illustrates the DPDCH fields that are available. As can be seen from the table, there is a range of possible data rates that can be supported. This is part of the variable data rate capability provided by the WCDMA system. In order to support the variable data rates, the spreading factor for the DPDCH fields is varied from 256 for the lowest data rate of 15 kb/s, down to a spreading factor of 4 for the data rate channel of 960 kb/s.

The selection of pilot TFCI, FBI and TPC bits varies, in order to meet different QoS objectives. It is the MAC and RRC layers that are responsible for defining and maintaining the selection of the transport format combination. The MAC selects the TFC from the combination defined by the RRC.

### Multicode uplink DPDCH

The number of data bits per frame varies according to the data rate requirements. It is also possible, on the uplink, to support multiple coded transmissions. In this case, there is a maximum of six simultaneous coded channels allowed on the uplink for the DPDCH as well as one for the DPCCH.

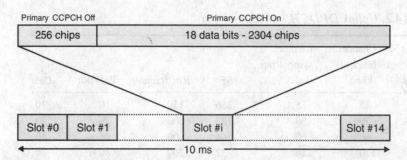

**Figure 4.6** Basic structure of the PCCPCH.

With a spreading factor of 4, there are potentially four I channels and four Q channels of the QPSK modulation scheme that can be utilised for this multicode transmission. However, one of these is already used by the DPCCH, leaving a maximum of seven. The spreading code from the part of the tree used by the DPCCH is not used in the I path and so only six codes are used for the DPDCH. Section 4.5 examines the structure for the spreading and scrambling functions.

## 4.4 Downlink channels

In this section we consider some of the many downlink channels that are illustrated in Figure 4.1. These channels carry higher layer information and are used by the UE to acquire and synchronise with the UTRAN. We start this section by considering the primary common control physical channel (PCCPCH), the physical channel that carries the broadcast information.

### 4.4.1 PCCPCH

The PCCPCH is used for transmission of the broadcast channel within the cell. Figure 4.6 illustrates its basic structure. One thing to note in this particular channel is that there are no pilot information bits present. It is intended that a UE that is receiving and demodulating this PCCPCH channel should use the primary common pilot channel (PCPICH) to provide the necessary channel estimation and timing information.

The PCCPCH is switched off for the duration of 256 chips at the start of each slot. As will be seen later, the switching off of this primary channel is to facilitate the transmission of the primary and secondary synchronisation channels (P-SCH and S-SCH) which occurs in this 256 chip part of the slot. In total, 18 data bits are transmitted within each slot of the PCCPCH and there are 15 slots per 10 ms radio frame. There are no TFCI bits, as the TF for the PCCPCH channel is fixed.

## 4.4 Downlink channels

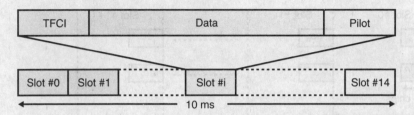

**Figure 4.7** The SCCPCH channel.

The PCCPCH is the physical channel that transports the broadcast transport channel (BCH). The PCCPCH uses the same spreading code in every cell and also the primary scrambling code for the cell.

The PCCPCH is used to provide system-wide cell broadcast information. The cell broadcast information is used by the UE when it is first trying to locate the UTRAN, and it decodes and uses the broadcast information in any subsequent communications with the UTRAN. Chapter 11 will consider the system broadcast information in more detail.

### 4.4.2 Secondary common control physical channel (SCCPCH)

The SCCPCH is used to carry two downlink transport channels, the forward access channel (FACH) and the paging channel (PCH). It comprises a number of TFCI bits, data bits and pilot bits, as illustrated in Figure 4.7. Each physical channel has a slotted structure, with a total of 15 slots per radio frame of duration 10 ms.

The FACH and PCH transport channels can be mapped onto either the same or separate SCCPCH channels. Multiple SCCPCH can be supported within the same cell. As mentioned previously, the SCCPCH can support variable transmission rates with the rate indicated through the TFCI field. The SCCPCH is not transmitted continuously, but only when there are data present to be transmitted.

### 4.4.3 P-SCH and S-SCH

Figure 4.8 illustrates the basic structure of the P-SCH and the S-SCH. For each slot, the first 256 chips are broadcast as P-SCH and S-SCH. The P-SCH uses a special waveform and the S-SCH a selection of special waveforms chosen on a slot by slot basis.

The P-SCH and the S-SCH are provided to assist the UE in tracking and locating cells in the network. The P-SCH and the S-SCH essentially provide a number of beacon signals from which the UE is able to derive frame and slot timing information, as well the cell code group. The P-SCH is a 256 chip hierarchical Golay sequence [7] waveform used for carrier location and frame timing extraction. All cells use the same P-SCH in the system and, as will be seen later, there is no scrambling code applied to the P-SCH;

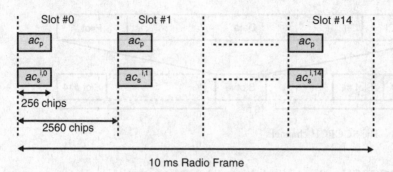

**Figure 4.8** Structure of the P-SCH and the S-SCH.

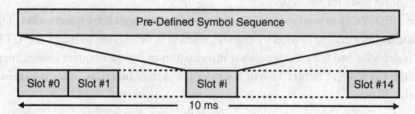

**Figure 4.9** Structure of P-CPICH and S-CPICH.

so the UE uses the P-SCH as a beacon signal from which it is able to determine the presence of a cell.

In addition to the spreading waveform, there is also a symbol defined as '$a$' which has the value of either $+1$ or $-1$, to indicate whether space time transmit diversity (STTD) encoding is present on the PCCPCH. If the value for '$a$' is equal to $+1$, then the PCCPCH is STTD encoded. If it is equal to $-1$, then the PCCPCH is not STTD encoded.

The S-SCH also uses a 256 chip waveform. The S-SCH is used, amongst other things, to define to which code group the specific cell belongs. This is covered in more detail in Section 6.7.

### 4.4.4 P-CPICH and S-CPICH

Figure 4.9 shows the structure for the P-CPICH and the S-CPICH. The structure within the P-CPICH and the S-CPICH consists of only a set of predefined symbols, which are broadcast continuously within the slot. In total there are 15 slots per frame.

The P-CPICH and S-CPICH are there to provide a cell phase reference. The P-CPICH is the main cell phase reference for most of the common and dedicated channels broadcast from the cell. This channel assists in the demodulation of some of the other downlink channels. In some circumstances, particularly in the use of adaptive antennas where traffic hot spots are present, it may be necessary to provide an additional phase reference. In this case, the S-CPICH can be used. There is only one P-CPICHs

## 4.4 Downlink channels

**Table 4.3.** *Mapping between PIs and PICH bits*

| Number of PIs per frame ($N_p$) | $P_q = 1$ | $P_q = 0$ |
|---|---|---|
| 18 | $\{b_{16q}, \ldots, b_{16q+15}\} = \{-1, -1, \ldots, -1\}$ | $\{b_{16q}, \ldots, b_{16q+15}\} = \{+1, +1, \ldots, +1\}$ |
| 36 | $\{b_{8q}, \ldots, b_{8q+7}\} = \{-1, -1, \ldots, -1\}$ | $\{b_{8q}, \ldots, b_{8q+7}\} = \{+1, +1, \ldots, +1\}$ |
| 72 | $\{b_{4q}, \ldots, b_{4q+3}\} = \{-1, -1, \ldots, -1\}$ | $\{b_{4q}, \ldots, b_{4q+3}\} = \{+1, +1, \ldots, +1\}$ |
| 144 | $\{b_{2q}, b_{2q+1}\} = \{-1, -1\}$ | $\{b_{2q}, b_{2q+1}\} = \{+1, +1\}$ |

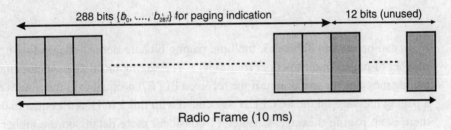

**Figure 4.10** Structure of the PICH.

per cell, broadcast over the entire cell coverage area. The P-CPICH provides the phase reference for the PCCPCH and the various indication channels.

The S-CPICH is used with an arbitrary channelisation code. There may be 0, 1, or several S-CPICHs per cell. It may be transmitted over only part of the cell using systems such as adaptive or smart antennas.

### 4.4.5 Paging indication channel (PICH)

Figure 4.10 illustrates the structure of the PICH. The paging indicators (PIs) are transmitted using 288 bits in a 10 ms radio frame. The 12 bits at the end of the frame are unused. Table 4.3 illustrates the correspondence between the number of PIs per frame and the use of the number of bits within the frame structure.

Within each PICH frame, $N_p$ PIs are transmitted and are given by $P_q$. The specific PI is related to the system frame number ($SFN$), the number of PIs per frame ($N_p$) and a quantity page indicator ($PI$), which is defined by the higher layers (see Chapter 11). In the paging indicator $P_q$, $q$ is given by:

$$q = (PI + \{18 + [SFN + (SFN)/8) + (SFN/64) + (SFN/512)]\} \\ \times \bmod 144 \times [N_p/144] \bmod N_p$$

Once $P_q$ is identified for a specific $SFN$, the paging bits shown in Figure 4.10 are given by Table 4.3.

## Physical layer

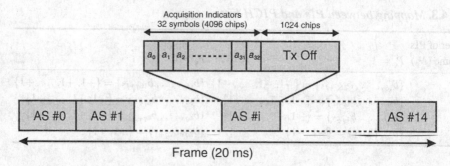

**Figure 4.11** Structure of the AICH.

As can be seen in Table 4.3, multiple paging bits are defined, depending upon the number of paging indicators per frame ($N_p$). The paging bits define whether there is a paging message present or not. If the received PI ($P_q$) is equal to 1, this shows there is a paging message on the SCCPCH associated with this PICH. A PI equal to 0 shows there is no paging message. Chapter 11 considers more details on the higher layers related to the paging process.

### 4.4.6 AICH

Figure 4.11 shows the structure of the AICH. The frame structure is based on two radio frames and has a duration of 20 ms. Within that, there are 15 access indication slots. These access slots are composed of acquisition indication bits as well as a number of unused bits. The acquisition indication bits correspond to 32 symbols spread with a spreading factor of 256, i.e. a total of 4096 chips per slot. The unused part of the slot corresponds to 4 symbols, or 1024 chips. No AICH is transmitted.

The AICH is used to respond to the PRACH. The waveform that is selected for use on the AICH is in direct response to the waveform detected as being used on the PRACH. When multiple users are being responded to within a given access slot, each of the orthogonal waveforms is summed after being multiplied by the correct acquisition indicator ($AI$) (+1 is ACK, −1 is NACK, and 0 is NULL).

The 16 signature patterns are summed after multiplication by the $AI$ for the specific PRACH signature to which it is responding. As the set of signatures are orthogonal, the UE can easily extract the correct signature.

### 4.4.7 AP-AICH and CD/CA-ICH

For the PCPCH channel, two additional indicators have been defined. The AP-AICH is used in response to an initial access attempt for the PCPCH channel, and CD/CA-ICH is used in response to the collision detection phase of the access attempt.

## 4.4 Downlink channels

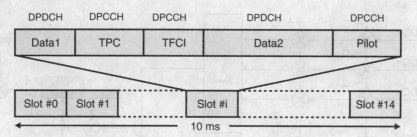

**Figure 4.12**  Slot and frame structure for the downlink DPCH.

Two methods of handling access to the PCPCH channels are defined: one is referred to as channel assignment (CA) not active, and the other as CA active.

The operation of the PCPCH is described in Chapter 8.

### 4.4.8  Physical downlink shared channel (PDSCH)

The PDSCH transports the downlink shared transport channel (DSCH). The PDSCH is always associated with a downlink DPDCH.

The PDSCH can support variable data rates, and is shared by a number of different users, with each user using a different channelisation code. The DPCCH is not transported by the PDSCH; instead, it is carried by the associated downlink DPCH.

The TFCI bits in the associated DPCH are used to indicate whether there are any data to decode on the PDSCH. The TFCI bits are also used to define the TFC used on the PDSCH.

### 4.4.9  Downlink DPCH

The downlink DPCH is used to carry the DPDCH and the DPCCH within the same burst structure. The DPDCH is used to transport the dedicated transport channel (DCH) and the DPCCH is used to transport the physical control information necessary to maintain the layer 1 link.

The DPDCH can support dynamically varying data rates, either directly through the use of TFCI bits contained within the burst or through methods employing blind-rate detection of the TFC. A variable number of pilot bits are provided for the DPCCH and this depends on the propagation conditions prevalent at the time. It is up to the UTRAN to define which of the many pilot combinations can be selected. Also within the pilot information, the patterns for the pilot bits define the frame synchronisation words that can be used by the UE to establish frame synchronisation.

The DPCH can support multicode transmission to a specific UE which consists of a DPCCH and multiple DPDCHs. In this case, only a single DPCCH is broadcast, and all the DPDCH channels for that UE use this for the physical layer control.

Figure 4.12 illustrates the slot and frame structure for the downlink DPCH. The DPDCH and the DPCCH are transmitted together, within the same slot but separated

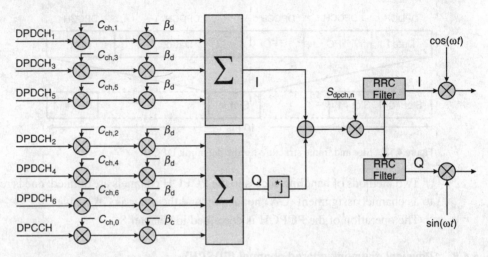

**Figure 4.13** Application of the spreading and scrambling codes to the uplink DPDCH and DPCCH.

in time. The DPDCH is divided into two parts, data 1 and data 2; the DPCCH into three parts, the TFCI bits, the power control bits and the in-band pilot bits. In total, there are 15 slots per radio frame of duration 10 ms.

## 4.5 Spreading and scrambling codes

### 4.5.1 Spreading and scrambling, general summary

In general, there are two different functions applied at the same chip rate on both the uplink and the downlink channels. These two separate functions are known as the spreading function and the scrambling function, considered in Chapter 3.

The spreading function is achieved using the OVSF waveforms that are used for the channelisation codes. On the uplink, these OVSF waveforms are used to separate different channels for a specific user. On the downlink, they are used to separate the different control channels and traffic channels for different users within a cell.

The scrambling codes may be either long or short codes on the uplink or a long code on the downlink. On the uplink, the scrambling codes are used to isolate different users accessing the same cell. On the downlink, scrambling codes are used to reduce the effect of co-channel interference from adjacent cells.

### 4.5.2 Uplink spreading and scrambling

Figure 4.13 illustrates the application of the spreading and scrambling codes to the uplink DPDCH and DPCCH. On the left hand side, it can be seen that there are a number of DPDCHs, numbered from 1 to 6, and a single DPCCH. If there is a single DPDCH, then this is transmitted in the I plane of the QPSK modulation scheme. The DPCCH is transmitted in the Q-plane of the QPSK modulation scheme. If there are multiple DPDCHs to be transmitted on the uplink, each DPDCH must be at the lowest possible spreading factor of 4.

In the multicode case on the uplink, the orthogonality of the modulation scheme is used in addition to the orthogonality of the OVSF codes. In theory, four codes of length 4 are used in the I-plane and four codes of length 4 are used for the Q plane. In practice, only three of the four codes in I and Q are used for the DPDCH.

After spreading, the DPDCH and the DPCCH are scaled by factors $\beta_d$ and $\beta_c$ respectively. The factor $\beta_d$ is constant for all DPDCHs and is used to vary the relative powers between the DPDCH and the DPCCH. The I and Q waveforms are then independently summed, and combined to form a complex signal that is multiplied by the complex scrambling waveform $C_{scramb}$. These waveforms are then filtered using RRC filters, before being modulated and passed through to the rest of the transmission chain, as described in Chapter 3.

As mentioned previously, one DPCCH, and up to six parallel DPDCHs are transmitted simultaneously. The channelisation codes are derived from the OVSF spreading codes. The DPCCH is always carried on the Q channel and the additional DPDCHs are alternately on the I and the Q channels, starting with the I channel. The gain factors are there to balance the QoS requirements for different channels. The criterion is that the peak gain factor should be equal to 1. After summation of the multicode DPDCH, the composite signalling has a higher peak-to-mean variation than would be obtained with a single DPDCH.

## Uplink channelisation codes

The channelisation codes on the uplink are defined as $C_{CH,SF,K}$ where $SF$ is the spreading factor for the channelisation code and $K$ is the code number. On the uplink, the DPCCH channel is always spread using the channelisation code $C_{CH,256,0}$. With only one DPDCH, $K$, i.e. the code number, is equal to the spreading factor divided by 4 ($SF/4$). When there is more than one DPDCH, the spreading factor is equal to 4 and the codes used are shown in Figure 4.13.

## Uplink scrambling codes

There are two types of scrambling codes that can be used on the uplink. The first, called the long code, is used for normal modes of operation. The second, called the short code, is used for interference cancellation receivers within the Node B.

In addition, the uplink scrambling sequences are defined in such a way as to produce a waveform that has very good peak-to-mean performance. This was considered in greater detail in Chapter 3, along with the structure for the uplink long and short scrambling code generators.

## PRACH message part scramble code

The PRACH message part scramble code is based on the long scramble code, but with a 4096 chip offset. The message part scramble code is 10 ms long. There are a total of 8192 different scramble codes defined.

The 8192 scramble codes are broken into 512 groups with 16 PRACH message scramble codes per group. The scramble code group for a cell is the same as the code set for the cell (code sets are defined in Section 4.5.8). The choice of which of the 16 available in the cell will be used is made by the higher layers based on broadcast information from the cell.

## PRACH preamble code

The PRACH preamble code comprises a preamble signature, a preamble scramble code and a complex vector which rotates 90° for each chip period. This composite structure leads to a complex waveform whose transitions are 90°, avoiding zero crossings in the complex plane. The scramble code that is chosen for the PRACH preamble code is the $C_{\text{long},1,n}$ scramble code. There are 8192 preamble scramble codes and the codes are paired with the scramble codes used by the PRACH message part. The preamble signature is one of 16 length-16 Hadamard vectors that are sequentially repeated to a total length of 4096 chips.

## PCPCH message part scramble code

The PCPCH message part scramble code is based on the uplink long scramble code, but starting from code number 8192 and going to code number 40 959. The message part scramble code is 10 ms long. In circumstances where the message part is in multiples of 10 ms, the scramble code is reset at the beginning of each 10 ms boundary.

There is a link between the scrambling code selected, the signature sequence and the access subchannel. There are 64 scrambling codes per group. There is a one-to-one link between the scrambling code groups and the primary downlink scrambling code.

## PCPCH AP/CD preamble code

The PCPCH AP/CD code is generated in a similar manner to the PRACH preamble code, using similar scrambling and spreading waveforms.

### 4.5.3 Downlink spreading and scrambling

Figure 4.14 illustrates the downlink spreading and scrambling processes. The DPCH channels are applied to a serial to parallel converters, which converts the incoming bit stream into two parallel bit streams that are multiplied by an OVSF channelisation code.

A complex scrambling code multiplies the complex sequence obtained from the channelisation codes. The output is then split into real and imaginary parts, before

## 4.5 Spreading and scrambling codes

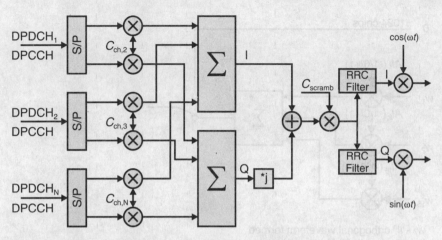

**Figure 4.14** Downlink spreading and scrambling processes.

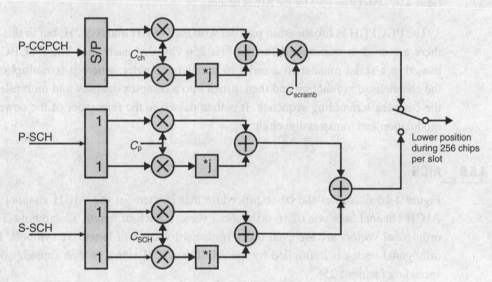

**Figure 4.15** Transmission scheme for PCCPCH, P-SCH and S-SCH.

being filtered, using an RRC filter. It is next passed on to the rest of the transmission and modulation chain.

### 4.5.4 PCCPCH, P-SCH and S-SCH

Figure 4.15 illustrates the transmission scheme that is proposed for the PCCPCH, the P-SCH and the S-SCH. Both the P-SCH and the S-SCH consist of a sequence of 1s that are multiplied by two real spreading functions. The P-SCH is multiplied by the spreading function $C_p$ and the S-SCH is multiplied by the spreading function $C_{sch}$. Chapter 6 discusses in more detail the structure and use of the P-SCH and S-SCH.

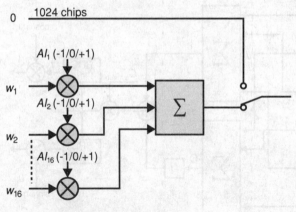

**Figure 4.16** Procedure used for the AICH channel.

The PCCPCH is broadcast in parallel with the P-SCH and S-SCH, but in this case there are no data transmitted for the first 256 chips of each slot. The PCCPCH is passed in a serial manner to a serial to parallel converter, where it is multiplied by the channelisation function and then turned into a complex quantity and multiplied by the complex scrambling sequence. It is then passed to the remainder of the downlink modulation and transmission chain.

### 4.5.5 AICH

Figure 4.16 illustrates the basic procedure that is used for the AICH channel. The AICH channel uses one of 16 orthogonal vectors, each of length 32 symbols. These orthogonal vectors are the even order Hadamard vectors of length 32 symbols. Each orthogonal vector is multiplied by the AI. The AICH signal is then spread, using a spreading factor of 256.

### 4.5.6 Downlink channelisation codes

The P-CPICH uses the channelisation code given as $C_{CH,256,0}$. The PCCPCH uses the spreading waveform denoted by $C_{CH,256,1}$. All other channelisation codes that are used on the downlink by the Node B equipment are assigned by the UTRAN.

### 4.5.7 Downlink scrambling codes

The downlink scrambling code is based on a 218-1 Gold code. This results in 262 143 scrambling codes, which are divided into 512 sets of 1 primary and 16 secondary scrambling codes. Thus the total number of codes used is 8192. Each cell is

## 4.5 Spreading and scrambling codes

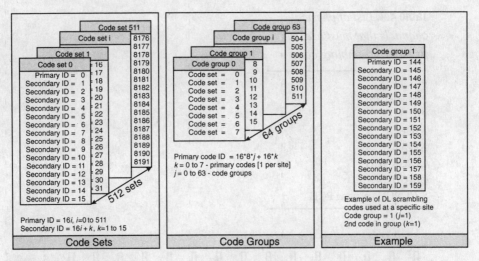

**Figure 4.17** Relationship between code sets and code groups.

allocated one, and only one, primary scrambling code. The reason for adopting a single primary scrambling code in the cell is to ease the burden placed on the UE when it comes to detecting and decoding the downlink broadcast information from the PCCPCH. The structure for the downlink scrambling code generator was presented in Chapter 3.

### Downlink SCH codes

The P-SCH and S-SCH are defined to have special properties. The P-SCH uses a generalised hierarchical Golay sequence which has good aperiodic autocorrelation properties. For the S-SCH, a set of codes $C_{CSCH,n}$ is defined using a combination of the hierarchical sequence and a Hadamard matrix.

Scrambling groups are defined where one from the set of codes is assigned for each of the 16 timeslots. By identifying the codes and their sequence it is possible to determine the scrambling group and hence the scrambling code for the cell.

### 4.5.8 Downlink code sets and code groups

Figure 4.17 illustrates the relationship between the downlink codes, code sets and code groups.

There are 8192 downlink scrambling codes defined for normal use on the downlink (there are some additional codes defined for use with compressed mode). The 8192 codes are collected into code sets with 16 codes per code set and a total of 512 code sets. Within each code set the 16 codes are collected as 1 primary scrambling code and 15 secondary scrambling codes, as shown in Figure 4.17. Each cell in the system is allocated to a code set, and through code planning it is intended that adjacent cells do not use the same code set.

**Table 4.4.** *List of physical channels that must use cell primary scrambling code*

PCCPCH
Primary CPICH
PICH
AICH
AP-AICH
CD/CA-ICH
CSICH
S-CCPCH carrying PCH

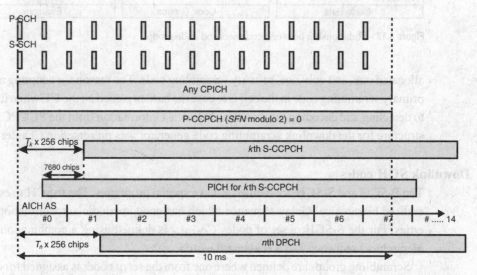

**Figure 4.18** Cell timing.

The code sets are collected into code groups. There are 64 code groups and 8 code sets per code group. Figure 4.17 illustrates the primary scrambling codes for each of the 62 code groups. The primary scrambling codes are used by specific physical channels in the cell. Table 4.4 identifies which physical channels in the cell must use the primary scrambling code.

## 4.6  Cell timing

Figure 4.18 illustrates the cell timing. The PCCPCH that transmits the SFN is used as the timing reference for all physical channels, directly for the downlink and indirectly for the uplink. The P-SCH, S-SCH, PDSCH and CPICH timing is the same as that

## 4.6 Cell timing

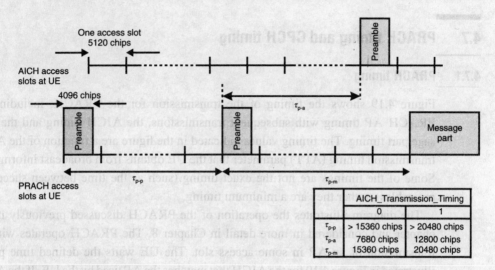

**Figure 4.19** PRACH timing.

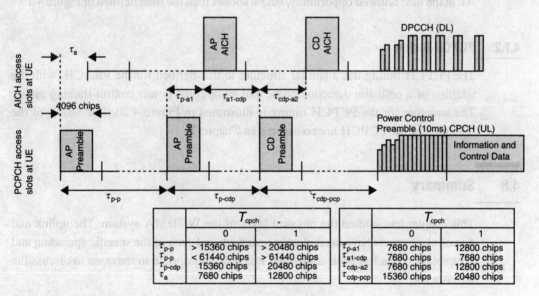

**Figure 4.20** PCPCH timing.

for PCCPCH. SCCPCH timing can be different for each SCCPCH, but offset from the PCCPCH is in multiples of 256 chips. PICH timing is variable, but occurs 7680 chips prior to the corresponding SCCPCH (this allows decoding of the PICH and then the decoding of the appropriate SCCPCH), which carries the PCH transport channel. AICH access slot #0 starts at the same time as the PCCPCH frame ($SFN$ modulo $2 = 0$). DPCH timing may be different for each DPCH, but the offset is a multiple of 256 chips from the PCCPCH.

## 4.7 PRACH timing and CPCH timing

### 4.7.1 PRACH timing

Figure 4.19 shows the timing of the transmission for the PRACH, including the PRACH AP timing with subsequent transmissions, the AICH timing and the message part timing. The timing values indicated in the figure are a function of the AICH transmission timing (ATT) parameter that the UE obtains from broadcast information. Some of the timings are not the exact timing (such as the time between successive preambles), rather they are a minimum timing.

The diagram illustrates the operation of the PRACH discussed previously in this chapter and considered in more detail in Chapter 8. The PRACH operates with the transmission of an AP in some access slot. The UE waits the defined time period illustrated in Figure 4.19 for the AICH that matches the AP used by the UE. If the AICH is detected, the UE transmits the message at the time shown, else it transmits the next AP at the next allowed opportunity, but no sooner than the time defined in Figure 4.19.

### 4.7.2 PCPCH timing

The PCPCH timing has a similar structure to that defined for the PRACH, with the addition of a collision detection phase and an optional power control-training phase. The structure for the PCPCH timing is illustrated in Figure 4.20. The details of the operation of the PCPCH are considered in Chapter 8.

## 4.8 Summary

This chapter has studied the physical layer of the WCDMA system. The uplink and downlink channels have been closely reviewed, and each of the specific spreading and scrambling codes has been addressed. It is now appropriate to move on to discuss the RE aspects.

# 5 RF aspects

In this chapter we consider the RF aspects for the WCDMA transceiver. We focus on the FDD mode, starting with a review of the basic transmitter specifications for the UE and the Node B. Following this, we introduce some terminology and parameters that define the receiver characteristics. Then, we examine the receiver specifications themselves with some comments on the likely design targets for the receiver. In the final section, we review elements of an example design, taking the design issues for a UE transceiver as the reference.

## 5.1 Frequency issues

### 5.1.1 UMTS frequency bands

Table 5.1 illustrates the 'current' proposed bands for the deployment of the UMTS system. Band I is the 'IMT-2000' band, Band II is the US personal communication system (PCS) band, and Band III is the digital cellular network at 1800 MHz (DCS 1800) band.

### 5.1.2 North American PCS bands

To understand the frequency allocations in the US PCS bands, it is useful to examine the band allocations for the US bands. Table 5.2 presents the US PCS bands. The 'C block' licences have been reauctioned to create a number of subbands as shown in the table. Figure 5.1 illustrates the band structure for the PCS bands, including the different band allocations for the 'C block'.

### 5.1.3 UTRA absolute radio frequency channel number (UARFCN)

The UARFCN is used to define the channel numbers for WCDMA carriers. For all bands the UARFCN is defined by the equations in Table 5.3 for uplink and downlink.

**Table 5.1.** *UMTS FDD mode proposed band structure*

|          | Uplink         | Downlink       |
|----------|----------------|----------------|
| Band I   | 1920–1980 MHz  | 2110–2170 MHz  |
| Band II  | 1850–1910 MHz  | 1930–1990 MHz  |
| Band III | 1710–1785 MHz  | 1805–1880 MHz  |

**Table 5.2.** *US PCS frequency band allocations*

| Licensed band | Bandwidth (MHz) | Starting frequency (MHz) Up | Starting frequency (MHz) Down | Centre frequency (MHz) Up | Centre frequency (MHz) Down |
|---|---|---|---|---|---|
| A  | 15  | 1850   | 1930   | 1862.5  | 1937.5  |
| D  | 5   | 1865   | 1945   | 1967.5  | 1947.5  |
| B  | 15  | 1870   | 1950   | 1877.5  | 1957.5  |
| E  | 5   | 1885   | 1965   | 1887.5  | 1967.5  |
| F  | 5   | 1890   | 1970   | 1890.5  | 1972.5  |
| C  | 15  | 1895   | 1975   | 1902.5  | 1982.5  |
| C1 | 7.5 | 1895   | 1975   | 1898.75 | 1978.75 |
| C2 | 7.5 | 1902.5 | 1982.5 | 1906.25 | 1986.25 |
| C3 | 5   | 1895   | 1975   | 1897.5  | 1977.5  |
| C4 | 5   | 1900   | 1980   | 1902.5  | 1982.5  |
| C5 | 5   | 1905   | 1985   | 1907.5  | 1987.5  |

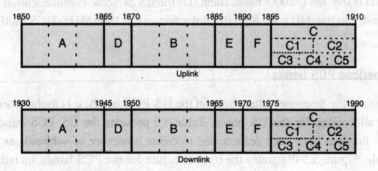

**Figure 5.1** US PCS band structure.

To manage the situation in the US PCS bands, and the fact they have a 100 kHz offset with respect to the IMT-2000 band allocation, an alternative channelisation scheme is defined for the case where the channel is 5 MHz wide and the main channelisation scheme does not work. This alternative scheme also ensures the maximum compatibility between the WCDMA system and the GSM system deployed in those bands.

## 5.1 Frequency issues

**Table 5.3.** *UMTS UARFCN numbering scheme for Bands I, II and III*

| | | |
|---|---|---|
| Uplink | $N_u = 5 * F_{uplink}$ | $0.0 \text{ MHz} < F_{uplink} < 3276.6 \text{ MHz}$ |
| Downlink | $N_d = 5 * F_{downlink}$ | $0.0 \text{ MHz} < F_{downlink} < 3276.6 \text{ MHz}$ |

**Table 5.4.** *UMTS US PCS additional channels UARFCN numbering scheme*

Band II – 12 additional channels

| | | | |
|---|---|---|---|
| Uplink $N_u = 5\,[(F_{uplink} - 100 \text{ kHz}) - 1850]$ | 1852.5, | 1857.5, | 1862.5, |
| | 1867.5, | 1872.5, | 1877.5, |
| | 1882.5, | 1887.5, | 1892.5, |
| | 1897.5; | 1902.5, | 1907.5 |
| Downlink $N_d = 5\,[(F_{downlink} - 100 \text{ kHz}) - 1850]$ | 1932.5, | 1937.5, | 1942.5, |
| | 1947.5, | 1952.5, | 1957.5, |
| | 1962.5, | 1967.5, | 1972.5, |
| | 1977.5, | 1982.5, | 1987.5 |

**Table 5.5.** *UMTS UARFCN channel numbers for different channels*

| Frequency band | Uplink UARFCN | Downlink UARFCN |
|---|---|---|
| Band I | 9612–9888 | 10 562–10 838 |
| Band II | 9262–9538 and 12, 37, 62, 87, 112, 137, 162, 187, 212, 237, 262, 287 | 9662–9938 and 412, 437, 462, 487, 512, 537, 562, 587, 612, 637, 662, 687 |
| Band III | 8562–8913 | 9037–9388 |

To achieve this, 12 additional duplex pairs are defined as shown in Table 5.4, but require a 100 kHz offset from the normal band plan. Table 5.5 illustrates the UARFCN for the uplink and downlink for the three different bands.

For Band II, there are also 12 'spot channels', which are specified from the 'low frequency range' to define the channel numbers in the case when a 5 MHz channel has to co-exist with GSM, or on its own (e.g. block D, E, F licence holders).

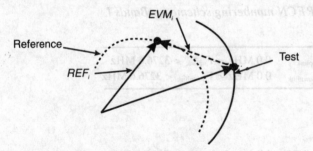

**Figure 5.2** UE EVM calculation.

## 5.2 UE transmitter specifications

### 5.2.1 General specifications

| | |
|---|---|
| Power level: | Class 3 +24 dBm |
| | Class 4 +21 dBm |
| Power control range: | Power level to −50 dBm |
| Power steps: | 1 dB, 2 dB and 3 dB |

### 5.2.2 EVM

The next performance characteristic of interest is the EVM that is obtained from the WCDMA transmitter. This is illustrated in Figure 5.2. The EVM is a 'time domain' measure of the transmitter performance. The EVM measures how closely the transmitter performs to an ideal transmitter. To do this the EVM is a vector difference measurement between the reference signal and the signal under test.

The EVM is a useful measure of how well the transmitted signal compares to the ideal signal.

$$EVM = \sqrt{\frac{\sum_i (EVM_i)^2}{\sum_i (REF_i)^2}}$$

$EVM \leq 17.5\%$

The equation above defines the EVM as the ratio of the sum of the squares of the individual error vector components to the sum of the squares of the reference signal. The summations are made across a WCDMA slot (2560 chips), and the WCDMA specifications state that the EVM has to be less than or equal to 17.5%.

## 5.2 UE transmitter specifications

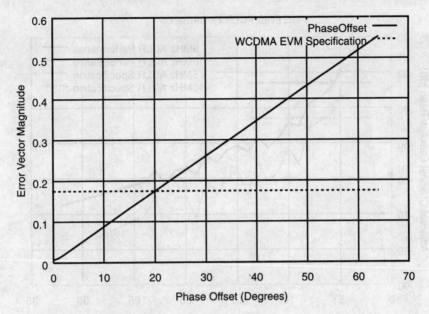

**Figure 5.3** EVM versus I/Q phase offset.

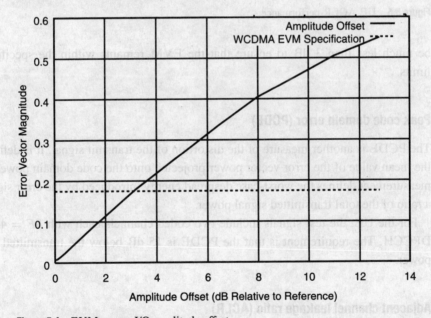

**Figure 5.4** EVM versus I/Q amplitude offset.

Figure 5.3 illustrates the effects that the I/Q phase offset error has in the contribution to the error vector. In this case, to achieve the specification level of 17.5%, the I/Q phase offset error should be much less than 20°.

Figure 5.4 illustrates the effects that the amplitude offset has in the contribution to the error vector. From the curves, it is clear that the I/Q amplitude balance should

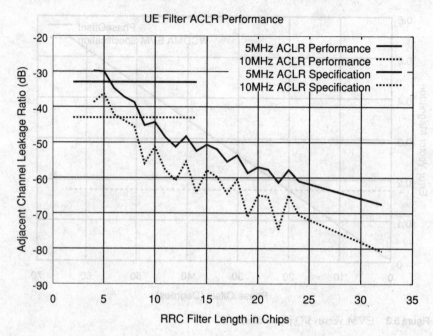

**Figure 5.5** UE ACLR performance.

be much less than 3 dB to ensure that the EVM remains within the specification limits.

### 5.2.3 Peak code domain error (PCDE)

The PCDE is another measure of the distortion of the transmit signal. It is defined as the mean value of the error vector power projected onto the code domain power. This measurement defines the worst-case despread energy introduced by the error signal as a ratio of the total transmitted signal power.

For the UE, the test signals include two coded channels each with $SF = 4$ and a DPCCH. The requirement is that the PCDE is 15 dB below the transmitted signal power.

### 5.2.4 Adjacent channel leakage ratio (ACLR)

The next transmitter performance measure of interest is the ACLR, which is a measure of how much adjacent power leaks from the wanted band into the adjacent band. The ACLR is measured with an RRC filter centred on the adjacent channel and used to estimate the average adjacent channel power as a ratio to the wanted channel power, both measured in RRC filter alpha $= 0.22$. The specification defines two ACLR performance levels of 33 dB for a 5 MHz offset and 43 dB for a 10 MHz offset.

## 5.3 Node B transmitter specifications

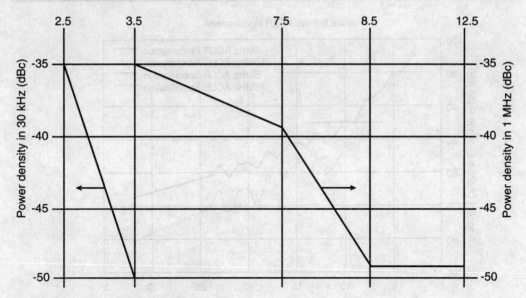

**Figure 5.6** UE spectrum emission mask.

Factors that affect the ACLR are issues such as the length of the RRC filter. Although out of the control of the RF designer, it is instructive to see what effect the RRC filter length has on the ACLR performance specification for the transmitter. Figure 5.5 illustrates the 5 MHz and 10 MHz ACLR results obtained from an RRC filter whose length is increased from 4 to 32 chips. The curves show a general trend of decreasing ACLR with increasing filter length. The RRC filter needs to be a minimum length of 6 chips and typically 12 chips to provide sufficient margin over the specification levels.

### 5.2.5 Spectrum emission mask

The next performance characteristic of interest is the spectrum emission mask for the WCDMA transmitter, illustrated in Figure 5.6. The mask is applied in addition to the ACLR performance. The measurements are based on either 30 kHz or 1 MHz measurement bandwidths, with the close-to-carrier measurements using 30 kHz and the away-from-carrier ones 1 MHz.

## 5.3 Node B transmitter specifications

The power of the transmitter is defined by the manufacturer but typically is in the 10–40 W range. There is downlink power control over at least 18 dB in steps of 1 dB or optionally 0.5 dB. The pilot channel broadcasts at a level within 2 dB of its announced value.

# RF aspects

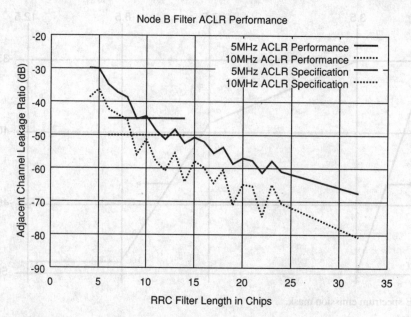

**Figure 5.7** Node B ACLR performance specifications.

## 5.3.1 EVM

The same requirements for the EVM apply to the Node B as apply to the UE.

## 5.3.2 PCDE

For the Node B, the requirement is that the PCDE is 33 dB below the transmitted signal power.

## 5.3.3 ACLR

The ACLR specification for the Node B is the same definition as for the UE, but with different performance levels. The specification defines two ACLR performance levels of 45 dB for a 5 MHz offset and 50 dB for a 10 MHz offset. Figure 5.7 illustrates the 5 MHz and 10 MHz ACLR results obtained from an RRC filter whose length is increased from 4 to 32 chips. The curves show a general trend of decreasing ACLR with increasing filter length. The RRC filter needs to be a minimum length of 10 chips and typically 16 chips provide sufficient margin over the specification levels.

## 5.3.4 Spectrum emission mask

The next performance characteristic of interest is the spectrum emission mask for the WCDMA transmitter as illustrated in Figure 5.8. The mask is applied in addition to

## 5.3 Node B transmitter specifications

**Table 5.6.** *Spurious emission specifications*

| Band (MHz) | Usage | Maximum level | Measurement bandwidth |
|---|---|---|---|
| 1920–1980 | UTRA Rx Band I | −96 dBm | 100 kHz |
| 1850–1910 | UTRA Rx Band II | −96 dBm | 100 kHz |
| 1710–1785 | UTRA Rx Band III | −96 dBm | 100 kHz |
| 921–960 | GSM 900 MS | −57 dBm | 100 kHz |
| 876–915 | GSM 900 BTS | −98 dBm | 100 kHz |
| 1805–1880 | DCS1800 MS | −47 dBm | 100 kHz |
| 1710–1785 | DCS1800 BTS | −98 dBm | 100 kHz |
| 1893.5–1919.6 | PHS | −41 dBm | 300 kHz |
| 1900–1920 | UTRA-TDD | −52 dBm | 1 MHz |
| 2010–2025 | UTRA-TDD | −52 dBm | 1 MHz |
| 1900–1920 | UTRA-TDD (co-located) | −86 dBm | 1 MHz |
| 2010–2025 | UTRA-TDD (co-located) | −86 dBm | 1 MHz |

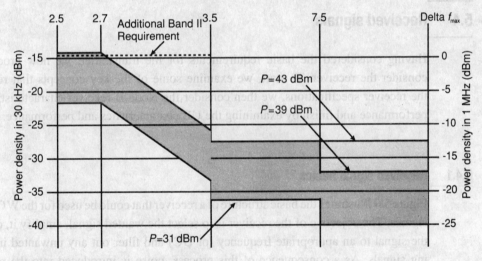

**Figure 5.8** Node B spectrum emission mask.

the ACLR performance. The mask is dependent on the carrier power, and the diagram illustrates the mask for powers from 31 dBm to greater than 43 dBm. The measurements are based on either 30 kHz or 1 MHz measurement bandwidths, with the close-to-carrier measurements using 30 kHz and the away-from-carrier ones 1 MHz.

### 5.3.5 Spurious emissions

Table 5.6 illustrates the spurious emission specifications to which the transmitter must conform.

# 146 RF aspects

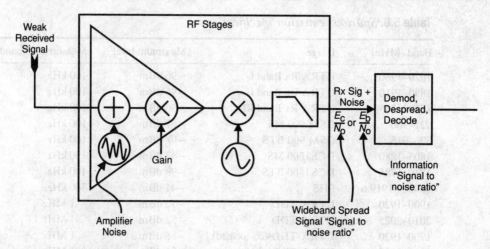

**Figure 5.9** Typical UMTS receiver block diagram.

## 5.4 Received signals

Having considered the basic requirements for the transmitter, we now proceed to consider the receiver. To start, we examine some of the key concepts that relate to the receiver specifications, we then consider the Node B receiver characteristics and performance and finish by examining the UE characteristics and performance.

### 5.4.1 Received signal basics

Figure 5.9 illustrates the basic structure of a receiver that could be used for the WCDMA signals. The objective of the receiver is to select the wanted signal, amplify it, convert the signal to an appropriate frequency (or DC) and filter out any unwanted interfering signals. As a consequence of this process, noise is introduced into the receiver signal by the components in the RF stages of the receiver. To help quantify the performance of the receiver we can use a number of terms or expressions as shown in the diagram.

In Figure 5.9 we can see that the receiver consists of what we could consider to be traditional RF functions (amplification and down-conversion) as well as traditional baseband functions (demodulation, despreading and encoding). To specify the characteristics and performance of the RF stages we need to define some terminology. The baseband stages are normally characterised in terms of BERs and BLERs for specific signal levels. The signal levels themselves are referred to as a ratio to the total receiver noise and are shown in the diagram as $E_b/N_o$ (bit energy per total noise power spectral density). Additionally, because we are considering a CDMA system, we also need to

## 5.4 Received signals

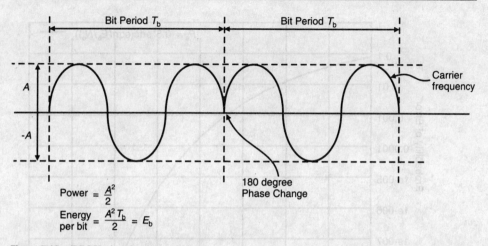

**Figure 5.10** BPSK signal representation.

examine the baseband performance in terms of its chip energy and consequently $E_c/N_o$ is also used.

In the following sections, we consider how we define these quantities and how we can relate them to the quantities that are of greater relevance to the RF designer, namely the noise figure and filter performance required from the receiver.

### 5.4.2 Bit energy

Figure 5.10 illustrates a simplistic representation of a binary phase shift keyed (BPSK) signal. With BPSK, the information is carried in the phase of a carrier waveform that has some desired frequency, e.g. the UMTS frequency. The information is carried by the change in the phase of the carrier as shown in the diagram. In this example there is a phase shift of 180° at the end of the first information bit.

We can define the power and energy of the RF signal as shown. The energy per information bit is defined as the energy carried by the RF carrier over an information bit period. The variable $E_b$ is used to represent this energy and Figure 5.10 illustrates how it relates to the RF carrier characteristics.

**Noise power**

The noise power is referred to as $N_o$, which represents the noise power spectral density in terms of watts per hertz. The equation below presents the definition of the noise power:

$$N_o = kT,$$

where $T$ is the temperature in Kelvins, and $k = 1.3803 \times 10^{-23}$ is Boltzmann's constant. At room temperature, $N_o = -174$ dBm/Hz.

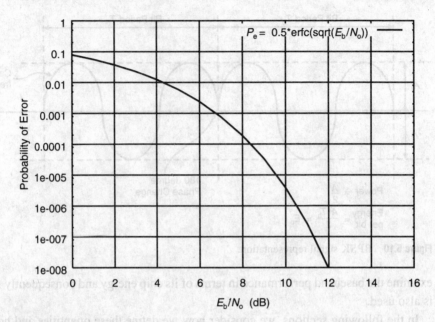

**Figure 5.11**  BPSK BER performance.

## BER

Figure 5.11 illustrates the basic relationship that we are interested in, i.e. the relationship between $E_b/N_o$ and the BER. The figure illustrates the equation and the curve for the probability of error or the BER for a BPSK/QPSK signal. The fact that the results for QPSK and BPSK are the same is a consequence of the orthogonal waveforms used for QPSK: the noise is uncorrelated for the two phases of the signal.

For WCDMA, the spread signals are based on BPSK/QPSK modulation, and so we should expect to obtain error probabilities of a similar order to these ideal curves, except that there will also be channel coding associated with the WCDMA case.

### 5.4.3 Energy per chip

Figure 5.12 illustrates the situation for the transmission of a CDMA spread spectrum signal. In this example there are a single information bit and eight chips (i.e. the spreading factor is 8). We can define a quantity known as the energy per chip as shown in the diagram. The energy per chip is related to the energy per information bit via the spreading factor.

### 5.4.4 Processing gain

Figure 5.13 illustrates the relationship between $E_c/N_o$, which is the wideband 'signal to noise ratio,' and $E_b/N_o$, which is the information signal 'signal to noise ratio'. The

## 5.4 Received signals

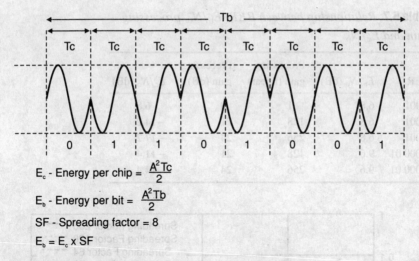

**Figure 5.12** Relationship between energy per chip and energy per bit.

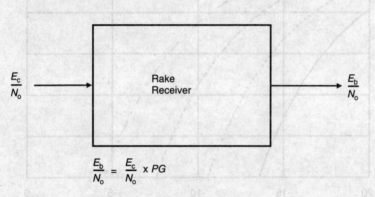

**Figure 5.13** Processing gain.

two are related through a quantity called the processing gain as shown in the diagram. The processing gain is the same as the spreading factor.

Table 5.7 illustrates the relationship between $E_b/N_o$, $E_c/N_o$ and the processing gain. In the first case, the processing gain is 1 (i.e. not spread) and consequently the $E_b/N_o$ and $E_c/N_o$ values are the same. For the other cases the processing gain is greater than 1 and consequently the required $E_c/N_o$ is negative. Conceptually, this means that the wanted signal is below the noise floor of the received signal.

Figure 5.14 illustrates the error probability for a BPSK signal that has been spread with differing degrees of spreading from 32 to 256. From these curves, it is clear that the greater the spreading factor the lower the energy per chip, potentially falling below the noise floor of the receiver. In all of these cases, however, the energy per bit remains the same.

## 150 RF aspects

**Table 5.7.** *Relationship between BER, $E_b/N_o$, processing gain and $E_c/N_o$*

| BER | $E_b/N_o$ (dB) | Processing gain (linear) | Processing gain (dB) | $E_c/N_o$ (dB) |
|---|---|---|---|---|
| 0.001 | 6.8 | 1 | 0 | 6.8 |
| 0.001 | 6.8 | 128 | 21 | −14.2 |
| 0.001 | 6.8 | 256 | 24 | −17.2 |
| 0.000 01 | 9.6 | 128 | 21 | −11.4 |
| 0.000 01 | 9.6 | 256 | 24 | −14.4 |

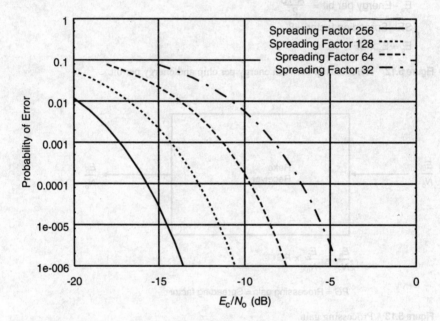

**Figure 5.14** Error probability as a function of $E_c/N_o$ and processing gain.

### 5.4.5 Channel coding

Figure 5.15 illustrates the typical performance of the three types of channel code that are employed in the WCDMA system. There is a convolutional code and a turbo code, with the convolutional code having two different rates: a half rate code and a third rate code. The turbo code has the best performance, with a BER of 0.1% at around 0.5 dB $E_b/N_o$.

### 5.4.6 Coding gain

The curves in Figure 5.16 illustrate the coding gains that are obtainable from the different types of codes used in the WCDMA system. As expected the turbo code exhibits

## 5.4 Received signals

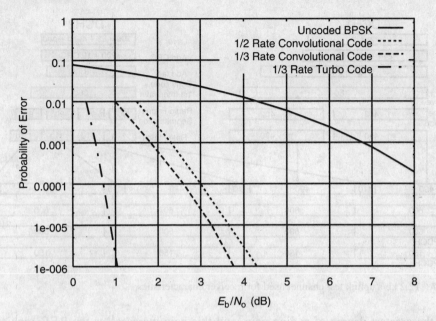

**Figure 5.15** Channel coding performance for WCDMA convolutional, and turbo codes.

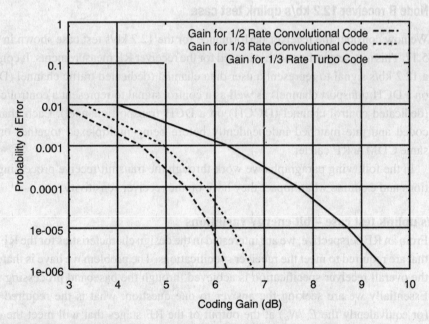

**Figure 5.16** Channel coding gain for WCDMA channel coding schemes.

# 152  RF aspects

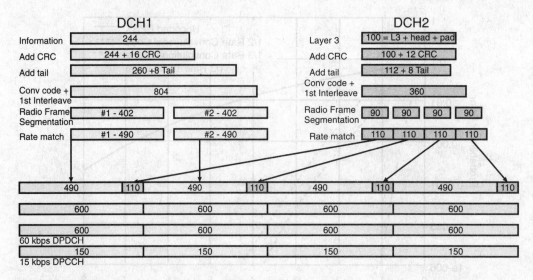

**Figure 5.17** 12.2 kb/s uplink test channel used for receiver characteristics.

the greatest degree of coding gain, and this gain increases as the BER requirement decreases.

### 5.4.7 Node B receiver 12.2 kb/s uplink test case

We now consider the receiver performance for the 12.2 kb/s test case shown in Figure 5.17. This test case is the one that is used for the receiver RF measurements. It comprises a 12.2 kb/s signal to represent a user data channel (dedicated traffic channel (DTCH) on a DCH transport channel) as well as a control signal to represent a control channel (dedicated control channel (DCCH) on a DCH transport channel). Each channel is coded and rate matched independently before being multiplexed together onto the same CDMA RF carrier.

In the following paragraphs we work through the transmit/receive processing functions and examine what impact they have on the receiver sensitivity.

**12.2 kb/s uplink test case – bit energy variations**

From an RF perspective, we are interested in the design characteristics for the RF stages that are required to meet the receiver specifications. The problem we have is that part of the overall receiver specification is achieved through the baseband processing stages. Essentially we are seeking the answer to one question: what is the required $E_b/N_o$ (or equivalently the $E_c/N_o$) at the output of the RF stages that will meet the overall receiver specification targets and from which we can estimate the RF specifications? One solution to this question is to simulate the entire performance of the baseband stages to arrive at specification targets at the output of the RF stages. While this is likely to be the solution adopted, we can lose some insight into how the specification targets are derived. Here, we examine what factors contribute to the desired receiver

## 5.4 Received signals

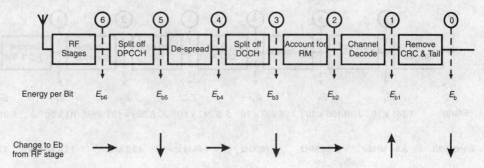

**Figure 5.18** Receiver energy variations.

performance based on analysis of the receiver chain, rather than through the use of simulation.

Consider the receiver chain from the RF stage through the baseband stages that are illustrated in Figure 5.18. The diagram shows how the energy per bit changes as it passes through the different stages. The energy per bit drops at stage 5; this is due to the removal of the DPCCH. At stage 4, after despreading the energy per bit remains the same; spreading and despreading have no effect on the energy per bit. At stage 3, after the removal of the DCCH that is carried through the second transport channel, the energy per bit has decreased. Some of the received bit energy is required for the DCCH and consequently the remaining bit energy is less. At stage 2, after rate matching, the energy per bit remains the same. This is the case when rate matching through repetition is used. If rate matching using puncturing were required, then the energy per bit would change. At stage 1, after the channel decoding, the effective energy per bit has increased due to the coding gain from the channel decoder. Finally, at stage 0, after the removal of the CRC bits and accounting for the tail bits, the energy per bit has reduced.

Figure 5.18, therefore, presents a qualitative description of the changes in energy per bit moving through the receiver chain. If we examine Figure 5.19, we can explore a quantitative performance of the receiver. Using Figure 5.19 we can estimate the energy per bit and hence the $E_b/N_o$ that we require at stage 6, just after the RF stages, and from which we can estimate the receiver noise figure. To do this, we traverse the receiver in the opposite direction starting at the output of the receiver on the right hand side. At the output of the receiver, for this specific test channel, we know from the specifications that the required bit error rate is $1 \times 10^{-3}$. From Table 5.7, we know that this corresponds to 6.8 dB $E_b/N_o$, or in linear terms 4.78. If we assume a certain value for the noise power spectral density ($N_o$), which (in general terms) remains constant through the receiver chain, we can estimate the received bit energy that we require at stage 0 ($E_b$ in Figure 5.19). In this case, we assume that the receiver noise figure is 5 dB, $kT$ is $-174$ dBm/Hz (hence $N_o$ is $1.2575 \times 10^{-20}$ J) and consequently $E_b$ can be estimated as being $6.011 \times 10^{-20}$ J. As we move towards the front of the receiver, the received energy changes (see $E_0$ to $E_6$), and the received bit energy ($E_b$ to $E_{b6}$) changes. These

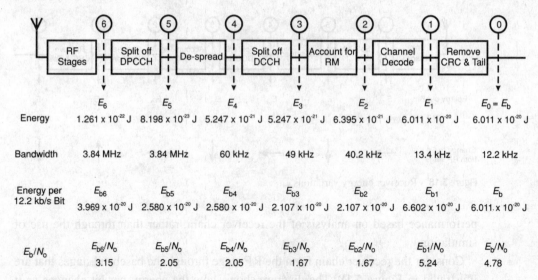

**Figure 5.19** Uplink receiver received energy variations.

changes are due to changes in the signal bandwidth and also changes in the received energy that were highlighted by Figure 5.18.

At stage 6, just after the RF stages, we find that we require a received energy per bit of $3.969 \times 10^{-20}$ J. Using the assumed value for $N_o$ from above, we find that $E_b/N_o$ is 3.15 (linear) or 5 dB. Using this value for $E_b/N_o$ we can now proceed in Section 5.5 to calculate the receiver sensitivity.

There is one final note of caution. The value of $E_b/N_o$ assumes ideal signal processing. In practice an implementation margin of around 2 dB is normally added to the calculated $E_b/N_o$ to account for real-world implementation issues.

## 5.5 Node B receiver characteristics

In this section, we explore the receiver RF characteristics for the Node B receiver. The characteristics for the receiver define the basic receiver performance for the Node B. In the section that follows this, we will also explore the Node B receiver performance specifications.

### 5.5.1 Receiver sensitivity

Here we are going to estimate the noise figure that is required to achieve the specified reference sensitivity for the WCDMA base station of $-121$ dBm for a BER of 0.1%.

Figure 5.20 illustrates the test system that will be used to estimate the reference sensitivity performance for the base station receiver. For WCDMA, it is assumed that the base station incorporates a diversity receiver as a minimum. For the reference

## 5.5 Node B receiver characteristics

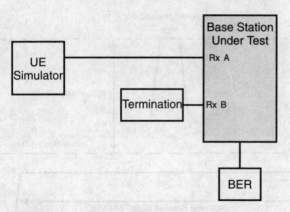

**Figure 5.20** Uplink receiver characteristics test system.

sensitivity tests, one of the receivers is terminated and the test signal applied to the other receiver and the BER measured for the required reference sensitivity of −121 dBm.

### 12.2 kb/s example – estimation of receiver noise figure

In Section 5.4.4 we saw that $E_b/N_o$ and $E_c/N_o$ are related through the quantity referred to as the processing gain ($PG$). This is illustrated in

$$\frac{E_b}{N_o} = \frac{E_c}{N_o} PG \tag{5.1}$$

We can express the noise power spectral density $N_o$ in terms of Boltzmann's constant ($k$), temperature ($T$) and the receiver noise figure ($NF$) according to

$$N_o = kTNF \tag{5.2}$$

The energy per chip ($E_c$) can be expressed in terms of the received signal power ($P_{Rx}$) and the chip duration according to

$$E_c = P_{Rx} T_c \tag{5.3}$$

Substituting (5.2) and (5.3) in (5.1) we obtain:

$$\frac{E_b}{N_o} = \frac{P_{Rx} T_c PG}{kTNF} \tag{5.4}$$

Expressing both sides of the equation in terms of decibels, we obtain

$$\left(\frac{E_b}{N_o}\right)_{dB} = P_{RxdB} + T_{cdB} + PG_{dB} - kT_{dB} - NF_{dB} \tag{5.5}$$

Rearranging (5.5) to obtain an expression for $NF_{db}$ we obtain

$$NF_{dB} = P_{RxdB} + T_{cdB} + PG_{dB} - kT_{dB} - \left(\frac{E_b}{N_o}\right)_{dB} \tag{5.6}$$

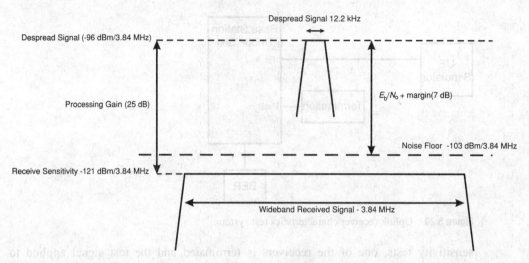

**Figure 5.21** Graphical representation of Node B receiver operation.

where $N_o$ is the effective input noise power spectral density including the effects of the receiver noise, $NF$ is the receiver noise, figure, $T_c$ is the chip period, and $E_b/N_o$ is the required energy per bit over the noise power spectral density.

Having established the input value of $E_b/N_o$, we can now proceed to estimate the input receiver sensitivity. To start we redefine the definitions for $E_b$ and $N_o$ in the terms that we used previously. Next we can derive an expression for the noise figure in terms of $E_b/N_o$. For the realistic case, the receive sensitivity is calculated including the implementation margin of around 2 dB.

Using (5.6) we can estimate the noise figure we require for the receiver. We know from the specifications that the received power in dBm ($P_{RxdB}$) is −121 dBm. We also know that the chip duration in decibels ($T_{cdB}$) is −65.8, the processing gain in decibels ($PG_{dB}$) is 25 and Boltzmann's constant times the temperature in kelvins ($kT_{dB}$) is −174 dBm. Following the discussion outlined in Section 5.4.7, we saw that the required $E_b/N_o$ expressed in decibels and including an implementation margin should be 7.0 dB. If we put these values into (5.6), then we can estimate the required noise figure in decibels ($NF_{dB}$). In this example the result is 5.2 dB.

Figure 5.21 illustrates physically what has happened in the receiver. The wideband signal (spread across 3.84 MHz) is despread by the receiver into a bandwidth of 12.2 kHz. This means that there is a processing gain of 25 dB that raises the level of the received signal spectral density from −121 dBm/3.84 MHz to a level of −96 dBm/ 3.84 MHz. The rise is due to the bandwidth reduction achieved through the despreading procedures. The despreading process has concentrated the received signal into a bandwidth equal to the original signal (12.2 kHz) and consequently, the received spectral density in the 12.2 kHz bandwidth is −121 dBm/12.2 kHz.

The received power spectral density should be greater than the noise floor by an amount equal to $E_b/N_o$. We can represent the noise floor in a 3.84 MHz bandwidth as

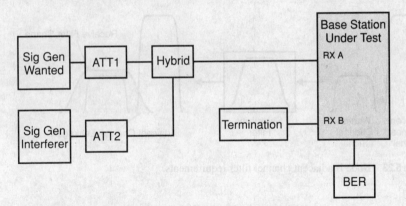

**Figure 5.22**  Node B adjacent channel selectivity test system.

shown in Figure 5.21 (−103 dBm/3.84 MHz), or we could represent it with a bandwidth of 12.2 kHz (−128 dBm/12.2 kHz). However, whichever representation we use, we need to keep a consistent set of units. In the case of the wideband representation shown in Figure 5.21, the received signal is $E_b/N_o$ greater than the wideband noise level.

What is also interesting is that the RF signal is 18 dB below the receiver noise floor of −103 dBm/3.84 MHz. This is one of the phenomenona of a CDMA signal: the received signal can lie below the noise floor of the receiver.

### 5.5.2 Receiver adjacent channel selectivity

In this section we are going to estimate the adjacent channel selectivity requirements for the receiver. The Node B receiver specifications [10] stipulate that the receiver should be able to receive a wanted signal at a level of −115 dBm with an average error rate of $1 \times 10^{-3}$ whilst there is an interfering signal at a level of −52 dBm in the adjacent channel. The two characteristics that we need to design for are the adjacent channel filter requirements and the receiver local oscillator phase noise requirements. In this section we consider each of these independently.

Figure 5.22 illustrates the test system that is used to estimate the receiver adjacent channel filtering requirements and the local oscillator phase noise requirements. A wanted signal is applied on-frequency along with an interfering signal that is 5 MHz offset. The wanted signal has a signal level of −115 dBm, whilst the interfering signal has a signal level of −52 dBm. The receiver has to be designed to meet a BER performance of 0.1% whilst subject to this high level of interference.

**Adjacent channel filter requirements**

Figure 5.23 illustrates the basic problem with the adjacent channel selectivity requirements. The receiver needs to detect a weak wanted signal in the presence of a strong adjacent channel signal. To achieve this, a selection filter is needed somewhere in the receiver architecture to reject the adjacent channel signal but allow the wanted signal.

## 158 RF aspects

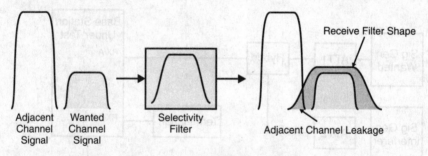

**Figure 5.23** Node B adjacent channel filter requirements.

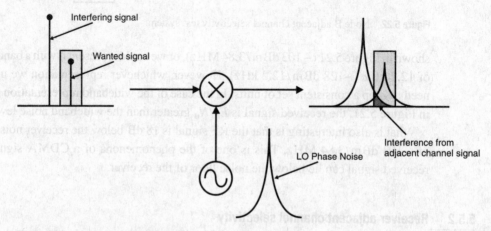

**Figure 5.24** Node B phase noise performance requirements.

It is unlikely that the selection filter is able to achieve this totally, and consequently some of the adjacent channel signal passes through the selection filter. The requirement, therefore, is to limit the adjacent channel signal to a sufficiently low level.

**Phase noise requirements**

Figure 5.24 illustrates the second problem that the receiver design has to overcome when considering the adjacent channel performance: this is the phase noise of the local oscillator. The local oscillator is not a perfect sine wave, and includes noise that is referred to as phase noise. These phase noise components mix with the interfering signal injecting them into the wanted band. By specifying the local oscillator phase noise appropriately, it is possible to control the degree of interference introduced into the wanted channel.

**Receiver adjacent channel selectivity (ACS) calculation**

Figure 5.25 illustrates the basic situation that is used to estimate the level of receive filtering required to meet the ACS. The interfering signal is attenuated by an amount

## 5.5 Node B receiver characteristics

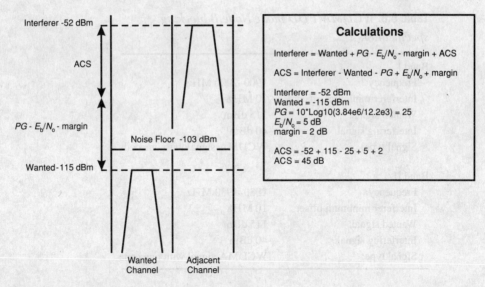

**Figure 5.25** Node B receiver ACS filtering estimation.

equal to the ACS plus the processing gain ($PG$) less $E_b/N_o$ and the margins required for the threshold sensitivity.

From an analysis of the situation, we write an equation for the interferer signal power as shown in Figure 5.25. From this, we can solve the equation for the ACS. Using the numbers presented and calculated previously, assuming an implementation margin of 2 dB, we end up with a minimum ACS of 45 dB for the receiver. This means that the receiver has to have a total average attenuation across the adjacent band of 45 dB. Depending upon the filter shape, this total attenuation can be converted into a stop band performance requirement.

### Local oscillator (LO) phase noise calculation

To estimate the phase noise requirements in the receiver, we calculate the maximum allowed phase noise from the local oscillator. We set the maximum phase noise equal to the wanted signal plus the processing gain less the required $E_b/N_o$. After accounting for the bandwidth differences (phase noise is normally expressed in dB/Hz and not per 3.84 MHz) we arrive at the following expression for the phase noise:

$$L(f) = DPCH\_E_c + PG_{dB} - \frac{E_b}{N_o} - I_o - 10\log_{10}(r_c) \tag{5.7}$$

where $r_c$ is the chip rate and $I_o$ is the interference. $L(f)$ is the required LO phase noise converted to dBc/Hz. When $DPCH\_E_c = -115$, $PG_{dB} = 25$, $E_b/N_o = 7.2$, $I_o = -52$ dBm, $r_c = 3.84 \times 10^6$, (5.7) gives $L(f) = -110.8$ dBc/Hz averaged across the adjacent channel.

From (5.7) it is possible to estimate the maximum value that can be accepted for the phase noise. In this case, a phase noise level better than $-110.8$ dBc/Hz is required

**Table 5.8.** *WCDMA FDD mode Node B blocking specification*

Band I
- Frequency: 1900–2000 MHz
- Interferer minimum offset: 10 MHz
- Wanted signal: −115 dBm
- Interfering signal: −40 dBm
- Signal type: WCDMA signal with one code

Band II
- Frequency: 1830–1930 MHz
- Interferer minimum offset: 10 MHz
- Wanted signal: −115 dBm
- Interfering signal: −40 dBm
- Signal type: WCDMA signal with one code

averaged across the adjacent channel. It should be noted that this is the maximum value, and that in reality the LO phase noise needs to be better than this.

## LO spurious signals

An additional source of interference introduced in the adjacent channel selectivity test are the local oscillator spurious signals. We need to estimate the requirements for the spurious signals. To do this we adopt the same approach as for the phase noise, except we assume that the source of interference is confined to a fixed number of discrete sources.

An equation similar to the phase noise equation can be presented:

$$S(f) = DPCH\_E_c + PG_{dB} - \frac{E_b}{N_o} - I_o - 10\log_{10}(n) \qquad (5.8)$$

where $n$ is the number of spurious components and $I_o$ is the interference. $S(f)$ is the maximum level for $n$ equal power spurious signals.

When $DPCH\_E_c = -115$, $PG_{dB} = 25$, $E_b/N_o = 7$ and $I_o = -52$ dBm, (5.8) gives $S(f) = -55$ dBc for each spurious signal, assuming $n = 10$.

From (5.8), it is possible to determine the level required per spurious signal assuming that there are $n$ spurious signals. Each of the spurious signals is assumed to have the same amplitude. In this example, assuming ten signals, each must be less than −55 dBc. The same comments about additional margins apply as in the case for the phase noise.

### 5.5.3 Receiver blocking

The data in Table 5.8 illustrate the requirements for the receiver blocking for two of the three defined bands. For the receiver blocking specifications, the error

## 5.5 Node B receiver characteristics

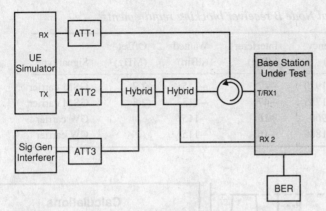

**Figure 5.26** Receiver blocking test system.

performance from the static reference sensitivity described in Section 5.5.1 is achieved for the receiver blocking test with the wanted signal and interfering signals defined according to Table 5.8. The blocking specification extends the ACS specification into the adjacent frequency bands. As a consequence, the interfering signal powers are greater than for the ACS specification.

**Receiver blocking specification**

Figure 5.26 illustrates the test system that is used to estimate the receiver blocking specification. The test setup is similar to that used for the ACS specification with the two test signals applied to one of the base station's receiver ports and the second port is terminated.

Figure 5.27 illustrates the physical processes in the receiver and the method used to estimate the receiver blocking performance. Similar to the ACS case seen previously, the blocking filter and the processing gain protect the receiver from the effects of the interference in the receiver. Using the equation shown in Figure 5.27 and the values specified and calculated previously, it is possible to estimate the blocking requirement. In this case, the receiver needs to be designed with a blocking requirement of 57 dB.

Table 5.9 illustrates the additional blocking performance required for a receiver. The first two cases apply to operation in the DCS1900/DCS1800 bands and the possibility of mobiles blocking the receiver using a GSM signal. The second two cases are for the blocking performance when the WCDMA system is co-located with an existing GSM system. In this case the interferer is from the GSM BTS transmitter.

### 5.5.4 Receiver dynamic range

Using the 12.2 kb/s measurement channel, a wanted signal of −91 dBm and an interfering AWGN signal of −73 dBm (in a 3.84 MHz bandwidth), the receiver needs to meet a BER of 0.1%.

**Table 5.9.** *Additional Node B receiver blocking requirements*

| Band | Frequency (MHz) | Interferer (dBm) | Wanted (dBm) | Offset (MHz) | Signal type |
|---|---|---|---|---|---|
| Band II | 1850–1910 | −47 | −115 | 2.7 | GSM carrier |
| Band III | 1710–1785 | −47 | −115 | 2.8 | GSM carrier |
| GSM | 921–960 | +16 | −115 | | CW carrier |
| DCS1800 | 1805–1880 | +16 | −115 | | CW carrier |

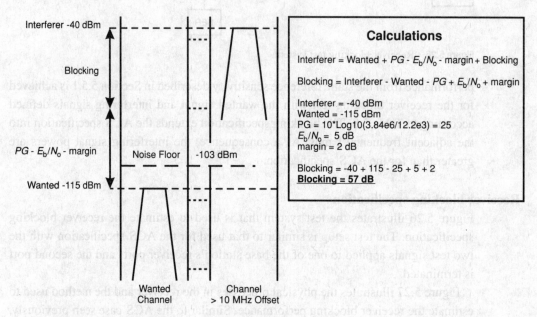

**Figure 5.27** Node B receiver blocking calculations.

Figure 5.28 illustrates the test case for the dynamic range test and also the physical interpretation of the dynamic range test. The interfering signal is attenuated by an amount equal to the processing gain. As long as the resultant noise signal is $E_b/N_o$ plus margin below the wanted signal, the BER of 0.1% can be achieved. This leads to a receiver dynamic range of 18 dB.

### 5.5.5 Receiver intermodulation specification

Table 5.10 illustrates the requirements for the receiver intermodulation performance. The receiver is subject to two interfering signals: the first (10 MHz offset) is a CW signal, the second (20 MHz offset) is a WCDMA signal. The receiver should be able to operate at the reference BER with these two signals present. Figure 5.29 illustrates the test configuration that is required to measure the receiver intermodulation performance.

## 5.5 Node B receiver characteristics

**Table 5.10.** *Node B intermodulation test requirements*

| | |
|---|---|
| Measurement channel | 12.2 kbps |
| Wanted signal | −115 dBm |
| Interfering CW signal | −48 dBm |
| Interfering CW signal offset | 10 MHz |
| Interfering WCDMA signal | −48 dBm |
| Interfering WCDMA signal offset | 20 MHz |
| Test BER | <0.1% |

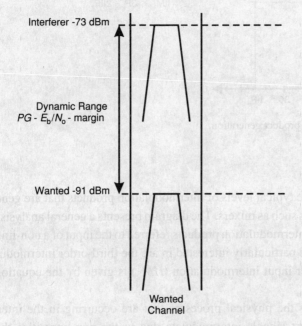

**Figure 5.28** Node B receiver dynamic range calculations.

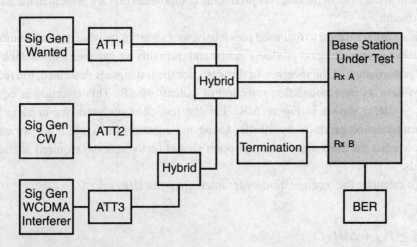

**Figure 5.29** Receiver intermodulation test system.

# 164 RF aspects

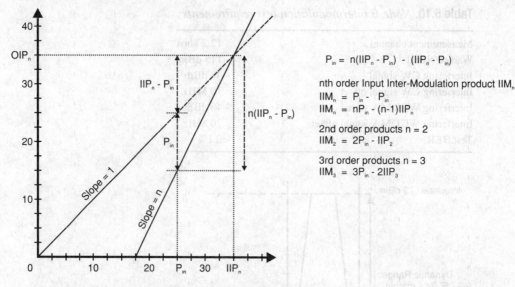

**Figure 5.30** Receiver intermodulation product generation.

## Intermodulation products

Figure 5.30 illustrates the typical levels of intermodulation products that are generated within non-linear devices such as mixers. The diagram presents a general analysis in the generation of $n$th-order intermodulation products referred to the input of a non-linearity. The products that we are particularly interested in are the third-order intermodulation products. The third-order input intermodulation ($IIM_3$) is given by the equation presented in Figure 5.30.

Figure 5.31 illustrates the physical processes that are occurring in the intermodulation test. The two test signals intermodulate due to the non-linearities that are present at the front of the receiver producing components that are present in the receiver bandwidth.

In a similar method to that used previously, we can estimate what the maximum level for these intermodulation products can be and from this we can then estimate the input $IP_3$ performance of the receiver. In this case, for the test signals described, the receiver must have an intermodulation rejection of at least 49 dB. This rejection is equal to ($P_{in} - IIM_3$) shown in Figure 5.30. For the test case analysed, we need to reject intermodulation products by 49 dB. Using the equation in Figure 5.30, we can estimate what the third-order intercept point should be to meet the required attenuation levels.

To calculate the receiver third-order intercept point ($IIP_3$):

$$IIP_3 \geq P_{sig} + \Delta IM_3/2$$

## 5.6 Node B receiver performance

**Table 5.11.** *Node B narrowband intermodulation test requirements*

| | |
|---|---|
| Measurement channel | 12.2 kb/s |
| Wanted signal | −115 dBm |
| Interfering CW signal | −47 dBm |
| Interfering CW signal offset | 3.5 MHz |
| Interfering WCDMA signal | −47 dBm |
| Interfering WCDMA signal offset | 5.9 MHz |
| Test BER | <0.1% |

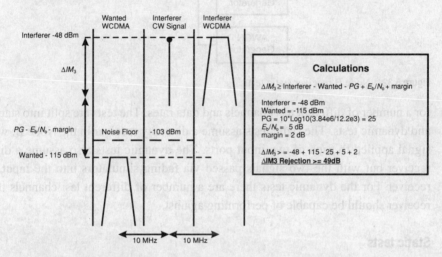

**Figure 5.31** Intermodulation generation processes.

where $P_{sig}$ the power of the interfering signal is −48 dBm, and $\Delta IM_3$ (the third-order product attenuation) is greater than −49 dB. Therefore:

$$IIP_3 \geq -48 + \frac{49}{2}$$

i.e.

$$IIP_3 \geq -23.5 \text{ dBm}$$

**Narrowband receiver intermodulation specification**

Table 5.11 illustrates the narrowband receiver intermodulation performance that needs to be achieved in cases when the receiver is operating in bands also used by GSM equipment (e.g. Band II and Band III).

## 5.6 Node B receiver performance

In this section we examine the Node B receiver performance. The previous section examined the receiver characteristics, i.e. how the receiver performed from an RF perspective. The receiver performance tests characterise the performance of the receiver

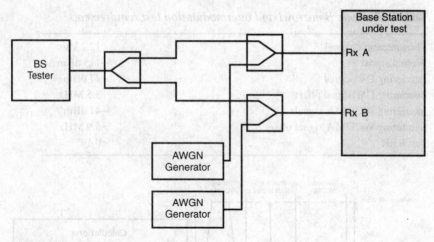

**Figure 5.32**  Node B receiver static tests.

for a number of different test channels and data rates. The tests are split into static tests and dynamic tests. The static tests assume a diversity receiver in the Node B, with the signal applied to each of two input ports. The dynamic tests also assume a diversity receiver but with the two signals passed via fading simulators into the input of the receiver. For the dynamic tests there are a number of different test channels that the receiver should be capable of performing against.

### 5.6.1 Static tests

Figure 5.32 illustrates the test methodology that is used for characterising the static performance characteristics of the Node B receiver. The test equipment consists of a base station (BS) tester to create the required signals, noise generators and the Node B receiver. In this and all performance tests the Node B receiver is a two-branch diversity receiver.

### 5.6.2 Dynamic multipath fading tests

Figure 5.33 illustrates the Node B receiver test configuration used to ascertain the receiver performance of the Node B in the case where a multipath channel is assumed. The Node B receiver is a diversity receiver and so two independently fading signals are used to feed each of the receivers.

Figure 5.34 illustrates the fading profile for a number of different test cases defined for the dynamic receiver performance tests. For the tests, the uplink power control is inactive.

**Case 1 fading profile**

In Case 1 (Figure 5.34(a)), there is a direct path and a single reflected path. The reflected path is 976 ns delayed with respect to the direct path. In both cases the signal paths are fading at a rate equivalent to a mobile velocity of 3 km/h.

## 5.6 Node B receiver performance

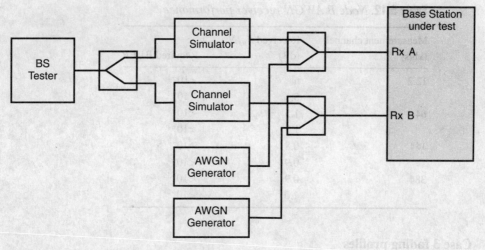

**Figure 5.33** Node B receiver dynamic tests.

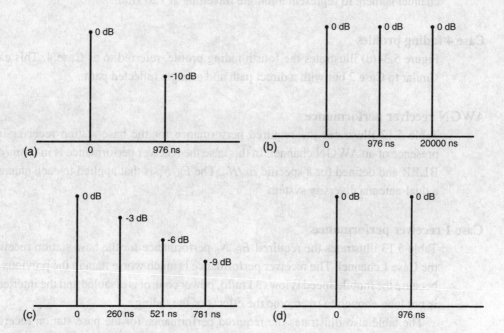

**Figure 5.34** Fading profiles: (a) Case 1: speed 3 km/h, classical Doppler spectrum; (b) Case 2: speed 3 km/h, classical Doppler spectrum; (c) Case 3: speed 120 km/h, classical Doppler spectrum; (d) Case 4: speed 3 km/h, classical Doppler spectrum.

### Case 2 fading profiles

Figure 5.34(b) illustrates the multipath channel parameters for the conditions referred to as Case 2. In this model, the simulated vehicle speed is 3 km/h and two reflected paths are included along with the direct path. In this case, one of the paths is of very high delay. All paths are of equal power level.

**Table 5.12.** *Node B AWGN receiver performance*

| Measurement channel (kb/s) | Received $E_b/N_o$ (dB) | Required BLER |
|---|---|---|
| 12.2 | na | $<10^{-1}$ |
|  | 5.1 | $<10^{-2}$ |
| 64 | 1.5 | $<10^{-1}$ |
|  | 1.7 | $<10^{-2}$ |
| 144 | 0.8 | $<10^{-1}$ |
|  | 0.9 | $<10^{-2}$ |
| 384 | 0.9 | $<10^{-1}$ |
|  | 1.0 | $<10^{-2}$ |

**Case 3 fading profiles**

Figure 5.34(c) illustrates the third multipath channel model, referred to as Case 3. This channel is there to represent a mobile travelling at 120 km/h.

**Case 4 fading profiles**

Figure 5.34(d) illustrates the fourth fading profile, referred to as Case 4. This case is similar to Case 2 but with a direct path and a single reflected path.

**AWGN receiver performance**

Table 5.12 illustrates the required performance for the base station receiver in the presence of an AWGN channel. In this case the receiver performance is in terms of the BLER and defined for a specific $E_b/N_o$. The $E_b/N_o$ is that applied to each antenna of a dual-antenna diversity system.

**Case 1 receiver performance**

Table 5.13 illustrates the required $E_b/N_o$ performance for the base station receiver in the Case 1 channel. The receiver performance is much worse than in the previous case, because the mobile speed is low (3 km/h), power control is disabled and the interleaving is not long enough to overcome the effects of the fading.

The table also illustrates the required performance for the base station receiver in the Case 2 channel. The main difference between this channel and the previous one is that there is a large number of equal power reflections that provide a large degree of diversity gain from the rake receiver.

In addition, the table illustrates the required performance for the base station receiver in the Case 3 channel. In this case the main difference is the change to the channel impulse profile and the change in the mobile speed to 120 km/h. This provides some improvement over what can be obtained in the previous two fading cases as the interleaving can now have some effect on the fading.

**Table 5.13.** *Node B dynamic receiver performance*

| Measurement channel (kb/s) | Required BLER | Case 1 received (dB) | Case 2 received (dB) | Case 3 received (dB) |
|---|---|---|---|---|
| 12.2 | $<10^{-1}$ | | | |
| | $<10^{-2}$ | 11.9 | 9.0 | 7.2 |
| | $<10^{-3}$ | | | 8.0 |
| 64 | $<10^{-1}$ | 6.2 | 4.3 | 3.4 |
| | $<10^{-2}$ | 9.2 | 6.4 | 3.8 |
| | $<10^{-3}$ | | | 4.1 |
| 144 | $<10^{-1}$ | 5.4 | 3.7 | 2.8 |
| | $<10^{-2}$ | 8.4 | 5.6 | 3.2 |
| | $<10^{-3}$ | | | 3.6 |
| 384 | $<10^{-1}$ | 5.8 | 4.1 | 3.2 |
| | $<10^{-2}$ | 8.8 | 6.1 | 3.6 |
| | $<10^{-3}$ | | | 4.2 |

## 5.7 UE receiver characteristics

Figure 5.35 illustrates one of the measurement channels that are used for the UE receiver tests. In this specific case, two logical channels, a traffic channel (DTCH) and a control channel (DCCH), are combined together and transmitted on the downlink the DPCCH. The DPCCH carries pilot, TFCI and other physical control bits.

### 5.7.1 Estimation of UE receiver sensitivity

In a manner identical to that used for the Node B receiver, we can calculate the receiver noise figure required for the UE receiver. The parameters that differ are the $E_b/N_o$ required (5.2 dB plus 2 dB implementation margin) and the receiver sensitivity of $-117$ dBm. Using these values in the expression in Section 5.5.1 we can calculate that the receiver noise figure for the UE is 9 dB. Figure 5.36 illustrates the test system that is used to estimate the reference sensitivity performance for the UE receiver.

### 5.7.2 UE receiver ACS

In this section, we are going to estimate the ACS requirements for the receiver for an input signal of $-103$ dBm, an interfering signal of $-52$ dBm and a BER of 0.1%.

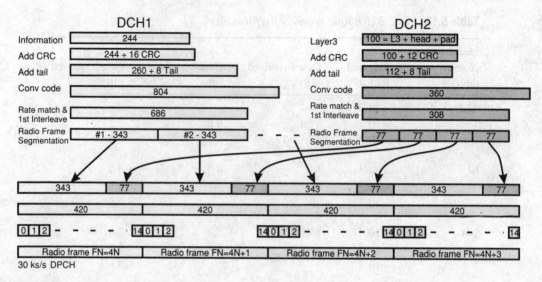

**Figure 5.35** UE measurement characteristics.

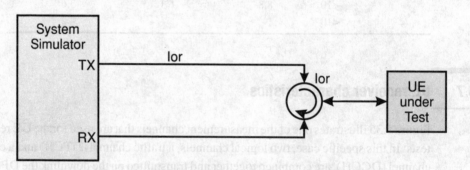

**Figure 5.36** UE receiver sensitivity.

Figure 5.37 illustrates the test system that is used. A wanted signal is applied on frequency along with an interfering signal 5 MHz offset. The wanted signal has a signal level of −103 dBm, whilst the interfering signal has a signal level of −52 dBm. The receiver has to be designed to meet a BER performance of 0.1% whilst subject to this high level of interference.

Using the numbers presented and the equations in Section 5.5.2 we end up with a minimum ACS of approximately 33 dB for the receiver. This means that the receiver has to have a total average attenuation across the adjacent band of approximately 33 dB. Depending upon the filter shape, this total attenuation can be converted into a stop band performance requirement.

### 5.7.3 Receiver blocking

Table 5.14 illustrates the requirements for the receiver blocking for the UE within the frequency band. The blocking specification extends the ACS specification into the

## 5.7 UE receiver characteristics

**Table 5.14.** *UE receiver ACS requirements*

| | |
|---|---|
| Measurement channel | 12.2 kb/s |
| Wanted signal | −103 dBm |
| Interfering signal | −52 dBm |
| Frequency offset | 5 MHz |
| Test BER | <0.1% |

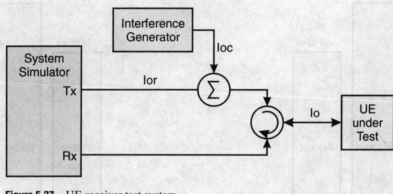

**Figure 5.37** UE receiver test system.

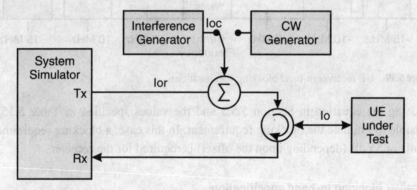

**Figure 5.38** UE receiver blocking test system.

adjacent frequency bands. As a consequence the interfering signal powers are now much greater than for the ACS specification.

### Receiver blocking specification

Figure 5.38 illustrates the test system that is used to estimate the receiver blocking specification. The test setup is similar to the ACS specification with the two test signals applied to the UE transmit/receive port. Similarly to the blocking case for the Node B, the blocking filter and the processing gain protect the receiver from the effects of the interference in the receiver.

**Table 5.15.** *UE receiver blocking requirements*

| Interferer minimum offset | 10 MHz | 15 MHz |
|---|---|---|
| Wanted signal | −114 dBm | −114 dBm |
| Interfering signal | −56 dBm | −44 dBm |
| Signal type | WCDMA signal with one code | WCDMA signal with one code |

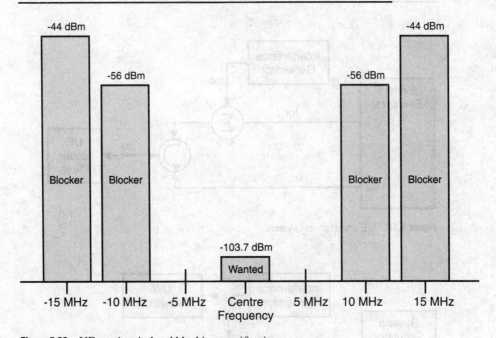

**Figure 5.39** UE receiver in-band blocking specifications.

Using the equation in Section 5.5.3 and the values specified in Table 5.15, it is possible to estimate the blocking requirement. In this case, a blocking requirement of 43 dB or 55 dB (depending upon the offset) is required for the receiver.

### 5.7.4 Receiver blocking in-band specification

Figure 5.39 summarises the in-band receiver blocking specifications.

### 5.7.5 Receiver blocking out-of-band specification

Figure 5.40 summarises the out-of-band receiver blocking specifications.

### 5.7.6 Receiver intermodulation specification

Let us consider the requirements for the receiver intermodulation performance. Table 5.16 summarises the performance requirements for the UE receiver intermodulation

## 5.7 UE receiver characteristics

**Table 5.16.** *UE intermodulation test requirements*

| | |
|---|---|
| Measurement channel | 12.2 kb/s |
| Wanted signal | −114 dBm |
| Interfering CW signal | −46 dBm |
| Interfering CW signal offset | 10 MHz |
| Interfering WCDMA signal | −46 dBm |
| Interfering WCDMA signal offset | 20 MHz |
| Test BER | <0.1% |

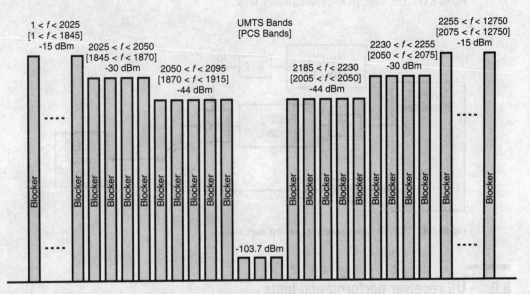

**Figure 5.40** UE receiver out-of-band blocking specifications.

test. The receiver is subject to two interfering signals: the first (10 MHz offset) is a CW signal, the second (20 MHz offset) is a WCDMA signal. The receiver should be able to operate at the reference BER with these two signals present. The two test signals intermodulate due to the non-linearities that are present at the front of the receiver producing components that are present in the receiver bandwidth.

In a similar method to that used previously for the Node B, we can estimate what the maximum level for these intermodulation products can be and from this we can then estimate the input $IP_3$ performance of the receiver. In this case, for the test signals described, the receiver must have an intermodulation rejection of at least 53 dB. Using the values estimated previously, and the test signals, we can estimate that the UE $IIP_3$ needs to be greater than or equal to −19.5 dBm.

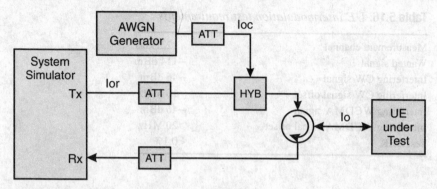

**Figure 5.41** UE static receiver performance tests.

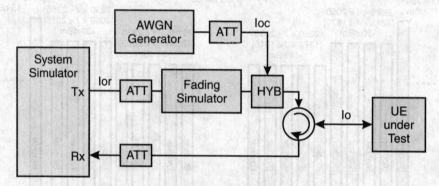

**Figure 5.42** UE dynamic receiver performance tests.

## 5.8 UE receiver performance tests

### 5.8.1 Static tests

Figure 5.41 illustrates the test methodology that is used for characterising the static performance characteristics of the UE receiver. The test apparatus consists of a system simulator to create the required signals, noise generators and the UE.

**Dynamic multipath fading tests**

Figure 5.42 illustrates the UE test configuration used to ascertain the receiver performance in the case where a multipath channel is assumed.

**Case 1–Case 4 fading profiles**

The same test cases as defined for the Node B are also used in the UE dynamic performance tests.

## 5.8 UE receiver performance tests

**Table 5.17.** *UE static AWGN receiver performance*

| Measurement channel | Received DPCH $E_c/I_{or}$ | Required BLER |
|---|---|---|
| 12.2 kb/s | na | $<10^{-1}$ |
|  | −16.6 dB | $<10^{-2}$ |
| 64 kb/s | −13.1 dB | $<10^{-1}$ |
|  | −12.8 dB | $<10^{-2}$ |
| 144 kb/s | −9.8 dB | $<10^{-1}$ |
|  | −9.8 dB | $<10^{-2}$ |
| 384 kb/s | −5.6 dB | $<10^{-1}$ |
|  | −5.5 dB | $<10^{-2}$ |

**Table 5.18.** *UE receiver performance for the Case 3 fading channel*

| Measurement channel kb/s | Received DPCH $E_c/I_{or}$ (dB) | Required BLER |
|---|---|---|
| 12.2 | na | $<10^{-1}$ |
|  | −11.8 | $<10^{-2}$ |
| 64 | −8.1 | $<10^{-1}$ |
|  | −7.4 | $<10^{-2}$ |
| 144 | −9.0 | $<10^{-1}$ |
|  | −8.5 | $<10^{-2}$ |
|  | −8.0 | $<10^{-3}$ |
| 384 | −5.9 | $<10^{-1}$ |
|  | −5.1 | $<10^{-2}$ |
|  | −4.4 | $<10^{-3}$ |

### 5.8.2 AWGN receiver performance

Table 5.17 illustrates the required performance for the UE in the presence of an AWGN channel. In this case the receiver performance is in terms of the BLER and is defined for a specific DPCH $E_c/I_{or}$.

**Case 3 receiver performance**

Table 5.18 illustrates the required performance for the UE in the Case 3 channel. In this case, the main difference is the change to the channel impulse profile and the change in the mobile speed to 120 km/h.

## 5.9 UMTS transceiver architecture study

For both the UE and the Node B, there are many possible architecture design options that can be made for both the transmitter and the receiver. For the transmitter we need to consider the basic up-conversion architecture (such as direct up-conversion or superheterodyne up-conversion). For the Node B and high complexity UEs the linearity of the power amplifier is important. For the receiver there are also architecture decisions to be made. There are many potential receiver architectures such as direct conversion, superheterodyne, wideband intermediate frequency (IF), direct IF-sampling, to name a few. In this section we aim to consider some aspects of these architecture design criteria. To simplify the considerations, however, we focus on one specific architecture structure for the UE.

### 5.9.1 General UMTS transceiver architecture

Let us consider an example architecture design for the UE transceiver. The design needs to include support for other 2G/3G systems such as GSM/GPRS, and so an architecture such as that illustrated in Figure 5.43 should be considered. Before considering the details of the architectural elements, let us examine the basic elements of the architecture.

Figure 5.43 illustrates a multimode design that is intended to encompass WCDMA and GSM/GPRS across a number of frequency bands. We are assuming that we have a single common antenna, and that only one mode is active at a given time. As a consequence, we have a single antenna and a mode-switch that can switch between the WCDMA mode of operation and the GSM/GPRS mode of operation. The GSM/GPRS elements in the architecture consist of multiband transmitter and receiver chains connected to the antenna via a multiway switch (this is assuming that GSM/GPRS is low data rate and does not require continuous transmission and reception).

The WCDMA part of the design consists of a duplex filter that connects to a transmit and receive path. In this example, the receive path is of a direct conversion design considered in more detail in Section 5.9.2, and the transmitter is a superheterodyne design considered in more detail in Section 5.9.3. WCDMA is a full duplex system and consequently requires a duplex filter to separate the transmit and receive paths. The duplex filter isolates the transmitter from the receiver and also provides some band filtering that will be useful in meeting both the transmitter and receiver specifications for frequencies far from the wanted band.

Figure 5.43 illustrates the basic components for the direct conversion receive path and which include filtration and amplification, down-conversion and then digitisation. There are many design decisions that need to be made concerning the location and extent of filtering and amplification across the receive chain.

## 5.9 UMTS transceiver architecture study

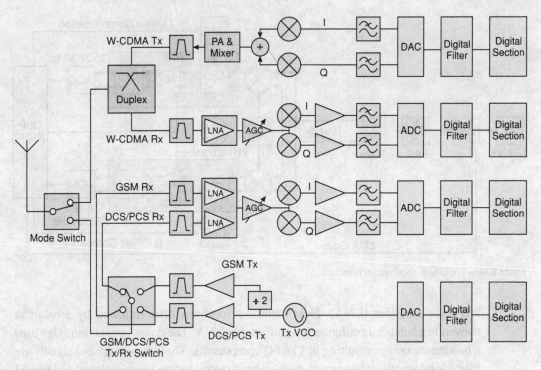

**Figure 5.43** General UMTS transceiver architecture.

For the transmit path, we are assuming a superheterodyne architecture that comprises up-conversion to IF, filtering and then up-conversion to the desired RF frequency. In this specific design we have a simple linear power amplifier. If our design was intended for high uplink data rates using multiple physical channels, we might expect an RF power amplifier that included some linearisation techniques to improve upon the basic linearity.

### 5.9.2 Receiver architecture

If we now go to consider the receiver. There are many options that could be selected for the WCDMA receiver architecture. A key consideration for the architecture components must be their suitability for integration. There are many potential architectures that could be used for the receiver, and there are many issues for using these differing architectures, some of which are considered in the literature [11, 12, 13]. In this section we consider the design issues related to a direct conversion design, starting with a brief overview, which is then followed by a more detailed consideration of the key components.

Figure 5.44 illustrates a potential architecture for the WCDMA receiver based on a direct conversion design. Starting at the duplex filter the signals arrive at a switched

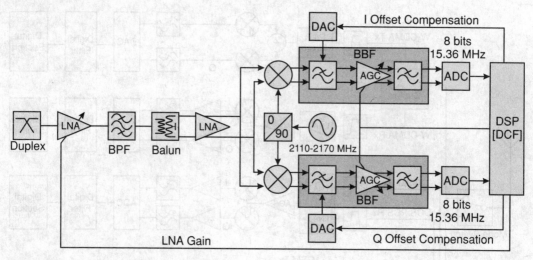

Figure 5.44  WCDMA receiver architecture.

low noise amplifier (LNA). Next there is a bandpass filter followed by a balun to convert to a balanced configuration and another LNA. The down-conversion stage uses a quadrature design resulting in I and Q components. These two baseband signals are filtered and amplified in an analogue baseband circuit before being digitised and passed to the digital signal processor (DSP), where the chip rate and symbol rate processing functions are implemented.

**Switched LNA**

The switched LNA is used when there are low level signals and is present to help overcome the losses associated with the mode switch, the duplex filter, the bandpass filter and the balun. As part of the automatic gain control (AGC) circuit, when the signals are at a high level the LNA can be switched to a low gain mode to help maintain the linearity of the receiver front-end. In a practical design, the LNA is most likely be part of an RF receiver application specific integrated circuit (ASIC). The typical design performance that we may require for the LNA when operating in its high sensitivity, high gain mode is a gain of 10 dB, a noise figure of 1.5 dB and an $IIP_3$ of 10 dBm.

**Bandpass filter (BPF) and balun**

The BPF is present to help reduce the effects of the out-of-band interference components received through the antenna, and also to attenuate the signals and noise leaking through the duplex filter from the transmitter. The balun is used to convert the unbalanced signal received from the antenna into a balanced signal. The benefits of using a balanced signal are the reduction in second order non-linearity effects, which for a direct conversion receiver can lead to significant problems. In the typical design considered here, the BPF and balun have an insertion loss of 3.5 dB and a noise figure of 3.5 dB. We can expect an attenuation of 9, 15 dB, 60–85 MHz offset from the carrier (respectively), and this will increase to better than 20 dB at 190 MHz offset from the carrier to provide sufficient transmitter attenuation. Both the BPF and the balun tend to be external RF components due to the low loss constraints on these components.

## Second LNA

The next LNA is included to further improve on the sensitivity of the receiver whilst minimising the degradation cause by non-linearities. The amplifier is balanced to help reduce the generation of second order non-linearities. In a typical design this LNA has a gain of 17 dB, a noise figure of 4.5 dB and an $IIP_3$ of $-3$ dBm. This component will be part of an RF receiver ASIC.

## Quadrature mixer

The quadrature mixer converts the RF signal to what is often referred to as a complex baseband signal. The complex baseband signal has an I and a Q component. To produce the baseband signal a quadrature synthesiser is required that generates two RF signals with a constant 90° phase offset between them. The typical performance for the mixer will be a gain of 10 dB, a 16 dB noise figure and a 10 dBm $IIP_3$. The mixer will form part of the RF receiver ASIC.

## Quadrature baseband filter (BBF)

The quadrature BBF is an analogue filter that also incorporates the main AGC amplification stage in the receiver design. The filtering is required to partially achieve the close-in filtering necessary to meet the ACS and blocking specifications and also to contribute to the out-of-band filtering requirements. An additional function for the quadrature BBF is to assist in the I/Q offset compensation. Digital to analogue converters (DACs) driven by the digital section are used to help null any DC offsets caused by amplitude and phase imbalances between the I and the Q paths. The analogue BBFs need to achieve a noise figure below 30 dB and an $IIP_3$ better than 15 dBm. The filtering from the BBF extends from adjacent channel filtering of around 18 dB, in-band blocking at 10 and 15 MHz offset of around 50 and 80 dB respectively, and an out-of-band filtering in excess of 80 dB, 85 MHz offset from the carrier frequency. The quadrature BBF most likely has a fifth-order butterworth shape, a cut-off at around 2.0 MHz. The filter is implemented as part of the RF receiver ASIC.

## ADC

The details of the design considerations for the ADC are presented in Chapter 6. To summarise, for this design, an 8 bit ADC sampling at four times the chip rate

**Table 5.19.** *Gain, noise figure and $IIP_3$ performance for the reference design*

| Parameter | Duplex + switch | LNA1 | BPF + Balun | LNA2 | Mixer | BBF | Combined |
|---|---|---|---|---|---|---|---|
| Gain (dB) | −3 | 10 | −3.5 | 17 | 10 | | |
| NF (dB) | 3 | 1.5 | 3.5 | 4.5 | 16 | 30 | 7.1 |
| $IIP_3$ (dBm) | | 10 | | −3 | 10 | 15 | −17 |

**Table 5.20.** *Summary of required filter specifications and target values achieved in the reference design*

| Offset (MHz) | Required attenuation (dB) | BPF + DPX (dB) | BBF (dB) | DCF (dB) | Margin (dB) |
|---|---|---|---|---|---|
| 5 | 33 | 0 | 18 | 18 | 3 |
| 10 | 45 | 0 | 51 | 18 | 24 |
| 15 | 55 | 0 | 68 | | 13 |
| 60 | 70 | 9 | 80 | | 19 |
| 85 | 85 | 15 | 80 | | 10 |

(15.36 MHz) is suggested. The sampling rate and ADC dynamic range were selected to allow the digital channel filter that follows to contribute to the overall receiver selectivity. The ADC forms part of the receiver RF ASIC.

**Digital channel filter (DCF)**

The final stage of interest in our design is the DCF. The DCF is implemented digitally either in a DSP or as part of a digital ASIC. The design details of the DCF are considered in Chapter 6. The DCF implements the required RRC filter shape, providing around 18 dB of selectivity for the adjacent channel and the first in-band blocking test.

**Summary of receiver architecture**

Tables 5.19 and 5.20 summarise the sensitivity, non-linearity and filter specifications achieved in our reference design and compare them with the design requirements outlined earlier in the chapter. Table 5.19 summarises the gain, noise figure and $IIP_3$ performance for the design presented in Figure 5.44. Table 5.20 summarises the filter performance for the design and compares it with the requirements outlined in the chapter; it also present a design margin that defines how closely the design meets the requirements.

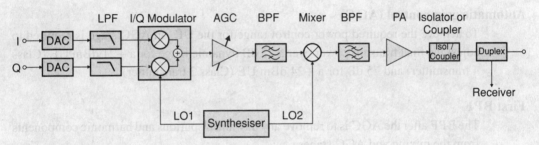

**Figure 5.45** Reference design transmit architecture.

### 5.9.3 Transmitter architecture

The proposed transmitter architecture is presented in Figure 5.45. This architecture utilises a superheterodyne structure with an IF frequency used for the quadrature upconversion, a second up-converter taking the transmitted signal to the required frequency and a power amplifier used to set the required transmit power. The amplified signal is then passed to the transmitter port of the duplex filter before being passed to the antenna. We can now examine some of the details of the proposed architecture starting with the DACs.

### DACs

We need to consider the requirements for the DAC located after the digital baseband section and before the analogue filter. There are a number of trade-offs that need to be considered and these are the sampling rate, the DAC dynamic range and the digital filter length. The longer the filter impulse response, the greater the sampling rate and the greater the number of DAC bits, the lower the ACLR. To meet the transmitter performance requirements of an ACLR 33 dB below the carrier power we require a 9 chip impulse response for the filter, a sampling rate four times the chip rate and an 8-bit DAC. The design and structure of the digital RRC filter are considered in Chapter 6.

### Low pass filter

The low pass filter that follows the DAC is used to remove the alias frequency components and the DAC noise. To meet the noise and ACLR specifications in our reference design a fourth order butterworth filter with 3 MHz cut-off and a stop-band attenuation of 20 dB to at least 10 MHz was selected.

### I/Q modulator

The I/Q modulator up-converts the baseband signals to an appropriate IF frequency. The choice of IF frequency depends upon an analysis of the transmitter spurious emissions. If we select a high side final up-conversion stage, the IF frequency needs to be greater than 250 MHz (typically a value of 380 MHz is selected) to avoid spurious components entering the receiver frequency band.

## Automatic gain control (AGC)

To achieve the required power control range for the UE, an AGC stage is required to adjust the final transmit power over a 73 dB dynamic range for a +21 dBm UE (Class 4 transmitter) and 75 dB for a +24 dBm UE (Class 3 transmitter).

## First BPF

The BPF after the AGC is to remove any unwanted spurious and harmonic components from the mixing and AGC stages.

## RF mixer

The RF mixer is a standard integrated component that converts the signal from the IF frequency to the RF frequency with the use of the appropriate synthesised local oscillator.

## Second BPF

The second BPF is present to remove the unwanted mixing products from the up-conversion stage. The filter most likely has an insertion loss in the region of 3.5 dB and will provide an attenuation of noise and interference components of around 20 dB in the receiver band.

## Power amplifier

The power amplifier needs to be a linear design. It must either be designed as a linear amplifier or use linearisation techniques to improve the linearity of the amplifier. For our reference design, we have selected an amplifier with an output power of +27 dBm, a gain of 36 dB and a noise floor of $<-130$ dBm/Hz to meet the receiver sensitivity requirements. The high transmit power is required for the Class 4 transmitter to overcome the losses in the isolator/coupler and the duplex filter.

## Isolator/coupler

The isolator/coupler provides two main functions in the reference design. First, any signals that are reflected back from the antenna due to a poor antenna load are passed to a dumping termination rather than back to the amplifier. This reduces the possibility of the transmitter generating unwanted out-of-band emissions. Second, with the coupler, a sample of the transmitted signal can be taken and this can be used as part of the power setting and control loop to ensure that the transmitter is operating at the correct point.

### 5.9.4 Transceiver architecture summary

Figures 5.44 and 5.45 present an architecture suitable for meeting the WCDMA FDD mode UE transceiver specifications. Many elements of the architecture can be integrated; some elements need to be discrete RF components to achieve the performance required. The architecture is proposed as an example reference architecture. Alternative architectural components were discussed; however, each variation in the architecture needs appraising in terms of its ability to satisfy the complete transceiver specifications. A consequence of the interaction of the different architecture elements is the need to consider the total transceiver architecture and not simply the component pieces in isolation.

# 6 Chip rate processing functions

## 6.1 Introduction

This chapter considers the chip rate processing functions that relate to the FDD mode receiver. When addressing specific implementation issues, we focus particularly on the UE aspects and, where reasonable to do so, make comments on the implications for the Node B. The chapter starts with a consideration of the receiver and the ADC – the link between the analogue and the digital domains. We continue with the reference architecture for the UE that was presented in Chapter 5. Following this, we then consider the other major chip rate functions in the receiver, including the rake receiver and the cell acquisition functions.

## 6.2 Analogue to digital converter (ADC)

### 6.2.1 Introduction

We start by considering the receiver reference architecture illustrated in Figure 5.44. We consider the chip rate processing functions in the receiver to start at the ADC. For the ADC it is necessary to consider criteria such as the sampling rate and the dynamic range; and there are different considerations for the Node B and the UE. We will also need to consider the location of selectivity filtering within the receiver. Some takes place at the RF/IF, some at the analogue baseband and some at the digital baseband. The amount of digital baseband filtering affects the ADC dynamic range requirements.

### 6.2.2 Sampling rate

An anti-aliasing filter acts as part of the channel selectivity filtering, in addition to rejecting aliasing components. The sampling rate should be as high as possible. As the sampling rate increases, the quantisation noise is spread across a wider bandwidth and therefore has a reduced effect. For the direct conversion architecture, the minimum sampling rate should be twice the chip rate (7.68 MHz/s). The higher the sampling

## 6.2 Analogue to digital converter

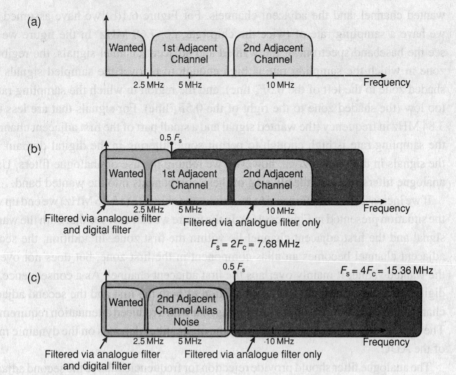

**Figure 6.1** Baseband received signal spectrum and zones of filter effectiveness: (a) wanted and two adjacent channel signals; (b) wanted and adjacent channel signals, also showing zones of filter effectiveness for two times over sampling (7.68 MHz); (c) wanted and adjacent channel signals and zones of filter effectiveness for sampling at four times over sampling (15.36 MHz).

rate, the lower the complexity of the analogue anti-aliasing filter; and also, the greater the cost and the power consumption of the converter.

To understand the various trade-offs between the sampling rate and the filter selectivity it is beneficial to review some of the basic principles of the sampling theorem and its effects on signal spectra. The sampling theorem, proposed by Nyquist, gave a theoretical definition for the minimum sampling rate required for a band-limited signal. If the sampling rate is greater than twice the maximum frequency component in the band-limited signal, the sampled signal is a faithful representation of the original signal. If the sampling rate is lower than the required minimum, then distortion products are introduced through alias components of the sampled signal. We may wish to design our transceiver and sampling rate to ensure that all the signals we need to consider are well within this sampling bandwidth. It turns out, however, that if we give careful consideration to the effects that the alias components have on our received WCDMA signal, we can simplify the design of both the ADC and the baseband filtering.

Consider Figure 6.1 which illustrates aspects of the sampling theorem and how they apply to the WCDMA receiver specifications. Figure 6.1(a) shows the spectrum of the wanted channel, and the adjacent channels. For Figure 6.1(b) we have assumed that we have a sampling rate of twice the chip rate, i.e. 7.68 MHz. In the figure we can see the baseband spectrum of the wanted and adjacent channel signals, the region or zone in which the sampling rate is high enough to recover the sampled signals (the shaded zone to the left of the $0.5F_s$ line), and the region in which the sampling rate is too low (the shaded zone to the right of the $0.5F_s$ line). For signals that are less than 3.84 MHz in frequency (the wanted signal and a small part of the first adjacent channel), the sampling rate is high enough to permit some filtering in the digital domain. For the signals in the second zone, however, we require the use of analogue filters. Using analogue filters prevents the aliasing of these components into the wanted band.

If we increase the sampling rate to four times the chip rate (15.36 MHz) we end up with the situation presented in Figure 6.1(c). In the figure we can now see that both the wanted signal and the first adjacent channel lie within the first zone. In addition, the second adjacent channel becomes an alias component in the first zone, but does not overlap the wanted signal, it mainly overlaps the first adjacent channel. As a consequence, the digital filter can contribute to the attenuation of both the first and the second adjacent channels. We still require analogue filtering, but with a reduced attenuation requirement. The degree of attenuation achieved from the digital filter depends on the dynamic range of the ADC.

The analogue filter should provide rejection for frequencies above the second adjacent channel (greater than 12.5 MHz approximately). It should also provide some, but not all, rejection for frequencies below the third adjacent channel.

### 6.2.3 ADC dynamic range

The dynamic range of the ADC is utilised for two purposes: first, to allow the sampling of the wanted signal with its amplitude fluctuations and also to allow for the amplitude fluctuations that are caused by other codes in the downlink; and second, to provide the dynamic range that is required for the filtering that we will be implementing in the digital filter. The dynamic range required to receive the wanted signal has been estimated through a number of simulations [20, 21] to be in the region of 4–6 bits.

Providing sufficient dynamic range for the attenuation of the other interference in the receiver depends upon the rejection of the analogue filter that precedes it. From Chapter 5 we can see that the specifications for the UE require an average rejection across the band of 33 dB, 43 dB and 55 dB for the 5 MHz ACS test and the 10 MHz and 15 MHz blocking tests respectively.

For the analogue filter that precedes the ADC (shown in Figure 5.44) we assume that we are using a low-order Butterworth filter. The Butterworth filter is a simple shape to implement and has a reduced effect on the signal phase and time-delay characteristics. Next we need to decide upon the filter order that is required. If we assume a Butterworth analogue lowpass filter and a linear rejection across the band of interest, then

**Table 6.1.** *Summary of Butterworth baseband filter performance*

| Filter order $n$ | Rejection 5 MHz offset (dB) | Additional rejection (dB) to meet specification | Rejection 10 MHz offset (dB) | Additional rejection (dB) to meet specification | Rejection 15 MHz offset (dB) | Additional rejection (dB) to meet specification |
|---|---|---|---|---|---|---|
| 2 | −12 | 21 | −27 | 16 | −35 | 20 |
| 3 | −16 | 17 | −40 | 3 | −52 | 3 |
| 4 | −20 | 13 | −53 | −10 | −69 | −14 |
| 5 | −24 | 9 | −65 | −22 | −62 | −31 |

Table 6.1 summarises the amount of rejection achieved for the 5 MHz, 10 MHz and 15 MHz offsets. In the simple analysis, it is assumed that the filter cut-off is 1.92 MHz and that the average filter attenuation is being measured in a bandwidth of 3.84 MHz at the offset frequency.

From Table 6.1 we can see that the critical frequency specification is the 5 MHz ACS specification. Assuming that we wish to perform most of the filtering in the analogue domain, Table 6.1 shows how much additional attenuation is required in the digital domain for the different filter orders.

If we were to use a fourth order analogue filter, we will be 13 dB out of specification for the 5 MHz offset requirement and if we use a fifth order analogue filter we will be 9 dB out of specification. In both cases, the higher offset frequency specification points are satisfactorily achieved with the analogue filter.

To meet the specification, we need to include some additional filtering in the digital domain. The amount of filtering and hence the additional dynamic range requirement is given by the additional rejection column. With a dynamic range extension of 6 dB per bit, we require at least three additional bits for the fourth order filter and two additional bits for the fifth order one.

To estimate the total ADC dynamic range in terms of number of bits, we add the additional bits required for filtering dynamic range to those that we require for signal detection; if we assume a fourth order Butterworth filter this results in a requirement of around 7–9 bits for the converter. For the purposes of further consideration we can assume that the digital converter is an 8 bit ADC with a sampling rate of 15.36 MHz.

## 6.3 Receive filtering

### 6.3.1 Analogue and digital filtering

Analogue filtering, as mentioned in the previous section, is required for anti-aliasing and also to provide the receiver selectivity. The following digital filter provides the

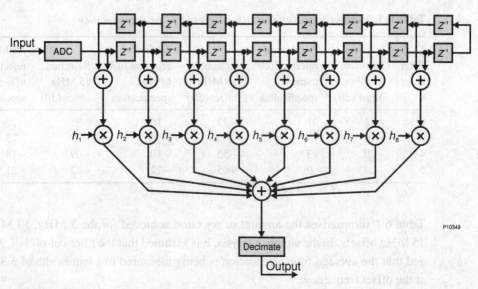

**Figure 6.2** Basic block diagram for the digital filter.

remainder of the receiver selectivity and the noise and receive pulse shaping. The details of this digital filter are considered in the next section.

### 6.3.2 Digital filter

The basic block diagram for the digital filter is illustrated in Figure 6.2, which shows the ADC, the RRC filter and a decimation (also known as down-sampling) filter. The impulse response of the RRC waveform is sampled at the required sampling rate and these samples become the coefficients $h_1$, $h_2$, etc. in the digital filter illustrated in Figure 6.2. In reality, the filter and the decimation stages are combined into one. The filter is usually an FIR filter called a matched filter because it is matched to the filter used in the transmitter. The filter length is selected to provide the attenuation needed to meet the selectivity, shaping and noise requirements, and is implemented using an input sampling rate four times the chip rate for reasons discussed previously.

The output of the filter is decimated by a factor of 4. At its simplest, the decimator simply passes every $n$th sample to the output to produce a change in the sampling rate. More complex decimation schemes can be considered in which the decimation stage is part of a polyphase filter, where the output is the sum of $n$ elements. Figure 6.3 shows this.

One of the benefits of the polyphase filter is that the use of four parallel filters of a quarter the length reduces the number of multiply and accumulate operations by four. One feature that is probably desirable is to be able to shift the timing of the receiver signal in steps of a quarter chip period.

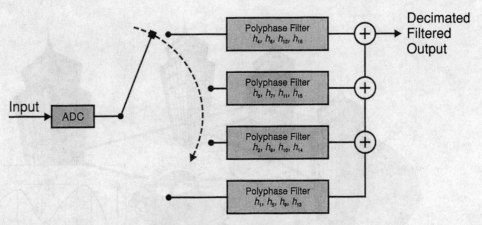

**Figure 6.3** Digital baseband polyphase decimation filter.

## 6.4 Rake receiver overview

Within the FDD mode WCDMA system the rake receiver is used in both the UE and the Node B to overcome the effects of multipath fading and to facilitate soft-handover. In this section we consider the design and operation of the rake receiver. We start with a consideration of the multipath channel and then move on to develop the basic structure and operation of the rake receiver.

### 6.4.1 Multipath channel

It can be shown that radio propagation in a mobile communication system is commonly via reflections and diffraction [22]. As a consequence, the signal arrives along with multiple reflections.

Figure 6.4 illustrates two specific points. First, the signals that arrive at the Node B (or the UE in the downlink) generally travel via multiple independent paths caused by reflections and diffraction from the buildings and terrain features in the coverage area. These multiple signals are referred to as multipath signals and their characteristics are well defined [22]. The second point illustrated by Figure 6.4 is that the reflected signals generally arrive at the receiver at different times, owing to the different path lengths that the signals take. Typically in an urban environment, these additional path delays can be as high as 5 µs and for a hilly area as high as 20 µs. Within Figure 6.4 we can see the consequence of these multiple reflected paths. If we imagine that we transmitted a single chip, at the receiver we receive not only this single chip, but also all of the reflections from this chip. The received signal becomes a time-smeared version of the transmitted signal as shown in the diagram. If the bandwidth of the system is small (typically less than 25 kHz) then to some extent we can ignore the multipath effects,

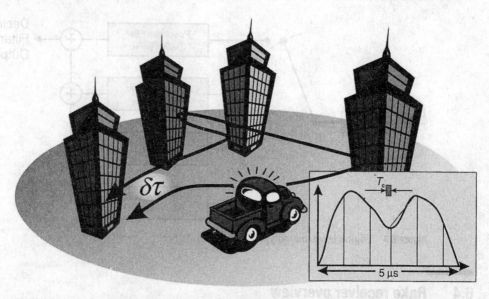

**Figure 6.4** Reception of a single transmitted impulse convolved with a typical receiver filter.

except for the fading effects considered previously. For our WCDMA system, however, with a bandwidth of 3.84 MHz we must consider these multipath delays and the effect that they have on the system performance.

In Chapter 5 we saw that the receiver performance characteristics are measured using multipath channels with a varying number of reflections. In these test channels, as in the real multipath channels, the location and the amplitude of the individual reflections changes with time. These effects need to be overcome by the receiver.

In summary, the multipath channel as described provides a challenge for the designer of a receiver for the CDMA system for a number of reasons:

- First, the delays in the multipath components are much greater than the chip period of the modulation (chip period is 0.260 μs) and multipath delays greater than 5 μs are not uncommon.
- Second, the location of the multipath components changes as the mobile moves through the environment.
- Third, the amplitudes of the multipath components are complex time varying quantities, which vary at a rate proportional to the Doppler shift of the radio channel.
- Fourth, all of the processing required to manage the issues described above needs to happen in real or near-real time.

### 6.4.2 Optimum receiver

This section develops the basic concepts and techniques leading to the design of the rake receiver. To start, the optimum receiver is presented. This receiver is optimum in the sense that it minimises the error rate of a signal that is detected in an additive white

## 6.4 Rake receiver overview

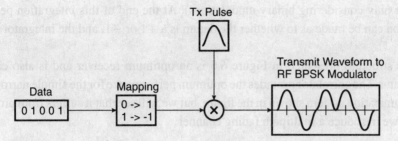

**Figure 6.5** Transmitter operating using a BPSK type of modulation.

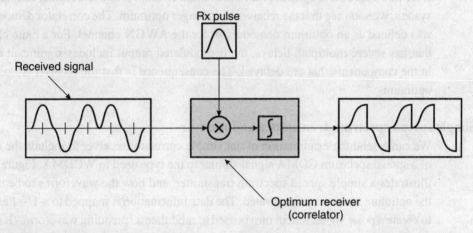

**Figure 6.6** Optimum receiver structure.

Gaussian noise (AWGN) channel. One method of implementing the optimum receiver is based on correlation and integration. The basics of this structure are examined.

Whilst most correlation based receivers are capable of operating efficiently with a signal whose amplitude varies with time (e.g. in flat Rayleigh fading), if multipath delay is present, the resulting intersymbol interference can degrade performance. The use of additional correlators is a means of overcoming the multipath propagation, leading to the signal combining and hence the complete rake receiver.

### Basic structure of optimum receiver

Figure 6.5 illustrates the basic structure for a simple narrow band transmitter operating using a BPSK type of modulation. The data after being mapped to $+1/-1$ are used to modulate a pulse that is then passed to the RF stages to be converted into a BPSK waveform.

Figure 6.6 shows the structure of the optimum receiver for detecting the transmitted signal, and illustrates its basic principles. The received signal is multiplied by a copy of the transmitting pulse. The result from this multiplication process is integrated in an integrator for the duration of the symbol period (the same as the bit period, as we are only considering binary modulations). At the end of this integration period, a decision can be made as to whether the datum is a +1 or −1, and the integrator can be reset.

The structure presented in Figure 6.6 is an optimum receiver and is also called a correlator. The correlator provides the optimum performance for the simple narrowband communication system shown in the figure, but we will see that it suffers from problems when we introduce a multipath fading channel.

### Effects of multipath channel on optimum receiver

If we now move on to consider the kind of radio channel found in a typical cellular radio system, we soon see that the receiver is no longer optimum. The correlator demodulator was defined as an optimum demodulator for the AWGN channel. For a radio channel that has severe multipath delays, the demodulated output includes significant energy in the components that are delayed. The consequence is that the receiver is no longer optimum.

### Simple spread spectrum transmitter and receiver

We can extend the examination of our simple correlator receiver to include the effects of a spread spectrum CDMA signal similar to the type used in WCDMA. Figure 6.7(a) illustrates a simple spread spectrum transmitter, and how the waveforms to be used in the optimum receiver are generated. The data information is mapped to $+1/-1$ as usual to create a polar signal that in turn is used to modulate a spreading waveform. (It should be noted that in this example, the spreading waveform is an arbitrary waveform and could be, for instance, a PN sequence.) In practice, in a system such as WCDMA this waveform would be an OVSF code combined with the scrambling code.

Figure 6.7(b) illustrates the procedures in the spread spectrum receiver. The received signal is multiplied (on a sample by sample basis) by a locally stored or generated despreading waveform. The resultant signal is integrated in an integrator for the duration of the transmitted symbol or bit period (which are the same due to the use of binary modulation). At the end of the integration period, the output is sampled and the integrator reset.

### 6.4.3 Operation of simple spread spectrum receiver

#### Operation of receiver with a single signal

Let us now address the principles of operation within our simple spread-spectrum receiver. Figure 6.8 illustrates the processes that occur in the radio channel and the receiver in terms of the time domain and the frequency domain. The upper part of the diagram symbolises that the radio channel has no additional multipath components other than the direct path, i.e. the received impulse response corresponds to the transmitted signal. The start time of the received data aligns with the time delay of the direct

## 6.4 Rake receiver overview

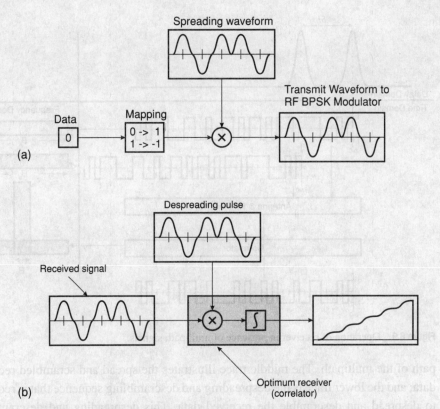

(a)

(b)

**Figure 6.7** (a) Transmitter, and how waveforms used in optimum receiver are generated. (b) Procedures in the spread spectrum receiver.

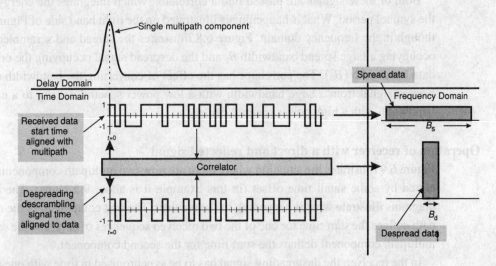

**Figure 6.8** Processes occurring in the radio channel and the receiver.

# 194 Chip rate processing functions

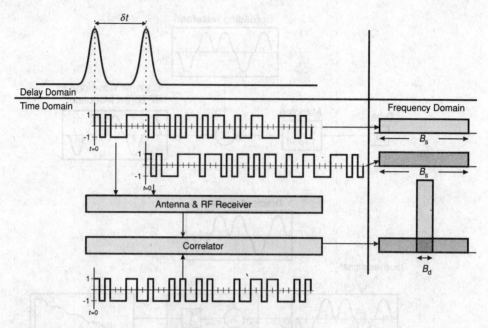

**Figure 6.9** Operation of receiver in presence of multipath signals.

path of the multipath. The middle trace illustrates the spread and scrambled received data; and the lower trace is the despreading and descrambling sequence that is required to despread and descramble the received data. This despreading and descrambling sequence is time-aligned with the received data (i.e. the time position of the multipath signal).

Both of these signals are passed into a correlator, which integrates the energy over the symbol period. What is happening is illustrated on the right hand side of Figure 6.8, though in the frequency domain. Figure 6.8 illustrates the spread and scrambled data occupying a large spread bandwidth $B_s$ and the despread signal occupying the original data bandwidth ($B_d$). The correlator has the effect of converting the bandwidth of the wanted signal from a large bandwidth with a low power spectral density to a narrow bandwidth with a high power spectral density.

## Operation of receiver with a direct and reflected signal

Figure 6.9 illustrates the situation where there are now two multipath components separated by some small time offset (in this example it is about 9.5 chips). The lower diagrams illustrate what is happening in the receiver. The first component of the multipath defines the start time for one of the two received sequences of data, and the second multipath component defines the start time for the second component.

In the receiver, the despreading signal has to be synchronised in time with one of the two received multipath components; in this example the first component is selected. At the output of the correlator, the result is that the first component is despread, but because

## 6.4 Rake receiver overview

the received signal arriving from the second component is not time synchronised with the first, it remains spread. (This is due to the autocorrelation characteristics of the spread and scrambled signal that were described in Chapter 3.)

The right hand side of Figure 6.9 illustrates what is happening, but in terms of the frequency domain. At the input to the receiver, there are two signals summed in the antenna, corresponding to the two components of the multipath. Both of these signals are spread and occupy some bandwidth $B_s$.

Let us now consider what is being shown in Figure 6.9. As we can see, there are two multipath signals entering the receiver. In a more realistic example, these two signals would have an arbitrary amplitude and phase, and they would be summed within the antenna prior to the receiver. In our example, the sum of these two signals enters the correlator with one of the signals delayed. The other input to the correlator is the combination of the despreading and descrambling code (the channelisation code and the scrambling code for the WCDMA system) that we assume is assigned to this particular radio link.

The first decision that needs to be made is which of the two received signals we intend to despread and descramble. In this example, both have equal amplitude (we can see this from the channel model) and so we make the decision arbitrarily and choose the first signal. The decision to despread and descramble the first received signal results in the time of the local despreading and descrambling codes being time-aligned with the first received signal as seen in the diagram.

If we now turn to the frequency domain, we can start to appreciate the processes that are occurring in the receiver. First, at the input to the correlator, both of the received signals are spread across the RF bandwidth defined as $B_s$ (this is 3.84 MHz for the WCDMA system). As a result of the correlator, however, the two input signals at the output manifest themselves differently. The signal that we have decided to despread and descramble (the signal that arrives first in this example) is returned back to its original data bandwidth (represented by $B_d$ here). The second delayed signal, however, remains spread across the RF bandwidth of 3.84 MHz, the reason for which is considered shortly. The integrator in the correlator acts as a filter and attenuates all frequency components that are greater than the data bandwidth $B_d$. At the output of the correlator we have all of the despread wanted signal and only that portion of the second delayed and spread signal that lies within the bandwidth of the wanted signal (this is shown as the shaded square in Figure 6.9). As a consequence of the operation of the correlator, we have recovered the signal that arrives first (we will call this the direct path), and only a small percentage of the reflected signal (the signal arriving second) is present to cause interference to the wanted signal. With this demodulation scheme we are able to overcome most of the effects of the multipath propagation and only suffer a marginal degradation in performance from the reflected signal.

One question concerning the difference in the outputs from the correlator remains. The correlator despreads the direct signal and leaves the reflected signal spread because

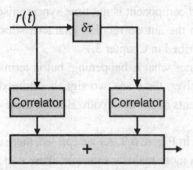

**Figure 6.10** Structure of receiver incorporating a second correlator.

of the autocorrelation characteristics of the spread and scrambled signal. In Chapter 3, when we looked at the scrambling code we saw that its autocorrelation had a large peak when the reference code and the test code were perfectly aligned, but it dropped to a very low level as soon as there was more than a single chip offset. If we return to our received signal and the correlator, we can see that the local despreading and descrambling code is perfectly aligned for the direct path (this was a requirement for the start time of the local despreading and descrambling codes), but is not aligned for the reflected path. The direct path therefore is correctly despread and descrambled, but the reflected path is not correctly descrambled, and cannot be correctly despread as a consequence.

The extent of the interference that the reflected signal introduces onto the despread direct signal depends upon their relative signal strength, and also the spreading factor that is used for the transmitted signals (in this context, the spreading factor is the ratio $B_s/B_d$). The higher the spreading factor the less the effects of the interference and conversely for lower spreading factors.

To summarise, we can effectively communicate with our simple CDMA system even in the presence of severe multipath reflections provided that the spreading factor is high enough, and the correlation of the scrambling codes ensures that the interference from the reflected signal is sufficiently small. The next step is to consider the addition of a second correlator into the receiver, which will be used to capture the energy from the reflected signal.

**Operation of receiver with a direct and reflected signal and two correlators**

In Figure 6.10, we now introduce a second correlator demodulator. Between the first and the second correlators there is a delay element with a magnitude $\delta t$. The magnitude of the signal delay between the correlators corresponds to the magnitude of the delay between the direct signal and the reflected signal. We use the two correlators as follows. In the first correlator, we arrange for the timing of the despreading and descrambling codes to despread and descramble the reflected signal. In the second correlator, we arrange for the timing of the despreading and descrambling codes to despread and

## 6.4 Rake receiver overview

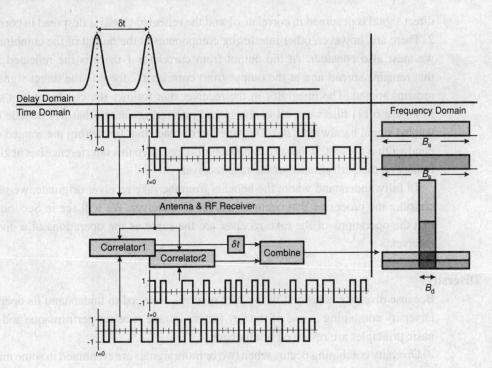

**Figure 6.11** Operation of dual correlator receiver in presence of multipath signals.

descramble the direct signal. At the output of the two correlators there will be the two despread and descrambled signals. The delay that is present between the first correlator and the second correlator ensures that we can add together the two outputs and in doing so we are able to correctly combine the output from the first correlator with the output from the second correlator. The important benefit in performing this combination is that the signal to noise ratio (SNR) of the combined signal will be better than the SNR of either of the two signals in isolation. The combination of the outputs from multiple correlators is the basis of the rake receiver.

For an explanation of what is happening with the two correlators, consider Figure 6.11. In a similar manner to the previous diagrams, Figure 6.11 illustrates the operation of the receiver in three domains. At the top there is the delay domain that illustrates the multipath delays between the direct and reflected signals, below there are the signals in the time domain, and to the right are the signals in the frequency domain.

In the time-domain representation we can see how the direct and reflected signals are summed by the antenna and passed to the two correlators. A combined despreading/descrambling code signal is applied to the two correlators with a time delay equal to the multipath delay. The outputs of the two correlators are synchronously combined (by including a delay equal to the multipath delay for the direct path).

In the frequency domain, at the input to the correlator both the direct and reflected signals are spread across the full RF bandwidth. At the output of the correlator, the direct signal is despread in correlator 1 and the reflected signal is despread in correlator 2. There are, however, other interfering components at the output of the combiner that we must also consider. At the output from correlator 1 there is the reflected signal that remains spread and at the output from correlator 2 there is the direct signal that remains spread. The integrator in the receiver that follows the combiner (not shown in Figure 6.11) filters out all of the spread interfering energy that lies outside of the wanted signal bandwidth ($B_d$). Only the interference that lies within the wanted signal bandwidth is present. If the spreading factor is large, then this interference has negligible impact on the performance of the rake receiver.

To fully understand where the benefits from the rake receiver originate, we need to consider the processes that occur in a diversity receiver. We will see in Section 6.4.5 that the operations of the rake receiver are the same as the operations of a diversity receiver.

## Diversity

Because diversity is present in the rake receiver, we need to understand its operation. Diversity combining is one of the key factors in rake receiver performance and so its basic principles are reviewed briefly here.

Diversity combining occurs when two or more signals are combined in some manner. The source of these two signals could be multiple antennas, different polarisation or, in the case of the rake receiver, different correlators in a rake receiver. The basic idea behind diversity combining in a diversity receiver is that the signals are added coherently (except for selection diversity) and the noise is added non-coherently. This results in a net power gain.

There are three basic techniques used for diversity combination: selection diversity; equal-gain combining; and maximum ratio combining.

### Selection diversity

Multiple signals are available in the receiver, the output of which is selected from one of the input signals. A control processor, or equivalent, selects one of the input signals based on some criteria. There are a number of selection strategies, and the choice depends on the nature of the signals. Generally, the selection diversity combiner has a benefit only if the received signal has a time varying amplitude or interference and the variation is decorrelated between the input signals, and so it is not commonly employed in a rake receiver.

### Equal-gain diversity combining

A typical block diagram for an equal-gain diversity combiner is shown in Figure 6.12(a). The multiple input signals are co-phased (for coherent combination) and then summed. As the name implies, the gains in the different branches of the combiner are the same. The performance of the equal-gain combiner is better than that of the selection diversity

## 6.4 Rake receiver overview

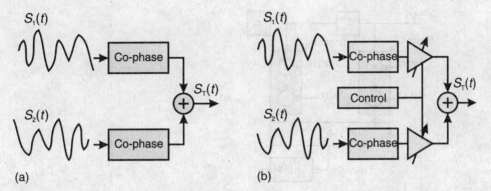

**Figure 6.12** Basic structure and operation of diversity receivers: (a) equal-gain combiner; (b) maximum ratio combiner.

combiner. The equal-gain combiner could be used in a rake receiver, but generally the technique of maximal ratio combining is preferred.

Consider the basic principles of what is happening in an equal-gain combiner in a rake receiver. Consider two signals (from two correlators say). The $SNR$ for these correlators is given for correlator 1 ($SNR_1$) and correlator 2 ($SNR_2$) by:

$$SNR_1 = \frac{A_1^2}{2N_1}$$

$$SNR_2 = \frac{A_2^2}{2N_2}$$

where $A_1$ and $A_2$ are the amplitudes of the two wanted signals being combined, and $N_1$ and $N_2$ are the powers of the noise present with the wanted signals. The combiner coherently adds the signal, i.e. the output signal ($A_c$) will be:

$$A_c = A_1 + A_2$$

But the noise power at the output ($N_c$) adds non-coherently due to the correlation statistics of the noise and interference at the output of the two correlators (this depends on the time delay separation of the correlators). The noise power of the output is given by

$$N_c = N_1 + N_2$$

If we assume that the input signal amplitudes are equal ($A_1 = A_2$) and the input noise and interference are equal ($N_1 = N_2$), then the combined $SNR$ at the output is given by

$$SNR_c = \frac{(A_1 + A_2)^2}{2(N_1 + N_2)} = \frac{(2A_1)^2}{2(2N_1)} = \frac{A_1^2}{N_1} = 2SNR_1$$

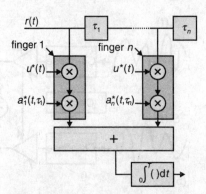

**Figure 6.13** Rake receiver structure.

That is, the combined $SNR$ at the output is twice the input $SNR$ (assuming both inputs have the same $SNR$).

**Maximal ratio combining**

The block diagram for a maximal ratio diversity combiner is shown in Figure 6.12(b). Multiple input signals are co-phased and then scaled according to their received signal quality (such as amplitude). The better the signal quality, the higher the scaling. The signals are then summed to produce a maximal ratio combined signal. Maximum ratio combining (MRC) is the optimum combining technique, as it performs better than both selection diversity and equal-gain combining. MRC is used in a rake receiver to combine the energy from the different rake fingers.

### 6.4.4 Rake receiver structure

Having considered the operation of the main components that make up a rake receiver we can now proceed to consider a more complete representation of the rake receiver. We start in this section by considering the rake architecture. In the following section we will then apply this architecture to the specifics of the WCDMA system and explore in a little more detail the mathematical operation of the rake.

Figure 6.13 illustrates a simplified structure for a rake receiver. At the top left the input signal is introduced into the rake receiver via a delay line. In a DSP implementation, this delay line is a data storage buffer, with the taps implemented as pointers to the appropriate buffer location. Attached to the taps at the appropriate points in the delay line structure there are multipliers that form part of the correlator. Each of these multiplier pairs is colloquially referred to as a finger of the rake receiver. The first multiplier multiplies the received signal by $u^*(t)$, which is the complex conjugate of the combination of the despreading and descrambling codes. Following the despreading and descrambling complex weight factors of $a_n^*(t, \tau_n)$ are applied. These are the complex conjugate of the channel impulse response that corresponds to that specific time and delay $(t, \tau_n)$.

In multiplying the received signal by this factor we perform two important functions which are prerequisite for MRC:
1. The signal is co-phased.
2. The signal is scaled according to the amplitude of the received signal ($a_n^*(t,\tau_n)$).

When we sum together each of the delayed and weighted taps, we are performing maximum ratio diversity combination of each of the multipath components.

The number of fingers used in the rake receiver depends upon the amount of multipath energy present in the received signal. Eventually, we reach a point of diminishing returns. For each additional finger we increase the total wanted signal energy but we also increase the noise and interference levels. If the wanted signal energy is low (a small multipath reflection) but the noise and interference energy are high, the net result is a degradation in the overall receiver SNR. Under these circumstances, it is better to exclude some components from the combiner.

### 6.4.5 Rake receiver in WCDMA

Having considered the architecture of the rake receiver, we can now apply this to the specific case of the WCDMA system. As we have seen, the objective for the rake receiver is to receive, despread and combine the different multipath components that are received via the multipath radio channel. To achieve this we utilise the fingers of the rake receiver.

The number of fingers required in the rake receiver is not specified explicitly in the WCDMA system specifications. There are many factors that affect the number of fingers that are required. In its use to collect multipath signals, the number depends upon design decisions and the searcher algorithm that is used to detect the multiple paths.

In its use as part of soft-handover, the specification [23] requires that the UE must be able to be in soft-handover with up to six cells at the same time, but it is not so definitive in terms of what the UE needs to do to be in soft-handover with that many cells. In [24], the specifications stipulate that the maximum number of cells in the active set is eight, which would imply that the UE needs to have up to eight fingers in the rake receiver.

For the Node B, the situation is even less well defined, but it can be expected that a similar degree of complexity is required in terms of the number of fingers needed per UE.

**Received signals**

The input to the rake receiver consists of a number of delayed, attenuated and phase shifted replicas of the transmitted signal. The received signal comprises the spread and scrambled information received via a number of multiple paths, with each of the paths introducing a time varying amplitude and phase variation. We can represent the received signal using the following equation:

$$r(k) = \sum_{p=1}^{N} a_p(k) s\left(k - \frac{\tau_p}{T_c}\right) + n(k)$$

where $a_p$ is the complex amplitude of the individual multipath components, $\tau_p$ is the relative delay for each of the multipath components, $T_c$ is the chip duration and $n(k)$ is the noise.

In the rake receiver we multiply a delayed version of the received signal by a despreading and descrambling signal that is correctly synchronised to the specific component of multipath that we wish to receive. We perform this operation for each rake finger and then sum the output from each finger scaled according to the channel amplitude for the specific path being collected by the rake.

The structure for the receiver is illustrated in Figure 6.13. The output of the rake receiver can be represented using the equation below:

$$r\left(k + \frac{\tau_i}{T_c}\right) a_i^*\left(k + \frac{\tau_i}{T_c}\right) = \alpha(k) s(k)$$
$$+ \sum_{p=1, p \neq i}^{N} a_p\left(k + \frac{\tau_p}{T_c}\right) a_i^*\left(k + \frac{\tau_i}{T_c}\right) s\left(k - \frac{\tau_p}{T_c} + \frac{\tau_i}{T_c}\right) + n\left(k + \frac{\tau_i}{T_c}\right) a_i^*\left(k + \frac{\tau_i}{T_c}\right)$$

The equation shows that the output of the rake combiner comprises three elements. The first is the desired detected signal scaled by the multipath component ($\alpha(k)$), the second is a received signal component that comprises reflections of the transmitted signal that are incorrectly despread and descrambled and which is referred to as interpath interference. The final component is from the noise in the receiver. The wanted signal is time varying as a consequence of the fading signal ($\alpha(k)$).

By combining the outputs from a number of different fingers in the rake, we can implement a diversity combination scheme. If the weighting of the combination is set appropriately (as is the case in the example shown), then the MRC of the different paths can be obtained.

There are two basic approaches to the location of the combining function within the rake: either we can perform symbol level combining or we can perform chip level combining.

Consider Figure 6.14(a), which illustrates an example of symbol level combining. The received signal is stored in a receiver buffer and, under the control of the searcher, a number of fingers are activated within the rake. With symbol level combining, the received signals are despread and descrambled prior to being combined in the maximum ratio combiner. The benefits of symbol level combining come from a complexity reduction. The multiplication of the signals by the channel impulse response estimates

## 6.4 Rake receiver overview

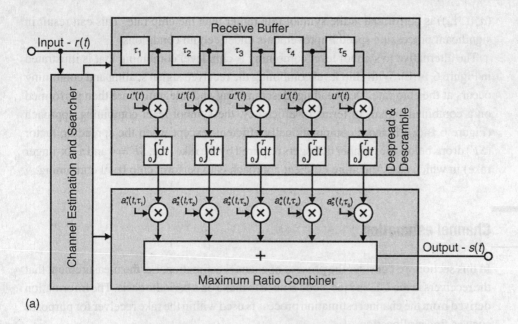

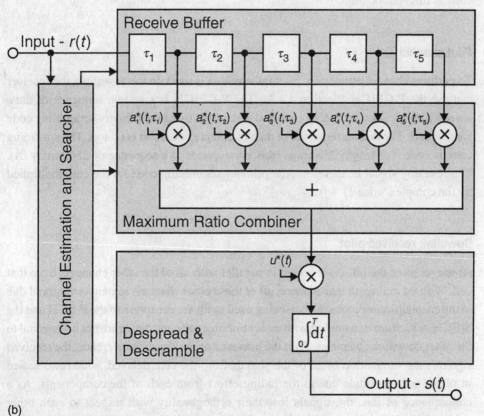

**Figure 6.14** (a) Symbol level combining in a rake receiver. (b) Chip level combining in a rake receiver.

($a_n^*(t, \tau_n)$) is performed at the symbol rate rather than the chip rate. This can result in significant processing speed improvements under certain conditions.

The alternative to symbol level combining is chip level combining that is illustrated in Figure 6.14(b). With chip level combining the received signal scaling and combining occurs at the chip rate. Despreading, descrambling and integration are then performed on a combined signal. In terms of efficiency, the symbol level combining approach (Figure 6.14(a)) is more computationally efficient, except when the spreading factor ($SF$) drops below the number of fingers required by the rake (e.g. $SF = 4$ and a six-finger rake) in which case, the more efficient approach is to perform chip level combining.

## 6.5 Channel estimation

In this section we consider the process of channel estimation, i.e. the measurement that the receivers in the UE and the Node B need to make on a periodic basis. The information derived from the channel estimation process is used within the rake receiver for purposes such as finger allocation.

### 6.5.1 Pilot channel

To perform channel estimation, pilot information is used. In the downlink, the receiver can use the P-CPICH (Section 4.4.4). The P-CPICH is a known sequence of data, spread with a known spreading code and scrambled using the primary scrambling code for that cell. The known sequence of data in complex notation is $(1 + j)$. The spreading code is code 0 of length 256 chips (this corresponds to a sequence of 256 binary 0s). The resulting signal is, therefore, the primary scrambling code for the cell, multiplied by the complex value $(1 + j)$.

### 6.5.2 Downlink received pilot

At the receiver, the pilot is received in parallel with all of the other channels from that cell. With no multipath interference, all of these other channels remain orthogonal due to the channelisation codes that are being used (with the exception of the P-SCH and the S-SCH, which are transmitted with no scrambling code and hence are not orthogonal to the other downlink channels). With the presence of multipath interference, the received signal is the summation of all of the signals from the cell, delayed, added and scaled in phase and amplitude due to the fading effect from each of the components. As a consequence of this, the signals lose their orthogonality with respect to each other due to the effects of the multipath signals. The interference caused by the multipath reflections is generally referred to as interpath interference which was considered in

## 6.5 Channel estimation

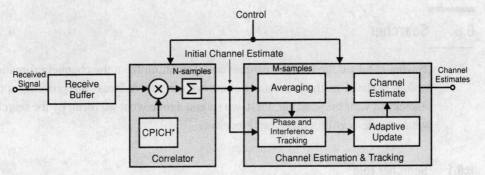

**Figure 6.15** Channel estimation architecture.

the previous section. When trying to perform the channel estimation, the effects of the interpath interference must be considered.

### 6.5.3 Architecture

The typical architecture for the channel estimation function is illustrated in Figure 6.15. In the figure, we can see that the received signal is despread and descrambled using the spreading and scrambling codes assigned to the pilot. This produces the required estimation for the channel impulse response. Next, using some form of adaptive technique such as the mean square error (MSE) or recursive least square (RLS) algorithms, we can track the changes to the channel as shown.

There are many methods that can be utilised to produce an initial estimate in the channel estimation process. In this section, we start by considering an example method, and then go on to consider some of the variations that may also be used for the channel estimation procedure. In this example, the basis for the initial channel estimation is correlation. Consider two real sampled data sequences $r(k)$ and $s(k)$. The correlation of the $m$th element between these two sequences is given by

$$C_{rs}(m) = \frac{1}{N} \sum_{K=0}^{N-1} r(k)s(k+m)$$

For the channel estimation procedure, the length of the correlation, $N$, is a trade-off between algorithm complexity, the ability of the algorithm to track the rapid changes in the channel impulse response and the accuracy of the channel estimation procedure.

In general, the received signal ($r(k)$) and the locally derived signal ($s(k)$) are complex quantities, and so the correlation to find the initial channel estimate is:

$$C_{rs}(m) = \frac{1}{N} \sum_{K=0}^{N-1} r(k)s^*(k+m)$$

In the following section we will consider the searcher and how it can utilise the channel estimation information to control the rate finger allocation.

## 6.6 Searcher

Having obtained the channel estimation information from the channel estimation procedure outlined in the previous section, we can now address the procedures that are associated with the searcher. First, we should define what we mean by the searcher and what its role is within the overall receiver.

### 6.6.1 Searcher role

The searcher is a device (an algorithm on a DSP or possibly on an ASIC) that is used to make decisions that are based on the measurements made in other parts of the receiver. Within the context of the rake receiver, the searcher is responsible for making decisions on the number and location of the rake fingers based on the measurements reported by the channel estimator.

The searcher should also be responsible for the control of the other measurements that the receiver should be making as part of its duties, including the monitoring of adjacent cells on the same and other frequencies. The details of the measurements that are to be made are considered in greater detail in Chapter 12.

### 6.6.2 Architecture

Figure 6.16 illustrates the typical architecture for the searcher, including the channel estimation procedures that were considered in the previous section. Here, we focus on the use of the searcher and its ability to perform finger allocation and control.

The typical searcher illustrated in Figure 6.16 has three components that are needed in addition to the channel estimation phase considered previously. These are the noise threshold calculation, the finger allocation and the finger cancellation stages. The use of each of these different elements is considered in more detail below.

**Channel estimation**

The details of the channel estimation were considered previously. The channel estimation element is responsible for providing periodic updates of the channel estimation. The rate of updates of the channel estimate can be dynamic, based on metrics such as mobile speed estimation and data rates. In addition, the window for the channel estimation can be variable based on specific circumstances of the mobile environment.

**Noise threshold calculation**

The noise threshold calculation algorithm is responsible for estimating and tracking changes in the correlation noise floor in the channel estimates. The procedure

## 6.6 Searcher

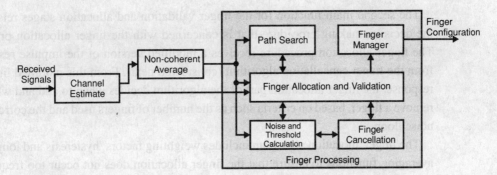

**Figure 6.16** Typical searcher architecture.

is important because it is used to identify when correlation peaks should be considered by the searcher and when they can be ignored as some spurious correlation component. The noise threshold calculation algorithm is an adaptive algorithm that identifies the underlying noise floor from the channel estimates. Changes in the channel estimation could also result in changes in the noise floor estimate, and so the algorithm needs to handle discontinuous changes in the channel estimates and still provide an accurate estimate for the channel impulse response.

### Finger validation and allocation

The finger allocation and validation algorithm is responsible for the allocation of fingers in the rake and also for validating the allocation prior to committing the rake to the allocation. One of the problems that occurs in a CDMA system is what is generically referred to as 'pilot pollution', when a strong adjacent cell using a different scrambling code is detected as part of the channel estimation procedure and appears as a component of the multipath. If a finger were allocated to that multipath component, it would not enhance the overall SNR of the receiver; rather, it would degrade the receiver due to the interference that is introduced by the adjacent channel signal. To guard against this eventuality, the finger allocation algorithm needs to validate that the specific component in the channel estimate being considered is from the wanted cell. The simplest method of achieving this is by comparing the signal quality of the combined despread and descrambled signal with and without the finger under test. If the signal is from a valid multipath component, the combined SNR will improve.

Various methods could be employed to assess the improvement or degradation in the SNR. The simplest is to measure the variance of the despread constellation points with and without the finger in question. If the signal is a wanted multipath component, the variance of the despread and descrambled data should decrease. If the signal is a strong interfering cell, the variance of the combined signal should increase due to an increase in the noise energy at the output of the receiver. To implement the finger validation stage may require access to the combined output from the receiver, unless a locally derived estimate for the combined signal can be obtained.

The second main function for the finger validation and allocation stages relates to the decision making procedure that is concerned with the finger allocation process. The finger allocation procedure receives a modified version of the impulse response from the finger cancellation algorithm (considered next). From this modified impulse response, the finger allocation part of the algorithm decides when to add and when to remove a finger, based on criteria such as the number of fingers used and the correlation noise floor.

The finger allocation algorithm includes weighting factors, hysteresis and long term averaging functions to ensure that the finger allocation does not occur too frequently and that the decisions that are made are good ones.

**Finger cancellation**

The finger cancellation stage, the final algorithm we are considering in the searcher, is responsible for tracking and modifying the channel impulse response estimate that comes from the channel estimation stage. When the finger allocation algorithm decides that a new finger is required (based on the modified channel impulse response), the finger cancellation algorithm can cancel the effects of the energy that would be extracted by the allocated finger of the rake from the impulse response. The most straightforward method is a simple subtraction of a weighted pulse shape from the magnitude squared impulse response. The modified impulse response that is produced is then passed back to the finger allocation algorithm for further processing. A more elaborate algorithm could apply averaging and smoothing filters to reduce the effects of temporal discontinuities in the cancellation procedure.

The result of the finger cancellation algorithm is a signal that in theory contains no useful multipath components (if there were any, they would be identified and a finger allocated). This residual signal could also be used by the noise threshold computation algorithm to improve its estimate of the correlation noise floor of the receiver.

## 6.7 Initial system acquisition

We now move on to consider the final chip rate processing functions of this chapter, namely the procedures involved in the initial system acquisition. We will start by considering the structure of the P-SCH and S-SCH channels.

### 6.7.1 P-SCH and S-SCH physical channels

The P-SCH and the S-SCH channels were described in Chapter 4. The P-SCH and S-SCH consist of a number of special waveforms that are designed to be efficiently detected by the receiver. A single waveform is used for the P-SCH and 16 waveforms are used for the S-SCH. The order in which the S-SCH waveforms are received can be used by the UE to decide the frame start time and the code group for the cell.

### 6.7.2 Three-stage cell acquisition process

The three-stage process is the name given to the procedure that the UE follows when it is first attempting to find and then synchronise to a cell.

**Stage 1 synchronisation**

The first stage of cell acquisition is to detect the P-SCH. To achieve this, the received signal is passed through a device such as a filter matched to the P-SCH waveform. The output of the matched filter is a series of pulses corresponding to the P-SCH at the start of each slot. It is possible that there are multiple pulses from multiple base stations. If this is the case, then the pulses with the greatest energy should be selected.

As well as indicating the presence of a base station, the detection of the P-SCH also provides an indication of the slot boundaries and a reasonable estimate for any gross frequency error.

**Stage 2 synchronisation**

Stage 2 synchronisation is where the mobile station attempts to determine the frame synchronisation and the scrambling code group. The scrambling code groups were discussed in Chapter 4.

The receiver uses the S-SCH to determine the frame synchronisation and scrambling code. As mentioned earlier, the start of every slot in each frame contains an S-SCH. Each slot transmits a 256 chip codeword selected from a family of 16 possible codewords. Each of these codewords is derived from a Hadamard matrix. The order and definition of these codewords defines the code group and the start of the frame. By looking for these codewords (using either a fast Hadamard transform or a matched filter) it is possible to identify the code sequence and hence the frame timing and code group.

Table 6.2 shows a segment of the table [7] that defines the relationship between the slot, the S-SCH waveform broadcast in that slot and the code group for the cell. The left hand column defines the cell code group (0–63), the remaining 15 columns define the S-SCH waveform used in slots #0 to #14 as a function of the code group. The UE can use the sequence of received S-SCHs to identify to which code group a specific Node B belongs.

**Stage 3 synchronisation**

The final stage of synchronisation requires the mobile to determine which of a possible eight primary scrambling codes is being used. By using the common pilot channel (CPICH) (which has a fixed spreading code) and a matched filter we can identify which of the eight scrambling codes is used in the cell by means of trial and error. Once

**Table 6.2.** *Relationships between code group (CG), slot number and S-SCH waveform in each slot*

| CG | \#0 | \#1 | \#2 | \#3 | \#4 | \#5 | \#6 | \#7 | \#8 | \#9 | \#10 | \#11 | \#12 | \#13 | \#14 |
|---|---|---|---|---|---|---|---|---|---|---|---|---|---|---|---|
| | | | | | | Slot number | | | | | | | | | |
| 0 | 1 | 1 | 2 | 8 | 9 | 10 | 15 | 8 | 10 | 16 | 2 | 7 | 15 | 7 | 16 |
| 1 | 1 | 1 | 5 | 16 | 7 | 3 | 14 | 16 | 3 | 10 | 5 | 12 | 14 | 12 | 10 |
| 2 | 1 | 2 | 1 | 15 | 5 | 5 | 12 | 16 | 6 | 11 | 2 | 16 | 11 | 15 | 12 |
| | | | | | Skipped rows: see [7] for details | | | | | | | | | | |
| 62 | 9 | 11 | 12 | 15 | 12 | 9 | 13 | 13 | 11 | 14 | 10 | 16 | 15 | 14 | 16 |
| 63 | 9 | 12 | 10 | 15 | 13 | 14 | 9 | 14 | 15 | 11 | 11 | 13 | 12 | 16 | 10 |

synchronisation is complete, the system and network data on the PCCPCH can be read. Having considered an outline of the cell acquisition procedure, we will now return to look at some aspects of the chip rate processing associated with cell acquisition.

### 6.7.3 Stage 1 – slot synchronisation

As described above, the first stage in the cell acquisition process is the slot synchronisation phase. To achieve slot synchronisation the UE uses the P-SCH. In this section we investigate how the special properties of the P-SCH can be exploited in a way that optimises the efficiency and reduces the complexity of the slot synchronisation process.

In Chapter 4, we described the structure of the P-SCH, and we review it briefly here. The P-SCH is a waveform, generated from a 256-chip-long hierarchical Golay sequence that is broadcast at the beginning of every slot on the downlink.

The objective for the slot synchronisation procedure is two-fold: first, it is to find the presence of a cell; and second to define the slot start time. To achieve both of these goals we expect to use some form of correlation in the receiver. The receiver correlates against a locally stored version of the P-SCH, with the objective of identifying its presence within a downlink signal.

The simplest method (simplest in terms of descriptive complexity, but not in terms of operational complexity) of detecting the P-SCH is to use a matched filter; we will see subsequently that there are other more computationally efficient approaches that exploit the special characteristic of the P-SCH. We start with the matched filter approach and then move on to consider two more efficient methods.

**Matched filter**

The basic block diagram of the matched filter is shown in Figure 6.17(a). The matched filter has a delay line structure with the input signal entering on the left, and the output correlation waveform exiting on the right. Within the matched filter, the received signal

## 6.7 Initial system acquisition

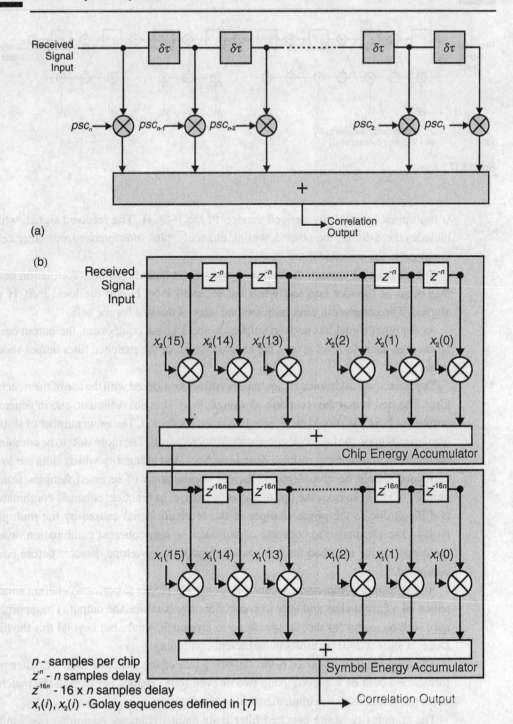

**Figure 6.17** (a) Matched filter for P-SCH. (b) Hierarchical matched filter for P-SCH. (c) EGC.

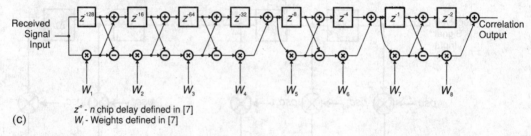

$z^n$ - $n$ chip delay defined in [7]
$W_i$ - Weights defined in [7]

(c)

**Figure 6.17** (*cont.*)

is multiplied by a locally derived version of the P-SCH. The received signal, which includes the S-SCH, the other downlink channels, plus interference from other cells, is passed to the matched filter.

If there is a cell present, the output signal should include periodic correlation peaks that occur at the slot rate and when the received P-SCH and the local P-SCH are aligned. The correlation peak indicates the start of the slot for the cell.

As the input signal has both an amplitude and a phase component, the output correlation does also. To resolve this, the absolute value of the matched filter output should be used.

There are some additional issues that should be considered with the use of the matched filter. The first is that the reception of a single P-SCH is not sufficient, and in general a number of P-SCHs should be received and then combined. The exact number of slots to use depends upon the false detection probability required. The more slots to be combined the lower the probability of false detection. Next, the manner in which slots are to be combined should be considered. Coherent combination of received samples from a number of slots provides the optimum performance. In practice, coherent combination is difficult due to the phase changes in the received signal caused by the multipath fading. The alternative to coherent combination is non-coherent combination, where the output of the matched filter is passed through an envelope detector before being combined.

An interesting compromise solution presented in [25] is to perform coherent combination of adjacent slots and then to non-coherently combine the output. This approach may well be useful for mobile speeds up to around 50 km/h, but beyond this the high Doppler shift at 2 GHz results in inefficient combining.

The final point to consider is the sampling rate of the matched filter. A better performance would be achieved using two or even four samples per chip in the matched filter, but this adds to the implementation complexity.

The complexity of the matched filter is its main weakness. Assuming one sample per chip, the matched filter will require 255 complex additions per chip, and with 256 chips per slot, this will result in 65 280 complex additions per slot.

## Hierarchical matched filter

As mentioned before, the main limitation of the matched filter is its algorithmic complexity; an alternative to using the type of matched filter described above is to exploit the hierarchical sequence properties of the P-SCH.

The P-SCH is created using a technique referred to as hierarchical sequences. This means that the same sequence $A = \{1, 1, 1, 1, 1, 1, -1, -1, 1, -1, 1, -1, 1, -1, -1, 1\}$ is sequentially repeated according to $P = \{A, A, A, -A, -A, A, -A, -A, A, A, A, -A, A, -A, A, A\}$ to produce the required 256 chips. This structure for the P-SCH allows the matched filtering to be done in two stages. The first stage is a matched filter correlating for the sequence $A$, and the second stage is another matched filter correlating against the sequence $P$. Figure 6.17(b) illustrates this algorithm.

The complexity of this alternative approach results in summation of 16 elements for sequence $A$, and 16 elements to obtain sequence $P$. The complexity is now $15 + 15 = 30$ complex additions. For a complete slot, this corresponds to 7680 complex additions, a reduction in complexity by a factor of around 8.

It should be noted that this alternative structure does not result in any loss in performance; it is achieved by exploitation of the hierarchical sequences used by the P-SCH.

## Efficient Golay correlator

The final approach to the slot synchronisation problem uses an additional feature of the P-SCH. The structure of the efficient Golay correlator (EGC) is illustrated in Figure 6.17(c) (originally posed in [26]).

The EGC is computationally more efficient than the previous two approaches. For every received signal sample 13 complex additions are required, leading to a total of 3328 complex additions for the full slot of P-SCH data. This is almost 20 times more efficient than the matched filter approach, and about twice as efficient as of the hierarchical matched filter

### 6.7.4 Stage 2 – frame synchronisation and code group

Once the UE has performed slot synchronisation, it can perform frame synchronisation using the S-SCH. In addition, the UE can estimate which code group a specific cell belongs to as part of this procedure.

The previous section showed how the S-SCH can be used to identify the frame synchronisation and the code group. It requires the UE to decode the sequence of waveforms (of which there are 16) and broadcast it across a frame of data, from which it can estimate the frame time and the code group as outlined previously. The S-SCH consists of the hierarchical sequence of length 256 chips modulated by a Hadamard waveform of length 256. The specific Hadamard waveform may be different in every

**Table 6.3.** *Calculation of FHT based on EGC correlation outputs*

| W | X | Y | Z |
|---|---|---|---|
| $W_0 = R_0 + R_1$ | $X_0 = W_0 + W_1$ | $Y_0 = X_0 + X_1$ | $Z_0 = Y_0 + Y_1$ |
| $W_1 = R_2 + R_3$ | $X_1 = W_2 + W_3$ | $Y_1 = X_2 + X_3$ | $Z_1 = Y_2 + Y_3$ |
| $W_2 = R_4 + R_5$ | $X_2 = W_4 + W_5$ | $Y_2 = X_4 + X_5$ | $Z_2 = Y_4 + Y_5$ |
| $W_3 = R_6 + R_7$ | $X_3 = W_6 + W_7$ | $Y_3 = X_6 + X_7$ | $Z_3 = Y_6 + Y_7$ |
| $W_4 = R_8 + R_9$ | $X_4 = W_8 + W_9$ | $Y_4 = X_8 + X_9$ | $Z_4 = Y_8 + Y_9$ |
| $W_5 = R_{10} + R_{11}$ | $X_5 = W_{10} + W_{11}$ | $Y_5 = X_{10} + X_{11}$ | $Z_5 = Y_{10} + Y_{11}$ |
| $W_6 = R_{12} + R_{13}$ | $X_6 = W_{12} + W_{13}$ | $Y_6 = X_{12} + X_{13}$ | $Z_6 = Y_{12} + Y_{13}$ |
| $W_7 = R_{14} + R_{15}$ | $X_7 = W_{14} + W_{15}$ | $Y_7 = X_{14} + X_{15}$ | $Z_7 = Y_{14} + Y_{15}$ |
| $W_8 = R_0 - R_1$ | $X_8 = W_0 - W_1$ | $Y_8 = X_0 - X_1$ | $Z_8 = Y_0 - Y_1$ |
| $W_9 = R_2 - R_3$ | $X_9 = W_2 - W_3$ | $Y_9 = X_2 - X_3$ | $Z_9 = Y_2 - Y_3$ |
| $W_{10} = R_4 - R_5$ | $X_{10} = W_4 - W_5$ | $Y_{10} = X_4 - X_5$ | $Z_{10} = Y_4 - Y_5$ |
| $W_{11} = R_6 - R_7$ | $X_{11} = W_6 - W_7$ | $Y_{11} = X_6 - X_7$ | $Z_{11} = Y_6 - Y_7$ |
| $W_{12} = R_8 - R_9$ | $X_{12} = W_8 - W_9$ | $Y_{12} = X_8 - X_9$ | $Z_{12} = Y_8 - Y_9$ |
| $W_{13} = R_{10} - R_{11}$ | $X_{13} = W_{10} - W_{11}$ | $Y_{13} = X_{10} - X_{11}$ | $Z_{13} = Y_{10} - Y_{11}$ |
| $W_{14} = R_{12} - R_{13}$ | $X_{14} = W_{12} - W_{13}$ | $Y_{14} = X_{12} - X_{13}$ | $Z_{14} = Y_{12} - Y_{13}$ |
| $W_{15} = R_{14} - R_{15}$ | $X_{15} = W_{14} - W_{15}$ | $Y_{15} = X_{14} - X_{15}$ | $Z_{15} = Y_{14} - Y_{15}$ |

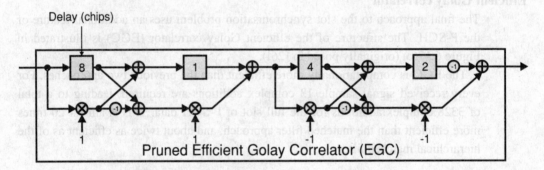

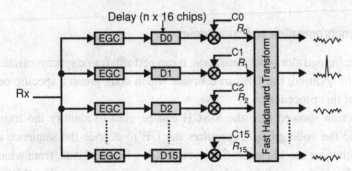

**Figure 6.18** Linkage between EGC and FHT for S-SCH detection.

## 6.7 Initial system acquisition

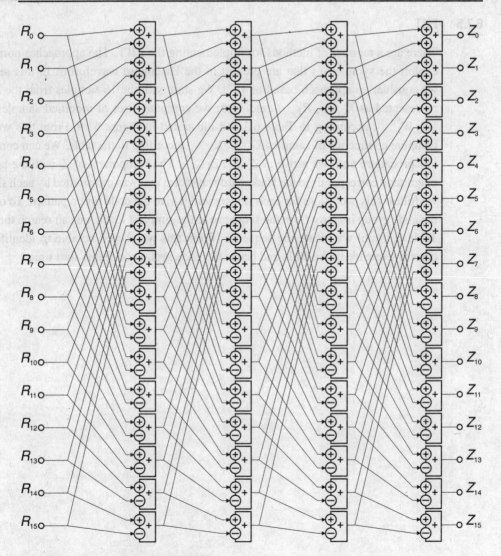

**Figure 6.19** Picture of operation of FHT.

slot. The UE needs to identify which waveform is used in each slot, and from this it is possible to deduce the code group and hence the frame start time.

To remove the hierarchical sequence from the S-SCH, the UE can use the arrangement illustrated in Figure 6.18. The EGC is similar to the one used for the P-SCH, and the output of the EGC is a signal that contains the symbols of a Hadamard sequence. The next stage is to identify which Hadamard waveform was used in a specific slot. To achieve this, the output from the EGC is input into a device called a fast Hadamard transform (FHT) that can take the 16 samples and produce an output that defines which of the 16 possible Hadamard waveforms was used. To proceed, we will consider the design of the FHT.

### 6.7.5 FHT

There are a number of methods for implementing the FHT. The approaches normally exploit the symmetries that are present in the Hadamard waveforms. In this section we consider an approach based on [27]. We start with the 16 samples from the EGC represented as $R_0, R_1, R_2, \ldots$ etc. Next we compute a set of modified samples $W_0, W_1, W_2 \ldots$ according to Table 6.3 and illustrated in Figure 6.19. From this we can compute a sequence of samples $X_0, X_1, X_2 \ldots$ according to the table. We can continue this procedure until we arrive at $Z_0, Z_1, Z_2 \ldots$. The component of $Z_i$ with the largest magnitude corresponds to the Hadamard waveform that was transmitted by each slot. If we follow this procedure for each slot we will derive the code group pattern. To obtain the code group identity from the sequence at the simplest level, we can search through Table 6.3 until we locate the sequence. From this we should be able to identify the correct sequence in the table and hence the code group and frame start time.

# 7 Symbol rate processing functions

## 7.1 WCDMA symbol rate transmission path

In this chapter we review the basic principles of what we call the symbol rate processing functions, but are also often referred to as bit rate processing functions. These functions apply to both transmit and receive paths of the WCDMA system. Additionally, we explore functions such as error protection coding, rate matching and the topic of transport channel combination. We start with a review of the uplink/downlink symbol rate transmission path (this is the same for the downlink/uplink receive path, but traversed in the reverse direction). Then we review some of the basic principles of convolutional error correcting codes, and finally we finish with an exploration of turbo codes, the turbo decoder and in particular the maximum a-posteriori probability (MAP) algorithm.

### 7.1.1 Coding introduction

Figure 7.1 presents the basic structure of the lower layers of the transmission link between the UE and the UTRAN. The diagram is a simplification of the processing stages that are considered in greater detail shortly. The diagram illustrates the basic principles in the operation of the symbol rate processing stages within the WCDMA system. Data are received via transport channels from the MAC. The data blocks may have cyclic redundancy check (CRC) bits appended for error detection purposes in the receiver. The data are then encoded using either a convolutional encoder or a turbo encoder. Next, interleaving is performed to 'shuffle' the data prior to transmission to help combat fading and interference across the radio link. Rate matching is applied to match the encoded data rate to the available data rate that has been selected for the combination of channels. The rate matching is performed on each channel individually, but with knowledge of the total incoming and outgoing data rates. The final stage of the symbol rate processing is the creation of what is referred to as a coded composite transport channel (CCTrCH) and a second stage of interleaving. The CCTrCH is a combination of all the transport channels to be multiplexed onto the physical channel. There are many restrictions associated with the creation of the CCTrCH, which are considered in greater detail later in this section.

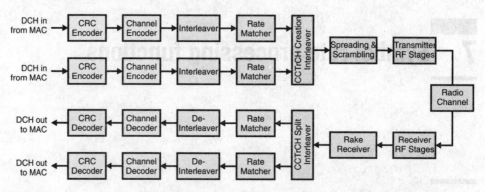

**Figure 7.1** Structure of the lower layers of the uplink of the WCDMA communications system.

After the symbol rate processing stages in the transmitter, there are the chip rate processing stages considered in Chapter 6 and the RF transmitter stages considered in Chapter 5. In the receiver, in the Node B, there are the RF receive stages and the chip rate receive stages (such as the rake receiver). After the receive chip rate processing stages, we come to the receive symbol rate processing stages, which in general, perform the reverse operation to the transmit symbol rate processing (deinterleaving, reverse rate matching, channel decoding and CRC decoding). From the symbol rate processing stages, the received transport blocks are passed via the configured transport channels to the MAC.

We start this chapter by examining the main processing stages performed by the symbol rate processing stages. Figure 7.2 illustrates the symbol rate transmission path for the uplink and the downlink in greater detail. The uplink diagram refers to the transmission in the UE and the reception functions in the Node B, and the downlink diagram relates to the transmission functions in the Node B and the reception functions in the UE. In this section we review each of the main processing stages presented in the diagram, starting at the top and working down through the different functions. We start this review by considering the definition and the structure of a CCTrCH.

### 7.1.2 CCTrCH

**What is a CCTrCH?**

The combining of the transport channels is one of the important functions undertaken by the physical layer. The combination process is illustrated in Figure 7.2. When the transport channels are combined they form what is known as a CCTrCH. There are some restrictions placed on the number and structure for the CCTrCHs that are considered in this section.

**Uplink restrictions on CCTrCH**

Various physical channels are used. For the DCHs, the number of transport blocks and the number of DPCHs that can be used is defined by the UE capability [14].

## 7.1 WCDMA symbol rate transmission path

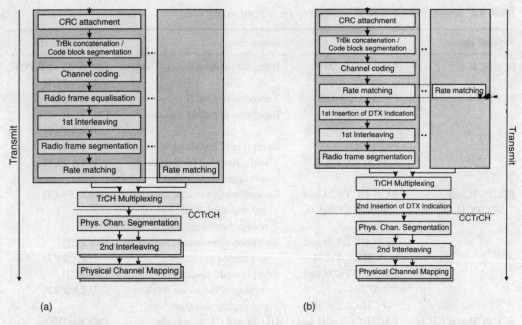

**Figure 7.2** Detailed representation of the symbol rate processing stages: (a) for the uplink transmission path and (b) for the downlink transmission path.

For the RACH, only one RACH transport channel can be in a RACH CCTrCH; for the CPCH, only one CPCH transport channel can be in a CPCH CCTrCH. For the DCH, a number of DCH transport channels can be combined into a single dedicated CCTrCH. At any time instant, only one CCTrCH is allowed on the uplink, either a dedicated CCTrCH or a common CCTrCH (RACH or CPCH).

**Downlink restrictions on CCTrCH**

Only transport channels with the same active set can be mapped onto the same dedicated CCTrCH. Different CCTrCHs cannot be mapped onto the same physical channel. One CCTrCH can be mapped onto one or more physical channels, but the spreading factor for the physical channels must be the same.

DCHs and common transport channels cannot be combined onto the same CCTrCH and hence the same physical channels. With the common transport channels, only the FACH and PCH may belong to the same CCTrCH. For the other common transport channels, there can be a CCTrCH for the downlink shared channel (DSCH), a CCTrCH for the broadcast channel (BCH) and a CCTrCH for the FACH/PCH. Only one CCTrCH of dedicated type is allowed for R99.

**Required downlink channel combinations for R99**

For the downlink, Table 7.1 presents the requirements for the UE to be able to receive various physical channels, the associated transport channels and the CCTrCHs that are required. To distinguish between the different CCTrCHs the associated transport channel prefixes the CCTrCH in the table.

**Table 7.1.** *Downlink CCTrCH options to be supported by UE*

| Physical channel | CCTrCH | Comment | Likely RRC state |
|---|---|---|---|
| PCCPCH | BCH-CCTrCH | Reception of broadcast messages | Idle mode, Cell_PCH, URA_PCH |
| SCCPCH | FACH-CCTrCH | Reception of FACH | Cell_FACH |
| | PCH-CCTrCH | Reception of paging messages | Cell_PCH, URA_PCH |
| | FACH+ PCH-CCTrCH | Reception of broadcast services while listening for paging messages | Cell_PCH, URA_PCH |
| PCCPCH and SCCPCH | BCH-CCTrCH and FACH-CCTrCH | Reception of broadcast messages and downlink FACH – not required continuously | CELL_FACH |
| | BCH-CCTrCH and PCH-CCTrCH | Reception of broadcast messages and paging messages | Cell_PCH, URA_PCH |
| | BCH-CCTrCH and FACH+ PCH-CCTrCH | Reception of broadcast messages, broadcast services and paging messages | Cell_PCH, URA_PCH |
| SCCPCH and AICH | FACH-CCTrCH and RACH-AICH | DL Rx and UL access via RACH | Cell_FACH |
| | FACH-CCTrCH and CPCH-AICH | DL Rx and UL access via CPCH | Cell_FACH |
| | PCH-CCTrCH and RACH-AICH | Reception of paging message and UL access via RACH | Move from Cell_PCH or URA_PCH to Cell_FACH |
| | PCH-CCTrCH and CPCH-AICH | Reception of paging message and UL access via CPCH | Move from Cell_PCH or URA_PCH to Cell_FACH |
| | FACH+PCH-CCTrCH and RACH-AICH | DL Rx (paging and broadcast services) and UL access via RACH | Move from Cell_PCH or URA_PCH to Cell_FACH |
| | FACH+ PCH-CCTrCH and CPCH-AICH | DL Rx (paging and broadcast services) and UL access via CPCH | Move from Cell_PCH or URA_PCH to Cell_FACH |
| SCCPCH and DPCCH [UE-RAC] | PCH-CCTrCH and CPCH-DPCCH | Reception of paging message and UL Tx via CPCH | Move from Cell_PCH or URA_PCH to Cell_FACH |
| | FACH-CCTrCH and CPCH-DPCCH | DL Rx and UL Tx via CPCH | Cell_FACH |
| | FACH+PCH-CCTrCH and CPCH-DPCCH | DL Rx (paging and broadcast services) and UL Tx via CPCH | Move from Cell_PCH or URA_PCH to Cell_FACH |
| PICH | na | | |
| DPCCH and DPDCH | DCH-CCTrCH | Multiple DCH on a single CCTrCH | Cell_DCH |

**Table 7.1.** (*cont.*)

| Physical channel | CCTrCH | Comment | Likely RRC state |
|---|---|---|---|
| DPCCH and multi-DPDCH [UE-RAC] | DCH-CCTrCH | Multiple DPDCH and multiple DCH using a single CCTrCH | Cell_DCH |
| Multi-PDSCH and DPCCH and multi-DPDCH [UE-RAC] | DCH-CCTrCH DSCH-CCTrCH | Multiple DPDCH and multiple DCH using a single CCTrCH and multiple PDSCH and multiple DSCH using a single CCTrCH | Cell_DCH |
| SCCPCH and DPCCH and multi-DPDCH [UE-RAC] | DCH-CCTrCH FACH-CCTrCH | Multiple DPDCH and multiple DCH using a single CCTrCH and single FACH using a single CCTrCH | Cell_DCH |
| SCCPCH and DPCCH and multi-DPDCH and multi-PDSCH [UE-RAC] | DCH-CCTrCH FACH-CCTrCH DSCH-CCTrCH | Multiple DPDCH and multiple DCH using a single CCTrCH and multiple PDSCH and multiple DSCH using a single CCTrCH and single FACH using a single CCTrCH | Cell_DCH |
| DPCCH and multi-DPDCH [UE-RAC] | Multi-DCH-CCTrCH | Multiple DCH coded onto multiple CCTrCH (Not R99) | Cell_DCH |
| PCCPCH DPCCH and multi-DPDCH and multi-PDSCH [UE-RAC] | BCH-CCTrCH DCH-CCTrCH DSCH-CCTrCH | Multiple DPDCH and multiple DCH using a single CCTrCH and multiple PDSCH and multiple DSCH using a single CCTrCH and monitoring of neighbour cell broadcasts | Cell_DCH |

In the table, the support for the different channel combinations will depend upon the UE radio access capabilities (UE-RAC). In some instances, the capabilities are not expected for R99. An example of this is the support for multiple CCTrCHs carrying dedicated TrCHs. For R99, only a single CCTrCH is supported.

**Required uplink channel combinations for R99**

Table 7.2 illustrates the uplink channel combinations required in R99. The uplink combinations are simpler than those used in the downlink.

### 7.1.3 Error detection coding and CRCs

The addition of the CRC bits in the WCDMA transmission path is for the purposes of error detection. At the transmitter, parity bits are appended to the data to be transmitted.

## Symbol rate processing functions

**Table 7.2.** *Uplink CCTrCH options to be supported by UE*

| Physical channel | CCTrCH | Comment | Likely RRC state |
|---|---|---|---|
| PRACH | RACH-CCTrCH | Single RACH and single CCTrCH | CELL_FACH, Idle Mode |
| PCPCH | CPCH-CCTrCH | Single CPCH and single CCTrCH | Cell_FACH |
| DPCCH and DPDCH | DCH-CCTrCH | Multiple DCH coded onto a single CCTrCH | Cell_DCH |
| DPCCH and multi-DPDCH | DCH-CCTrCH | Multiple DCH coded onto a single CCTrCH | Cell_DCH |

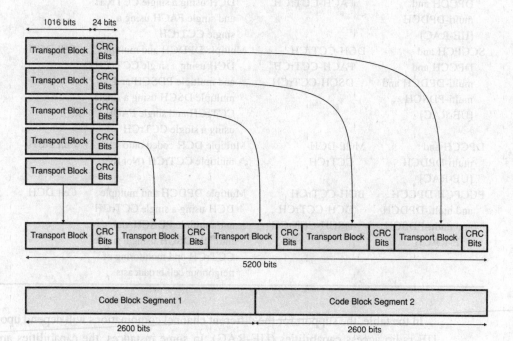

**Figure 7.3** Example CRC attachment and code block segmentation.

This type of codeword is referred to as a systematic code whereby the information and the parity bits are separated.

Figure 7.3 illustrates an example of data blocks having CRC bits appended to them. In the context of the WCDMA system, these CRC bits are of defined length and the data blocks are the transport blocks that arrive in the physical layer from the MAC.

Table 7.3 defines the different lengths of CRC bits that can be used for the WCDMA system, and the cyclic generator polynomials that can be used to create the parity check bits. Apart from the no-CRC case, four different length CRCs can be used in the WCDMA system. The choice of the length will be made by the UTRAN when

## 7.1 WCDMA symbol rate transmission path

**Table 7.3.** *WCDMA CRC cyclic generator polynomials*

| CRC length | Cyclic generator polynomial |
|---|---|
| 8  | $D^8 + D^7 + D^4 + D^3 + D + 1$ |
| 12 | $D^{12} + D^{11} + D^3 + D^2 + D + 1$ |
| 16 | $D^{16} + D^{12} + D^5 + 1$ |
| 24 | $D^{24} + D^{23} + D^6 + D^5 + D + 1$ |

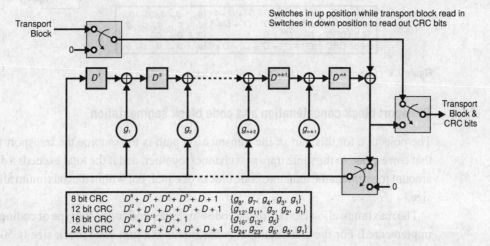

**Figure 7.4** General form for the encoder.

the specific transport channels are being configured. It is most probable that for a set of different services, while in the Cell-DCH state, there will be a collection of CRCs of different lengths being combined onto the same CCTrCH and hence physical channel. The parity bits are generated using a CRC encoder using the cyclic generator polynomials defined in Table 7.3.

The general form for the encoder is shown in Figure 7.4, which includes the generator tap locations for the different CRC lengths to be used in the WCDMA system. The CRC decoder operates by dividing the received codewords by the same polynomial as used in the transmitter, using a shift register structure similar to that used in the transmitter.

Figure 7.5 illustrates the general structure for the CRC decoder including the generator tap locations. The decoder operates by clocking the received data through the shift register. When all of the data have been passed into the shift register, then the contents of the shift register can be read. The contents of the shift register after processing the received data are called the syndrome. The syndrome can then be examined: if the syndrome vector is non-zero, errors must be present in the received data.

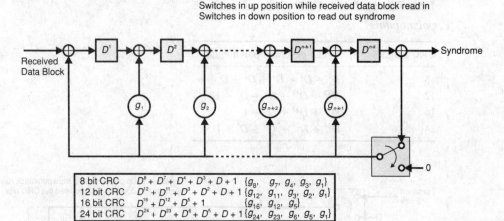

**Figure 7.5** General structure for the CRC decoder.

### 7.1.4 Transport block concatenation and code block segmentation

The objective for this part of the transmission path is to combine the transport blocks that correspond to the same transport channel together, and if the total exceeds a defined amount to segment the total into code blocks of equal, but within the maximum allowed, size.

The maximum allowed size for the code blocks depends upon the type of coding being implemented. For the convolutional codes, the maximum code block size is 504 bits, and for the turbo code it is 5114 bits. To ensure that the code blocks are the same size, filler bits are added to the beginning of the first transport block.

Figure 7.3 summarises the procedure. The data arrive from a specific transport channel to be transmitted within a defined transmission time interval (TTI). The CRCs are appended to the individual transport blocks. Next, the transport block concatenation occurs, and in the example shown, we produce a total of 5200 bits, which exceeds the capability of the turbo code. Code block segmentation now occurs. In this example, filler bits are not required, as we have an even number of bits and we only need two code blocks. Two code blocks each of length 2600 bits are created.

### 7.1.5 Channel coding

Channel coding is applied to the transport channels after the code block segmentation. The BCH, PCH and RACH transport channels use a 1/2-rate convolutional code. The CPCH, DCH, DSCH and FACH transport channels can use a 1/2-rate or 1/3-rate convolutional code or a 1/3-rate turbo code. For the convolutional codes, eight tail bits are added prior to the encoding. For the turbo codes, four tails bits are added through the encoding.

In the following sections we examine some of the details of the convolutional and turbo encoding and decoding schemes.

## 7.1.6 Radio frame equalisation

On the uplink, radio frame equalisation is applied after the channel encoding. The objective of the radio frame equalisation is to ensure that the number of data available for transmission within the required number of radio frames is equal for each radio frame.

The parameter that defines the amount of equalisation required is the TTI value that is set for the specific transport channel. If, for example, the TTI is set by higher layers to be 40 ms, this means that it will require four radio frames to transmit the data. The data must, therefore, be divisible by 4 in this example. If it is not divisible into equal-sized radio frames, padding bits are inserted at the end of the data block to ensure that it can be divided into equal-sized radio frames.

## 7.1.7 First interleaving

First interleaving is applied after the radio frame equalisation on the uplink, and after rate matching on the downlink. The first interleaving is a block interleaving that includes intercolumn switching. Depending upon the TTI, a matrix is defined with 1, 2, 4 or 8 columns for TTIs of 10 ms, 20 ms, 40 ms and 80 ms respectively. The data are read into the matrix structure in rows. Once the data are read into the matrix, then the columns are swapped according to a specific algorithm [15]. For TTIs of 10 ms and 20 ms, columns are not swapped, but for TTIs of 40 ms and 80 ms the columns are swapped. Once the column swapping has occurred the data are read from the matrix in columns starting at the position in the top left. For the 10 ms TTI, it turns out that the first interleaving has no effect on the data, as there is only one column available.

Figure 7.6 illustrates the principles behind the first interleaving for an example where the TTI is set to 40 ms, and consequently four columns are created with the swapping occurring between columns 1 and 2.

## 7.1.8 Radio frame segmentation

The next stage of the uplink is radio frame segmentation. On the downlink after rate matching and on the uplink after radio frame equalisation, the block sizes can be divided equally into the required number of radio frames based on the TTI for that transport channel. Radio frame segmentation then divides the data block into $n$ equal sized radio frames, where $n = \{1,2,4,8\}$ for TTI = 10 ms, 20 ms, 40 ms and 80 ms respectively.

## 7.1.9 Rate matching

On the uplink, after radio frame segmentation comes rate matching. On the downlink, rate matching occurs after the channel coding stage. The objective for rate matching is

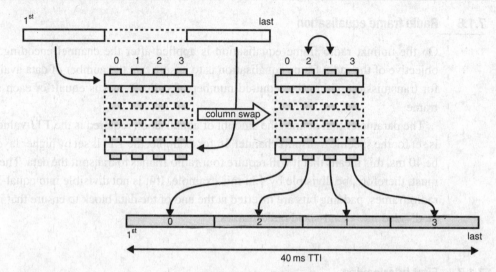

**Figure 7.6** Principles behind the first interleaving.

to match the amount of data to be transmitted to the available capacity of the different physical channels available.

Rate matching uses three basic mechanisms to adjust the data rate for the different channels. Repetition is used when there are insufficient data for the capacity of the physical channel; puncturing is used when there are too many data for the capacity of the physical channel; and DTX bits are used on the downlink when there are too few data.

In an example presented in Chapter 8, we will see an example on the uplink, where both repetition and puncturing occur within the same CCTrCH. Some of the transport channels combined onto the CCTrCH use repetition whilst other channels use puncturing. Whether to use repetition or puncturing depends upon the physical channel capacity, the number of transport channels to be combined and the rate matching parameter.

## Uplink rate matching

Starting with the uplink we can consider the basic principles behind the operation of the rate matching algorithm. One key difference between the uplink and the downlink is that the uplink spreading factor (and hence data rate) can be selected on a frame-by-frame basis. The downlink data rate, however, is fixed unless changed via higher layer signalling or through the use of compressed mode. The rate matching algorithm operates by either repeating or puncturing bits. The bits that are repeated or punctured are defined deterministically using algorithms defined in [15] once we know what spreading factor we can use and the number of physical channels, which, in turn, are defined by the number of data to be transmitted $N_{data}$. The first step, therefore, in rate matching is to establish the spreading factor and the number of physical channels to be used. To do this we need to estimate the value for $N_{data}$, which is the number of data bits to be transmitted for the TTI defined for that transport channel. Once we have selected

## 7.1 WCDMA symbol rate transmission path

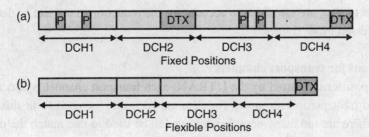

**Figure 7.7** Differences between (a) fixed and (b) flexible positions for transport channels.

$N_{\text{data}}$, we know how many bits we start with, therefore, the amount of puncturing or repetition is known and using the algorithm defined in [15], we can proceed to either puncture or repeat the data. To estimate $N_{\text{data}}$ we define three data sets SET0, SET1 and SET2, and we proceed as follows:

- The first step is to determine the SET0 that defines the possible allowed values for $N_{\text{data}}$, based on UTRAN and UE capability issues (data to be transmitted). Within SET0, SET1 is created by selecting the spreading factor that allows all data to be combined on one physical channel, without having to use puncturing. If SET1 is not empty, the element of SET1 that uses the largest spreading factor is chosen for $N_{\text{data}}$.
- If SET1 is empty, SET2 is formed from the allowed elements from SET0 that can be combined on one or more physical channels using the maximum allowed puncturing.
- The element from SET2 is selected as the lowest spreading factor that can be found in SET2 whilst using the minimum number of physical channels. This procedure ensures that the lowest number of physical channels possible is selected, and that the minimum amount of puncturing is performed.

### Downlink rate matching

In the downlink, rate matching is done to support the variable data rates introduced by combining different services and also to support the compressed mode. For the downlink rate matching there are two basic approaches to the rate matching. The first uses fixed positions for the transport channels, and the second uses flexible positions for the transport channels. The differences are highlighted in Figure 7.7.

When the downlink is configured to use fixed positions, the transmission resource is allocated per transport channel with rate matching applied per transport channel. With flexible positions, a more efficient use of the transmission resource can be made; if a transport channel does not need its full share of the downlink resource, it can be used by other transport channels.

### Downlink rate matcher operation

To define the number of bits available in all the physical channels in one radio frame $N_{\text{data}}$ must be defined. If in compressed mode with compression performed via puncturing, modifications to $N_{data}$ are estimated to account for the reduction in the size of data bits in the compressed mode frames.

**Fixed positions for transport channels**

When fixed positions are used by the UTRAN, each transport channel is allocated a specific fixed transmission resource. The data are then rate matched to fit this fixed resource. If there are too many data, puncturing can be used to rate match the data to the allocated resource. If there are too few data, discontinuous transmission (DTX) bits can be used (DTX bits are bits that are not transmitted). Figure 7.7(a) illustrates an example of the use of fixed positions for four transport channels. At the instance depicted in the diagram, transport channels DCH1 and DCH3 have too many data and are punctured, whereas transport channels DCH2 and DCH4 have too few and so DTX bits are used. It should be noted that this figure is prior to interleaving which will spread the DTX and punctured bits more evenly through the radio frame.

**Flexible positions for transport channels**

Figure 7.7(b) illustrates the use of flexible positions for rate matching on the downlink. With flexible positions, the decision on whether to puncture or use DTX bits is only made once all of the data to be transmitted within a radio frame are assembled. The benefit in this approach is that any spare capacity for one transport channel can be shared with transport channels that require additional capacity. In the example illustrated in Figure 7.7(b) we can see that DTX bits are being used, indicating that there are fewer data to transmit than the physical channel can manage. In this instance, the DTX bits are placed at the end of the transmission block, and again interleaving spreads these DTX bits across the radio frame.

**Rate matching**

An algorithm specified in [15] defines where the punctured bits and the DTX bits should be located based on the number of data to be transmitted, and the physical channel that is being selected.

### 7.1.10 Transport channel multiplexing

Once the transport channels have been rate matched, they can be multiplexed to form a CCTrCH. To achieve this, serial multiplexing is performed for the different transport channels.

### 7.1.11 Physical channel segmentation

Once the CCTrCH is formed, it can be segmented into the different physical channels, depending upon the number of physical channels that were configured by the higher layers. The data are then split between the different physical channel blocks.

### 7.1.12 Second interleaving and physical channel mapping

The second interleaving stage shuffles the data that are to be transmitted in the same physical channel. In this case, the interleaving period is a radio frame of 10 ms. The interleaver is a block interleaver that also includes intercolumn permutations. After completing the interleaving operation, the data are mapped onto the physical channels that were defined for the combined data.

As mentioned previously, the data from a single CCTrCH can be sent via multiple physical channels, but the spreading factors for the different physical channels must be the same (this is true for both the uplink and the downlink).

The data from multiple CCTrCHs for the same user (only allowed on the downlink) must be sent via different physical channels. For R99, the only allowed case for multiple CCTrCHs comes from the mixing of common CCTrCHs and dedicated CCTrCHs. This mixing results in different physical channels: a common physical channel for the common CCTrCH and a dedicated physical channel for a dedicated CCTrCH.

## 7.2 Convolutional error correction codes

### 7.2.1 Introduction to convolutional codes

Convolutional codes were first presented in 1955 by P. Elias in a paper entitled 'Coding for noisy channels' [16]. Convolutional codes are different from the type of block codes seen in the last section that are used for the generation of CRC bits. Whereas block codes operate independently on segments of data, convolutional codes operate continuously on streams of data.

The encoder for a convolutional code is based on simple sequential logic and shift register sequences. The optimum decoder (i.e. the one which minimises the sequence error rate) is the maximum likelihood sequence estimator (MLSE). The MLSE algorithm, however, suffers from a high associated complexity. The alternative often employed is the Viterbi algorithm. The Viterbi algorithm is often associated with a convolutional decoder, and is an efficient solution to the MLSE decoding problem.

### 7.2.2 Convolutional encoder

**Code rate**

A general form for the convolutional encoder is presented in Figure 7.8. The diagram shows the case where the input data symbols are $k$ bits in size, and the output codewords are $n$ bits long. We can define a quantity known as the code rate, where the code rate is given by $k/n$. For binary systems it is normal to assume that $k = 1$, leading to code rates of $1/n$: for WCDMA, $n = 2$ or 3.

The constraint length $L$ is related to the memory storage of the encoder. For a binary code ($k = 1$) with a constraint length of $L$, there are $L - 1$ bits of memory in the encoder.

## Symbol rate processing functions

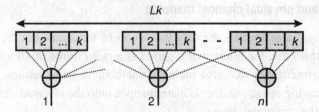

**Figure 7.8** General form for the convolutional encoder.

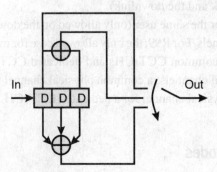

**Figure 7.9** Basic structure for a rate 1/3, $L = 3$ encoder.

To specify the convolutional encoder, a generator vector is used to specify the location of the taps in the modulo $-2$ adder.

**Example encoder**

Figure 7.9 shows the basic structure for a rate 1/3, $L = 3$ encoder. The modulo 2 taps are defined by the binary vectors for the generator polynomial $g$, which are usually specified as an octal number as shown. The basic operation of the example encoder can be considered through the use of state and trellis diagrams. The output from the encoder is given by the current state of the shift registers and the modulo $-2$ adders.

**State diagram**

Figure 7.10 shows the states for the (3,1,3) code considered earlier, and illustrates the basic operation of the encoder, indicating the incoming bits on the transitions, as well as the output bits in the form $X/YYY$, where $X$ is the incoming bit and $YYY$ is the three-digit output sequence for the rate 1/3 code. There are a number of interesting points that can be made about the state diagram. First, not all state transitions exist. For instance, from state S1 the only transitions are back to state S0, or to state S2. Next, from each state there is one transition corresponding to an input binary 0, and one for an input binary 1. Finally, the transitions into a state are either all binary 0 or all binary 1.

## 7.2 Convolutional error correction codes

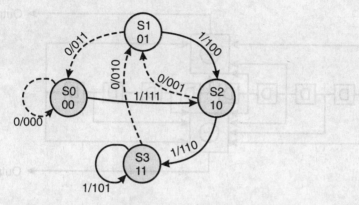

**Figure 7.10** States for the (3,1,3) code and the operation of the encoder.

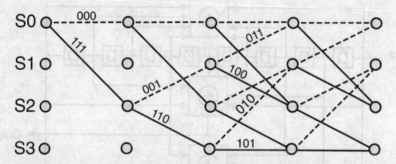

**Figure 7.11** States of the encoder represented using a trellis diagram.

### Trellis diagram

An alternative representation to the state diagram is the trellis diagram. The trellis diagram is illustrated in Figure 7.11 for the simple encoder considered in Figure 7.9. The trellis diagram represents the state transitions as an unfolding sequence of time steps. In this example (as is usually the case), the encoder starts in the 'all-zero' state S0. As the data enter the encoder, the different possible transitions result as shown in the diagram. In Figure 7.11, the dashed lines correspond to an input bit of binary 0, and the solid lines an input bit of binary 1. The numbers above the transitions represent the output codeword that is associated with each of the input data bits, but dependent upon the current state of the encoder.

### Tail bits

Because convolutional encoding is a continuous process, it is necessary to force the encoder into a known end state, otherwise data may be lost within the memory of the encoder. To achieve this it is customary to append $L-1$ zero bits to the end of the data sequence that is to be encoded. This data sequence forces the trellis (or state diagram)

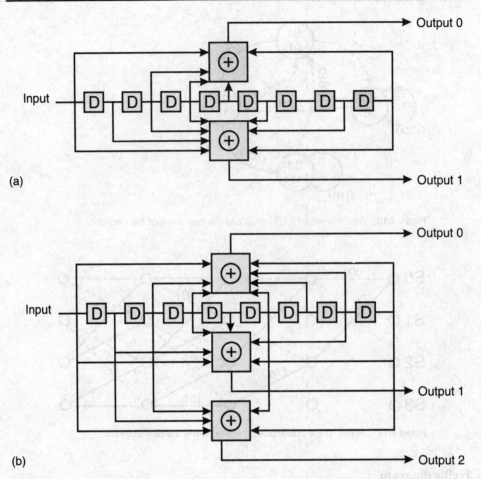

**Figure 7.12** (a) WCDMA 1/2-rate convolutional encoder; (b) WCDMA 1/3-rate convolutional encoder.

to finish in a known state (S0 in this case). The fact that the code finishes in a known state is useful information that can be used in the decoding process.

### 7.2.3 WCDMA convolutional codes

Figure 7.12(a) illustrates the 1/2-rate convolutional encoder and Figure 7.12(b) the 1/3-rate convolutional encoder used in WCDMA. The 1/2-rate convolutional encoder has a constraint length of 9 and comprises an eight-element shift register. For every input bit there are two output bits. The 1/2-rate shift register also requires eight tail bits to ensure that all of the data bits are correctly passed through the encoder.

The 1/3-rate convolutional encoder also has a constraint length of 9 and includes an eight-element shift register. With the 1/3-rate encoder there are three output bits for every input bit. Similarly to the 1/2-rate encoder, the 1/3-rate encoder requires eight tail bits to flush the encoded data out.

### 7.2.4 Convolutional code decoding

The process of convolutional code decoding is different from the CRC block code decoding considered earlier. Generally, convolutional code decoders are based on MLSE. This is an optimum decoding technique described later. The Viterbi algorithm is an efficient implementation of the MLSE decoding process – it too is covered later. Hard decision decoding is where a decision between a 1 or 0 is made by the demodulator. For soft decision decoding, the demodulator output is a quantised word.

**MLSE**

MLSE operates by considering all the possible transmitted sequences and comparing them with the actual received sequence. For each of the transmitted sequences the number of similarities with the received sequence is estimated. The transmitted sequence that has the most similarities with the received sequence is chosen as the most likely transmitted sequence. Having selected the most likely transmitted sequence, the receiver can derive the input data that were at the transmitter. This procedure corrects some, if not all, errors in the received data stream. The problem with MLSE occurs if we are transmitting a very large number of bits. If the input data sequence was for 100 bits, say, the output coded sequence would be 300 bits (1/3-rate code and ignoring the tail bits), and so the number of possible transmitted sequences is exponential to the length of the input data and for this simple example could lead to a total number of potential transmitted sequences in excess of $1 \times 10^{30}$. The estimation of the correct sequence can become a computationally burdened problem.

**Viterbi algorithm**

MLSE is the optimum method of decoding a sequence of received bits. As we have seen, the issue with MLSE, however, is that it is a computationally expensive method of performing the decoding. To reduce the amount of computation, a simplifying step in the decoding algorithm can be implemented. This simplification is known as the Viterbi algorithm – named after the person who first proposed it [17].

To understand the operation of the Viterbi algorithm, consider the situation shown in Figure 7.13, where two paths in the trellis meet at a state (e.g. A and B meet at A'). Assuming that we are performing some form of MLSE where we are tracking all possible transitions between the states at each time step (this is referred to as a path), we can calculate what is known as a transition metric. This metric is the measure of how closely a particular transition compares with the actual received sequence for that

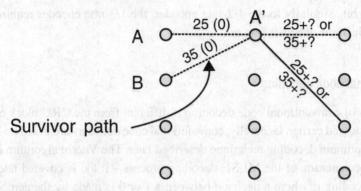

**Figure 7.13**  Viterbi decoding step.

particular time step. We return to Figure 7.13, and assume that the metric sum for the two paths arriving at some state A′ is either 25 or 35, where 25 implies that there were 25 similarities between what was received and the test path taken, whilst 35 represents 35 similarities between the received signal and the alternative test path. In addition, the number of similarities (metric increments) for the transition from A′ to A is 0 and from B to A′ is 0. These two paths emerge from A′ as four potential paths (two from A via A′ – assuming that transmitted datum was an 0 or a 1; and two from B via A′ – assuming that the datum was a 0 or a 1). From this point on, it is impossible to distinguish between the two paths entering A′. For this reason, only the path with the largest metric is allowed to continue (the 'survivor path'). Allowing both paths to continue makes no sense as there is always a constant difference in the metric; the path with the highest metric is always chosen.

The benefit of this algorithm is that a decision is taken at each step, significantly reducing the number of potential paths through the trellis, hence reducing the computational effort considerably.

**Soft decisions**

So far, we have considered a decoding strategy that is based on the reception of bits having one of two possible values. In a real system, the received signals may have a range of values, depending upon the degree of quantisation used within the receiver.

The decoding scheme can exploit the multivalued nature of the received signal using what is known as soft decision decoding. With soft decision decoding, soft data are passed by the demodulator to the decoder. The soft data can be derived from a number of sources, but are essentially a measure of the energy of the digital received data.

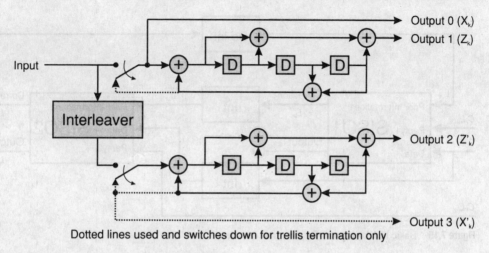

**Figure 7.14** PCCC proposed for the WCDMA system.

In using soft decision decoding, the performance of the channel decoding algorithm can be improved above that obtained with hard decision decoding algorithms.

## 7.3 Turbo codes as used in WCDMA

### 7.3.1 Turbo code origins

Turbo codes were discovered by a team of researchers investigating the soft output Viterbi algorithm (SOVA) and its implementation in VLSI [18]. Based on their intuition, they were trying to replicate from electronics the improvements obtained in an amplifier that used feedback. They started by investigating coding techniques that utilised feedback [19]. Through a process of investigation, the choice of encoder was made, the SOVA was replaced by the MAP algorithm and the decoders were iteratively connected. The rest, as they say, is history.

### 7.3.2 Turbo coding principles

The main parts of a turbo encoder are two encoders and an interleaver. The encoders are systematic, meaning that the input data is transmitted in parallel with the derived parity information. There are a number of different structures that can be used to implement the turbo encoder. For the WCDMA application, a parallel concatenated recursive systematic convolutional encoder is used. Figure 7.14 illustrates the parallel concatenated convolutional code (PCCC) that is proposed for the WCDMA system. The encoder consists of two recursive systematic convolutional encoders separated by

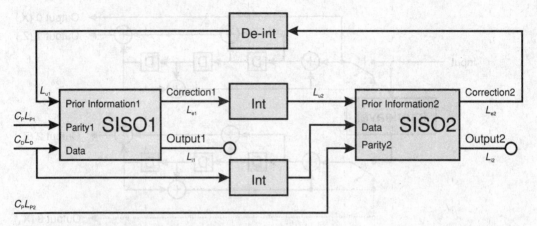

**Figure 7.15** Basic block diagram for the turbo decoder.

an interleaver. The turbo encoder uses a forced trellis termination technique to ensure that the encoder ends in a known state.

### Turbo interleaver

The design of the interleaver is crucial to the performance of the turbo code. Ideally, the optimum performance can be achieved by using a very long random interleaver. Such a scheme would result in a performance which approximates that of long random codes. For the WCDMA system, the interleaver performs intrarow and interrow permutations of a rectangular matrix based on prime number considerations.

### 7.3.3  Turbo decoder

#### Basic structure and operation

Figure 7.15 illustrates the basic structure of a turbo decoder. In this section we explore this structure and its operation, and in the following sections examine in more detail its component parts.

The turbo decoder consists of two soft-in-soft-out decoders (SISO1 and SISO2 in the figure) that are connected together via interleavers and deinterleavers (Int and De-int in the figure). The turbo decoder receives three types of input information: uncoded user data (Data in the figure); and two types of parity information (Parity1 for SISO1 and Parity2 for SISO2).

At the output of SISO1 there is the decoded data output (Output1 in the figure) and a second output (referred to as Correction1 in the figure, but later we will see it is more correctly called extrinsic information). Correction1 is passed from SISO1 to SISO2 via an interleaver where it becomes an input into SISO2 (referred to as Prior Information2 in the figure). The other inputs are the received data (passed via an interleaver) and the Parity2 information. At the output of SISO2 there is the decoded data (Output2 in the figure) and also a second correction term (Correction2 in the figure). This correction term is passed back to the Prior Information1 input of SISO1 via a deinterleaver.

The turbo decoder operates iteratively as follows. A block of received data consisting of Data bits, Parity1 bits and Parity2 bits is passed to the turbo decoder. In the first part of the first iteration the Data and Parity1 are passed to SISO1. Using the Data and Parity1 bits the algorithms within SISO1 attempt to correct for errors in the received data and additionally derive a correction term. Output1 is the decoded data with some errors corrected through the use of Parity1. Correction1 is a second output derived mainly from the parity bits. Correction1 is applied to SISO2 via an interleaver where it becomes Prior Information2. The interleaver is required because an interleaver is used in the transmitter. The other inputs to SISO2 are the Data (via an interleaver) and Parity2 bits.

With the additional information available to SISO2 (Correction1 and Parity2), it should be able to correct additional errors in the received data which is Output2. SISO2 also generates a correction term (Correction2) that is passed back to the input of SISO1 as Prior Information1.

On the second iteration of the decoder, SISO1 uses the same information for the Data and Parity1, but now has new information from SISO2 that can be used to correct for additional errors in the received signal. The decoder passes through a number of iterations. The exact number required depends upon the target performance required.

## SISO decoder – basic structure

The SISO decoder is a type of optimum receiver. There are two well-known types of optimum receiver, MLSE and the MAP summarized by (7.1) and (7.2).

MLSE: $m' = \max_m P[m|r(t)]$  (7.1)

MAP: $m'_i = \max_{m_i} P[m_i|r(t)]$  (7.2)

MLSE, given by (7.1), defines the most likely received sequence as the sequence that maximises the conditional probability of receiving the sequence given some received signal. MAP, given by (7.2), however, defines the most likely received symbol as the one that maximises the conditional probability of receiving the symbol given some received signal.

Both approaches result in an optimum receiver. The MAP algorithm minimises the symbol error rate and MLSE minimises the sequence error rate. For turbo codes, the MAP algorithm is used most in the SISO decoder, even though it is more complex than the equivalent based on MLSE. In this chapter we will only consider the MAP algorithm and its derivatives.

## SISO decoder – soft information

Soft information can take the form of either soft inputs or soft outputs, and for the MAP algorithm can be generated from (7.2). We start by estimating the probabilities that the most likely symbol is a binary 1 and the probability that the most likely symbol was a binary 0. We assume at this stage that we have some methods by which we can estimate these probabilities, and in Section 7.3.4 we will review one such approach. If we then form the ratio of these two probabilities and take the natural logarithm, we have what is referred to as the log likelihood ratio (LLR). The LLR is the soft information at the output of the decoder and is given as $L_i$ in (7.3). In this example $r(t)$ is the input signal that is being decoded. We assume that this signal is a soft information input signal.

$$L_i = \ln \frac{P[m_i = 1 | r(t)]}{P[m_i = 0 | r(t)]} \tag{7.3}$$

A decoder of this form satisfies the basic requirement for a decoder that can be used as the turbo decoder, i.e. it takes soft information at the input and delivers soft information to the output, hence it is a SISO decoder.

Equation (7.3) represents a general form for the estimation of soft output information based on soft input information. In its present form, however, this equation is not suitable for use within the turbo decoder. Below we will modify the equation that is presented, introducing the concept of extrinsic information (the correction term) that was mentioned previously.

## SISO decoder – extrinsic information (the correction term)

Extrinsic information is part of the reason for the performance of turbo codes. Extrinsic information is additional relevant information (in this case prior information) that is used by the turbo decoder. To establish what is extrinsic information, it is beneficial to return to the definition of the LLR and through the use of a mathematical tool referred to as Bayes theorem, modify the form of the LLR.

The LLR expressed by (7.3) is not in a form suitable for application in a practical decoder. The conditional probabilities defined by the equation assume that we have knowledge of the probabilities of $m_i$ given a specific input signal $r(t)$. In reality, the calculation of this probability is not straightforward. It transpires, that using Bayes theorem, we can modify the structure of (7.3) into a form that is easier to implement in a practical decoder. First we will consider Bayes theorem.

## Bayes theorem and log a-posteriori probability (LAPP)

Bayes theorem relates the posterior probabilities (i.e. those observed after some event) to their prior probabilities, i.e. before the event. Using Bayes theorem, it is possible to develop solutions to the optimum receiver problem. First consider Bayes theorem in a form suitable for the optimum receiver problem, this is shown in

$$P(m_i | r(t)) = \frac{P[r(t) | m_i] P(m_i)}{P[r(t)]} \tag{7.4}$$

## 7.3 Turbo codes used in WCDMA

We can apply Bayes theorem to the conditional probabilities defined in (7.3), expressed as a ratio and taking the natural logarithm to obtain an alternative expression for the LLR. The result of this procedure can be seen in (7.5). The first term in (7.5) is referred to as the likelihood, and the second term is referred to as the prior information. In a non-turbo decoder environment, we would likely assume that the prior information as expressed in (7.5) was zero as we would assume that a bit equal to a binary one is as likely as a bit equal to a binary zero. In our turbo decoder, however, we have access to this additional prior information and consequently this second term remains.

$$L_i = \frac{P[r(t)|m_i = 1]}{P[r(t)|m_i = 0]} + \ln \frac{P(m_i = 1)}{P(m_i = 0)} \tag{7.5}$$

Next, we can represent the logarithm of the ratios of the probabilities in terms of received signals obtained by the receiver. It can be shown [42] that the quantities defined in (7.5) can be replaced directly by the received signals and the prior information obtained (in the case of the turbo decoder) from a second decoder.

Equation (7.6) presents the LAPP in terms of the received signals that were illustrated in Figure 7.15 excluding the parity information that will be considered next:

$$L_i = C_D L_D + L_u \tag{7.6}$$

where $C_D L_D$ is the likelihood (channel reliability × received signal) and $L_u$ is the prior information.

**Derivation of extrinsic information**

Now we can address how the extrinsic information is defined. It has been shown that the LAPP output $L_i$ [42] for an iterative decoder of the form shown in Figure 7.15 can be expressed in a slightly modified form:

$$L_i = C_D L_D + L_u + L_e \tag{7.7}$$

The new term $L_e$ is referred to as the extrinsic information. It is additional reliability information that is obtained from within the decoder and is a component part of $L_i$. $L_e$ is found by solving (7.7):

$$L_e = L_i - C_D L_D - L_u \tag{7.8}$$

Referring to Figure 7.15, $L_e$ is the correction term that is passed to the following decoder (via an interleaver) where it becomes the prior information ($L_u$). The decoder operates, therefore, by decoding the data and deriving $L_i$. From $L_i$ and the other inputs $C_D L_D$ and $L_u$, $L_e$ is estimated. $L_e$ is then passed to the following decoder.

The final thing that we need to consider, is how we estimate $L_i$ in the first place. To do this, we will consider the operation of the MAP algorithm in the next section.

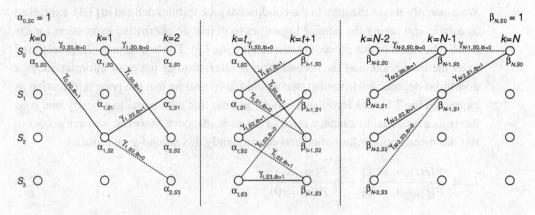

**Figure 7.16** Trellis diagram for a simple turbo encoder.

### 7.3.4 MAP algorithm symbol by symbol decoding

To generate the soft output information, we consider the operation of the MAP algorithm in some detail. To do so, we need to understand the operations on the trellis that define the operation of the encoder. In this example, we consider a simple trellis with only four states. For WCDMA, the turbo encoder is more complex, but the analysis can be easily extended to include the additional states.

Based on forward and backward analysis of the trellis, we need to define three quantities: $\alpha$, $\beta$ and $\gamma$. For each point in the trellis, $\alpha$ and $\beta$ have a defined value. For a specific state (e.g. $S_0$) for a specific time step (e.g. $k = 3$) $\alpha$ has a value that defines the probability that the coder is in that state at that point in time. Therefore, $\alpha$ defines the forward path encoder state probabilities and it can be calculated recursively, starting at the beginning of the trellis moving forward in time. $\beta$ is also an encoder state probability, but is defined from the end of the trellis and estimated working backwards in time. $\gamma$ is defined as the state transition probability, i.e. the probability that the coder will move from some state at time $k = t$ to some other allowed state at time $k = t + 1$.

**Recursive calculation of $\alpha$**

Having defined $\alpha$, $\beta$ and $\gamma$, we now wish to proceed to calculate them. First, for $\alpha$, we start at state $S_0$ where $\alpha_{0S_0}$ the probability of the encoder (this is $\alpha$ at time step $k = 0$ being in state $S_0$) is known to equal 1. We know that the probability is 1 as the encoder started in this state.

From Figure 7.16 we can see that there are two non-zero values for $\alpha$ at time $k = 1$, and these can be calculated from:

$$\alpha_{1,S_0} = \alpha_{0,S_0} \cdot \gamma_{0,S_0,Ib=0}$$

$$\alpha_{1,S_2} = \alpha_{0,S_0} \cdot \gamma_{0,S_0,Ib=1}$$

That is, the probability of the coder being in some state at time $k = 1$ is the probability that the coder was in the previous state at time $k = 0$ multiplied by the probability of the transition between these states for the specific input bit ($Ib$; with $Ib = 0$ or $Ib = 1$). Once the trellis is fully expanded, the expression for $\alpha$ needs to take into account both the transitions arriving at a given state that correspond to an input bit being a 1 or a 0.

Observing the trellis in Figure 7.16, we see that the general expression for $\alpha$ at state $S_0$ becomes:

$$\alpha_{t,S_0} = \alpha_{t-1,S_0} \cdot \gamma_{t-1,S_0,Ib=0} + \alpha_{t-1,S_1} \cdot \gamma_{t-1,S_1,Ib=1} \tag{7.9}$$

Following this simple forward recursion formula, the probability that the coder is in a specific state can be calculated for all states in the trellis based on the state transition probabilities $\gamma$.

## Recursive calculation of $\beta$

Start at state $S_0$, where $\beta_{N,S_0}$ is known to equal 1. At time $k = N - 1$, there are two non-zero values for $\beta$ and these can be calculated from:

$$\beta_{N-1,S_0} = \beta_{N,S_0} \cdot \gamma_{N-1,S_0 Ib=0}$$

$$\beta_{N-1,S_1} = \beta_{N,S_0} \cdot \gamma_{N-1,S_1 Ib=1}$$

That is the probability of the coder being in some state at time $k = N - 1$ is the probability that the coder was in the next state at time $k = N$, multiplied by the probability of the transition between these states for the specific input bit. Following this simple backwards recursion formula, the probability that the coder is in a specific state can be calculated for all states in the trellis, based on the state transition probabilities $\gamma$.

Observing the trellis in Figure 7.16, we see that the general expression for $\beta$, including both trellis transitions for $k = t$ at state $S_0$ can be found. The general expression for Beta becomes:

$$\beta_{t,S_0} = \beta_{t+1,S_0} \cdot \gamma_{t,S_0,Ib=0} + \beta_{t+1,S_2} \cdot \gamma_{t,S_0 Ib=1} \tag{7.10}$$

Following this simple backward recursion formula, the probability that the coder is in a specific state can be calculated for all states in the trellis based on the state transition probabilities $\gamma$.

### 7.3.5 Probabilities

**State probability**

Using the recursive equations (7.9) and (7.10), two definitions for the probability of the encoder being in a specific state are derived. The state probability estimate based on $\alpha$ uses the received information (obtained through $\gamma$) in the first part of the trellis, while the probability based on $\beta$ uses the received information in the later part of the trellis.

## Bit probability

For any specific state in the trellis (state $S_i$ at time $k = t$), we can now define the probability that the input bit was a 1 and also the probability that the input bit was a 0. Starting with the probability that the input bit was a 1:

$$P(Ib = 1)_{t, S_i}$$

We can define this probability as the probability of being in that state multiplied by the transition probability from that state (with $Ib = 1$) multiplied by the probability of being in the state at the end of the transition (calculated from the end of the trellis)

$$P(Ib = 1)_{t, S_i} = \alpha_{t, S_i} \cdot \gamma_{t, S_i, Ib=1} \cdot \beta_{t+1, (S_i | Ib=1)} \quad (7.11)$$

In (7.11) $(S_i | Ib = 1)$ means the state that the encoder would be in if it started from state $S_i$ and the transition corresponded to an input bit being a binary 1. Similarly, for the same state the probability that the input bit was 0 can be estimated from:

$$P(Ib = 0)_{t, S_i} = \alpha_{t, S_i} \cdot \gamma_{t, S_i, Ib=0} \cdot \beta_{t+1, (S_i | Ib=0)} \quad (7.12)$$

For a specific time step ($k = t$) we can now estimate the probability that the input bit was a 1 at that time step by summing the individual state probabilities corresponding to a transition with an input bit equal to 1. Similarly, the same can be done for the state probabilities corresponding to an input bit equal to 0.

We can define these two probabilities as:

$$P_{t, Ib=1}$$

and

$$P_{t, Ib=0}$$

The LLR is defined as the logarithm of the ratio of these two quantities:

$$L_{i,k} = \log \left[ \frac{P(Ib = 1)_k}{P(Ib = 0)_k} \right] \quad (7.13)$$

This represents the soft output information from the decoder at trellis step $k$.

## Return to turbo decoder

We now need to tie the MAP algorithm into the turbo decoder. We start by defining $\gamma$, the transition probability or branch metric, as:

$$\gamma_{k, S_i, Ib} = \exp \left[ \frac{1}{2} (L_{u,k} Ib'_k + C_D L_{D,k} + C_P L_{P,k} Ob'_k) \right] \quad (7.14)$$

where $L_{u,k}$ is the prior information at time step $k$ input into the decoder (this is either $L_{u_1}$ or $L_{u_2}$ in Figure 7.15), $Ib'_k$ is the input bit at time step $k$ converted into a polar signal ($Ib'_k = 1$ if $Ib = 0$ and $Ib'_k = -1$ if $Ib = 1$). $C_P$ and $C_D$ are the channel reliability

## 7.3 Turbo codes used in WCDMA

information given by $E_b/N_o$. The received datum at time step $k$, is $L_{D,k}$, the received parity information at time step $k$ (Parity1 or Parity2 shown in Figure 7.15, depending upon which SISO is being considered) is $L_{P,k}$ and the output bit that would result from the input bit at time $k$ is $Ob'_k$.

### Extrinsic information

In Section 7.3.3 (7.7) gave an expression for $L_i$ in terms of the received data and the extrinsic information $L_e$. In (7.8) we defined an expression for $L_e$. To obtain $L_e$ and hence the prior information for the next decoder we must first obtain $L_i$. We can obtain $L_i$ from (7.13) by performing the trellis processing steps outlined above:

$$L_{i,k} = L_{u,k} + C_D I_{D,k} + L_{e,k}$$

where $L_{e,k}$ is the extrinsic information from the current coder that is derived from the parity information. $L_{e,k}$ is passed through to the second encoder where it becomes $L_{u,k}$. $L_{e,k}$ is found from:

$$L_{e,k} = L_{i,k} - C_D L_{D,k} - L_{u,k}$$

### Turbo coder operation

With reference to Figure 7.15, the turbo decoder can now operate as follows:
1. Data, Parity1 and Parity2 blocks are received by the turbo decoder.
2. Data and Parity1 are passed into SISO1, prior information $L_{u1}$ set initially to zero.
3. $L_i$ are calculated in SISO1 for each received bit using (7.13) with the trellis processing stages as described.
4. For all the bits received for Data and Parity1, $L_{e1}$ is calculated using equation (7.8).
5. $L_{e1}$ is passed through the interleaver and to SISO2 where it becomes the Prior Information2 ($L_{u2}$).
6. Parity2 bits and Data (via the interleaver) are also passed into SISO2.
7. $L_i$ for SISO2 is calculated using (7.13) and $L_{e2}$ calculated from $L_i$ using (7.8).
8. $L_{e2}$ is passed through a deinterleaver where it becomes $L_{u1}$, and is passed into SISO1 as Prior Information1 along with the Data and Parity1 that were used in the previous iteration.
9. The process can now repeat iteratively until halted at which point the decoded data can be recovered.

### 7.3.6 Reduced complexity algorithms

The main issue with the MAP algorithm is its algorithmic complexity. The trellis processing steps require arithmetic calculations involving the exponentials of some received quantity and some stored quantities. These processing steps introduce numerical precision issues. For fixed-point implementation, it is likely that the precision

# Symbol rate processing functions

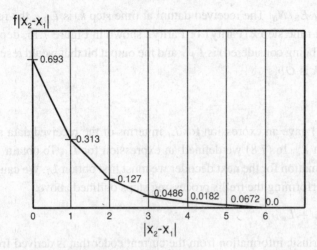

**Figure 7.17** Log MAP correction term.

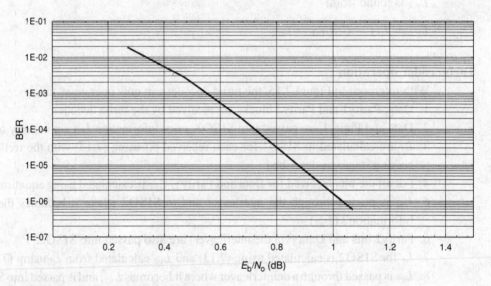

**Figure 7.18** BER for a 128-kb/s data signal.

will be an issue that needs addressing. To help overcome these implementation concerns, two reduced complexity algorithms are considered that bypass the calculations of those exponentials and lead to computationally efficient methods of implementing the decoder. The first algorithm to be considered is called the max-log MAP algorithm and the second the log MAP algorithm.

## 7.3.7 Log MAP and max-log MAP

When calculating $L_i$ the trellis calculations will involve expressions of the type $\log(e^{x_1} + e^{x_2})$. From an implementation perspective, evaluating expressions of this type

## 7.3 Turbo codes used in WCDMA

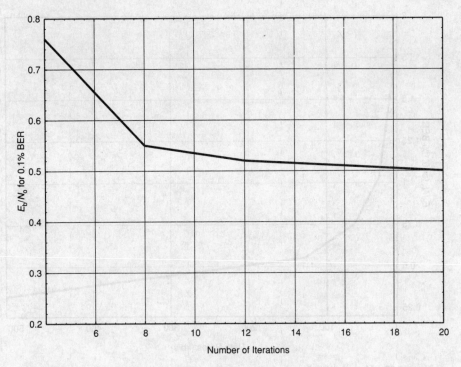

**Figure 7.19** $E_b/N_o$ required for a BER of 0.1% for variations in the number of decoder iterations.

consumes a large amount of processing time and also can lead to stability problems with the algorithms.

To overcome these limitations a number of approximations can be used that simplify the arithmetic but at the expense of accuracy in the final result. The two approximations that we will examine here are the max-log MAP algorithm and the log MAP algorithm. To examine these algorithms we will start by expressing the log of the sum of two exponentials in a slightly different form:

$$\log(e^{x_1} + e^{x_2}) = \max(x_1, x_2) + \log(1 + e^{-|x_1 - x_2|}) \quad (7.15)$$

The first approximation is for the max-log MAP algorithm, which reduces (7.15) to an approximate form represented by.

$$\log(e^{x_1} + e^{x_2}) \approx \max(x_1, x_2) \quad (7.16)$$

The second approximation is referred to as the log-MAP algorithm which improves on the approximation by adding a correction term:

$$\log(e^{x_1} + e^{x_2}) \approx \max(x_1, x_2) + f(|x_1 - x_2|) \quad (7.17)$$

The function in (7.17) can be implemented through the use of a look-up table. Figure 7.17 presents the seven values that would be required in such a look-up table.

**246** Symbol rate processing functions

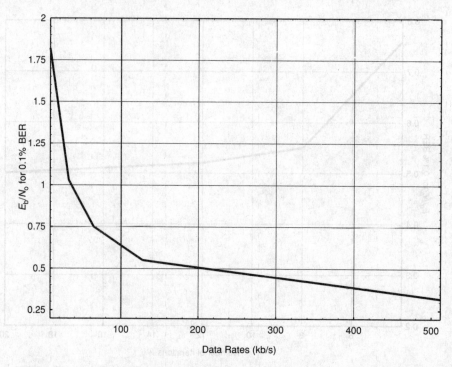

**Figure 7.20** How the $E_b/N_o$ required to achieve a BER of 0.1% changes as the data rate coming out of the encoder changes.

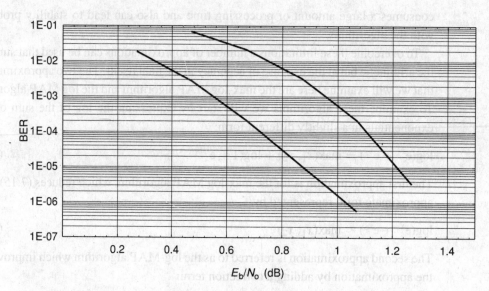

**Figure 7.21** How the max-log MAP algorithm compares with the MAP algorithm: on the left the MAP algorithm; on the right the max-log MAP algorithm.

## 7.4 The performance of the WCDMA turbo code

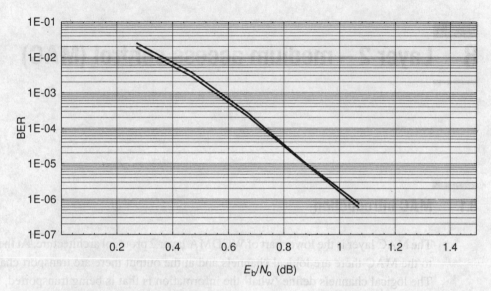

**Figure 7.22** How the log-MAP algorithm compares with the MAP algorithm: on the left the MAP algorithm; on the right the log-MAP algorithm.

## 7.4 The performance of the WCDMA turbo code via examples

### 7.4.1 MAP turbo decoder performance

Figure 7.18 shows the BER for a 128 kb/s data signal coming out of the decoder with a 10 ms interleaving depth and using the MAP algorithm. Figure 7.19 shows the $E_b/N_o$ required for a BER of 0.1% with a data signal coming out of the encoder at 128 kb/s, a 10 ms interleaving depth and the MAP algorithm. The $x$ axis shows the number of iterations. Figure 7.20 shows how the $E_b/N_o$ required to achieve a BER of 0.1% changes as the data rate coming out of the encoder changes. The decoder used the MAP algorithm.

**MAP and max-log MAP turbo decoder performance**

Figure 7.21 shows how the max-log MAP algorithm (upper right curve) compares with the MAP algorithm (lower left) for a data rate of 128 kb/s and a 10 ms interleaving depth. From the curves, we can see a slight degradation in performance for the max-log MAP compared with the MAP algorithm.

Figure 7.22 shows the how log MAP algorithm (upper right curve) compares with the MAP algorithm (lower left) for a data rate of 128 kb/s and a 10 ms interleaving depth. From the curves, we can see that there is little degradation using the log MAP algorithm when compared with the MAP algorithm.

# 8 Layer 2 – medium access control (MAC)

## 8.1 MAC introduction

The MAC layer is the lower part of WCDMA layer 2 protocol architecture. At the input to the MAC there are logical channels and at the output there are transport channels. The logical channels define 'what' the information is that is being transported, whilst the transport channels define 'how' the information is transported. One of the prime functions of the MAC, therefore, is to map a specific logical channel onto the appropriate transport channel. We will see later that this mapping can change dynamically as the characteristics of the network or the user vary (for instance due to a rise or fall of the loading in the network).

This chapter starts by considering the logical channels that enter the MAC, their structure and uses, and also transport channels that exit the MAC, their structure and uses. When we consider the architecture of the MAC, we see that it comprises a number of component parts to reflect its distributed nature within the UTRAN. The functions and services provided by the MAC include items such as random access procedure and transport format selection control, and the mapping and switching between logical and transport channels. To fully understand the function of the MAC we then explore in greater detail some of the key operations it provides, including the random access procedure, the control of CPCH and the TFC selection in the uplink within the UE. We start this first section with an introduction to the logical channels that the MAC provides for the transportation of higher layer data, and review the transport channels considered in earlier chapters.

### 8.1.1 Logical channels

Figure 8.1 illustrates the different types of logical channels that are defined to operate across the radio interface of the WCDMA system. In general, the logical channels are separated in two distinct ways. First, there is the differentiation as to whether the logical channel is utilised by many mobiles (referred to as common logical channels) or whether it is utilised by a single mobile (a dedicated logical channel). In general in a cell, there are relatively few common logical channels shared by the users in the cells,

## 8.1 MAC introduction

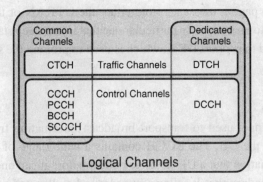

**Figure 8.1** Definition of different types of logical channel.

but there are a large number of dedicated logical channels, typically more than one per user that is active in the cell.

In addition to the definition of logical channels in terms of the likely number of users, there is also differentiation of the channels in terms of what the channels are used for. In general, the logical channels are divided into traffic channels and control channels. Traffic channels are logical channels that are used for the transfer of user plane information such as speech or packet data (for instance from an Internet connection). Control channels are the common or dedicated logical channels used for the transfer of signalling or control messages. These control channels are used to control the radio interface or user services that are transported across the radio interface. Examining the different types of channel in Figure 8.1, we see that there are common traffic channels as well as dedicated traffic channels and equally there are common control channels as well as dedicated control channels.

### 8.1.2 Common control logical channels

**Common control channel (CCCH)**

CCCH is the name given to a general purpose logical channel used by mobiles in a cell when they do not have any cell specific logical channels assigned to them. The common control logical channel is a bi-directional channel, i.e. it is used by both the UE and the UTRAN.

The CCCH is used, for instance, when a UE is requesting the establishment of an RRC connection between the UE and the UTRAN or alternatively as part of a cell update procedure when a UE that does not have a dedicated physical connection moves into a new cell and needs to perform a cell update procedure.

**Paging control channel (PCCH)**

PCCH is the name given to the channel that carries the paging messages from the UTRAN to the UE. The PCCH is a uni-directional channel, i.e. it is only available on the downlink.

The PCCH is used to carry paging messages from either the UTRAN or the CN. The paging messages are used to alert a UE, in particular one that is in a sleep state. The PCCH operation was designed to support DRX, which is used to prolong the standby time of a UE.

**Broadcast control channel (BCCH)**

The BCCH is the logical channel used to transport broadcast information from the network (UTRAN or CN) to the UE. The BCCH contains a wide range of system information such as the information that a UE needs to collect before it can undertake the initial random access procedure that it follows in order to be allocated an RRC connection.

**Shared channel control channel (SHCCH)**

The SCCCH is a shared bi-directional logical channel used to transport control information between the UE and the network. The SHCCH is only used within the TDD mode of operation.

### 8.1.3 Dedicated control logical channels

**DCCH**

The DCCH is the logical channel assigned to a UE when that UE is allocated an RRC connection. The DCCH is a point-to-point bi-directional service allocated on both the uplink and the downlink. DCCHs are assigned as part of the initial random access process that results in the establishment of the RRC connection. They are used to carry all of the signalling messages between the UE and the network (either UTRAN or CN). These signalling exchanges are for situations such as the establishment of a NAS service, or alternatively, for use by the RRC, e.g. to put a UE into soft-handover.

We will see later in this section that a UE is typically allocated four DCCHs. Two of them are used to carry UTRAN specific signalling messages (part of SRB1 and SRB2) and two are used to transport CN specific signalling messages (part of SRB3 and SRB4).

### 8.1.4 Common traffic logical channels

**Common traffic channel (CTCH)**

Currently, there is only one logical channel transporting common traffic information, which is the CTCH. The CTCH is a uni-directional channel (downlink) used for broadcast type services, where the information is destined for a number of UEs in a cell (point-to-multipoint). These broadcast services could be for cell broadcast services, or alternatively they could be used for services such as IP-Multicast.

### 8.1.5 Dedicated traffic logical channels

**DTCH**

The DTCH is probably the main logical channel as far as a specific UE is concerned. The DTCH is a point-to-point bi-directional logical channel used to transport all of the user information such as speech, video and data. A UE can have a number of DTCHs active at the same time, one for speech, one for video and one for a packet switched connection to the Internet, for example. Each of the DTCHs. are established with characteristics (QoS) that match the specific requirements of that service.

### 8.1.6 Transport channels

The transport channels have been considered to some degree in Chapters 4 and 7. Like the logical channels, the transport channels can be considered as being of common type, shared type or dedicated type.

There are a number of common transport channels. The CPCH is an uplink channel used to carry control and user plane data. The RACH is another uplink transport channel used to carry control and user plane data. The FACH is a downlink channel used to carry control and user plane data, and the PCH is a downlink transport channel used to carry paging messages on the downlink. The broadcast channel (BCH) is a downlink channel used to carry broadcast messages. The shared transport channels include the DSCH used to carry data and control, and which is shared between users in a cell. The uplink shared channel (USCH) is an uplink channel used in the TDD mode only to carry data and control. Finally, there is the DCH, which carries data and control in both the uplink and the downlink. Chapter 4 describes how these different transport channels relate to the physical channels.

## 8.2 MAC architecture

The MAC comprises a number of different elements needed to manage the different types of transport channel. The basic architecture for the MAC-c/sh and MAC-d in the UE is shown in Figure 8.2 and that for the UTRAN in Figure 8.3. The MAC also includes elements to manage broadcast functions (seen in Figure 8.11). Additionally, there is a fourth MAC element, the MAC-hs, which is defined to manage the high speed downlink packet access (HSDPA) functions that are defined as part of R5.

The MAC shown in both Figure 8.2 and Figure 8.3 illustrates how it is divided into two main functions: the MAC-c/sh (MAC – common or shared) and the MAC-d (MAC – dedicated). The MAC-c/sh in both the UE and the UTRAN is there to manage the operation of the MAC for the common transport channels. The MAC-d in both the UE and the UTRAN is there to manage the operation of the MAC for the

## 252 Layer 2 – medium access control (MAC)

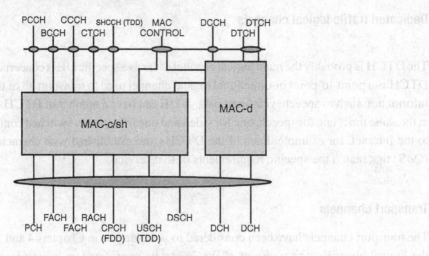

**Figure 8.2** MAC architecture on UE side.

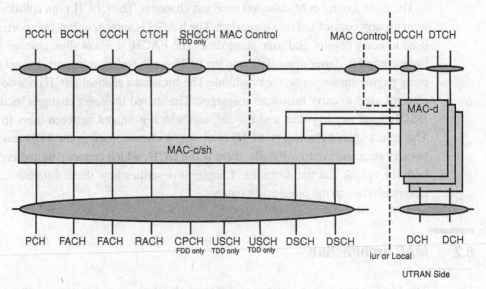

**Figure 8.3** MAC architecture on UTRAN side.

dedicated transport channels. As we will see, part of the interaction between the MAC-c/sh and the MAC-d is to allow dedicated logical channels (e.g. DTCH or DCCH) to be transported using a common transport channel (e.g. RACH in UL or FACH in DL). To facilitate this type of operation, the MAC-c/sh and the MAC-d are interconnected, either directly as is the case in the UE, or possibly via the interface between RNCs (the Iur interface) in the UTRAN.

In the following sections we examine the structure of the MAC-c/sh and the MAC-d in both the UE and the UTRAN.

## 8.2 MAC architecture

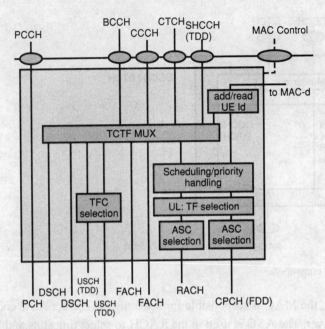

**Figure 8.4** MAC-c/sh structure in UE.

### 8.2.1 MAC-c/sh structure on the UE side

Figure 8.4 illustrates the MAC-c/sh entity on the UE side. The target channel type field (TCTF) multiplex (MUX) entity shown in the diagram is responsible for managing the TCTF. The TCTF entity is responsible for detecting, inserting and deleting the TCTF in the MAC protocol data unit (PDU) header. The TCTF field defines the mapping between the logical and transport channels for a specific PDU. A specific transport SAP (FACH for instance) has bits in the TCTF field to indicate whether the PDU is going to (or from) a BCCH, CCCH, CTCH or DCCH/DTCH logical channel SAP.

Figure 8.5 illustrates the mapping applied for different logical channels to transport channel mappings. The length of the TCTF bits depends upon the first two bits. TCTF equal to 00 or 11 indicates a mapping between BCCH or DCCH/DTCH and the FACH respectively. The longer TCTFs starting 01 and 10 indicate a mapping to the CCCH and CTCH respectively. For the uplink (via the RACH), TCTF 00 corresponds to a CCCH mapping and 01 a DCCH/DTCH mapping.

The MAC-c/sh entity in the UE uses the TCTF bits to identify which logical channel is being carried on the received FACH transport channel and also to indicate to the UTRAN which logical channel is being transported on the RACH transport channel. The transport format (TF) selection entity shown in Figure 8.4 is responsible for selecting the TF for the RACH and CPCH transport channels. For the CPCH channel, the TF is selected from those indicated to be available in the CSICH.

## Layer 2 – medium access control (MAC)

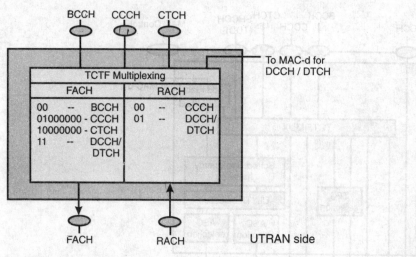

**Figure 8.5** TCTF field contents.

In ASC selection, the MAC is responsible for selecting the correct ASC and notifying the physical layer. The ASC is used in the RACH to select timeslots and access preambles for use in the random access procedure. The scheduling and priority handling function refers to the control of the flow of data through the MAC-c/sh to ensure that the relative priority of the services is observed and services with higher priority given preferential access to the transmission resources of the uplink common channels.

### 8.2.2 MAC-c/sh structure on the UTRAN side

Figure 8.6 illustrates the MAC-c/sh entity on the UTRAN side. In a similar way to the UE, the TCTF mapping function in the UTRAN is responsible for indicating the association between transport channels (e.g. FACH) and logical channels (e.g. BCCH, CCCH and CTCH). The scheduling and priority handling functions are used to manage the use of limited resources between users or by a single user.

TFC selection requires the MAC to choose the appropriate TFC for the channel in question. The downlink code allocation is to indicate the spreading code for the downlink-shared channel that is associated with a specific DCH. The flow control and link to the MAC-d entity is there to manage the data exchanges between the MAC-c/sh and MAC-d. The flow control is required in the UTRAN because the MAC-d and MAC-c/sh could be in different RNCs. The flow control function ensures that the data flow does not cause buffer overflow.

### 8.2.3 MAC-d structure on the UE side

Figure 8.7 illustrates the MAC-d structure on the UE side. The channel switching function provides a dynamic transport channel switching based on decisions made by

## 8.2 MAC architecture

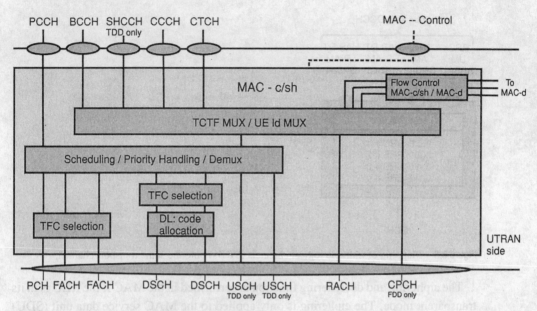

**Figure 8.6** MAC-c/sh architecture in the UTRAN.

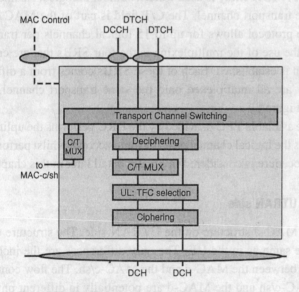

**Figure 8.7** MAC-d structure on the UE side.

the RRC layer. This function allows the RNC to change the transport channel (and hence the physical channel) currently used by the logical channel to a different transport channel. This switching is most likely to occur as a consequence of a change in the loading on a specific cell in the network. The MAC is there to manage the change, applying new TFs defined for the new transport channels.

## Layer 2 – medium access control (MAC)

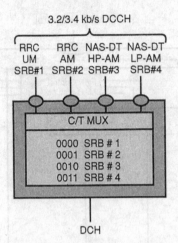

**Figure 8.8** Mapping of multiple logical channels (SRBs) onto the same transport channel.

The ciphering and deciphering function is performed in the MAC if the RLC is in its transparent mode. The ciphering is only applied to the MAC service data unit (SDU) (i.e. the RLC PDU), the MAC header is not ciphered.

The control/traffic (C/T) MUX entity is used when a number of logical channels are multiplexed onto a single transport channel. The C/T field is part of the MAC PDU header (4 bits long). The protocol allows for up to 15 logical channels per transport channel. An example of the use of the multiplexing is the four SRBs that are created when an RRC connection is established. Each of these SRBs comes from a different logical channel, but they are all multiplexed onto the same transport channel. This situation is illustrated in Figure 8.8.

The MAC based on the available TFC provided by the RRC performs the uplink TF selection. The MAC takes the logical channel priorities into account whilst performing the TF selection. This procedure is considered in greater detail later in this chapter.

### 8.2.4 MAC-d structure on the UTRAN side

Figure 8.9 illustrates the MAC-d structure on the UTRAN side. The structure of the MAC-d is essentially the same as in the UE. The only differences are the inclusion of flow control functions between the MAC-d and the MAC-c/sh. The flow control is required because the MAC-c/sh and the MAC-d are potentially in different physical entities connected via an asynchronous transfer mode (ATM) link. The flow control regulates the flow of data between the MAC-c/sh and the MAC-d entities.

### 8.2.5 MAC-b structure in the UTRAN or UE

Figure 8.10 illustrates the structure of the MAC-b function within the UE and the UTRAN. Unlike the other MAC functions (except the MAC-hs) the MAC-b function is located in the Node-B.

## 8.3 MAC functions and services

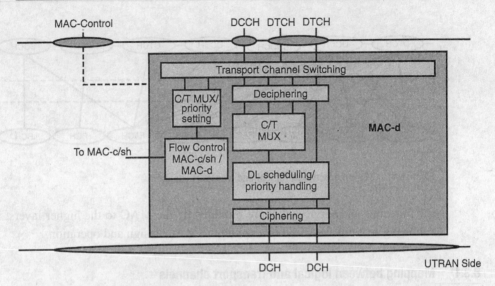

**Figure 8.9** MAC-d structure on the UTRAN side.

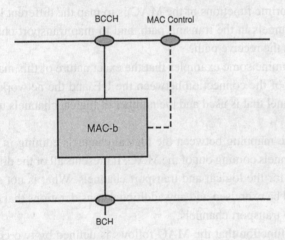

**Figure 8.10** MAC-b structure in the UTRAN or UE for BCCH.

The BCCH broadcasts a number of system information messages, some of which can be changing rapidly based on measurements in the Node B. By placing the broadcast functions in the Node B the MAC-b layer can respond rapidly to these dynamic changes.

## 8.3 MAC functions and services

As we saw in the introduction, the MAC performs the mapping between the logical and the transport channels. In addition, the MAC also provides for functions such as TFS and the priority handling of data flows. In this section we examine in greater detail

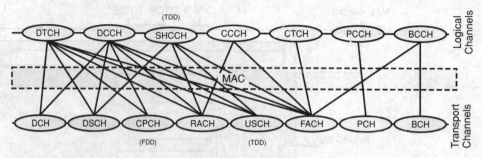

**Figure 8.11** Mapping between logical and transport channels.

the functions and services that are provided by the MAC to the higher layers. In the following sections we will then explore the MAC design and operation.

### 8.3.1 Mapping between logical and transport channels

In Section 8.2 we introduced the different logical channels that are defined for the radio interface. One of the prime functions of the MAC is to map the different logical channels onto transport channels in the transmit path, and to map transport channels onto the logical channels in the receive path.

We will see when we examine some examples that the exact nature of this mapping depends upon the specifics of the connection between the UE and the network, such as the type of physical channel that is used and the number of logical channels using a specific transport channel.

Figure 8.11 illustrates the mapping between the logical channels coming in to the MAC and the transport channels coming out of the MAC. It presents all of the different mapping options that exist for the logical and transport channels. What is not shown is when such options would be exercised and any additional considerations that define the mapping of logical onto transport channels.

In general, the mapping function that the MAC follows is defined by two criteria: the type of physical connection present and the type of information to be transported across the radio interface. If the UE has no active physical connection (CELL_PCH state, URA_PCH state and idle mode) it is still required to monitor paging and broadcast channels. When the UE has an active physical connection it can be one of two general types: a connection via a common physical channel (CELL_FACH state) or a connection via a dedicated physical channel (CELL_DCH state). If the connection is via a common physical channel this restricts the transport channels to those that use the common physical channels, with a similar consideration for the dedicated physical connection. The type of information defines what type of logical channel is to be used. User data travel via a DTCH, dedicated control information via a DCCH and common control information via a CCCH. A UE that is in a low power state and using DRX receives broadcast information via the BCCH and paging requests via the PCCH. We

## 8.3 MAC functions and services

will see later that the UE can have multiple DCCHs and DTCHs depending upon the services that the UE has active at the time.

### 8.3.2 TFS

The next major function provided by the MAC is that of TFS. The main objective of TFS is to match the offered traffic in the MAC to the radio resources available. TFS requires the MAC to continuously examine the data that is being offered by the higher layers, selecting the data based on the capacity of the physical resources that have been allocated to the UE. To achieve this, the MAC is constantly told how many data are available in the RLC, and it also has a set of rules which it uses to decide how many data to take from each logical channel (the rules are part of the TFC procedures signalled from the UTRAN to the UE). In this way, the MAC ensures that the physical resources available via the physical layer are used as efficiently as possible.

### 8.3.3 Priority handling of data flows

As part of the TFS procedure described above, the MAC needs to respond to the different priorities of the services that are being multiplexed onto the physical layer. To facilitate this, each of the services transported via the MAC has a priority (MAC logical channel priority), and the MAC utilises this priority information when deciding how many data to take from each of the services.

In total eight priorities are defined for the MAC services. It can be expected that real-time services such as speech and video will be allocated a high priority (1 is the highest priority), whilst services such as packet data will be allocated a lower priority.

### 8.3.4 UE identification management

A UE transmitting or receiving a dedicated transport channel is identified through the physical layer (the spreading and scrambling codes used by that UE). Common transport channels such as RACH/CPCH in the uplink or FACH in downlink, however, are shared by a number of UEs within a specific cell. Because these channels are transmitted to a number of UEs at the same time, it is necessary to indicate which UE is the intended recipient of a specific message on the downlink and the originator on the uplink. To achieve this, UE identification information is inserted by the MAC into the transport channel that is sent to a group of UEs. On the downlink the MAC needs to decode the UE identification to ascertain whether that specific message is intended for that UE, and on the uplink the MAC inserts the appropriate identification in the MAC PDU.

### 8.3.5 Measurements

To ensure that the operation of the radio interface remains within defined limits the MAC reports to the RRC on traffic volume estimates made for the different services that may be active at the time. The MAC in the UE can achieve this by monitoring the RLC buffer occupancy, compiling statistics and reporting these statistics back to the UTRAN.

If buffers in the RLC start to fill, this implies that there is insufficient capacity for the specific RBs assigned for these services. Once the RRC has reported on the traffic measurements (either using a periodic measurement or via an event triggered measurement report) then the UTRAN can respond by modifying the characteristics of specific RBs.

### 8.3.6 Ciphering (transparent RLC mode (TM RLC) only)

Ciphering or encryption is one of the procedures performed in the MAC, but only under a specific circumstance, i.e. when the RLC is operating in the transparent mode. Encryption is performed on individual logical channels, and so the MAC will only encrypt the logical channels that correspond to a TM RLC. The encryption is only applied to the MAC SDU (i.e. the RLC PDU) and so the MAC header information is not encrypted.

### 8.3.7 ASC selection

On the uplink the UE can access the network via the random access channels. By random access we mean that there can be a number of UEs attempting to make simultaneous access attempts using the channels. A consequence of these simultaneous access attempts is collisions, which result in a reduction in the capacity of the channels.

To attempt to reduce the collisions, the system structures how and how often a specific UE can use these RACHs. The quantity that defines how the UE can access the RACH transport channels is called the ASC. The MAC, therefore, is responsible for controlling the process that the UE follows to select the ASC and hence how the UE accesses the network via these RACHs. The exception to this is the initial random access used as part of an RRC connection establishment procedure. In this case the ASC is defined by the RRC based on broadcast information in that cell.

### 8.3.8 Transport channel switching

This final function of the MAC requires the MAC to control the switching of the logical channels between different transport channels. As the load on the system varies, and

## 8.4 MAC PDUs and primitives

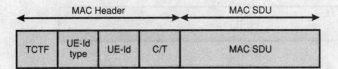

**Figure 8.12** MAC PDU structure.

as the load within the UE varies, the optimum selection of the transport channel (and hence physical channel) may change. Under the instruction of the network, therefore, the MAC can change from using common transport channels (CELL_FACH state) to using dedicated transport channels (CELL_DCH state). The MAC is responsible for administering this change and the changes that are required within the transport channel combining functions to facilitate the mapping of the logical channel from one transport channel (e.g. a common transport channel) to a different transport channel type (e.g. a dedicated transport channel).

## 8.4 MAC PDUs and primitives

There is only one MAC PDU defined and it is illustrated in Figure 8.12. In the diagram it is shown that the MAC PDU comprises a MAC SDU and the MAC header. The MAC SDU is the RLC PDU to be transported within the transport channel. The MAC header has three main fields: the TCTF, the UE-Id field and the C/T field.

The TCTF is used to define the logical channel type that is carried within the MAC SDU. The TCTF field is used to help route the PDUs between the correct SAPs. The TCTF provides the mapping function between the logical channels and common transport channels.

The C/T field defines the logical channel Id within a specific transport channel. Through the use of the C/T field, up to 15 logical channels can be multiplexed onto the same transport channel. An example of this was shown in Section 8.2.3.

The UE-Id is the identity of the UE. There are two types of identity that might be assigned: a c-RNTI and a UTRAN radio network identity (u-RNTI). The c-RNTI is normally associated with connections to a controlling RNC (CRNC) and the u-RNTI with connections to the serving RNC (SRNC).

### 8.4.1 MAC primitives

Figure 8.13 illustrates the elements of the layer to layer protocol associated with the MAC. These are the MAC primitives. There are MAC primitives between the MAC and the RRC layer that are used for configuration and control of the MAC. There are also primitives between the MAC and the RLC layer.

## Layer 2 – medium access control (MAC)

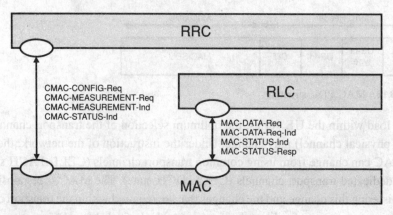

**Figure 8.13** MAC layer to layer protocol.

In Tables 8.1 and 8.2, we examine the primitives between the MAC and the RLC and the MAC and the RRC. Table 8.1 defines the primitives between the MAC and the RLC. In general the MAC-DATA-Req/Ind primitives are exchanged per TTI per logical channel. The idea of these primitives is to allow the data transmitted/received to be passed between the layers. The primitives include information on the current buffer occupancy in the RLC and the amount of data passed to the physical layer by the MAC.

The MAC-STATUS-Ind/Resp primitives are used for passing status information. The MAC uses the MAC-STATUS-Ind primitive to request defined numbers of data from the RLC per logical channel and per TTI. The MAC-STATUS-Resp is used by the RLC to report on buffer occupancy levels and the RLC configuration information in the case when there are no data to be sent for a logical channel for a specific TTI.

Table 8.2 defines the primitives between the MAC and the RRC. The primitives are used for configuration and status reporting. The CMAC-CONFIG-Req contains the information that is used to configure the MAC for the different logical channels and transport channels. The information changes as the configuration of the logical channels and transport channels changes. The information that is included is:

- UE information (s-RNTI, SRNC identity, c-RNTI and activation time).
- RB information (RB multiplexing information such as transport channel identity, logical channel identity, MAC logical channel priority).
- Transport channel information (TFCS).
- RACH transmission information (set of ASC parameters, maximum number of preamble ramping cycles, Min and Max NBO1, ASC for RRC CONNECTION REQUEST).
- Ciphering information (ciphering mode, ciphering key, ciphering sequence number).
- CPCH transmission information (CPCH persistency value, maximum preamble of ramping cycles, maximum number of Tx frames, backoff control timer, TFS, channel assignment active indication, initial priority delays).

## 8.4 MAC PDUs and primitives

**Table 8.1.** *RLC to MAC primitives*

| Primitive | Direction | Parameters | Comments |
| --- | --- | --- | --- |
| MAC-DATA-Req | RLC to MAC | Data, buffer occupancy, UE-ID type indicator, RLC entity information | Used by RLC to request transfer of RLC PDU. Includes buffer occupancy per logical channel. Includes the UE ID type – common transport channels. Includes RLC configuration information per channel |
| MAC-DATA-Ind | MAC to RLC | Data, numbers of TBs transmitted, error indication | Used by MAC to notify RLC of arrival of RLC PDU. Includes number of blocks received (based on TFI). Includes error indication (CRC failures). |
| MAC-STATUS-Ind | MAC to RLC | Numbers of PDUs per TTI, PDU size, Tx status | Used by MAC per logical channel to request amount of data from RLC and size of PDUs. Used by MAC to report success of data transmission. |
| MAC-STATUS-Resp | RLC to MAC | Buffer occupancy, RLC entity information | Indicates no data to send this TTI but includes the buffer occupancy and RLC configuration per channel. |

**Table 8.2.** *RRC to MAC control primitives*

| Primitive | Direction | Parameters | Comments |
| --- | --- | --- | --- |
| CMAC-CONFIG-Req | RRC to MAC | UE IEs, RB IEs, TrCH IEs, RACH Tx control IEs, ciphering IEs, CPCH Tx control IEs, | Configuration information from RRC to MAC |
| CMAC-MEASUREMENT-Req | RRC to MAC | Measurement IEs | Measurement configuration |
| CMAC-MEASUREMENT-Ind | MAC to RRC | Measurement results | Measurement results relating to RLC buffer occupancy |
| CMAC-STATUS-Ind | MAC to RRC | Status information | Status information to RRC |

For the measurements, the measurement information configures the traffic volume measurements to be made by the MAC. The measurement result is the reporting quantity to be reported.

The status information is used by the MAC to indicate to the RRC the success or otherwise for the transmission of TM-RLC PDUs. By transmission it means that the PDU was successfully submitted to the lower layers for transmission.

## 8.5 MAC operation

The MAC is located in the centre of the radio interface protocol architecture and is therefore involved in many of the procedures and processes that require dynamic control. These procedures require information from higher layers (e.g. the RRC) to configure and setup the operation of the procedure, and equally require the MAC to control the operation of the physical layer.

In this section, therefore, we examine some of the procedures that are specifically controlled by the MAC. As a consequence of their multilayer extent described above, these procedures involve the layers both above and below the MAC. Rather than just focussing on the specific MAC related elements of the procedure, we also involve elements of the upper and lower layers to provide a complete description of the procedure and its operation. The intention is not to go into very great detail; for instance, for an appreciation of the details of the physical channels used in the random access procedure we recommend that Chapter 4 is consulted.

### 8.5.1 Traffic volume measurements

The traffic volume measurements are performed by the MAC in the UE. The MAC is responsible for monitoring the RLC buffer levels and for computing the current buffer occupancy, the average buffer occupancy and the variance of the buffer occupancy. This information is collated into a measurement report that is sent from the UE to the RNC. The RNC requests this information either periodically or through the enabling of triggers that generate status reports. The purpose of the information is to allow the RNC to monitor and control the buffer levels in the RLC, possibly by changing the state, from CELL_FACH to CELL_DCH for instance.

More details on traffic volume measurements can be found in Chapter 12.

## 8.6 Random access procedure

### 8.6.1 Basic principles of random access procedure

In this section we examine the basic principles of the random access procedure. We start by reviewing the likely uses for the random access procedure and then look at the different channels involved in the random access procedure from the higher layers down to the lower layers of the protocol architecture. Next we discuss the basic principles of the random access process and finally we look at the different physical channels involved and the timing issues related to these different physical channels.

## 8.6 Random access procedure

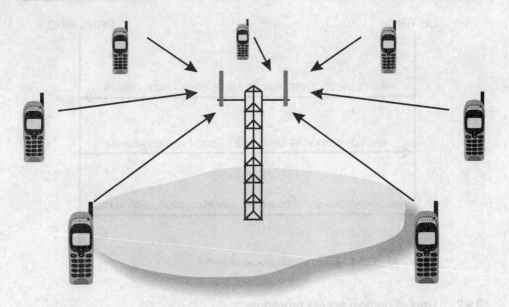

**Figure 8.14** Mobiles making random access attempts.

### What is random access?

The first questions that we may wish to consider are what is random access and why do we need it. If we consider a scenario such as that shown in Figure 8.14, we see one of the first problems that we meet in the design of the radio system. With a large number of mobiles in the system, there is a very high likelihood that some of these mobiles will need to establish a connection to the radio system at roughly the same time. In the radio system we only have a finite resource available and so we can expect that there will be a number of mobiles requesting access to this resource at the same time. The consequence of this high level of activity is collisions in the messages from the different mobiles. In addition to this, mobiles that are some distance from the base station will have a greater difficulty in making a connection due to the differences in the power levels between the distant mobiles and those that are close to the base station.

The random access process itself provides the solution to these and a number of other problems. It is the procedure that is followed by mobiles within a specific cell. The objective of the procedure is to control the access attempts made by a group of mobiles to ensure that the likelihood of collisions can be reduced, ideally to zero.

The basic concept behind random access is that it provides a mobile with relatively quick access to the resources of the cell. In a subsequent stage, the device that controls the resources of that cell (the RNC in WCDMA) can allocate additional resources to be used by the mobile. In the next section we examine some examples of the use of the random access process in the WCDMA system.

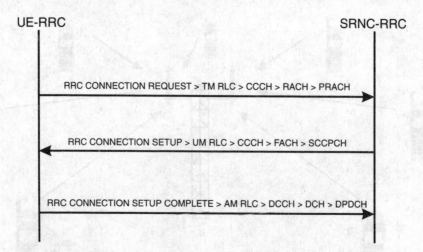

**Figure 8.15** Initial random access procedure by UE.

### 8.6.2 Uses of random access procedure

The random access process is used in a number of likely scenarios in the WCDMA system. The obvious ones are:
- initial random access by a UE;
- cell update;
- transfer of small PS data messages.

**Initial random access**

The first example of the use of the random access channel is in the initial random access procedure that a UE uses when it is first accessing the network. The messages that comprise this procedure are illustrated in Figure 8.15. It is the first message that is passed via the random access channel; however, we shall consider all of the messages for completeness.

The first message is the 'RRC CONNECTION REQUEST' message. This message is the layer 3 RRC message that indicates that the UE wants to establish a radio connection. The message is generated within the UE. The diagram illustrates the transmission path that the message takes as it passes through the layers.

The RRC layer uses the TM RLC to send the initial message. The TM RLC uses a CCCH which is mapped to the RACH transport channel and then to the PRACH, which is the name for the physical channel.

This initial message indicates to the RNC that there is a UE in the coverage area of the cell and that the UE wishes to establish a radio connection (the details and rationale behind these layer 3 message exchanges are covered in detail in Chapter 11). The key element as far as we are concerned here is that the message went via the random access channel and used the random access process.

## 8.6 Random access procedure

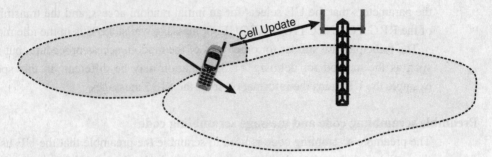

**Figure 8.16**  UE performing cell update procedure.

The response from the RNC shown in Figure 8.15 (RRC CONNECTION SETUP) indicates to the UE what resources are being assigned to it in that cell (in this case they are DPCH resources). This response is transported via the FACH and the SCCPCH.

The acknowledgement from the UE to the RNC uses the DCCH, the DCH and the DPDCH.

**Cell update**

A second potential use of the random access procedure occurs when a UE moves from one cell to another, whilst it is not using DPCHs and may even be using DRX (idle mode, CELL_PCH or URA_PCH state). When the UE notices that it has moved into a new cell it will perform a cell update to the new cell; see Figure 8.16. When the UE performs a cell update, it is telling the RNC that it is available in a new cell. When it enters into the cell in which it makes the update, it does not have any resources assigned in that cell and, therefore, it needs to use the RACH in that cell to perform the update. In this example, the UE uses the same RLC mode, logical, transport and physical channels as the UE used in the previous example for the initial random access.

**Transfer of small PS data message**

The final example of the use of the RACH is when a UE has a data connection active, and it needs to transfer some packet data between the UE and the network. In this example, the UE has previously been configured to use the RACH in place of a dedicated channel (this happens in the RADIO BEARER SETUP message, which is considered in Chapter 11).

When the UE uses the RACH, the difference between this and the previous example is that here the UE is probably using an AM RLC connection and a DTCH. The UE is still using the RACH and the PRACH.

### 8.6.3  Layer 1 information required for random access

In this section we review the layer 1 information that the UE needs to obtain prior to making a random access attempt. We focus here on the FDD mode of operation and the parameters that the UE selects for an initial random access, and the transmission of the RRC CONNECTION REQUEST message whilst a UE is in the idle mode.

The principles are similar for other uses of the random access procedure, but items such as the method for defining TF information may be different. In this specific example the UE gains the information from the SIB5 messages.

### Preamble scrambling code and message scrambling code

The preamble scrambling code is used to scramble the preamble that the UE uses on the uplink. It is based on the long scrambling code that is defined for use by the UE on the uplink. The scrambling code uses what is called a code offset to the normal scrambling codes used by the UE for the dedicated channels.

The UE selects the scrambling code number based on two quantities. First it takes the code number for the downlink primary scrambling code and it adds to that a number (the preamble scrambling code number) that the UE has received from the SIB5. The preamble scrambling code number is a number ranging from 0 to 15.

When the UE has identified the code number to use for the preamble scramble code, it will be the same number as that used for the message scramble code. It should be pointed out, however, that although the message scramble and the preamble scramble codes use the same number there are code phase offsets in the generation of the codes that will ensure that they are physically different.

### Preamble signature

The preamble signature is the short waveform (16 chips) that is repeated sequentially 256 times to form the access preamble (total length 4096 chips). There are 16 access preambles based on 16 signatures. The UE uses a number of stages to identify what signature to use as follows:

- Identify which of the 16 signatures are available in the cell by listening to the SIB5 message.
- Identify which ASC the UE belongs to.
- Select a subset of the available signatures based on the ASC.
- Randomly select one of the subset of signatures for use in a specified access slot.

The objective of this procedure is to reduce the likelihood that two UEs select the same signature, and to also reserve some signatures for use by UEs with higher priority.

### Message length TTI

The message part of the random access process can be either 10 ms or 20 ms in length. The length of the message defines the TTI that is to be used for the message. The UE listens to the broadcast channel to identify what TTI is to be used in that particular cell.

If a 10 ms and a 20 ms TTI are allowed in the same cell, then the UE selects the shorter TTI (10 ms) as long as the UE has sufficient transmit power to transmit the access preamble and the message. Sufficient power is defined as being less than 6 dB from the maximum power that the UE can use in the cell. If the UE does not have sufficient power, then the 20 ms TTI is selected with a consequent lowering of the transmit power requirements.

There may be a number of PRACHs within the cell that the UE can use (the different PRACHs are separated by different scrambling codes) and the UE randomly selects from this set.

**Available spreading factor**

The UE needs to know what the minimum spreading factor is for the message part on the RACH. The spreading factor is defined by the specifications to be in the range 32–256. The broadcast channel (SIB5) defines what is the minimum value to be used, but the UE can select a higher spreading factor if the number of data to send in the message part permits this.

**AICH transmission timing parameter**

The AICH transmission timing parameter is used to define the timing for the access preambles and the message with respect to the perceived timing of the AICH. The UE obtains this parameter from the downlink SIB5 broadcast channel. The parameter can take two values, 0 or 1.

**ASC**

The ASC is an integer in the range 0–7. The ASC is used in the random access procedure to define both the access preamble signature and which access slot the UE should use. The ASC essentially defines a priority with ASC = 0 being the highest priority.

For the transmission of the RRC CONNECTION REQUEST message, the UE will obtain the ASC in a two-stage process. First the UE must identify which access class (AC) applies to the call (the AC is defined on the USIM). Secondly, the UE then performs a mapping of AC to ASC using information that is sent to the UE on the SIB5 broadcast channel.

**Subchannels**

Subchannels are used in the WCDMA system to allow the UE to define in which access slot they should make their access attempt. The subchannels and supporting information are sent in broadcast messages to the UE. For the purposes of the initial random access, this is carried in the SIB5 broadcast message. There are 12 subchannels defined, and the UE uses the subchannel along with the SFN to define which access slot to use in a specific access frame.

To understand the operation of subchannels, we should consider Figure 8.17. First, the broadcast message (SIB5) broadcasts a list of available subchannels in this specific cell. This information is conveyed as a bit string denoted as $B_{11}B_{10}B_9 \ldots$ etc. in the

## From SIB5
Available Subchannel Number: $B_{11}B_{10}B_9B_8B_7B_6B_5B_4B_3B_2B_1B_0$
Assigned Subchannel Number: $b_3b_2b_1b_0$ [ASC=x]

Available Subchannel (ASC=x, ATT=0) = $(b_2b_1b_0b_2b_1\,b_0b_2b_1\,b_0b_2b_1b_0)$ AND $(B_{11}B_{10}B_9B_8B_7B_6B_5B_4B_3B_2B_1B_0)$
Available Subchannel (ASC=x, ATT=1) = $(b_3b_2b_1b_0b_3b_2b_1b_0b_3b_2b_1b_0)$ AND $(B_{11}B_{10}B_9B_8B_7B_6B_5B_4B_3B_2B_1B_0)$

## Example
Available Subchannel Number: 1 1 1 1 0 0 1 1 1 1 0 1
Assigned Subchannel Number: 1 1 0 1 [ASC=1]

ATT = 0
Available Subchannel = (1 0 1 1 0 1 1 0 1 1 0 1) AND (1 1 1 1 0 0 1 1 1 1 0 1)
Available Subchannel = (1 0 1 1 0 0 1 0 1 1 0 1)

ATT = 1
Available Subchannel = (1 1 0 1 1 1 0 1 1 1 0 1) AND (1 1 1 1 0 0 1 1 1 1 0 1)
Available Subchannel = (1 1 0 1 0 0 0 1 1 1 0 1)

**Figure 8.17** Derivation of subchannel from SIB5 message.

diagram. The bit string denotes the presence or absence of a specific subchannel. For instance $B_0 = 1$ indicates that subchannel 0 is available in the cell, but $B_4 = 0$ indicates that subchannel 4 is not available in the cell. In addition to the available subchannel number, there is a quantity called the assigned subchannel number. There are a number of assigned subchannels, one for each ASC. The assigned subchannel numbers are indicated by four bits $b_3b_2b_1b_0$ defined for each of the ASCs (although default values could also be defined).

In the next step, the UE needs to identify which ATT parameter is defined for that cell. If ATT = 0 for a specific ASC, then the UE constructs a string of 12 bits using the assigned subchannel number as shown in the diagram. The string consists of four repetitions of the assigned subchannel with $b_3$ removed.

In the last stage, the UE logically ANDs the available subchannel number and the bit string obtained from the assigned subchannel. This procedure produces a 12 bit string that defines which subchannels are available for that UE in that specific cell. An example of this procedure is illustrated in Figure 8.17. If the cell parameter ATT is set to 1, then the bit string is constructed from all the assigned subchannel bits repeated three times in this case. The remainder of the procedure is the same. The bit string derived from the assigned subchannel logically ANDs the available subchannel bit string to define the subchannel available for the UE in that cell. Figure 8.17 illustrates an example.

Having defined the available subchannels for that UE in that cell, the UE now needs to convert this to an access slot. To do this the UE uses Table 8.3. The UE selects the next complete access slot set (as illustrated in Figure 4.2) available to the UE. Essentially, this means the next radio frame. The UE calculates the SFN for the next access slot set modulo 8, which determines the row to select in Table 8.1. Next the UE selects the subchannels that are available for that UE (the result of the calculation described previously). The available subchannels are used by the UE to select those

## 8.6 Random access procedure

**Table 8.3.** *Subchannel number to access slot mapping*

| SFN mod 8 | Subchannel number | | | | | | | | | | | |
|---|---|---|---|---|---|---|---|---|---|---|---|---|
| | 0 | 1 | 2 | 3 | 4 | 5 | 6 | 7 | 8 | 9 | 10 | 11 |
| 0 | 0 | 1 | 2 | 3 | 4 | 5 | 6 | 7 | | | | |
| 1 | 12 | 13 | 14 | | | | | | 8 | 9 | 10 | 11 |
| 2 | | | | 0 | 1 | 2 | 3 | 4 | 5 | 6 | 7 | |
| 3 | 9 | 10 | 11 | 12 | 13 | 14 | | | | | | 8 |
| 4 | 6 | 7 | | | | 0 | 1 | 2 | 3 | 4 | 5 | |
| 5 | | | 8 | 9 | 10 | 11 | 12 | 13 | 14 | | | |
| 6 | 3 | 4 | 5 | 6 | 7 | | | | | 0 | 1 | 2 |
| 7 | | | | | | 8 | 9 | 10 | 11 | 12 | 13 | 14 |

columns in the table. The intersection of the SFN and the available subchannels defines the access slots that the UE can use for the random access attempt.

The UE then randomly selects one of the access slots from those available in the access slot set. If there are no access slots available in this access slot set, the UE waits for the next access slot set and repeats the procedure described above.

Consider the example with ATT = 0 shown in Figure 8.17. If we assume that SFN modulo 8 is 3, then the available access slots are: $\{8, 9, 11, 12, 14\}$. None of these access slots are in the first access slot set, and so the UE has to wait for the second access slot set to make the access attempts. The UE then randomly selects one of the access slots from those available and transmits the access preamble as described in the following sections.

### Power ramp step

The power ramp step (used only for the FDD mode) defines the amount by which the access preamble transmit power is increased in the case when the UE does not receive an acknowledgement from the UTRAN. The power ramp step is signalled to the UE in the broadcast channel SIB5 and can take an integer value between 1 dB and 8 dB, but typically takes a value of 3 dB. Each subsequent access preamble transmission attempt results in an increase in the access preamble transmit power by this amount.

### Preamble retrans max

Preamble retrans max is a counter that defines the number of times that a UE can transmit an access preamble (without being acknowledged) before it has to abort the procedure and obey the higher layer protocols. The preamble retrans max parameter is signalled to the UE on the broadcast channel SIB5 and has the range 1–64, but typically takes a value of 2.

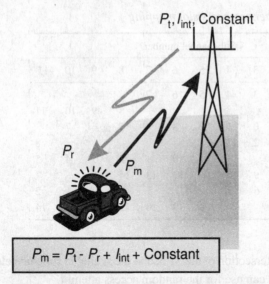

**Figure 8.18** Uplink PRACH open loop power $P_m$ based on the received power $P_r$, Node B transmitted power $P_t$, Node B receiver interference levels $I_{int}$ and a constant value.

## Initial preamble power

The initial preamble power is the initial transmit power that the UE uses for the transmission of the first access preamble. It is defined using what is called the open loop power control procedure. The basic principles of this procedure are illustrated in Figure 8.18. The UE receives the power level for the P-CPICH that is broadcast on the downlink in the SIB5 message. The UE also receives the value of the parameter referred to as the 'constant value' as well as the interference levels at the Node B on the uplink. The constant value is sent as part of the SIB5 message and the interference levels are part of the SIB7 message.

Next, the UE measures the receive signal code power (RSCP) for the UE and from this estimates the path loss between the UE and the Node B. The initial preamble transmit power is defined as:

$$\text{Init\_Pream\_Power} = \text{PCPICH\_Tx\_Power} - \text{PCPICH\_RSCP} \\ + \text{Interference\_Levels} + \text{Constant\_Value}$$

## Power offset ($P_{p-m}$)

The power offset is the difference between the power of the preamble that is successfully acknowledged and the power of the message. The UE receives the value of the power offset from a SIB5 message. The power offset can be in the range −5 to 10 dB.

## Message data-control power offset

The power for the data channel and the power of the control channel within the message part of the random access channel are different in general. This is because the spreading factor may be different in which case more or less power is needed for the data part with respect to the control part.

The UE defines the power level it needs based on the values for the gain factors $\beta_c$ and $\beta_d$ that are signalled to the UE as part of the SIB5 broadcast message. Either the gain factors can be signalled explicitly or the UE can be told to calculate them based on a reference value that is signalled to the UE.

**TF parameters**

The TF parameters tell the UE which TFs can be applied for the RACH. The UE obtains the RACH parameters from the SIB5 broadcast channel. As mentioned earlier, the UE selects the RACH based on the TF parameters that are available for a specific RACH, and on the transmit power requirements for that RACH.

**AICH spreading code**

The AICH is the channel that is used to acknowledge the receipt of the access preamble. The AICH is a downlink channel that is spread and scrambled in a similar way to the other downlink channels. The spreading code that is selected for the AICH is chosen by the RNC and the UE is told which spreading code is used in the downlink channel SIB5. The AICH always uses a spreading factor of 256.

**AICH scrambling code**

The AICH is scrambled as well as being spread and so the UE needs to identify which scrambling code is used by this channel. The scrambling code used by the AICH is the same as the primary downlink scrambling code that is used by the PCPICH and the P-CCPCH.

### 8.6.4 Layer 2 information required for random access

The UE also needs to collect a range of layer 2 information that the MAC uses to control the random access process. The MAC receives this information from the RRC layer via a CMAC-CONFIG-Req primitive. The RRC layer receives the information from broadcast channels and, in the case of an initial random access considered here, it is the SIB5 message carried on the BCCH. This information is considered in the following section.

**Set of ASC parameters**

The ASC parameters include the PRACH partition information that is used to select the access preamble signature and the subchannel that was described in the layer 1 information section previously. One of these ASC parameters is the persistence scaling factor $s_i$ that the UE needs to collect for each defined ASC. The UE uses this scaling factor to define the persistence value $P_i$ that is used to define how often and when the

## Layer 2 – medium access control (MAC)

**Table 8.4.** *Persistence value calculations for different ASCs*

| ASC#$i$ | 0 | 1 | 2 | 3 | 4 | 5 | 6 | 7 |
|---|---|---|---|---|---|---|---|---|
| $P_i$ | 1 | $P(N)$ | $s_2 P(N)$ | $s_3 P(N)$ | $s_4 P(N)$ | $s_5 P(N)$ | $s_6 P(N)$ | $s_7 P(N)$ |

UE can make a random access attempt. The persistence scaling factor $s_i$ can take a value between 0.9 and 0.2 in steps of 0.1 for ASC#2–ASC#7 and has a value set to 1 for ASC#0 and ASC#1.

### Dynamic persistence level

To calculate the persistence value $P_i$, the UE needs to obtain a value known as the dynamic persistence level as well as the persistence scaling factor $s_i$ described previously.

The dynamic persistence level $N$ takes a value between 1 and 8, and the UE obtains this value from the SIB7 broadcast message. SIB7 is the set of broadcast messages that change rapidly and as the dynamic persistence value is there to control the number of uplink access attempts, it also needs to change dynamically. From the dynamic persistence level $N$, a quantity $P(N)$ (the dynamic persistence value) is derived as follows:

$$P(N) = 2^{-(N-1)}$$

For each ASC we can then calculate the persistence value ($P_i$) as shown in Table 8.4. For ASC#0, the persistence value is fixed at 1, for ASC#1, only the dynamic persistence value is used, and for ASC#2–ASC#7, a variable persistence value is calculated based on the dynamic persistence value and the persistence scaling factor.

We will see in Section 8.6.5 that the closer the persistence value is to 1, the more likely the UE will be to make an access attempt at any given time instant.

### Number of preamble ramping cycles $M_{max}$

The quantity $M_{max}$ defines the number of times that the UE can pass to the layer 1 part of the access procedure (i.e. where the UE can transmit a sequence of access preambles). The UE obtains the value of $M_{max}$ from the SIB5 message; it can take a value between 1 and 32.

A typical value of $M_{max}$ would be 2; this would mean that the UE can have two attempts at the layer 1 transmission phase. If it is unsuccessful, it needs to stop the procedure and report to the higher layers that the access attempt was unsuccessful.

### $N_{BO1min}$ and $N_{BO1max}$

The parameters $N_{BO1min}$ and $N_{BO1max}$ define the lower and upper bounds on a scaling factor for a back off timer that is used to control a subsequent attempt at the random access procedure in the event that the UE is given a NACK to its level 1 preamble

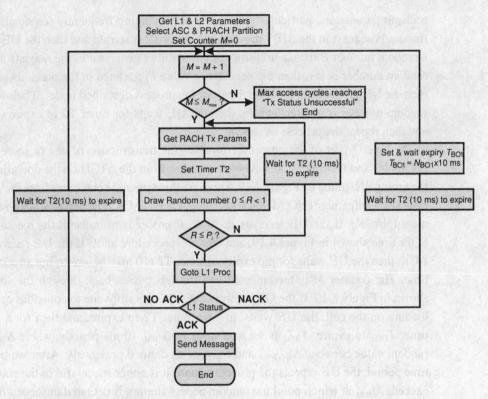

**Figure 8.19** Layer 2 random access procedure.

access attempt. The objective of the back off timer is to ensure that two mobiles that collided with a first set of preamble transmissions do not collide on a second occasion.

$N_{BO1min}$ is the lower value for the back off time and can take a value between 0 and 50; $N_{BO1max}$ is the upper value for the back off time and can also take a value between 0 and 50. From $N_{BO1min}$ and $N_{BO1max}$, the UE randomly selects a value $N_{BO1}$ that is between $N_{BO1min}$ and $N_{BO1max}$ such that $N_{BO1}$ has the value:

$$0 \leq N_{BO1min} \leq N_{BO1} \leq N_{BO1min}$$

The UE obtains the values of $N_{BO1min}$ and $N_{BO1max}$ from the SIB5 message but typical values are $N_{BO1min} = 3$ and $N_{BO1max} = 10$.

### 8.6.5 Layer 2 random access procedure

The flow chart that summarises the layer 2 part of the random access procedure can be seen in Figure 8.19. The procedure starts with the UE collecting the necessary information from broadcast channels and setting a loop counter $M$ to zero. $M$ is incremented and a test made to see if it exceeds the value $M_{max}$ defined from a broadcast message. If it does, the procedure terminates, else it continues. Next, the UE updates the RACH transmit parameters, particularly those that are changing frequently (especially those that are broadcast in the SIB7 message). A timer T2 is started and then the UE selects a random number $R$ from a uniform random number generator in the range 0–1. If the random number is less than the persistence value $P_i$ (defined in the previous section), then the UE proceeds to the layer 1 part of the process described in detail below. If the random number is greater than $P_i$, then the UE waits for timer T2 to expire (10 ms) and then repeat the process as shown.

The layer 1 part of the procedure involves the transmission of one or more access preambles and waiting for an acknowledgement from the AICH on the downlink with the timing illustrated by Figure 4.19. There are three possible outcomes from the AICH: no acknowledgement (NO ACK), acknowledgement (ACK) and negative acknowledgement (NACK). If the UE receives an ACK, it proceeds to transmit the message part at the time shown in Figure 4.19, and then the procedure ends. If the UE receives NO ACK, then the UE waits for the expiry of timer T2 (10 ms), before trying an additional time. The counter $M$ is incremented and the UE passes back through the stages as shown in Figure 8.19. If the UE receives a NACK, possibly due to collisions or uplink loading on the cell, the UE waits, first for timer T2 to expire, and then for a second timer $T_{BO1}$ to expire. $T_{BO1}$ is set to a value of $N_{BO1}$ 10 ms periods, where $N_{BO1}$ is a random value between $N_{BO1min}$ and $N_{BO1max}$ as defined previously. After waiting this time period, the UE repeats the procedure until it is either successful or the counter $M$ exceeds $M_{max}$ at which point the random access attempt is declared unsuccessful.

### 8.6.6 Layer 1 random access procedure

The layer 1 procedure is followed once the UE reaches the appropriate point in the layer 2 procedure described above. Figure 8.20 illustrates the layer 1 random access procedure. This starts with the UE selecting an access slot based on those available in the first access slot set (this could be access slot set 1 or access slot set 2). The procedure to select the access slot was defined above.

Next, the UE selects a preamble signature from the ones available for the UE in the cell based on the ASC for the UE. The UE then sets the preamble retransmission counter to the maximum value, and transmits the access preamble in the selected access slot using the transmit power defined by the initial preamble power.

The UE then waits for a response on the AICH on the downlink. If the UE receives an ACK, it transmits the message with a power offset defined by the parameter $P_{p-m}$, and the procedure ends. If the UE receives a NACK, the UE reverts to the layer 2 procedure and follows whatever commands are defined there. Alternatively the UE may receive no acknowledgement in which case it increases the transmit power, selects a new preamble signature and transmits the new signature in the next available access slot with the increased power and waits for a response on the AICH channel as before.

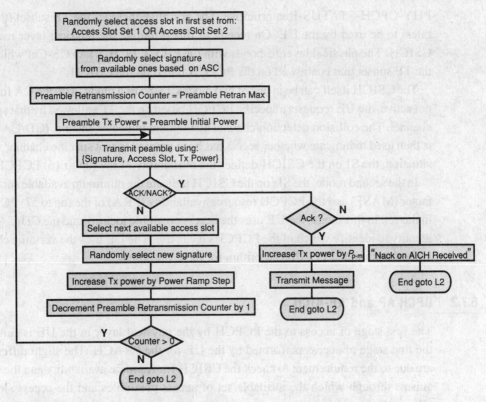

**Figure 8.20** Layer 1 random access procedure.

## 8.7 Control of CPCH

In this section we explore the basic principles of the control and operation of the CPCH. The CPCH is similar to the RACH considered previously. It is a channel that is only available on the uplink, but it is designed to transport higher rate data than the RACH. The CPCH is more complex than the RACH in operation; it includes a collision detection phase in addition to a random access phase. The CPCH has associated with it a number of indication channels, the purpose of which we cover in this section. There are a number of CPCHs available within a CPCH set, and there are a number of CPCH sets that may be available in a cell (dependent upon operator provisioning). The RRC layer selects a CPCH set, and the procedures that follow in this section relate to that set.

### 8.7.1 CSICH

The CSICH is a physical channel that carries the PCPCH status indicators (SIs). The detection of this channel by the physical layer is requested by the MAC using the

PHY-CPCH-STATUS-Req primitive. The primitive includes the TF subset (the data rates) to be used by the UE. On receipt of this primitive the physical layer reads the CSICH. The physical layer responds with the PHY-CPCH-STATUS-Cnf which lists the TF subset that is allowed on the PCPCH based on the CSICH.

The CSICH itself can be in one of two modes. In the first mode, with the CA function not active, the UE requests a specific PCPCH based on the TFs allowed by that specific channel. The collision detection/channel assignment indicator channel (CD/CA-ICH) is then used to indicate whether access has been granted to that specific channel. In this situation, the SI on the CSICH defines the availability of the (up to 16) PCPCHs.

In the second mode, the SIs on the CSICH indicate the minimum available spreading factor (MASF) and the PCPCH resource availability (PRA) of the (up to 57) PCPCHs in the set. In this mode, the UE uses the access preamble identity and the CD/CA-ICH identity to identify which of the PCPCHs it can use. The UE uses the versatile channel assignment mode (VCAM) algorithm for these purposes.

### 8.7.2 CPCH AP and AP-AICH

The first stage of access to the PCPCH by the physical layer in the UE is similar to the first stage of access performed by the UE for the PRACH. The slight differences are due to the requirement to check the CSICH for resource availability and the mechanisms through which the available set of access preambles and the access slots are allocated.

On the uplink, the UE transmits a CPCH-AP signature in a specific access slot. The procedure to define which signature and which slot is based on those signatures and slots that are available to the UE, which, in turn, depends upon whether CA is active or inactive. When the CPCH-AP signature and access slot are selected, the UE (after testing that downlink resources are still available by detecting the CSICH) transmits the signature in the slot. The UE then detects the CSICH again to ensure that the resources that are being requested are still available. If the resources are available, the UE then detects the AP-AICH on the downlink. If the AP-AICH is an ACK the UE can proceed to the collision detection phase; if it is an NACK it aborts the procedure and returns to the MAC procedure; if there is no acknowledgement the UE will select the next slot and retransmit the preamble.

### 8.7.3 CD/CA-ICH

The CD and CA phase for the CPCH access procedure is included to reduce the probability that two UEs are granted access to the same resource and, in the case that CA is active, it is used to allocate the resources on the downlink. The UE randomly selects a CD-AP from the available signature set and a CD access slot from those available but with some constraints dependent upon the usage of the PRACH. The UE then transmits the CD-AP in the access slot and waits for the CD/CA-ICH on the downlink. If no CD/CA-ICH is detected then the procedure aborts and a failure message is sent to the MAC. The CD/CA-ICH is a physical channel that carries the CD information and optionally the CA information. If CA is not active, the CD/CA-ICH indicates whether the CD phase was successful or not. It carries the CD signature (CDI) which is used in response to the CD access preamble (CD-AP). If the CDI matches the CD-AP then this indicates success. If the CA is active, then the CD/CA-ICH indicates whether the access attempt was successful and simultaneously a CA indicator (CAI). CD/CA-ICH carries 16 CDIs (8 odd signatures using $+1$ and $-1$) and also carries 16 CAIs (8 even signatures using $+1$ and $-1$). The CDI acknowledges CD-AP and the CAI assigns the channel. The original CPCH-AP used by the UE together with the CAI can be used to indicate which PCPCH the UE should use based on the VCAM algorithm.

## 8.7.4 VCAM

VCAM is used to identify the PCPCH that is assigned to a specific UE. VCAM uses the number of available CPCH-AP signatures ($S_r$), the number of PCPCHs in the CPCH set ($P_r$), the selected CPCH-AP, the signature ($i$), the received CAI ($j$) and an integer $n$ that is derived from $i, j$ and $S_r$. The algorithm uses the CPCH-AP and CAI and maps them to the assigned PCPCH ($k$), where $k$ is given by:

$$k = \{[(i + n) \bmod S_r] + j \times S_r\} \bmod P_r \tag{8.1}$$

Using this information and the algorithm for $k$ illustrated in the equation, the UE can identify which of the PCPCHs have been assigned to that specific UE. Once the PCPCHs have been assigned, the UE can then proceed to recheck the availability of the assigned PCPCHs from the CSICH before transmitting the message part of the PCPCH.

## 8.7.5 PCPCH information

Table 8.5 illustrates some of the PCPCH physical layer information that the UE needs to identify prior to accessing the PCPCH. The table presents the typical ranges for these quantities and the possible source of the information. The selection of the access slot from the subchannel group uses a procedure that is very similar to that used on the PRACH.

Table 8.6 illustrates some additional information that relates to the layer 2 part of the PCPCH access. This information includes the timers that are used as part of the PCPCH access, some of the counters that define the maximum number of frames and the end of transmission counter.

**Table 8.5.** *PCPCH information*

| Parameter | Range | Comment |
|---|---|---|
| UL AP scrambling code | {0...79} | From SIB or DCCH |
| UL AP signature set | {0...15}[16] | From SIB or DCCH |
| AP subchannel groups | {0...11}[12] | Same use as in PRACH |
| AP-AICH spread code | {0...255} | From SIB or DCCH |
| UL CD scrambling code | {0...79} | From SIB or DCCH |
| UL CD signature set | {0...15}[16] | From SIB or DCCH |
| CD subchannel groups | {0...11}[12] | Same use as in PRACH |
| CD-AICH spread code | {0...255} | From SIB or DCCH |
| CPCH UL scrambling code | {0...79} | Used for message |
| DPCCH DL spread code | {0...511} | Used for power control |

UL = uplink; DL = downlink.

**Table 8.6.** *PCPCH timer and counter information*

| Parameter | Range | Comment |
|---|---|---|
| NF_bo_all busy | {0...31} | SIB8 – number of frames |
| NF_bo_mismatch | {0...127} | SIB8 – number of frames |
| EOT | {0...7} | End of transmission from SIB8 & DCCH |
| NF_max | {1...64} | Max nos of data frames from SIB8 & DCCH |
| $T_{BOC1}$ | {0...31} | Random time between 0 and NF_bo_all busy |
| $T_{BOC2}$ | {0...31} | Taken from NF_bo_busy from SIB8 |
| $T_{BOC3}$ | {0...31} | Taken from NF_bo_no_AICH from SIB8 |
| $T_{BOC4}$ | {0...127} | Random time between 0 and NF_bo_mismatch |

### 8.7.6 CPCH transmission

In this section we consider the MAC aspects of the CPCH transmission procedure. Figure 8.21 illustrates the first part of the procedure, which is the access stage. The second part is the message transfer stage. The procedure starts with the collection of the necessary information and the reading of the CSICH by the physical layer. The response from the physical layer (PHY-CPCH-STATUS-Cnf) is used by the MAC to build a busy table that defines the availability of the different PCPCHs in the CPCH set.

Next the MAC selects a CPCH TF based on the busy table, and other CPCH configuration parameters. The UE selects a random number $R$. If $R$ is less than the persistence value associated with the selected CPCH TF, then the UE proceeds to the L1 part by sending the PHY-ACCESS-Req primitive.

The PHY-ACCESS-Cnf primitive from L1 indicates five possible outcomes as shown. Depending upon the outcome the UE either proceeds to the sending of the message stage, or follows the specific retry procedure illustrated.

## 8.7 Control of CPCH

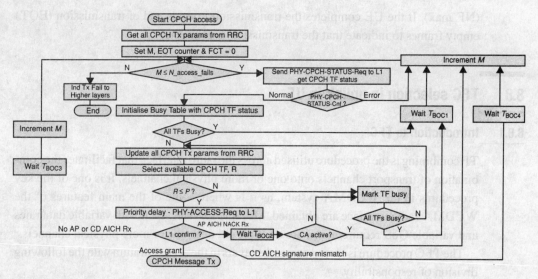

**Figure 8.21** CPCH access stage flow diagram.

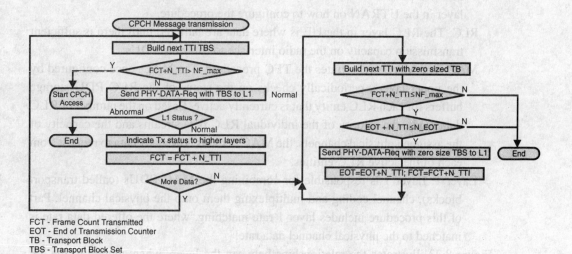

FCT - Frame Count Transmitted
EOT - End of Transmission Counter
TB - Transport Block
TBS - Transport Block Set

**Figure 8.22** CPCH message transmission flow diagram.

Figure 8.22 illustrates the next stage, which corresponds to the transmission of the CPCH message. The procedure consists of the transmission of N_TTI frames of data (the number of frames per TTI) and then a check on the L1 status from the DL-DPCCH, which is passed to the MAC via the PHY-STATUS-Ind primitive. If L1 indicates abnormal activities (emergency stop received, start of message indicator not received on DL DPCCH, L1 failure or out of synchronisation), then the UE aborts the transmission. The UE continues until all of the data are transmitted or the frame count transmitted (FCT) exceeds the maximum number of frames to be transmitted

(NF_max). If the UE completes the transmission it sends end of transmission (EOT) empty frames to indicate that the transmission is complete.

## 8.8 TFC selection in uplink in UE

### 8.8.1 Introduction to TFC

TF combining is the procedure utilised across the radio interface that facilitates the combination of transport channels onto one or more physical channels. It is one of the key procedures in the WCDMA system, as it is where some of the main features of the WCDMA radio interface are obtained. Multiple services, each with variable data rates and variable QoS requirements, are multiplexed onto one or more physical channels.

The TFC procedure is spread across the layers in the access stratum with the following division of responsibility.

- RRC: The RRC layer is responsible for configuring and monitoring the operation of the TFC process. The RRC layer in the UE receives information from the RRC layer in the UTRAN on how to configure the procedure.
- RLC: The RLC layer in the UE is where data are buffered until there is sufficient transmission capacity on the radio interface to carry the PDUs.
- MAC: The MAC orchestrates the TFC procedure. The MAC, once configured by the upper layers, periodically checks on the status of the RLC PDU storage buffers for each RLC entity that is currently active. Based on the number of RLC data ready, the priority of the individual RLC data streams and the capacity of the assigned physical channels, the MAC decides how many data to request from each of the active RLC entities.
- Layer 1: Layer 1 is responsible for combining the MAC PDUs (called transport blocks), channel coding and multiplexing them onto the physical channel. Part of this procedure includes layer 1 rate matching, where the offered data rate is matched to the physical channel data rate.

Figure 8.23 illustrates the relationships between the layers when applied to the TF combining procedure. As we can see, the TF combining involves layers both above and below the MAC. For this reason, we consider the operation of the TF combining procedure across all of the layers concerned.

### 8.8.2 TF combining configuration

The TF combining procedure is configured when an RB or an SRB is established, for instance as part of the radio bearer setup procedure. There is a range of information that the UE needs before it can perform the TF combining procedure. Logical channels, transport channels and physical channels need to be configured. The TFs for the new

## 8.8 TFC selection in uplink in UE

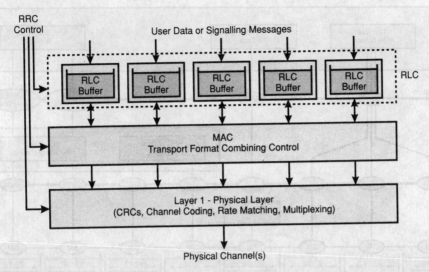

**Figure 8.23** Layer interactions for TF combining.

transport channels and the TFC set (TFCS) for the combination of the transport channels need to be defined. All of this information can be contained within the RADIO BEARER SETUP message that the UE receives as part of the establishment of a new RB to carry, for instance, a speech or PS data service.

Before considering the details of the TFC, we first consider what information that UE has received and how it uses this information. To do this we consider an example.

### 8.8.3 Introduction to example set of services

In this example we assume that there are a number of RBs active as shown in Figure 8.24. The RBs are for a speech signal adaptive multirate (AMR) codec at 12.2 kb/s, a packet data signal with a data rate up to 64 kb/s and a number of SRBs used to carry control information (e.g. SRB1, SRB2, SRB3 and SRB4 described previously). We will consider the MAC functions in the UE for the uplink only.

**12.2 kb/s AMR speech service**

From Figure 8.24 it is clear that the speech service actually comprises a number of data streams referred to as subflows. Three subflows are used to transport the speech data because of the different importances of the speech bits and the different susceptibilities that these bits have to errors across the radio channel.

The speech codec categorises the speech bits as high importance, medium importance and low importance and these are referred to as class A, B and C bits respectively. Each of these subflows requires a different amount of error protection across the radio interface and consequently they are treated as separate RBs as far as the lower layers

## 284 Layer 2 – medium access control (MAC)

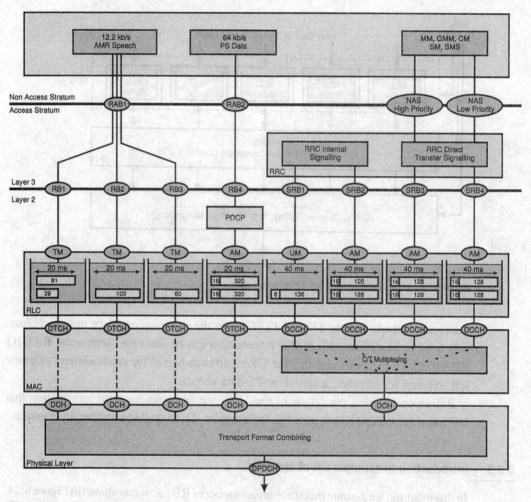

**Figure 8.24** Combination of two services and SRBs.

are concerned. Each of the streams uses the TM RLC, with the RLC providing a buffer for speech data until the MAC requests the RLC PDUs. We will see in detail later that the data rate offered by the speech codec is variable and so the RLC deals with different sized SDUs in its buffers, as shown diagrammatically in Figure 8.24.

In addition, there is a time period associated with the delivering of data from the RLC to the MAC. This time period is the TTI and is a semi-static parameter for the FDD mode, but can vary dynamically in the TDD mode. In Section 8.8.5 we will explore the values that the TTI can take and its effect on the operation of the different layers within the AS. The TTI values are used by the MAC and L1. As mentioned, the TTI is the rate at which data are pulled from the RLC by the MAC and passed to L1, and also it is the time it takes for L1 to transmit data via the physical channels. Figure 8.24 shows the TTI values in the RLC purely for diagrammatic convenience.

### 64 kb/s PS domain data service

Adjacent to the 12.2 kb/s AMR speech service we can see the 64 kb/s packet data service. The data are delivered to the RLC layer from the higher layers (via the PDCP layer). In this example, the PS data service is utilising the AM RLC to ensure high reliability for the transfer of data across the radio interface. One of the conditions applied to the RLC is that the PDU sizes for the AM RLC are fixed. In this example they are fixed at 336 bits in length.

Within the RLC layer, the RLC can perform segmentation and concatenation functions as described in Chapter 9. To support the segmentation, concatenation and AM operation, the RLC adds an RLC header to the segmented or concatenated blocks of data. In this example, the header is a minimum of 16 bits, but could be larger if one or more 'length indicators' are required as a consequence of the segmentation or concatenation procedures (in this case the data part (RLC SDU) would be smaller). Later, we will also see that the data rate provided by the RLC varies, and as a consequence the number of blocks leaving the RLC per TTI also varies.

### SRBs

Figure 8.24 illustrates the four SRBs that are established as part of the RRC connection establishment procedure. Two of the SRBs are used within the RRC for RRC signalling, one using an unacknowledged mode (UM) RLC and the other an AM RLC. The other two signalling channels are from the NAS. One is a high priority signalling channel and the other is a low priority signalling channel.

Within the RLC, it is worth pointing out the difference in the typical RLC header size. In this example, the UM RLC uses an 8 bit header and the AM RLC a 16 bit header. The same comment applies here as to the 64 kb/s PS data AM RLC service, i.e. segmentation and concatenation can cause this header size to change with the consequence that the number of data transferred may reduce.

Within the MAC, the four SRBs are multiplexed onto a single transport channel as described earlier when considering the operation of the MAC. The multiplexing is through a selection procedure. Every TTI (40 ms in this example for the signalling channels), the MAC selects data from the SRBs that have the highest logical channel priority.

Once the MAC has selected which of the four SRBs will be transported across the radio interface for a specific time period, it adds a 4 bit MAC header (C/T bits) to indicate which of the four logical channels is multiplexed onto the transport channel.

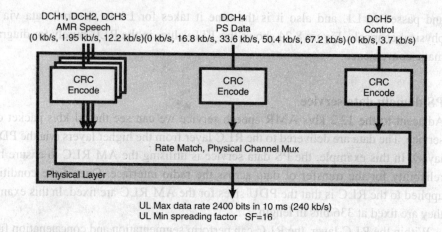

**Figure 8.25** Example combination of three transport channels.

This procedure increases the length of the MAC PDU (called a transport block) by 4 bits, and so the transport block size will be increased to 148 bits.

### 8.8.4 Data rates for example set of uplink services

Moving down through the MAC to the input to the physical layer, we now want to consider the data rates that the two services and the SRBs require. The data rates that we review here include any of the overheads that are introduced by the MAC and the RLC as the data passes through those layers.

Figure 8.25 illustrates the variation in the data rates that we are considering in this example. We have already seen that the data coming from the speech codec should be considered as being in three classes with each class provisioned with a different QoS.

#### 12.2 kb/s AMR codec speech data rates

For the speech service the variable data rate is needed to support the source controlled rate (SCR) operation of the UMTS AMR speech codec. With the SCR, the speech codec uses voice activity detection (VAD) to decide whether any speech information is being generated by the user. If the user stops speaking for a period of time the AMR codec can stop submitting speech data for transmission across the radio interface (this is often referred to as DTX).

Prior to the cessation of transmission of the speech data, however, the speech codec characterises the background noise signal that is present when the user has stopped speaking. This background noise signal is referred to as comfort noise and is sent using silence descriptor (SID) frames of data. The rationale for this process is that it is disturbing at the receiving end to have complete silence once the user has stopped talking, and so the background noise signal is 'played' at the receiver. A consequence of this procedure is that the speech codec has three effective data rates: 12.2 kb/s for normal speech transmission, 0 kb/s when the user has stopped speaking, and 1.95 kb/s for the transmission of the SID frames that carry the comfort noise signal from the transmitter to the receiver.

## 64 kb/s PS data rates

The fourth transport channel (DCH4) is carrying a packet switched data service (e.g. Internet access). This specific service has a variable data rate to support the QoS profile that was requested for the service. In this example the data rate varies from 0 kb/s to 64 kb/s in steps of 16 kb/s. The RLC adds a minimum of a 16 bit header (as seen in Figure 8.24) and consequently the data rates vary from 0 kb/s to 67.2 kb/s as shown in Figure 8.25. The RLC-PDU size is fixed at 336 bits per 20 ms TTI. For the four PS data rates (excluding the RLC header) going from 16 kb/s to 64 kb/s either 1, 2, 3 or 4 RLC PDUs are required per TTI. If, as a result of segmentation for instance, the RLC AM header needs to be greater than 16 bits, then the capacity available for the PS data is reduced as the RLC-PDU size is fixed.

## SRB data rates

The final transport channel (DCH5) carries the logical control channels (DCCHs). In this example the SRBs have three data rates: 0 kb/s when there is no data, 3.4 kb/s for the RRC signalling service using the UM RLC and 3.2 kb/s for the three SRBs using the RLC AM service. In addition, we need to add the 8 bit RLC header for the UM RLC and the 16 bit header for the AM-RLC leading to an RLC data rate of 3.6 kb/s for all SRBs. The MAC adds a 4 bit header for the C/T bits leading to a transport channel data rate of 3.7 kb/s.

### 8.8.5 Transport channel configuration for example set of uplink services

In this section we now want to consider some of the configuration parameters that are defined for the RBs when the RBs are established. This information is passed to the UE in the RRC CONNECTION SETUP message for the SRBs and in RADIO BEARER SETUP messages for the speech and PS data service and optionally for the SRBs, for instance if they are being reconfigured.

In general, each RB is established individually, but in doing so the effects of a new RB on the combination of RBs need to be considered as well.

To start, we consider the TF, then the TFS that is defined for each transport channel, and finally we will consider the TFC and TFCS.

## TF

For each transport channel there is an associated TF. The TF is information that is used by the MAC and the physical layer to define how the transport channels should be

## Layer 2 – medium access control (MAC)

**Table 8.7.** *Semi-static transport format information*

| Element | Description |
| --- | --- |
| TTI | The period that defines the transmission time for the transport blocks associated to a specific transport channel. Allowed values are:<br>• 10 ms<br>• 20 ms<br>• 40 ms<br>• 80 ms<br>• dynamic (TDD mode only) |
| Type of channel coding | This defines what kind of channel coding should be used on the radio interface for the specific transport channel. Allowed types are:<br>• convolutional coding<br>• turbo coding |
| Coding rate | This defines the code rate for the selected code. The code rate defines how many coded bits are sent per information bit. Allowed values are:<br>• 1/2-rate (convolutional code only)<br>• 1/3-rate (convolutional or turbo code) |
| CRC size | This defines how many CRC bits are appended per transport block. The CRCs are used in the receiver to detect whether there is an error in the transport block. Allowed values are:<br>• 0 bits<br>• 8 bits<br>• 12 bits<br>• 16 bits<br>• 24 bits |
| Rate matching attribute | The rate matching attribute acts as a weighting factor in the rate matching process to define how many bits should be repeated or deleted from a specific transport channel. The higher the attribute, the less likely that data bits will be deleted. Allowed values are: integers from 1 to 256. |

structured and subsequently combined. For any given transport channel there are two parts to the TF, the semi-static TF information and the dynamic TF information.

### Semi-static TF information

The semi-static transport format information defines the parameters for a specific transport channel that apply at least until they are modified by a subsequent higher layer message. The term semi-static is used to imply that it is not likely to change but that there is the possibility that it will change.

Table 8.7 shows the different elements of semi-static information. Each transport channel has semi-static transport information defined as part of the configuration procedure. For our example set of RBs the semi-static transport information is defined in Table 8.8. From the table we can see that the speech bits with the different degrees of importance have different levels of protection. Class A bits, which are the most

## 8.8 TFC selection in uplink in UE

**Table 8.8.** *Semi-static TF information for example services*

| Service | Transport channel | TTI | Channel coding | Coding rate | CRC size | Rate match |
|---|---|---|---|---|---|---|
| 12.2 kb/s speech | DCH1 | 20 ms | Conv. | 1/3 | 12 bit | 180 |
| 12.2 kb/s speech | DCH2 | 20 ms | Conv. | 1/3 | 0 bit | 170 |
| 12.2 kb/s speech | DCH3 | 20 ms | Conv. | 1/2 | 0 bit | 215 |
| 64 kb/s PS data | DCH4 | 20 ms | Turbo | 1/3 | 16 bit | 130 |
| SRBs | DCH5 | 40 ms | Conv. | 1/3 | 16 bit | 155 |

important bits, go via DCH1, which has a 1/3-rate convolutional code as well as CRC bits to detect errors. Class B bits also use a 1/3-rate convolutional code, but no CRC bits and class C bits (the least important) use a 1/2-rate convolutional code and no CRC bits. All of the speech transport channels have the same TTI of 20 ms (this is not surprising as the speech bits are all coming from the same speech codec).

The 64 kb/s packet data channel uses a 1/3-rate turbo code and also includes a 16 bit CRC for error detection purposes (this will be used by the AM RLC entity to decide whether or not to request the retransmission of a specific RLC PDU). The PS data channel also uses a 20 ms TTI.

The SRBs use a 1/3-rate convolutional code and a 16 bit CRC. The CRC is required for the SRBs that use an AM RLC link for retransmission requests. The TTI for the SRBs is longer than for the speech and PS data channels, being set at 40 ms.

### Dynamic TF information

The dynamic TF information defines the attributes for a specific transport channel that can change dynamically. For the FDD mode of operation, the dynamic information defines the instantaneous data rate that is offered by a specific transport channel to the layer 1 transport channel combining procedure. For the TDD mode, the dynamic TF information has the additional capability that allows the TTI to vary dynamically. There are two elements to the dynamic TF information: the transport block size and the number of transport blocks.

Each transport channel has dynamic TF information assigned to it. In addition, there can be a number of dynamic TFs configured for a specific transport channel and which can be applied at different time instants. This collection of dynamic TF information is referred to as a TFS.

### TFS

The TFS is a collection of dynamic TF information that applies to a specific transport channel. This collection of dynamic TF information allows the data rate for the transport channel to be changed, potentially every TTI. To illustrate what the TFS is we return to our example collection of services on the uplink, starting with the speech service.

**Table 8.9.** *TFS for speech bits carried on DCH1*

| TF | TBS | No. TBS | Comment |
|---|---|---|---|
| 0 | 81 | 0 | No blocks of size 81 bits. This TF contributes to the data rate of 0 kb/s. |
| 0 | 0 | 1 | 1 block of size 0 bits. This is an alternative to the previous case, the reasons for which are discussed in the text. This TF contributes to the data rate of 0 kb/s and 1.95 kb/s. |
| 1 | 39 | 1 | This TF is used to transfer the SID frames which have an effective data rate of 1.95 kb/s |
| 2 | 81 | 1 | This TF is used to transfer the class A speech bits which contribute to the speech codec rate of 12.2 kb/s. |

**Table 8.10.** *TFS for speech bits carried on DCH2*

| TF | TBS | No. TBS | Comment |
|---|---|---|---|
| 0 | 103 | 0 | No blocks of size 103 bits. This TF contributes to the data rate of 0 kb/s. |
| 1 | 103 | 1 | This TF is used to transfer class B speech bits which contribute to the speech codec rate of 12.2 kb/s. |

## TFS for speech example service

The speech service comprises three transport channels and consequently there are three TFSs. We consider the three TFs in turn starting with the high importance, class A bits that are transported via the DCH1 transport channel.

In Tables 8.9–8.13, TBS means transport block size and No. TBS is the number of transport blocks per TTI. Table 8.9 illustrates the TFS that is defined for the first transport channel DCH1 that carries the class A speech bits. The first thing to note is the three data rates that are facilitated by this set of TFs. In addition there is an option on how the data rate of 0 kb/s can be indicated. The two possibilities for 0 kb/s are no blocks of some size (81 bits in this example), or 1 block of no size. Although this would appear to be a pedantic way of saying the same thing (i.e. no bits) in reality it has an impact on what follows in the physical layer. 'No blocks of some size' results in no data being selected for that transport channel. One block of zero size, however, results in CRC bits being attached to that zero length block (assuming that CRC bits are applied to the transport channel, which they are in this case). By attaching CRC bits we are reliably indicating that there are no bits. We are also ensuring that there are some data to transmit for that period. So, for the first transport channel, there are three TFs that define three data rates for that transport channel.

Table 8.10 and Table 8.11 illustrate the TFSs that are defined for the class B and class C bits respectively. In both of these cases, there are only two TFs per TFS. The two TFs contribute to the three codec rates of 0 kb/s, 1.95 kb/s or 12.2 kb/s.

**Table 8.11.** *TFS for speech bits carried on DCH3*

| TF | TBS | No. TBS | Comment |
|---|---|---|---|
| 0 | 60 | 0 | No blocks of size 60 bits. This TF contributes to the data rate of 0 kb/s and 1.95 kb/s. |
| 1 | 60 | 1 | This TF is used to transfer class C speech bits which contribute to the speech codec rate of 12.2 kb/s. |

**Table 8.12.** *TFS for PS data bits carried on DCH4*

| TF | TBS | No. TBS | Comment |
|---|---|---|---|
| 0 | 336 | 0 | No blocks of size 336 bits corresponds to a PS data rate of 0 kb/s. |
| 1 | 336 | 1 | 1 block of size 336 bits corresponds to a PS data rate of 16 kb/s, but also includes RLC AM overheads. |
| 2 | 336 | 2 | 2 blocks of size 336 bits corresponds to a PS data rate of 32 kb/s, but also includes RLC AM overheads. |
| 3 | 336 | 3 | 3 blocks of size 336 bits corresponds to a PS data rate of 48 kb/s, but also includes RLC AM overheads. |
| 4 | 336 | 4 | 4 blocks of size 336 bits corresponds to a PS data rate of 64 kb/s, but also includes RLC AM overheads. |

**TFS for PS data example service**

Table 8.12 defines the TFS that is used by the 64 kb/s PS data service. Five TFs are defined within the TFS. The five TFs correspond to data rates of 0 kb/s, 16 kb/s, 32 kb/s, 48 kb/s and 64 kb/s. The PS data service is using an AM RLC entity and, therefore, the block size is fixed at 336 bits. The block size is 16 bits larger than required (320 bits for 16 kb/s in 20 ms TTI) and this is due to the 16 bit RLC header that is added to each RLC PDU by the RLC AM entity. As we mentioned previously, on occasions this 16 bit header may be larger, due, for instance, to segmentation or concatenation within the RLC. If the RLC header is larger, the data that can be transported are reduced to ensure a fixed block size of 336 bits.

**TFS for SRB example service**

Table 8.13 defines the TFS used by the SRBs that are multiplexed by the MAC onto the transport channel DCH5. There are two transport formats for DCH5, which will enable two sets of data rates. The SRB can have a data rate of 0 kb/s using TF0, or alternatively it can have a data rate of 3.2 kb/s or 3.4 kb/s. The reason for the differences in the SRB data rate is due to the two different RLC modes supported by this specific service. The UM RLC entity used by SRB1 has only an 8 bit header, whilst the AM RLC used by SRB2, SRB3 and SRB4 uses an RLC header of 16 bits. In addition, there is also a 4 bit MAC header within each transport block to account for the MAC C/T multiplexing.

**Table 8.13.** *TFS for SRB bits carried on DCH5*

| TF | TBS | No. TBS | Comment |
|----|-----|---------|---------|
| 0  | 148 | 0       | No blocks of size 148 bits. This TF contributes to the SRB data rate of 0 kb/s. |
| 1  | 148 | 1       | 1 block of size 148 bits. This TF contributes to the SRB data rate of 3.2 kb/s or 3.4 kb/s. |

The data rate for the TF1 transport channel, therefore, is 3.7 kb/s, independently of which SRB is using the transport channel.

### 8.8.6  TFC

Having considered the different TFSs that are defined for each of the transport channels that are active, we now turn to address how we may wish to combine these different transport channels together. To define how the different transport channels are combined, the TFC is used. The TFC specifies which of the different TFs from each of the transport channels can be combined together (combination occurs within the physical layer). It is obvious that with more than one TF for each transport format, that there must be more than one TFC. Therefore, we next consider the TFCS.

### 8.8.7  TFCS

The TFCS can be signalled to the UE on a number of different occasions, for instance as part of a transport channel reconfiguration (TRANSPORT CHANNEL RECONFIGURATION message) or when an RB is setup (RADIO BEARER SETUP message).

The TFCS is a table of allowed TFCs. By defining the combinations in this way, it is possible to very precisely control the peak data rates that would result across the radio interface. The RNC is responsible for defining the TFCS and can exclude combinations that result in high peak data rates if it needs to do so from a resource allocation or QoS perspective.

The TFCS is signalled to the UE using the calculated transport format combination (CTFC) procedure outlined in Chapter 11.

**TFCS for example services**

Table 8.14 illustrates an example TFCS for the set of services defined in the example we are considering. Each TFC is identified by a quantity referred to as the TFCI. The TFCI bits are transmitted in the DPCCH that is transporting the combined transport channels. In this specific example for the TFCS, we have all of the allowed data rates available in the table. Examples of combinations that are not present in the table are the ones that come from invalid combinations of the speech transport channels (e.g. there

## 8.8 TFC selection in uplink in UE

**Table 8.14.** *TFCS for example services*

| TFCI | TF-DCH1 | TF-DCH2 | TF-DCH3 | TF-DCH4 | TF-DCH5 |
|---|---|---|---|---|---|
| 0 | 0 | 0 | 0 | 0 | 0 |
| 1 | 1 | 0 | 0 | 0 | 0 |
| 2 | 2 | 1 | 1 | 0 | 0 |
| 3 | 0 | 0 | 0 | 1 | 0 |
| 4 | 1 | 0 | 0 | 1 | 0 |
| 5 | 2 | 1 | 1 | 1 | 0 |
| 6 | 0 | 0 | 0 | 2 | 0 |
| 7 | 1 | 0 | 0 | 2 | 0 |
| 8 | 2 | 1 | 1 | 2 | 0 |
| 9 | 0 | 0 | 0 | 3 | 0 |
| 10 | 1 | 0 | 0 | 3 | 0 |
| 11 | 2 | 1 | 1 | 3 | 0 |
| 12 | 0 | 0 | 0 | 4 | 0 |
| 13 | 1 | 0 | 0 | 4 | 0 |
| 14 | 2 | 1 | 1 | 4 | 0 |
| 15 | 0 | 0 | 0 | 0 | 1 |
| 16 | 1 | 0 | 0 | 0 | 1 |
| 17 | 2 | 1 | 1 | 0 | 1 |
| 18 | 0 | 0 | 0 | 1 | 1 |
| 19 | 1 | 0 | 0 | 1 | 1 |
| 20 | 2 | 1 | 1 | 1 | 1 |
| 21 | 0 | 0 | 0 | 2 | 1 |
| 22 | 1 | 0 | 0 | 2 | 1 |
| 23 | 2 | 1 | 1 | 2 | 1 |
| 24 | 0 | 0 | 0 | 3 | 1 |
| 25 | 1 | 0 | 0 | 3 | 1 |
| 26 | 2 | 1 | 1 | 3 | 1 |
| 27 | 0 | 0 | 0 | 4 | 1 |
| 28 | 1 | 0 | 0 | 4 | 1 |
| 29 | 2 | 1 | 1 | 4 | 1 |

are no combinations TF1-DCH1, TF1-DCH2 and TF1-DCH3 as these are not valid speech data rates).

In the following section we will explore how the MAC utilises this TFCS table to control the transport combining procedure.

### 8.8.8 TF combining for example set of services

There are two main aspects for the TF combining process. First, there is the procedure in the MAC that makes the decisions on which TFC to select based on how many data are available in the RLC and the relative priorities of these data. Second, there is the procedure in the physical layer that performs the physical combination of the transport channels and multiplexes them onto one or more physical channels.

We start by considering the MAC procedures. Then we move on to examine the physical layer procedures to round off the discussion on the TF combining process. The first stage of deciding upon the TFC within the MAC requires that the MAC knows how many data are available in the buffers of the different RLC entities present in the RLC layer. When an RLC entity delivers some data to the MAC for transportation in the next TTI, it does so using the MAC-DATA-Req primitive as illustrated in Figure 8.26. The MAC-DATA-Req primitive contains the data to be transmitted and also contains information on the RLC buffer occupancy (BO) in kilobytes of information (the primitive also contains other information not listed here). Alternatively, if there are no data to be transmitted, the RLC can use the MAC-STATUS-Res primitive to provide information on the BO for that specific logical channel. From the BO information the MAC knows how many data are available for each of the RLC entities active in the UE. The next stage is for the MAC to select the data it needs from each of the logical channels. To do this the MAC uses the TFCS information, the MAC priority information and a simple prioritisation algorithm that is outlined below.

### MAC TFC selection algorithm

The TFC selection occurs at a rate equal to, and coincident with, the shortest TTI within the transport format set. To start the MAC has to generate a set of valid TFCs. Invalid TFCs such as those requiring padding in the RLC layer, those in the blocked state (see below for a definition) and those not compatible with the current RLC configuration are excluded. Next, the MAC selects the TFC that allows the transmission of the highest priority data starting from the highest priority logical channel and then iteratively selecting the most data from the next lower priority logical channel consistent with the available TFCs. The final check is that there is no other TFC that transmits the same amount of data, but using a lower bit rate. On a specific TTI boundary, once the TFC has been selected, the MAC requests from each RLC entity the number of RLC PDUs and the size of RLC PDUs defined by the selected TFC. To do this, the MAC uses the MAC-STATUS-Ind primitive as illustrated in Figure 8.26.

### MAC TFC blocked and excess power states

There are two possible reasons why some TFCs may not be available for selection. Before considering the details behind MAC TFC blocked and excess power states a note on the relationship between TFCs and transit power is required.

In the WCDMA system, as the data rate changes (due to either changes in the spreading codes or procedures in the rate matching algorithms), the amount of transmit power required to achieve a specific QoS (e.g. BER) also changes. In selecting different TFCs from the TFCS, we are essentially changing the data rate and consequently the

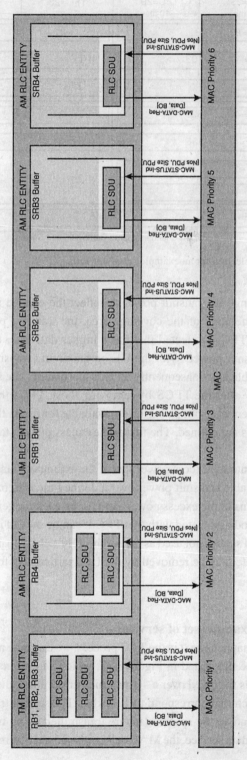

**Figure 8.26** RLC buffer status exchanges.

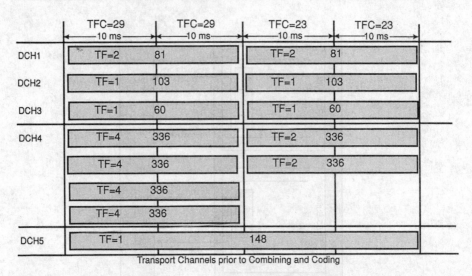

**Figure 8.27**  TFC selection for an example set of services.

UE needs to change its transmit power to reflect the change in TFC. Now, as a UE moves towards the edge of the coverage area, the transmit power tends towards its upper limit. If a TFC is selected that has a higher data rate, this results in a request for a higher transmit power. With the UE on maximum transmit power, this increase may not be possible and consequently there is a problem. The solution to this problem is to dynamically alter the TFCS by removing those TFCs that cannot be supported based on the current transmit power levels available for use in the cell. To facilitate this process two states are defined. The first is the excess power state and the second is the blocked state.

A TFC is put into the excess power state if the estimated transmit power for the TFC exceeds the available transmit power over a defined measurement period. In addition, if the TFC remains in the excess power state for an even longer duration relative to the longest time period associated with that TFC it can be moved into the blocked state. A TFC in a blocked state is not used as part of the TFC selection procedure. A TFC in the excess-power state may be removed by the UE, particularly for sensitive applications such as speech.

### TFC selection for an example set of services

To explore what happens next in the TFC process, we examine a specific TFC selection. Consider the case illustrated in Figure 8.27, which shows the TFC selection process that the MAC has followed over a 40 ms time period. First, the MAC has selected the TFs for the speech that correspond to 12.2 kb/s. The speech transport channels have a TTI of 20 ms, and so the speech data are transmitted using two 10 ms radio frames. Next, for the PS data service, the MAC selects the TF that corresponds to the maximum

## 8.8 TFC selection in uplink in UE

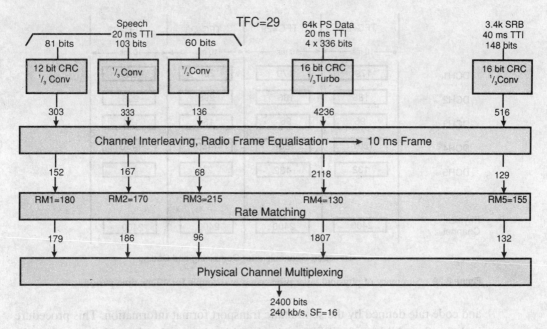

**Figure 8.28** Physical layer processing of example TFC = 29.

data rate of 64 kb/s. And finally for the SRB transport channel, the TF with the data rate of 3.7 kb/s is selected.

Referring back to the TFCS, we can see that this combination of TFs corresponds to TFC = 29. Because the PS data and speech have a 20 ms TTI and the SRB has a 40 ms TTI, the MAC needs to reselect the TFC during the transmission of the SRB data. In this case, the TFC selected needs to ensure that the TFC for DCH5 is present (in this example this means that only those TFC that include TF = 1 for DCH5 can be selected for the second 20 ms period).

In this example, we assume that the data available for the PS data transport channel drop and that only two transport blocks are available. In this instance the MAC selects TFC = 23 (a valid combination) for the next 20 ms.

**Physical layer procedures for TF combining for example set of services**

Having selected a specific TFC for the channels to be combined, we briefly examine the procedures that occur within the physical layer to combine the different transport channels onto the same physical channel. A more detailed explanation of the physical layer part of the transport format combination process was presented in Chapter 7.

Figure 8.28 summarises the physical layer procedures that occur to combine the transport channels with the specific TFs that are part of TFC = 29 onto a single physical channel. At the top, we have the bits that have been selected by the MAC from the different logical channels. In the first part of the physical layer processing, the CRC bits are added (where appropriate) and the data are encoded using the code

## 298 Layer 2 – medium access control (MAC)

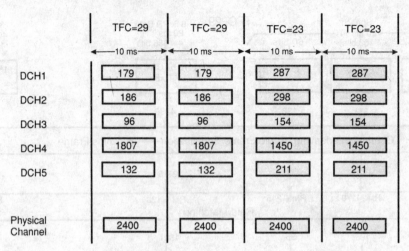

Figure 8.29  Summary of physical layer processing in example for 40 ms time period.

and code rate defined by the semi-static transport format information. This procedure includes the addition of tail bits, redundant bits that need to be added to ensure the channel coding schemes work efficiently.

Next, the coded data from the different transport channels are interleaved and the blocks sizes equalised to the radio frame size. The radio frame equalisation ensures that the coded data blocks can be equally divided into the appropriate number of radio frames based on the TTI that is valid for that channel. So, for example, a channel with a 20 ms TTI needs to be divisible by 2, and a channel with a 40 ms TTI needs to be divisible by 4.

The data are then segmented into blocks to be transmitted in specific radio frames. For a channel with a 20 ms TTI, two segmented blocks are created: the first block is transmitted in the first 10 ms of the 20 ms TTI and the second block in the second 10 ms.

Once the data are segmented into radio frames, the blocks of data from each transport channel to be broadcast in the first 10 ms radio frame are brought together into the rate matcher. The rate matcher either adds bits (through repetition) or deletes bits (through puncturing) according to the total number of data, the physical channel capacity and the rate match attribute that is defined in the semi-static TF information for each transport channel.

The rate match attribute acts like a weighting factor. The higher the attribute relative to the attribute for all of the other transport channels, the more likely data will be repeated. The lower the attribute, the more likely data will be punctured. Whatever the rate matching attribute, however, the algorithm is designed always to end up with exactly the right number of data for the physical channel data rate.

The final line in Figure 8.29 shows how many data there are after rate matching for the different transport channels. If we sum up all the data we find that we have exactly

2400 bits, which is exactly the capacity that we have for the 240 kb/s physical channel for that 10 ms radio frame.

Figure 8.29 summarises the physical layer processing for the 40 ms time period being considered. When the PS data channel changes from 64 kb/s to 32 kb/s the total number of transmitted bits remains the same at 2400 per 10 ms radio frame. As a consequence, all of the transport channels use repetition in the rate matching function to ensure that the data when all of the transport channels are combined add to exactly 2400 bits.

# 9 Layer 2 – RLC

## 9.1 Introduction

The RLC protocol is a layer 2 protocol that provides a range of transport services between an RLC entity in the UE and a peer RLC entity in the RNC. There can be multiple instances of the RLC protocol active depending upon the number of services and control channels that are required for a specific UE. In this chapter we consider the structure and operation of the RLC. The configuration and operation of the RLC in the UE is well defined within the RLC specification [28]. The configuration and operation of the RLC within the UTRAN is less well defined in [28] and to some extent is open to proprietary implementation. In this chapter, therefore, we focus on the aspects that are well defined, namely those pertaining to the UE, and where possible we address issues relevant to the UTRAN.

Three transport modes can be provided by the RLC layer: the TM, UM and the AM. In the following sections we examine the details for each of the different modes, their likely use and some aspects of their protocols.

## 9.2 TM

### 9.2.1 Introduction

The TM is a single direction protocol that operates between a TM entity in the UE and a TM entity in the RNC. The TM entity can be characterised as shown in Table 9.1.

### 9.2.2 TM structure

The structures of the transmitting and receiving TM entities are shown in Figure 9.1. As can be seen in the figure, the TM has limited functionality, providing a transmission buffer and some segmentation functions, the operational details of which are considered in the following section.

## 9.2 TM

**Table 9.1.** *TM characteristics*

| | |
|---|---|
| Transparent | The TM passes transparently through the RLC layer, meaning that no header information is added by the TM entity. |
| Single direction | The TM operates in a single direction, i.e. it is a connectionless protocol. |
| Segmentation | The TM can provide some limited segmentation functions where higher layer SDUs are broken into a number of RLC PDUs. |
| Buffering and discard | The TM can provide buffering and discard functions depending upon its configuration from higher layers. |

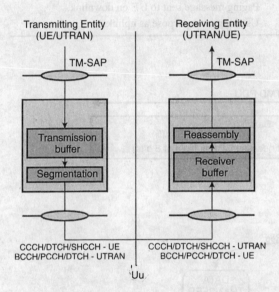

**Figure 9.1** Structures of RLC TM entities.

The RLC PDUs are sent via a number of logical channels to the MAC layer below. The selection of logical channel depends on the context of the message to be sent and the direction. The relationships between logical channel, direction and context are explained in greater detail in Table 9.2.

### 9.2.3 TM PDU

Figure 9.2 illustrates the structure of the RLC transparent mode data (TMD) PDU. One of the key features of the TMD PDU is that the length is not constrained to be an integer number of 8 bits (as is the case for the UM and AM PDUs as we will see later). The TMD PDU is constrained to have a defined size(s) depending upon the data that are to be transmitted using the TMD PDU.

**Table 9.2.** *RLC TM to logical channel mapping and usage*

| Logical channel | Direction | Typical usage |
|---|---|---|
| CCCH | UE to UTRAN | Used on the uplink for initial random access, cell updates, URA updates etc. |
| DTCH | UE to UTRAN | Used for services where delay is more important than reliability, e.g. speech and video services. |
| SHCCH | UE to UTRAN | (TDD mode only) Used by UE to request capacity on the PUSCH. |
| BCCH | UTRAN to UE | Used to transport system information in a cell. |
| PCCH | UTRAN to UE | Paging message sent to UE on downlink. |
| DTCH | UTRAN to UE | Used for same purpose as uplink. |

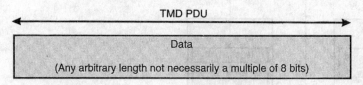

**Figure 9.2** TMD PDU.

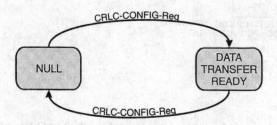

**Figure 9.3** RLC TM state diagram.

### 9.2.4 TM state diagram

Figure 9.3 illustrates the state diagram that is defined for the operation of the RLC TM in either the transmitter or the receiver. The CRLC-CONFIG-Req primitive is used to change the RLC TM state between NULL and DATA TRANSFER READY state. The CRLC-CONFIG-Req is used by the upper layers to control the operation of the RLC TM, i.e. to establish, reestablish and modify the RLC TM connection. The CRLC-CONFIG-Req primitive contains the TM parameters defined to control the TM, such as segmentation information, information on timers to be used for the RLC TM and information on how to handle erroneous PDUs.

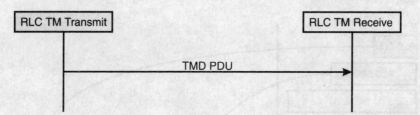

**Figure 9.4** RLC TM operation.

## 9.2.5 TM operation

Figure 9.4 shows the operation of the RLC TM. The RLC TM is a connectionless type protocol and so we can consider the operation of the transmit part and the receive part separately, starting with the transmit part.

### RLC TM entity creation

First we need to consider the creation and definition of an RLC TM entity. The RLC TM entity is defined and then created by the upper layers. The RLC layer knows that a TM RLC entity is being created when it receives the CRLC-CONFIG-Req primitive. First, we examine what may cause the establishment of an RLC TM entity such as:

1. A HTM entity created for SRB0 to allow the transmission of RRC messages such as the RRC CONNECTION REQUEST message. The structure for SRB0 is predefined in [24].
2. The UE needs a TM connection for a DTCH logical channel. In this case the TM RLC parameters for the RB are defined in the RLC INFO message, which is part of the RADIO BEARER SETUP message.

So we can assume that the higher layers of the UE have received the necessary parameters for the configuration of the TM RLC connection, and that now the specific RLC TM connection is in the DATA TRANSFER READY state ready to transmit (or receive) RLC PDUs using the TM operation.

Next we examine some of the operational aspects of the RLC TM, starting with the segmentation functions.

### Segmentation of RLC TM PDUs

Segmentation for the TM is configured by the upper layers and notified to the RLC entity via the parameters of the CRLC-CONFIG-Req primitive. For SRB0 (hard defined in [24]), segmentation is not configured; for other logical channels, for instance a speech or video channel using the DTCH, the upper layers are informed as to whether segmentation is required in the RLC INFO part of the RADIO BEARER SETUP message.

Let us assume that segmentation is configured and a large RLC SDU arrives that is larger than the TMD PDU. In this case all of the TMD PDUs that make up that

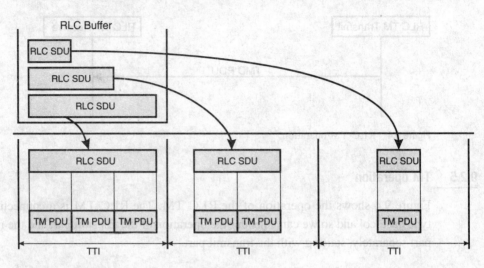

**Figure 9.5** Buffering and segmentation in RLC TM.

particular RLC SDU are sent in the same TTI and no other TMD PDUs can be sent from that specific RLC entity.

In Figure 9.5 we see the operation of the buffering and segmentation functions for a TM RLC assuming that both are configured within that specific RLC entity. The diagram illustrates the transmission of TM PDUs in successive TTI intervals. We assume that a number of RLC SDUs have arrived from the higher layers for this specific RLC entity (remember that there could also be a number of other RLC TM entities supporting other services, but operating in parallel).

The TM RLC entity extracts the data from the buffer in the order it arrives and segments each RLC SDU into the correct number of TM PDUs. It is a requirement, therefore, that the size of the RLC SDU is predefined and must be an integer multiple of a valid TM PDU size. Unlike the AM, however, there can be a number of different sized TM PDUs used for the same RLC TM entity.

Once the segmentation function is complete, the resulting TM PDUs are sent to the lower layers via the appropriate logical channel.

## Buffering and discarding of RLC TM PDUs

As part of the configuration of the RLC TM by the upper layers, the buffering and consequent discarding functions can be configured. For SRB0 (TM RLC on UL), the RLC SDU discard function is hard defined in [24] as not being configured. For any other RBs the buffering and discard functions are defined, for instance, as part of the RB setup procedure.

First, we consider the case when the discard function is not configured (as is the case for SRB0, for instance). In this situation, on receipt of a new RLC SDU, the TM RLC

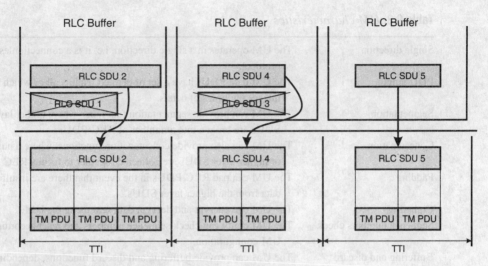

**Figure 9.6** RLC TM buffering with discard not configured.

entity discards all RLC SDUs received in previous TTIs and delivers the new RLC SDU for transmission by the lower layers in the next available TTI. Only the most recent RLC SDU per TTI is transmitted, as only one RLC SDU can be transmitted per TTI. This situation is illustrated by Figure 9.6. The alternative is that the RLC SDU discard procedure is configured through the use of upper layer signalling (e.g. on receipt of a RADIO BEARER SETUP message).

For the RLC TM, only 'timer based discard without explicit signalling' is available. The operation of the procedure is based on a timer (the timer is called TIMER_DISCARD, it is defined by the higher layers and takes a value from 10 ms to 100 ms in steps of 10 ms). The timers are configured when the RBs are defined (e.g. as part of the RADIO BEARER SETUP message for the RBs).

For the TM, the RLC TM entity starts TIMER_DISCARD when the RLC SDU enters the TM buffer. If the timer expires before the RLC SDU is submitted to the lower layers for transmission, the transmitter discards the SDU without any signalling to the peer RLC entity in the receiver.

**Operation at the receiver**

The operation of the receive part of the RLC TM entity depends upon the configuration of the segmentation function and whether erroneous SDUs should be delivered to higher layers (defined as part of the QoS requirement for a specific RAB). If segmentation is configured, the TM RLC entity reassembles the RLC SDU from the TMD PDUs received in a TTI; if it is not configured, it treats each TMD PDU as an RLC SDU.

The delivery of erroneous SDUs depends upon whether it is requested and whether segmentation is configured.

**Table 9.3.** *UM characteristics*

| | |
|---|---|
| Single direction | The UM operates in a single direction, i.e. it is a connectionless protocol. |
| PDU length | The UM data PDUs have a set of defined lengths all of which are integer multiples of octets. |
| Segmentation | The UM can provide segmentation functions, where higher layer SDUs are broken into a number of RLC PDUs. |
| Concatenation | The UM can also provide concatenation functions, where a number of higher layer SDUs are collected together to form a RLC PDU. |
| Padding | The UM can pad RLC PDUs in the event that there are insufficient data from the higher layer SDUs. |
| Ciphering | The UM can encrypt and decrypt data sent using the UM. |
| Sequence numbers check | The UM entity can check sequence numbers and react according the UM configuration. |
| Buffering and discard | The UM can provide buffering and discard functions, depending upon its configuration from higher layers. |

## 9.3 UM

### 9.3.1 Introduction

The UM is a single direction protocol that operates between a UM entity in the UE and a UM entity in the RNC (and vice-versa). The characteristics of a UM entity are given in Table 9.3.

### 9.3.2 UM structure

The structures of the transmitting and receiving UM entities are shown in Figure 9.7. The UM has greater functionality than the TM, but is still a connectionless protocol. Any erroneous SDUs received by the UM require higher layer intervention. The RLC PDUs are sent via a number of logical channels to the MAC layer below. The selection of logical channel depends on the context of the message to be sent and the direction. The relationships between logical channel, direction and context are explained in greater detail in Table 9.4.

### 9.3.3 UM PDU

Figure 9.8 illustrates the structure of the RLC UM data (UMD) PDU. One of the key features of the UMD PDU is that the length is constrained to be an integer number of 8 bits. The UMD PDU can have a range of sizes that are defined as part of the transport channel configuration process. We will see a little later that the MAC is responsible

## 9.3 UM

**Table 9.4.** *RLC UM to logical channel mapping and usage*

| Logical channel | Direction | Typical usage |
|---|---|---|
| DCCH | UE to UTRAN | Used on the uplink for measurement reports and acknowledging the release of an RRC connection. |
| DTCH | UE to UTRAN | Used for services where delay is more important than reliability. |
| SHCCH | UTRAN to UE | Used to notify UEs in the CELL_FACH state that system information is about to change. |
| CCCH | UTRAN to UE | Used to inform UEs of RRC connection setup information, successful URA/cell update, RRC connection release/reject. |
| DCCH | UTRAN to UE | Used for a range of purposes such as RB setup/reconfiguration, active set update, transport channel reconfiguration. |
| DTCH | UTRAN to UE | Used for same reasons as uplink. |
| CTCH | UTRAN to UE | Common traffic channels on downlink, e.g. cell broadcast information. |

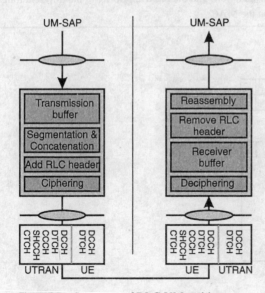

**Figure 9.7** Structures of RLC UM entities.

for informing the UM entity of what UMD PDU size should be used for a specific TTI. The UM entity is then responsible for segmentation and concatenation to ensure that the UMD PDU is filled as efficiently as possible. The meaning of the individual elements in the UMD PDU is defined in Table 9.5.

### 9.3.4 UM state diagram

Figure 9.9 illustrates the state diagram that is defined for the operation of the RLC UM in either the transmitter or the receiver and Table 9.6 defines each of the states.

## Layer 2 – RLC

**Table 9.5.** *Definitions of the contents of the UMD PDU*

| Element | Description |
| --- | --- |
| Sequence number | The sequence number is a 7 bit binary number that defines the sequence of an UMD PDU and is used for reassembly. |
| E bit | Extension bit is a single bit to indicate what is the following field ($E = 1$ a length indicator, $E = 0$ data or padding). |
| Length indicator | For UM, the length indicator can be either 7 bits (UMD PDU $\leq 125$ Octets) or 15 bits depending on the largest UMD PDU size. The length indicator is used to define the last octet of each RLC SDU ending within the UMD PDU. |
| Data | RLC SDU must be a multiple of 8 bits and concatenation should be used if possible to avoid the use of padding. |
| Padding | Padding (when used) is present to fill any unused space in an UMD PDU. A special length indicator is used to indicate the presence of padding. The total length of the UMD PDU is defined by the MAC based on TFCS considerations and the relative priority of the different transport channels in the MAC. |

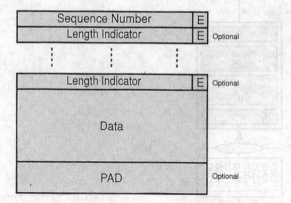

**Figure 9.8** UMD PDU.

The CRLC-CONFIG-Req primitive is used to change the RLC UM state between NULL, DATA TRANSFER READY and LOCAL SUSPEND states. It is used by the upper layers to control the operation of the RLC UM, i.e. to establish, reestablish and modify the RLC UM connection. The CRLC-CONFIG-Req primitive contains the UM parameters that are defined to control the UM such as information on timers to be used for the RLC UM, information on how to handle erroneous PDUs and the largest UMD PDU size currently being used by that UM entity.

CRLC-SUSPEND-Req and CRLC-RESUME-Req are used to suspend and then resume the operation of the UM data transfer service. Higher layers use these primitives

## 9.3 UM

**Table 9.6.** *State definitions for RLC UM entity*

| UM state | Description |
| --- | --- |
| NULL | In this state the RLC UM entity cannot transfer data. The RLC entity needs to be created through the use of an CRLC-CONFIG-Req primitive. |
| DATA TRANSFER READY | In this state the RLC UM entity can transfer data to its peer entity. From this state the RLC UM entity can move back to the NULL state or the LOCAL SUSPEND state. |
| LOCAL SUSPEND | In this state the RLC UM entity is suspended. |

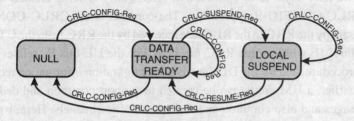

**Figure 9.9** RLC UM state diagram.

to control the operation of the UM entity. CRLC-SUSPEND-Req contains a parameter (*N*) that is used to define the maximum number of UMD PDUs that should be sent before suspending the operation of the UM data transfer service.

### 9.3.5 UM state variables

Within the RLC (for both UM and AM) there are state variables. The state variables are used internally within the different RLC entities to track the current state of the entity. Within the UM, VT(US) is the state variable that defines the sequence number of the next UMD PDU to be transmitted. Under normal operation, the state variable is incremented by 1 and starts with an initial value of 0. Under some error conditions (such as buffer discard), the state variable can be incremented by 2.

In the receiver, VR(US) is the state variable that defines the sequence number of the UMD PDU that it expects to receive next. VR(US) is set to equal the last received UMD PDU sequence number plus 1.

### 9.3.6 UM operation

Figure 9.10 illustrates the operation of the RLC UM. We start by considering the operation of the transmit part of the RLC UM entity.

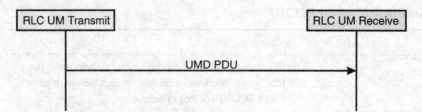

**Figure 9.10** RLC UM operation.

### Establishment of the RLC UM entity

In a similar way to the TM, the RLC UM entity is defined and then created by the upper layers. The RLC layer knows that a UM RLC entity is being created when it receives the CRLC-CONFIG-Req primitive. The contents of the CRLC-CONFIG-Req primitive (sent by the RRC to the RLC) are received by the RRC from the UTRAN via the 'RLC Info' IE within some RRC message. For the UL this IE relates to the PDU discard procedure, and for the DL there is no configuration information required.

As we saw earlier, a UM entity can be created to support common and dedicated signalling messages and also common and dedicated traffic channels. Here, we want to examine some of the configuration parameters that define these types of UM entity and how the UE receives this configuration information.

### Configuration of downlink SRB0 RLC UM entity parameters DL

As part of the establishment of an RRC connection, the UE needs to receive signalling messages on the downlink. To receive these messages (e.g. RRC CONNECTION SETUP) the UE needs to receive a CCCH logical channel that is defined to send data to an RLC UM entity. This signalling connection is referred to as SRB0, and is the downlink partner of SRB0, defined for the uplink. The RLC parameters for the UM entity to support SRB0 are hard defined in [24], remembering for the downlink, however, that it contains no RLC Info configuration parameters.

### Configuration of uplink and downlink SRB1 RLC UM entity parameters

Another example of the configuration of an UM RLC entity is a second SRB (SRB1), which is defined in the RRC CONNECTION SETUP message sent when the UE requests an RRC connection. Again, for the downlink there is no RLC Info for the UM RLC entity. For the uplink, the configuration of the discard procedure needs to be defined. There are two options for the configuring of the uplink discard procedure – 'SDU discard not configured' and 'Timer based discard, without explicit signalling'. A typical choice for SRB1 would be the latter case, and a typical value for the only configuration parameter would be 'Timer_Discard' set to 50 ms.

Next we can examine some of the operational aspects of the RLC UM, starting with the segmentation functions.

## 9.3 UM

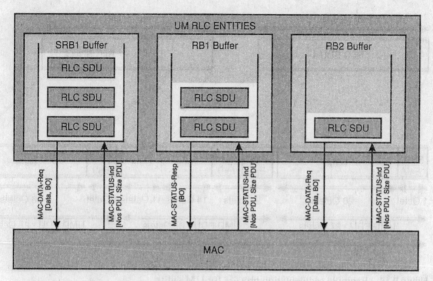

**Figure 9.11** Relationship between RLC UM and MAC.

### Preliminary processing prior to segmentation of RLC UM PDUs

To understand how the segmentation process works with RLC UM entities, we first need to understand the relationship between the MAC and the RLC UM entities.

As we saw previously, the RLC UM entities have a buffer that can store any RLC SDUs that are to be sent via an RLC UM link. Each RLC UM entity has a separate buffer space to store RLC SDUs until they can be transmitted across the radio interface. Figure 9.11 illustrates such a situation. In the diagram we have three UM entities. The first is SRB1, which is used for dedicated control messages, and the second and third are RB1 and RB2 respectively, which we assume are for dedicated traffic channels.

In the UM, the size of the UMD PDUs can vary depending on the allowed TFCs that have been configured for that specific UM link. The entity that controls the selection process for the size of the UMD PDUs to use for a specific TTI is the MAC. The MAC, therefore, needs to ascertain how many data are available in each of the UM buffers and also information that is used as part of the TFC selection process. To do this, a number of MAC primitives are defined. One primitive reports the RLC status (MAC-STATUS-Resp) and another delivers the data down to the MAC from the RLC layer (MAC-DATA-Req), and also includes the RLC buffer status.

Both of the primitives from the RLC UM to the MAC include buffer occupancy (the number of data in bytes) as well as parameters that affect the TFC selection process in the MAC. Whenever the UM entity receives an RLC SDU for transmission, it notifies the MAC indicating the current level of buffer occupancy. Upon receipt of this information from all of the RLC UM entities and similar information from any other RLC entities that are active, the MAC selects the appropriate TFC and hence the appropriate UMD PDU size as well as the number of UMD PDUs that the RLC should deliver to the

## 312 Layer 2 – RLC

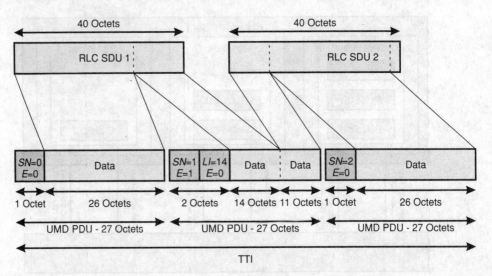

**Figure 9.12** Example segmentation process for UM entity.

MAC for transmission in the next available TTI. This information is sent from the MAC to the RLC UM entities in a MAC_STATUS-Ind primitive as illustrated figuratively in Figure 9.11. With this information, the RLC UM can now proceed with the segmentation process.

### Segmentation of RLC UM PDUs

One of the procedures for the UM entity is the segmenting of RLC SDUs and packing UMD PDUs. Figure 9.12 gives an example of the operation of the UM entity segmentation and concatenation procedure, in which we assume that the RLC UM entity has told the MAC that there are two RLC SDUs, each of length 40 octets, in the RLC UM entity buffer. For the next TTI, we assume that the MAC (based on priority and other factors) has requested the RLC UM entity to provide three PDUs each of length 27 octets. We also assume that the maximum PDU length at this time is such that a 7-bit length indicator can be used in place of the longer, 15-bit, length indicator.

The first UMD PDU is completely filled with the first 26 octets of the first RLC SDU. In addition, a UMD header of length 1 octet is added. The header defines the sequence number of the UMD PDU (set to zero in this case) and the E bit (set to zero indicating that data follows).

In the second UMD PDU, the remaining data from the first RLC SDU are put into the UMD PDU along with the beginning of the second RLC SDU. In this case, the first RLC SDU is completed in this UMD PDU and so a length indicator is added to the header. The length indicator defines the number of octets between the end of the RLC header and the last octet of the RLC SDU it refers to, including the last octet. In this example, there are 14 octets in the last segment of the RLC SDU and so the length indicator is set to 14. The header for this second UMD PDU has two octets, one for the sequence number (set to 1, i.e. incremented by 1 from the previous UMD PDU sequence number) and an E bit (set to 1 indicating a length indicator follows), next there is a second octet for the 7-bit length indicator, which also includes an E bit (set to 0 indicating data field next).

In the third and final UMD PDU that the RLC has to supply to the MAC, 26 octets of data from the second RLC SDU are added to the UMD PDU along with a 1 octet RLC header. The RLC header has a sequence number (set to 2) and an E bit (set to 0 indicating data field next). The remaining RLC SDU segments (three octets in total) can remain in the RLC UM entities buffer to be transmitted in the next TTI, concatenated with additional RLC SDUs as they arrive in the buffer, or alternatively follow any discard procedures configured for this RLC UM entity.

Although the use of padding is avoided where possible, the UMD PDU can be padded out to fill the available length in which case special length indicators are used to indicate the presence of padding. There are also a number of other special length indicators to cope with some less common situations where there is no room for a length indicator in the current UMD PDU and so these special length indicators are defined and included in subsequent UMD PDUs.

## Buffering and discarding of RLC UM PDUs

As we have seen above, the RLC UM entity uses a buffer to support its operation. There are discard functions that can be configured for use on the uplink. Like the TM entity, there are two possible settings for the configuration of this discard function. The first is 'SDU discard not configured', and the second is 'SDU discard without explicit signalling'. We review both of these procedures, starting with the first.

## 'SDU discard not configured'

In this mode of operation, the higher layers in the UE did not receive a configuration for the discard function. In the UM entity, RLC SDUs in the transmitter are not discarded unless the transmit buffer is full. The UM entity only discards the SDUs whilst the buffer is full, and it starts with the RLC SDUs that are at the front of a buffer (i.e. next to be transmitted). If the transmit buffer is full, there are two possible courses of action depending on what state the segmentation process is in:

- If no segments from the RLC SDU have passed to the MAC for transmission, the UM entity can discard the RLC SDU.
- If segments of the RLC SDU have been passed to the MAC for transmission, the transmit UM entity discards the remaining segments of the RLC SDU. Next, the sequence number is incremented by 2, and the first segment of the next RLC SDU is sent. A length indicator (length indication set to zero) is added to the first segment of the next SDU to indicate to the receive UM entity that the previous PDU contained the last segment of the previous SDU. The receive UM entity in reality did not actually receive this previous PDU and discards all segments of the SDU that it has

received. This course of action will fool the receiver into thinking that a complete RLC SDU was transmitted, but as it did not receive all of the segments, it will discard the RLC SDU (which is the desired result).

**'SDU discard without explicit signalling'**

This is the alternative buffer management procedure operated by the UM entity. In this mode of operation the UM entity starts a timer (Timer_Discard) when the RLC SDU is received from the higher layers for transmission. If the timer Timer_Discard expires, then the UM entity discards the RLC SDU. If segments of the RLC_SDU have already been transmitted, then the UM entity follows the procedure that we saw previously for the 'SDU discard not configured' mode. This causes the receiver to think that the RLC SDU was transmitted but received in error and it discards the segments of RLC SDU that it has already received.

**Summary of operation of the RLC UM entity**

The objective for the transmitter part of the RLC UM entity is to prepare RLC SDUs for transmission by the lower layers. To do this it has a buffer into which the RLC SDUs arrive. The RLC UM entity notifies the MAC of its current buffer occupancy (see Figure 9.11) and the MAC responds by requesting the UMD PDU size and the number of UMD PDUs to be used for the next TTI. The UMD SDU then segments and concatenates RLC SDUs into the required number of UMD PDUs as shown in Figure 9.12, and in doing so attempts not to use padding.

The RLC UM entity has a buffer management process that defines how to manage RLC SDUs that have been in a buffer too long and how to manage the buffer due to overflows.

**Operation at the receiver**

The objective of the receiver is to reassemble the RLC SDUs from the segments arriving in the UMD PDUs. If the received sequence number (VR(US)) increments by more than 1, it indicates that a UMD PDU was not received (either deliberately by the transmitter or due to noise and interference across the radio interface) and the RLC SDU that has some segments missing is discarded.

## 9.4 AM

### 9.4.1 Introduction

The AM is a bi-directional protocol that operates between an AM entity in the UE and an AM entity in the RNC. The characteristics of the AM entity are shown in Table 9.7.

## 9.4 AM

**Table 9.7.** *AM characteristics*

| | |
|---|---|
| Bi-direction | The AM operates in two directions and has a transmit and receive function in both of its peer entities. |
| PDU length | The AM PDUs have a length which is an integer multiple of octets, and the length of the AM PDUs is fixed for a specific entity. |
| Segmentation | The AM can provide segmentation functions where higher layer SDUs are broken into a number of AMD PDUs. |
| Concatenation | The AM can also provide concatenation functions where higher layer SDUs are collected together to form an AMD PDU. |
| Padding | The AM can pad AMD PDUs in the event that there are insufficient data from the higher layer SDUs. However, one of the criteria that the MAC follows when selecting the number of AMD PDUs is to minimise the use of padding wherever possible. |
| Ciphering | The AM can cipher and decipher data sent using the AM. |
| Sequence numbers check | The AM entity can check sequence numbers and react according to the AM configuration. |
| Buffering and discard | The AM can provide buffering and discard functions depending upon its configuration from higher layers. |
| In-sequence delivery | The AM can ensure that RLC SDUs are delivered in sequence to the higher layers if requested. |
| Duplicate detection | The AM can detect duplicate RLC SDUs, possibly received due to the retransmission process. |
| Flow control | The AM can control the flow of the data through the AM entity to prevent buffer overflow. |
| Error correction | Because the AM provides full retransmission procedures, any AMD PDUs received in error can be retransmitted to correct any errors at the receiver. |
| Protocol error detection and recovery | The AM provides facilities to detect whether the peer entities of the AM link protocol have lost synchronisation, and it also has procedures to recover from this. |

### 9.4.2 AM structure

The structure of the AM entity is shown in Figure 9.13.

**Number of logical channels**

The AM can operate using either one logical channel (shown in Figure 9.13 by a solid line) or alternatively using two logical channels (shown by the dashed lines). For the AM entity, there are two types of PDUs: AM data (AMD) PDUs and the AM control PDUs. The AMD PDUs are used to carry the data to be acknowledged and these data could be user traffic information (such as packet data destined for the Internet) or alternatively they could be user control information (such as layer 3 RRC control messages). The AM control PDUs are used to carry the AM entity control PDUs such as the STATUS PDU. These PDUs are used to control the AM connection.

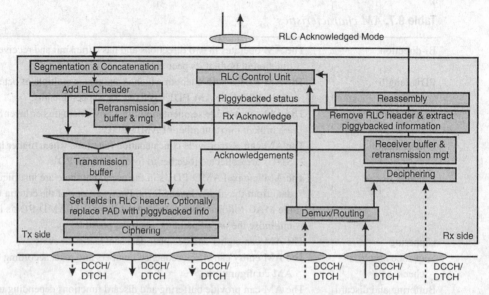

**Figure 9.13** Structures of RLC AM entity.

If the AM is using a single logical channel, the AMD PDUs and the AM control PDUs both go via the same logical channel and consequently have the same fixed PDU size. On the uplink, if the AM entity is configured with two logical channels, the first logical channel carries the AMD PDUs and the second the AM control PDUs. On the downlink, if the AM entity is configured with two logical channels, either logical channel can be used for AMD PDUs and similarly for the AM control PDUs.

**Explanation of structure**

The objectives for the AM entity are to transmit RLC SDUs reliably across the radio interface. To achieve this goal the AM entity uses a feedback technique whereby the receiving end of the link reports to the transmitting end how successfully it received the AMD PDUs. If there are missing or corrupted AMD PDUs the transmitter can arrange to retransmit these AMD PDUs at some time in the future.

To support these objectives, therefore, the AM entity is composed of two parts: a transmit part that is responsible for controlling segmentation and concatenation of RLC SDUs into AMD PDUs as well as the scheduling of AMD PDUs to be transmitted; and a receive part that is responsible for receiving AMD PDUs from the peer AM entity on the other side of the radio link, as well as detecting the feedback information on missing or corrupted PDUs received from the peer entity. The transmit part, therefore, provides functions to segment and concatenate RLC SDUs – it is used in a similar way to the UM entity. It adds AMD PDU header information, it has transmission and retransmission buffers and manages the encryption procedure. The receive part receives PDUs, which could be either AMD PDUs or AM control PDUs, and routes these PDUs to the appropriate entity. If the received PDU is a STATUS or PIGGYBACKED STATUS

## 9.4 AM

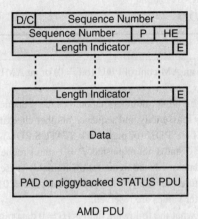

**Figure 9.14** AMD PDU structure.

PDU (that is one that is attached to the end of an AMD PDU), this information may result in the transmit entity retransmitting missing or corrupted AMD PDUs if required.

The following subsection now explores in greater detail the structure and use of the different AM PDUs.

### 9.4.3 PDUs used in AM

**AMD PDU structure**

Figure 9.14 illustrates the structure of the RLC AMD PDU. The meanings of some of the individual elements in the AMD PDU are very similar to those presented earlier for the UMD PDU and are given in Table 9.8.

**STATUS PDU**

Figure 9.15 illustrates the structure of the STATUS PDU. The STATUS PDU comprises a number of super fields (SUFIs). The SUFIs that are used in a specific STATUS message are dependent upon the implementation.

The SUFIs are used, in general, to allow the reporting of errors in the receiver, or for the transmitter to request the movement of the receive window (possibly due to RLC SDU discard in the transmitter). The general structure of the SUFI is a 4 bit type indicator, an optional 4 bit SUFI length indicator and optional (variable length) SUFI data.

Figure 9.16 shows the SUFI values that have been defined for the AMD STATUS PDUs. There are a number of different field types – for a brief description for each of the types and their use, see Table 9.9.

**Table 9.8.** *Definitions of the contents of the AMD PDU*

| Element | Description |
|---|---|
| D/C bit | This indicates if AM PDU is an AM control PDU (bit = 0) or an AMD PDU (bit = 1). |
| Sequence number | The sequence number is a 12 bit binary number that defines the sequence of an AMD PDU and is used for reassembly and sequence number checking. |
| P bit | This is used to request a STATUS PDU (or piggyback STATUS PDU) from the peer receive entity: $P = 0$ status not requested, $P = 1$ status requested. |
| Header extension (HE) bits | These are used to indicate if data or LIs are in the following field. These also include reserved bits for future use: $HE = 00$ next field data, $HE = 01$ next field is LI and E bit. All other values are reserved. |
| E bit | This is a single bit to indicate what the following field is ($E = 0$ data piggyback STATUS or padding, $E = 1$ another LI). |
| LI | For AM, the LI can be either 7 bits (AMD PDU $\leq$ 126 Octets) or 15 bits depending on the AMD PDU size. The LI is used to define the last octet of each RLC SDU ending within the AMD PDU. |
| Data | AMD PDU must be a multiple of 8 bits and concatenation should be used if possible to avoid the use of padding. |
| Padding | Padding (when used) is present to fill any unused space in an AMD PDU. A special LI is used to indicate the presence of padding. The size of the AMD PDUs is configured by the RRC, the number of AMD PDUs is selected by the MAC based on the TFCS and the relative priority of the logical channels entering the MAC. The use of padding, therefore, is selected by the MAC. |
| Piggybacked STATUS | Piggybacked STATUS PDU can be included to improve the efficiency of the AM link if capacity is available in a specific AMD PDU. |

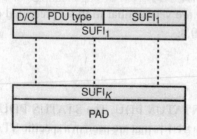

**Figure 9.15** STATUS PDU structure.

## Piggybacked STATUS PDU

Figure 9.17 illustrates the structure of a piggybacked STATUS PDU that can be sent attached to an AMD PDU as an alternative to sending a full STATUS PDU. In Figure 9.17, all the fields in the piggybacked STATUS PDU, except for the 'R2' field are the same as in the STATUS PDU that was considered previously. The R2 field is 1 bit long and set to binary 0. It is present to ensure that the piggybacked STATUS PDU is an integer number of octets in length.

## 9.4 AM

**Table 9.9.** *Definition of the different SUFIs used in the STATUS PDU*

| Field | Description |
| --- | --- |
| NO_MORE | The NO_MORE SUFI indicates that this is the end of the data part of a STATUS PDU. It is always the last SUFI in a STATUS PDU and any subsequent data will be padding bits. |
| WINDOW | The WINDOW SUFI is used by the receiver to change the transmit window size during a connection as long as it stays within the minimum and maximum values defined by higher layers. By changing the transmit window size, the receiver can control the flow of data from the transmitter to the receiver. The parameter window size number (WSN) defines the window size to be used. |
| ACK | The ACK field is used as part of the acknowledgement process from the receiver to the transmitter. The receiver uses the last sequence number (LSN) to indicate how many PDUs are received.<br><br>The transmitter can only move its transmit window on the basis of either an ACK or an MRW_ACK. There are a number of possible uses of the ACK field that will determine whether all AMD PDUs detected as being in error are reported in the same STATUS PDU or in additional STATUS PDUs.<br><br>The ACK is the last field in the STATUS PDU and so the NO_MORE field is not required, and any data following are treated as padding. |
| LIST | The LIST field defines a list of AMD PDUs received in error. The sequence number (SN) and length (L) define the number of the errored PDUs and how many subsequent PDUs are received in error. |
| BITMAP | The BITMAP field defines the errored PDUs in the form of a bitmap (0 indicates that the SN of the PDU at that bit position was not received correctly and 1 indicates that the PDU with SN defined by the bitmap position was received correctly). The bitmap is defined relative to FSN, which defines the SN of the first PDU in the bitmap. LENGTH defines the length of the bitmap in octets. This leads to a maximum bitmap size of 16 octets or 128 AMD PDUs.<br><br>This method of counting errors is good if there tend to be a lot of errors, and the errors tend to occur in bursts. |
| RLIST | RLIST defines a relative list SUFI. Starting from the sequence number of the first PDU received in error (first sequence number (FSN)), a code word (cw) is used to encode a number that defines how many PDUs have been received in error from the one defined by FSN up to the next PDU received in error. |
| MRW | Move receive window (MRW) is used to move the receive window and can be used to define the RLC SDUs that were discarded by the transmitter. LENGTH defines the number of SN_MRW fields in the SUFI and the SN_MRWs can be used to define the SNs of the RLC_SDUs that were discarded, if that is configured for the RLC AM entity.<br><br>NLENGTH is used to indicate the end of the last RLC SDU to be discarded. |
| MRW_ACK | MRW_ACK is the move receive window acknowledgement SUFI. It is used to acknowledge the receipt of the MRW SUFI and to update the receive window position. |

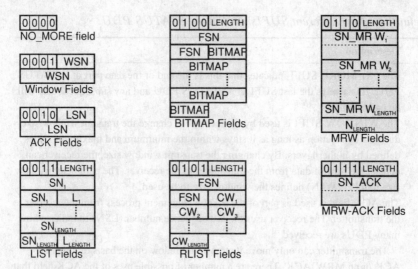

**Figure 9.16** STATUS PDU SUFI field definitions.

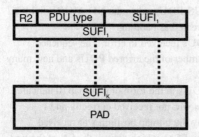

**Figure 9.17** Piggybacked STATUS PDU.

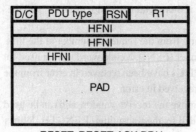

**Figure 9.18** RESET and RESET ACK PDU.

## RESET and RESET ACK PDU

Figure 9.18 illustrates the structure of the RESET AND RESET ACK PDUs. These PDUs are used as part of the AM reset procedure that is required when the peer RLC entities lose synchronisation. RSN is the reset sequence number and is a 1 bit number. R1 is a 3 bit field (coded as '000') and is used to ensure that the PDU lies on an octet

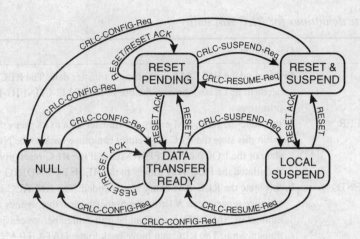

**Figure 9.19** RLC AM state diagram.

boundary. HFNI is the 20 bit hyperframe number indicator that is passed to the peer AM RLC entity. The passing of the HFNI ensures that the HFN in the UE and the UTRAN are synchronised.

### 9.4.4 AM state diagram

Figure 9.19 illustrates the state diagram that is defined for the operation of the RLC AM in either the transmitter or the receiver, and Table 9.10 defines each of these states. The CRLC-CONFIG-Req primitive is used to change the RLC AM states from the NULL state to the DATA TRANSFER READY state, or from any AM state back to the NULL state. It is used by the upper layers to control the operation of the RLC AM, i.e. to establish, reestablish and modify the RLC AM connection. The CRLC-CONFIG-Req primitive contains the AM configuration parameters such as the largest AMD PDU size currently being used by that AM entity.

CRLC-SUSPEND-Req and CRLC-RESUME-Req are used to suspend and then resume the operation of the AM data transfer service. Higher layers use these primitives to control the operation of the AM entity. CRLC-SUSPEND-Req contains a parameter ($N$) that is used to define the maximum number of UMD PDUs that should be sent before suspending the operation of the UM data transfer service.

Also shown in Figure 9.19 are the RESET and RESET ACK messages that can cause the change of state. The RESET message is used to reset the RLC AM entity in cases when the AM RLC looses synchronisation.

### 9.4.5 AM state variables

The AM state variables are used to define the current state of the AM entity. In total, there are 13 AM state variables and these are summarised in Table 9.11.

**Table 9.10.** *State definitions for RLC AM entity*

| UM state | Description |
|---|---|
| NULL | In this state the RLC AM entity cannot transfer data. The RLC entity needs to be created through the use of an CRLC-CONFIG-Req primitive. |
| DATA TRANSFER READY | In this state the RLC AM entity can transfer data to its peer entity. From this state the RLC AM entity can move back to the NULL state or the LOCAL SUSPEND state. If the RLC reset procedure is initiated the RLC AM moves to the RESET PENDING state. |
| LOCAL SUSPEND | In this state the RLC AM entity is suspended. The AM RLC is supplied a parameter $N$ (from the higher layers) that defines the number of AMD PDUs that can be transmitted before ceasing transmission. The RLC can move back to the DATA TRANSFER READY state on reception of a CRLC-RESUME-Req primitive. If an RLC reset procedure is initiated, the AM RLC moves to the RESET & SUSPENDED state. |
| RESET PENDING | No data are exchanged in the RESET PENDING state. |
| RESET & SUSPEND | Used as part of the reset procedure, the RESET & SUSPEND state means the entity cannot send or receive data and must wait for further commands. |

### 9.4.6 AM RLC configuration

The RLC can be configured at a number of instances such as the establishment of a radio connection, or the establishment of an RB.

Table 9.12 presents the configuration information that is within the 'RLC Info' part of the RRC message that is used to configure the RLC. The message is generic and can be applied to the TM, UM and AM RLC configuration. Here, we are focussing specifically on the configuration of an AM RLC entity. There are a number of messages that can contain the 'RLC Info' configuration information. The main one is the RADIO BEARER SETUP message, which is used to establish the RBs and SRBs that will use the AM RLC entity.

### 9.4.7 Types of AM polling

AM polling is the mechanism through which a status report from the receiver can be requested. In this section we review the different polling types that can be configured for a specific RLC AM mode entity. The different polling types are not exclusive and a number of different polling types can be active at the same time. These RLC AM entity parameters are configured by higher layers (RRC layer in the RNC) using the layer 3 RRC protocol, and in particular the 'RLC Info' information element that is sent to the UE and they are reviewed in Table 9.12. In this table under the heading 'Polling Info' a number of different configurable polling triggers are presented. We now review these polling triggers.

**Table 9.11.** *AM transmit and receive state variables*

|  | Variable | Description |
|---|---|---|
| Transmit state variable | VT(S) | Contains the sequence number of the next PDU to be transmitted (excluding PDUs that are to be retransmitted). |
|  | VT(A) | First PDU that needs acknowledging, lower edge of transmission window. |
|  | VT(DAT) | Counts the number of times each PDU has been transmitted. |
|  | VT(MS) | First PDU that should not be transmitted. Upper edge of transmit window. |
|  | VT(PDU) | Transmitted PDU counter. Used as part of the polling based on the number of transmitted PDUs. Reset when a poll is requested. |
|  | VT(SDU) | Transmitted RLC SDU counter. Used as part of the polling based on the number of transmitted RLC SDUs. Reset when a poll is requested. |
|  | VT(RST) | Counts the number of times the RESET PDU is sent. The variable is reset when the RESET ACK PDU is received. |
|  | VT(MRW) | Counts the number of times the MRW command is sent. Reset when the received window is moved. |
|  | VT(WS) | Defines the transmission window size. Initially the configured window size is used, but can be changed dynamically by the receiver using the STATUS PDU. |
| Receive state variable | VR(R) | Counts the received AMD PDUs. It is equal to the last received PDU plus one. |
|  | VR(H) | Defines the highest received sequence number, but must be less than the maximum acceptable sequence number. |
|  | VR(MR) | Maximum acceptable sequence number. This is the first AMD PDU that can be rejected and is at the end of the receive window. |
|  | VR(EP) | Counts the number of AMD PDUs expected to be retransmitted. |

### Last transmission PDU in buffer

The last transmission PDU poll refers to a poll that is triggered by the transmission of the last PDU within a specific RLC AM entity. The trigger occurs when the last PDU that has not yet been transmitted is sent to the lower layers for transmission. The RLC AM entity sets the poll bit in the AMD PDU header for the PDU prior to sending it to the lower layers. Table 9.12 shows this configurable trigger as being either active (True) or inactive (False).

### Last retransmission PDU poll

This is similar to the previous poll trigger except that it refers to the retransmission buffer rather than the transmission buffer. The trigger occurs when the last PDU in the retransmission buffer is sent to the lower layers for transmission. The RLC AM entity sets the poll bit in the AMD PU header for this PDU prior to sending it to the lower layers.

**Table 9.12.** *Typical configuration of AM RLC for PS domain*

| Message element | Values or range of values | Comment |
| --- | --- | --- |
| Choice RLC Info type | RLC Info or same as RB | Defines whether the RLC Info is defined explicitly or copied from an existing RB. We will examine the explicit mode |
| Choice UL RLC mode | | Defines the uplink RLC mode being configured |
| AM RLC | | We will consider the options for the AM RLC |
|   Transmission RLC discard | | A category that defines the different methods for discarding RLC SDUs. The options are: timer based explicit, timer based no explicit, Max_Dat retransmissions and no discard |
|     Timer based explicit | | Method that uses the MRW procedure to facilitate RLC SDU discard |
|       Timer_MRW | [50–900 ms] | Timer used to trigger the retransmission of a STATUS PDU containing MRW super field |
|       Timer_discard | [100–7500 ms] | Time before the SDU should be discarded |
|       MaxMRW | [1–32] | Maximum number of times that the MRW SUFI can be sent before a reset needs to occur. |
|     Timer based no explicit | | Method that uses a timer before the RLC SDU is discarded |
|       Timer_Discard | [10–100 ms] | Elapsed time before the RLC SDU is discarded |
|     Max_Dat retransmissions | | A discard mode that is based on the number of times that an AMD PDU can be retransmitted. It uses explicit signaling via the MRW SUFI to indicate the discard of an RLC SDU |
|       Max_Dat | [1–40] | Defines the maximum number of retransmissions before the RLC SDU is discarded |
|       Timer_MRW | [50–900 ms] | Timer used to trigger the retransmission of the STATUS PDU that contains the MRW superfield |
|       MaxMRW | [1–32] | Defines the maximum number of times the MRW procedure can occur |
|     No discard | | Used to reset the RLC entity if an AMD PDU is transmitted more than MAX_DAT times. |
|       Max_Dat | [1–40] | Defines the maximum number of times that an AMD PDU can be transmitted before an RLC reset will occur |
|   Transmission window size | [1–4095] | Defines the maximum number of AMD PDUs that can be transmitted before an acknowledgement is received. This will be based on the receive window size |
|     Timer_RST | [50–1000 ms] | Timer used to trigger RESET PDU retransmission |
|     Max_RST | [1–32] | Defines the maximum number of resets that can be performed. If the AM protocol exceeds this value, the RLC must indicate that there is an unrecoverable error to the upper layers |

**Table 9.12.** (*cont.*)

| Message element | Values or range of values | Comment |
| --- | --- | --- |
| Polling Info | | Defines information and parameters that configure and control the AM polling procedure |
| Timer_Poll_Prohibit | [10–1000 ms] | Parameter that defines the minimum time between polls. It is used to restrict additional polls being generated whilst the timer is active |
| Timer_Poll | [10–1000 ms] | A timer that is used to retransmit an existing poll if no STATUS PDU is received by the time the timer has expired |
| Poll_PDU | [1–128] | Defines how many PDUs are sent before setting the poll bit |
| Poll_SDU | [1–64] | Defines how many SDUs are sent before setting the poll bit |
| Last retransmission PDU poll | True/False | This causes the last PDU in transmission buffer to trigger a poll |
| Last transmission PDU poll | True/False | This will cause the last PDU in retransmission buffer to trigger a poll |
| Poll_Window | [50–99%] | Defines a threshold (expressed as a percentage of the transmission window) above which the poll bit will be set |
| Timer_Poll_Periodic | [100–200 ms] | Poll bit set via a periodic timer |
| Choice DL RLC mode | | Defines that the downlink RLC mode is being configured |
| AM RLC | | We will consider the options for the AM RLC |
| In-sequence delivery | [True/False] | Defines whether the RLC SDUs are passed to the higher layer in the receiver in order |
| Receiving window size | [1–4095] | Defines the maximum number of RLC PDUs allowed to be received |
| Downlink RLC status info | | |
| Timer_Status_Prohibit | [10–1000 ms] | A timer that blocks the transmission of status reports whilst the timer is active |
| Timer_EPC | [50–900 ms] | A timer that is used by the receiver to decide whether a PDU has arrived or not. It is used as part of the estimated PDU counter procedure |
| Missing PDU indicator | [True/False] | Defines whether the receiver should send a status report for each missing AMD PDU |
| Timer_Status_Periodic | [100–2000 ms] | Defines the period for periodic status reports to be sent from the receiver to the transmitter |

**Poll timer**

If configured by the higher layers (via the 'RLC Info' information element), the poll timer option can be used to set the poll bit, based on the expiry of the timer 'Timer_Poll'. It is optional whether or not the higher layers define the poll timer. If it is not defined, then the UE does not need to consider the use of the poll timer. The value for the timer can be set between 10 ms and 1 s and is configured by the higher layers. The operation of the poll timer is considered in more detail in Section 9.7. A typical value for this timer would be 200 ms.

**Every Poll_PDU PDU**

This is an optional polling function that is defined by the higher layers. If the option is configured, the RLC AM transmitter sets the polling bit every time Poll_PDU AMD PDUs have been transmitted or retransmitted (both types of PDUs will be counted together). The higher layers may define the parameter Poll_PDU. If the parameter is present in the higher layer message it indicates that this polling trigger is activated. It is an enumerated number in the range 1–128 PDUs.

**Every Poll_SDU SDU**

This polling trigger is similar to the Poll_PDU trigger, except that it considers the transmission of RLC SDUs in place of AMD PDUs. The poll bit is set when the last segment of the $n$th RLC SDU is transmitted in a specific AMD PDU. The $n$th RLC SDU is defined by the parameter Poll_SDU which is set by the higher layers with a range between 1 and 64 SDUs.

**Window based polling**

Window based polling is another optional trigger that can be used to set the poll bit in the RLC AMD PDU header. With window based polling, the poll bit is set whenever the transmitted PDUs expressed as a percentage of the transmission window is greater than the parameter Poll_Window that is set by the higher layers. The value for Poll_Window is 50–99% that of the transmit window.

**Timer based polling**

Timer based polling is another optional polling technique. With timer based polling, the poll bit is set periodically on the expiration of the timer (Timer_Poll_Periodic). The timer can be configured to take a value between 100 ms and 2 s.

### 9.4.8 Timers and timing constraints on RLC AM operation

A number of timers are associated with the operation of the AM RLC entity. In this section, we review each of these timers to define their purpose and also examine some of the time specific elements of the operation of the RLC AM.

## 9.4 AM

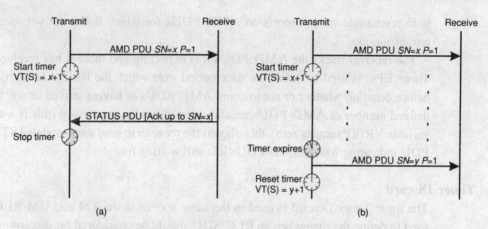

**Figure 9.20** Use of Timer_Poll timer in AM RLC entity: (a) successful case; (b) timer expires.

### Timer_Poll

The higher layers indicate whether or not to use this timer. The timer is used to control the timing of successive polls. We have seen in Table 9.12 that Timer_Poll takes a value between 10 ms and 1 s.

The basic principles of use for the Timer_Poll timer are illustrated in Figure 9.20 for two cases: (a) when the poll request is successful and (b) when the poll request is not successful. When the higher layers configure the use of this timer, it is started when an AMD PDU (with $SN = x$) is transmitted which has the poll bit set. In the successful case (a), the timer is stopped when the STATUS PDU is received which acknowledges the receipt of AMD PDUs with $SN = x$. If the timer expires before the receipt of the STATUS PDU (case (b)), another AMD PDU is sent with the poll bit set (in this case the sequence number is now $y$), the timer is reset and the variable VT(S) is set to the value of the next PDU to send ($y + 1$). If a new poll is sent whilst the timer is active, the timer will need to be reset and the variable VT(S) updated.

### Timer_Poll_Prohibit

Timer_Poll_Prohibit is a timer that is used to prohibit the transmission of additional polls. The timer is configured by the higher layers and we have seen in Table 9.12 that Timer_Poll_Prohibit takes a value between 10 ms and 1 s. If the timer is configured, it is started when an AMD PDU with the poll bit set is transmitted. Once active, no further polls can be transmitted until the timer has expired. Once the timer has expired, a single poll is transmitted if one or more polls have been requested.

### Timer_EPC

The use of this timer is configured by the higher layers. Timer_EPC takes values between 50 ms and 900 ms. The timer is used in conjunction with the estimated PDU counter (EPC) procedure available in the AM. The objective of the EPC procedure is to reschedule status reports on AMD PDUs for which it has not yet received a retransmission.

The receiver tracks the AMD PDUs it is expecting and those it has received. The Timer_EPC is used to define the time period over which the RLC AM entity waits before deciding whether or not to count AMD PDUs as having arrived or not. Once a defined number of AMD PDUs are indicated as not having arrived (this is when the variable VR(EP) equals zero), this triggers the receiver to send a new updated STATUS PDU indicating which AMD PDUs it is still waiting for.

**Timer_Discard**

The timer Timer_Discard is used in the same way as in the TM and UM RLC. It is used to define the time when an RLC SDU should be considered for discarding.

**Timer_Poll_Periodic**

This timer is configured by the higher layers and is used to define whether a periodic poll is to be sent. The timer is started when the RLC entity is created and restarted upon its expiration. When the timer expires the poll bit is set in the next AMD PDU to be sent to the receiver.

**Timer_Status_Prohibit**

The higher layers configure this timer. It is used to control the flow of status reports from the receiver to the transmitter. If the timer is active, status reports cannot be transmitted until the timer has expired. The timer is started when a STATUS PDU is sent to the transmitter.

**Timer_Status_Periodic**

The timer and its use are configured by the higher layers. This timer is used when periodic status reports are to be sent by the receiver. If the timer is configured, the receiver sends status reports periodically to the transmitter whenever the timer expires. Once the timer does expire it is restarted.

**Timer_RST**

This timer is used to control the transmission of additional RESET PDUs in the event that a RESET PDU is not acknowledged with a RESET ACK PDU. The timer is configured from the higher layers and started when a RESET PDU is transmitted. If no RESET ACK PDU is received by the time the timer has expired, then another RESET PDU is sent.

**Timer_MRW**

This timer is used to control the potential retransmission of the MRW SUFI. It is started when a STATUS PDU containing a MRW SUFI is transmitted. If the timer

## 9.4 AM

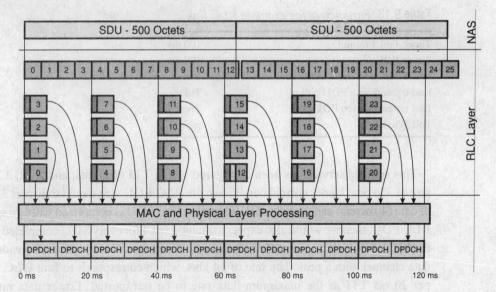

**Figure 9.21** Example of RLC AMD PDU creation – no retransmission required.

expires, an additional STATUS report is sent. The timer is stopped once the SDU discard procedure has been successfully completed.

### 9.4.9 AM operation

The AM data (AMD) transfer procedure is used to transfer data reliably between two peer AM entities. This transfer procedure is only available to entities that are in either the DATA TRANSFER READY state or the LOCAL SUSPEND state, which were described previously. The PDUs are transmitted on the DCCH for control messages and the DTCH logical channel for user plane messages.

The RLC AM procedure utilises the same segmentation procedure as used by the UM, except that the PDU size for the AMD PDU is fixed for a specific RLC AM connection.

**Simple example of RLC AMD PDU transmission**

Figure 9.21 illustrates the basic processing stages that occur within the operation of the AM RLC. In this example, we are considering the transmission of a number of higher layer SDUs. Here we see two SDUs of length 500 octets, but in general, the SDUs are an integer number of octets in length with a maximum size of 1502 octets for a PPP connection and 1500 octets for all other cases but excluding any additional header information introduced by the PDCP layer.

For this specific example, we saw previously that the RB is setup without the presence of the PDCP layer (i.e. header compression is not performed) and so we are looking at the higher layer SDUs arriving directly into the RLC.

**Table 9.13.** *Parameters for example RLC link*

| | |
|---|---|
| Timer_Poll_Prohibit | 100 ms |
| Timer_Poll | 100 ms |
| Poll_SDU | 1 |
| Last transmission PDU poll | True |
| Last retransmission PDU poll | True |
| Poll_Window | 99% |

This specific service has been configured with a TTI of 20 ms, and an RLC PDU size of 336 bits. Under normal conditions, the RLC PDU size would transport 320 bits of data (40 octets) and 2 octets of RLC header. If an SDU is completed within a specific RLC PDU, then we would also expect to include an LI for every SDU completed within the PDU, in which case the RLC header would increase. Finally, we are considering a data channel with a peak data rate of 64 kb/s, which corresponds to four RLC PDUs per 20 ms TTI as the maximum data rate to be transported. Lower data rates are possible (typically 0, 1, 2 or 3 PDUs per TTI), but are not shown in this example.

Figure 9.21 illustrates the segmentation of the SDU into a number of RLC PDUs, the inclusion of the RLC header and the combination and transmission via two 10 ms DPDCH radio frames per 20 ms TTI. The first SDU is converted into 13 RLC PDUs. The last part of the SDU is contained within the thirteenth PDU and consequently an LI is included in the RLC header to indicate where that SDU was completed. The following SDU also starts in the thirteenth RLC PDU, but due to the inclusion of the LI, the capacity of the thirteenth RLC PDU ($SN = 12$) is reduced to 39 octets.

The diagram assumes that all of the PDUs are received correctly and no retransmission is required. Below we will examine the detailed operation of the RLC, including the potential requirement to retransmit some of the RLC PDUs.

### Use of RLC retransmission

To view the operation of the RLC AM retransmission procedure, we consider another specific example. The AM RLC is quite complex in terms of the different configuration options that can be defined for its operation. Here, we focus on a specific example and examine the parts of the AM that feature in the example. For a detailed general understanding of the AM, the reader is referred to the RLC specification [28]. The configuration of the AM that we consider is outlined in Table 9.13. To consider the operation we consider Table 9.14, Figure 9.22, Figure 9.23, Figure 9.24 and Figure 9.25.

### Structure of the MAC

In the example we are assuming the same layer 1 and layer 2 structure as in the previous example. We have an AM link with a TTI of 20 ms, a peak data rate of 64 kb/s and

**Table 9.14.** *Example operation of RLC AM*

| SN Tx | P bit | VT (S) | VT (A) | VT (DAT) | VT (MS) | VT (SDU) | Timer_Poll | Timer_Poll_Prohibit |
|---|---|---|---|---|---|---|---|---|
| 0 | 0 | 0 | 0 | 0[0] | 32 | 0 | Inactive | Inactive |
| 1 | 0 | 1 | 0 | 1[0] | 32 | 0 | Inactive | Inactive |
| 2 | 0 | 2 | 0 | 2[0] | 32 | 0 | Inactive | Inactive |
| 3 | 0 | 3 | 0 | 3[0] | 32 | 0 | Inactive | Inactive |
| | | | | | | | | 20 ms |
| 4 | 0 | 4 | 0 | 4[0] | 32 | 0 | Inactive | Inactive |
| 5 | 0 | 5 | 0 | 5[0] | 32 | 0 | Inactive | Inactive |
| 6 | 0 | 6 | 0 | 6[0] | 32 | 0 | Inactive | Inactive |
| 7 | 0 | 7 | 0 | 7[0] | 32 | 0 | Inactive | Inactive |
| | | | | | | | | 40 ms |
| 8 | 0 | 8 | 0 | 8[0] | 32 | 0 | Inactive | Inactive |
| 9 | 0 | 9 | 0 | 9[0] | 32 | 0 | Inactive | Inactive |
| 10 | 0 | 10 | 0 | 10[0] | 32 | 0 | Inactive | Inactive |
| 11 | 0 | 11 | 0 | 11[0] | 32 | 0 | Inactive | Inactive |
| | | | | | | | | 60 ms |
| 12 | 1 | 12 | 0 | 12[0] | 32 | 1 | Inactive | Inactive |
| 13 | 0 | 13 | 0 | 13[0] | 32 | 0 | Inactive | Inactive |
| 14 | 0 | 14 | 0 | 14[0] | 32 | 0 | Inactive | Inactive |
| 15 | 0 | 15 | 0 | 15[0] | 32 | 0 | Inactive | Inactive |
| | | | | | | | | 80 ms |
| 16 | 0 | 16 | 0 | 16[0] | 32 | 0 | 0 ms | 0 ms |
| 17 | 0 | 17 | 0 | 17[0] | 32 | 0 | 0 ms | 0 ms |
| 18 | 0 | 18 | 0 | 18[0] | 32 | 0 | 0 ms | 0 ms |
| 19 | 0 | 19 | 0 | 19[0] | 32 | 0 | 0 ms | 0 ms |
| | | | | | | | | 100 ms |
| 20 | 0 | 20 | 0 | 20[0] | 32 | 0 | 20 ms | 20 ms |
| 21 | 0 | 21 | 0 | 21[0] | 32 | 0 | 20 ms | 20 ms |
| 22 | 0 | 22 | 0 | 22[0] | 32 | 0 | 20 ms | 20 ms |
| 23 | 0 | 23 | 0 | 23[0] | 32 | 0 | 20 ms | 20 ms |
| | | | | | | | | 120 ms |
| 24 | 0 | 24 | 0 | 24[0] | 32 | 1 | 40 ms | 40 ms |
| 25 | 0 | 25 | 0 | 25[0] | 32 | 0 | 40 ms | 40 ms |
| 26 | 0 | 26 | 0 | 26[0] | 32 | 0 | 40 ms | 40 ms |
| 27 | 0 | 27 | 0 | 27[0] | 32 | 0 | 40 ms | 40 ms |
| | | | | | | | | 140 ms |
| 6 | 1 | 28 | 6 | 6[1] | 38 | 0 | Inactive | Inactive |
| 8 | 0 | 28 | 6 | 8[1] | 38 | 0 | Inactive | Inactive |
| 28 | 0 | 28 | 6 | 28[0] | 38 | 0 | Inactive | Inactive |
| 29 | 0 | 25 | 6 | 29[0] | 38 | 0 | Inactive | Inactive |
| | | | | | | | | 160 ms |

a transport block size of 336 bits. The physical layer combines the transport blocks together, interleaves them and then transmits them via two radio frames. The consequence of this from the RLC perspective is that the RLC PDUs in the receiver cannot be recovered until all of the data from that 20 ms TTI has been received. Subsequent to that, the physical layer performs the necessary deinterleaving and decoding (see Chapter 7) to provide the MAC and hence the RLC with the recovered data.

As a consequence of the physical layer processing, there is a delay from when the RLC PDU is delivered to the lower layers for transmission to when the received RLC PDU is passed to the peer RLC entity in the receiver. The elements of this delay are the transmission time (defined by the TTI and set to 20 ms in this example), the propagation delay (e.g. 10 µs for a 3 km propagation path) and the receiver processing, consisting of the rake processing, deinterleaving and channel decoding (assumed to be around 20 ms total delay based on delay estimates presented in [29]). When we consider the operation of the RLC AM protocol, we need to consider the effects of the delay.

As a consequence of the 20 ms TTI, we now review the status and operation of the RLC AM entity in blocks of 20 ms.

### Details of the operation of RLC AM retransmission protocol

We now consider the detail of the operation of the RLC AM retransmission protocol in 20 ms time segments. The operation of the protocol is summarised in Table 9.14. The table presents some of the key RLC state variables and timers for each of the 20 ms time sequences considered here. The operation is also shown in Figures 9.22–9.25. The state variables and timers shown in the columns in Table 9.14 are defined in Tables 9.11 and 9.12.

### First 20 ms of operation

We are assuming that the RLC has received two large SDUs from the NAS, as shown in Figure 9.21. In this example, however, there is some retransmission of the data. Table 9.14 illustrates the operation of the RLC AM in terms of the various timers and state variables, as well as some of the values set in the RLC header.

With a 20 ms TTI, four AMD PDUs can be transmitted using two radio frames. In Table 9.14 this is illustrated by the value of the transmit sequence number (SN Tx) incrementing from 0 to 3. The P bit is set to zero, indicating that no poll is requested. VT(S) is initially set to zero, but is incremented by 1 for each AMD PDU, so after the fourth PDU has been passed to the lower layers for transmission, it has a value of 4. VT(A) defines the SN of the first PDU to be acknowledged, and in this case it is zero. VT(DAT) defines the number of occasions that each of the PDUs has been transmitted, and in all cases it is the first time and so they are all set to zero. VT(MS) is the first AMD PDU that cannot be transmitted, and in this example the window size is set to 32 PDUs and so VT(MS) is the PDU with $SN = 32$. VT(SDU) counts the number of complete SDUs transmitted so far. Currently, no complete SDUs have been transmitted

## 9.4 AM

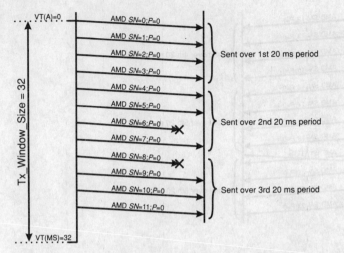

**Figure 9.22** UE transmits data, some data packets received corrupted.

and so this has a value of zero. The two timers, Timer_Poll and Timer_Poll_Prohibit are inactive and are not activated until a poll has been transmitted to the receiver.

### 40–60 ms of operation
The AMD procedure continues in a similar way for the transmission of PDUs with SN from 4 through to and including 11. Table 9.14 illustrates how the different state variables change. Figure 9.22 illustrates the transmission of the AMD PDUs from the transmitter to the receiver across the first 60 ms of operation. Indicated in the figure is the fact that some of the PDUs are not correctly received. In this example the PDUs with SNs of 6 and 8 are received in error.

### 60–120 ms of operation
The AMD PDU with $SN = 12$ includes the end of the first SDU and so the state variable VT(SDU) increments by 1. The parameter Poll_SDU has been configured to have a value of 1, and so the maximum number of SDUs have been transmitted before a poll needs to be transmitted and so the poll bit is set to 1 in AMD PDU with $SN = 12$. This indicates to the receiver that a STATUS PDU is requested.

The AMD PDU with $SN = 13$ is transmitted with the poll bit reset. The state variable VT(SDU) is reset as a consequence of the poll bit being set in the AMD PDU with $SN = 12$. The timers are inactive and they are not activated until the lower layers have indicated that the PDU has either successfully or unsuccessfully transmitted in the appropriate radio frames.

It takes over 20 ms for the radio frames that carry RLC PDUs 12–15 to be transmitted to the receiver. In addition, there is a processing period in the region of 20 ms in the receiver before the peer RLC entity in the UTRAN is able to detect the presence of the poll bit. In the interim, the transmitter continues to transmit RLC PDUs (at least

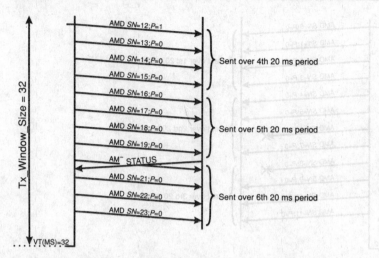

**Figure 9.23** UTRAN sends STATUS PDU on receipt of poll.

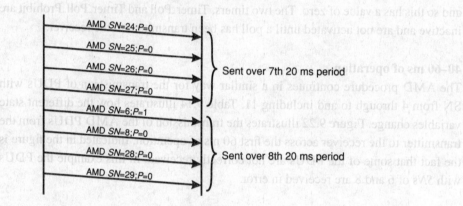

**Figure 9.24** UE resends missing PDUs.

until it reaches the end of the transmit window). The receiver transmits the STATUS PDU when it has successfully decoded the RLC PDUs. In this case it is transmitted in the downlink in radio frames 10 and 11 (100–120 ms from the start).

This phase of the procedure is illustrated in Figure 9.23.

### 120–160 ms of operation

In the period after the STATUS message is received, the UE processes the STATUS message, which again may take in the region of 20 ms. In the meantime, the end of the next SDU will occur in the AMD PDU with $SN = 24$. At this point in time, Timer_Poll_Prohibit is still active and therefore a new poll cannot be sent until either it expires or the STATUS message is received.

In the following 20 ms time period, we assume that the STATUS message has been decoded and passed to the RLC entity. The STATUS message uses the bitmap SUFI that indicates that AMD PDUs with $SN = 6$ and $SN = 8$ need retransmitting.

## 9.5 Summary

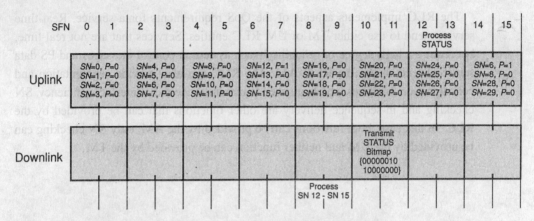

**Figure 9.25** Operation of RLC AM viewed in terms of radio frame transmission.

Additionally, the receipt of the STATUS PDU deactivates the Timer_Poll and the Timer_Poll_Prohibit timers.

In the following 20 ms period (140–160 ms – radio frames 14 and 15) the AMD PDUs with $SN = 6$ and $SN = 8$ are retransmitted. As the Timer Poll_Prohibit timer is now no longer active, the poll can be requested along with the retransmission of the PDU with $SN = 6$. This period is illustrated in Figure 9.24 and the complete process summarised in Figure 9.25.

### 9.4.10 AM RLC reset procedures

There are a number of occasions when it is necessary to reset the RLC AM entities. Examples of these occasions include:
- AMD PDU is retransmitted the maximum number of times defined by the configurable parameter (Max_Dat).
- The number of times that the receive window has been moved exceeds some defined maximum amount (MaxMRW).
- A STATUS PDU is received with an erroneous sequence number.

In any of these cases, the RLC AM is reset by transmitting a RESET PDU with the procedure completed on reception of the RESET ACK PDU. As part of the reset procedure, the HFN used by the encryption process is reset. The reset procedure resets the RLC AM to conditions similar to the initial conditions.

## 9.5 Summary

In this chapter we have reviewed the structure and operation of the RLC. All services passing through the RLC have an RLC entity assigned to them. These entities can offer a variety of transfer services such as the TM, the UM and the AM transfer service.

The RLC implements aspects of the QoS requirements for a service. Real-time services tend to use either TM or UM RLC entities. Services that are not real-time, but require a high degree of reliability (such as certain control messages and PS data services) tend to use the RLC AM. The RLC can also provide segmentation and concatenation functions which are used to maximise the radio interface efficiency. SN checking and in-sequence delivery are other functions that can be provided by the RLC. In this case, both functions can be provided by the AM, only SN checking can be provided by the UM and neither function can be provided by the TM.

# 10 PDCP and BMC protocols

In this chapter we consider two final layer 2 protocols, the PDCP and the BMC protocols. The PDCP provides the NAS with packet data transfer services such as header compression and lossless SRNS relocation. The BMC, for R99, provides cell broadcast services.

## 10.1 PDCP architecture and operation

Figure 10.1 illustrates the basic structure of the PDCP architecture. The PDCP is defined for the PS domain only. At the inputs to the PDCP layer are the PDCP SAPs. The RB is defined as the input to the PDCP layer for the PS domain. An RB is the transfer service that layer 2 provides to the higher layer protocols. The RAB is defined as the input to the AS and is a transfer service that the AS provides for NAS. An RAB is made from an RB and an Iu bearer between the RNC and CN.

The PDCP layer for R99 provides HC functions and support for lossless SRNS relocation. HC is defined using standard algorithms of which RFC 2507 is the one used for R99 [30].

### 10.1.1 PDCP PDUs

Figure 10.2 illustrates the three different PDUs that are defined for the PDCP layer operation: one when HC is not utilised, one when HC is utilised and one that also includes the SNs used as part of the PDCP SN synchronisation. The PDCP SeqNum PDU is used only when the PDCP layer entities lose synchronisation, e.g. as part of the RLC reset procedure.

### 10.1.2 HC

One of the functions of the PDCP layer is to support what is known as header compression (HC). With HC, TCP/IP, RTP/UDP/IP and other protocol headers can

# PDCP and BMC protocols

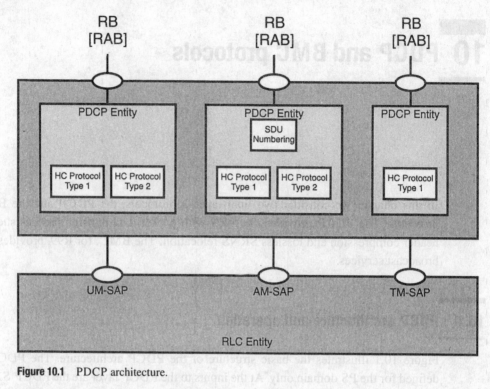

**Figure 10.1** PDCP architecture.

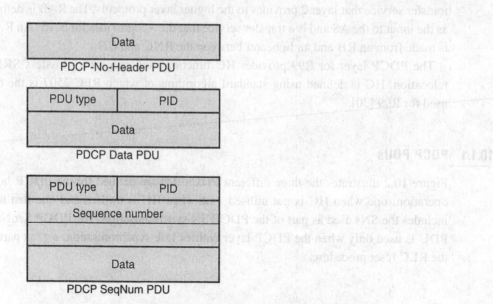

**Figure 10.2** PDUs that are defined for the PDCP layer operation.

## 10.1 PDCP architecture and operation

**Table 10.1.** *PDCP packet types used in support of HC*

| PDCP *PID* | Packet type | Comment |
|---|---|---|
| 0 | No header compression | PDCP PDU that is not compressed. |
| 1 | Full header | PDCP PDU whose header is not compressed; includes context ID (CID); includes a generation (a number related to context for non-TCP). |
| 2 | Compressed TCP | PDCP PDU with compressed TCP header; includes CID; includes flag indicating TCP fields that have changed. |
| 3 | Compressed TCP non-delta | PDCP PDU with compressed TCP header; includes CID; TCP headers elements normally sent in full rather than as delta. |
| 4 | Compressed non-TCP | PDCP PDU with compressed non-TCP header; includes CID and generation (for CID check) includes randomly changing fields from header. |
| 5 | Context state | List from decompressor to compressor containing CIDs which are out of synchronisation; sent without IP header across link. |

be compressed. For R99, HC is based on the Internet standard header compression algorithm defined in [30].

The basic idea of this algorithm is to send an uncompressed header which defines a context along with a context identifier (CID). Then the compressor can send the differences from the base context that can be used in the decompressor to rebuild the uncompressed header. The compression algorithm may need to be reinitialised (for instance after corruptions or after SRNS relocations).

Future versions of the standard will support additional HC protocols, for instance the robust header compression (ROHC) protocol [31] that is being defined for use with voice data across a noisy radio link.

Table 10.1 presents some of the packet identifier (PID) types that are defined for use with the header compression algorithms. The PID defines the type of packet that is being sent. The first option ($PID = 0$) is for a data packet where no HC is supported. The other *PID* values map onto the HC packet types defined by [30] with the details as they are laid out in the table.

### 10.1.3 Layer to layer protocol

Table 10.2 defines the primitives and the parameters that form the layer to layer protocol between the PDCP layer and the user of the PDCP services such as the NAS layer for data transfer or the RRC layer for configuration. There are two types of PDCP primitives defined: those used for the transfer of data, and those used for the configuration and control of the PDCP layer.

**Table 10.2.** *PDCP interlayer primitives*

| Primitive | Direction | Parameters | Comment |
|---|---|---|---|
| PDCP-DATA-Req | User to PDCP | | Used by upper layer to transfer upper layer PDU. |
| PDCP-DATA-Ind | PDCP to User | | Used to transfer received SDU to upper layers. |
| CPDCP-CONFIG-Req | RRC to PDCP | | Used to configure and reconfigure PDCP and assign it to an RB. |
| | | PDCP-Info | Parameters on HC to be used by PDCP entity. |
| | | RLC-SAP | RLC SAP to use for RLC (TM/UM/AM). |
| | | SN_Sync | Tells PDCP to start SN synchronisation procedure. |
| | | R/I | Tells PDCP to reinitialise/ initialise HC protocols. |
| CPDCP-RELEASE-Req | RRC to PDCP | RLC-SAP | Used by RRC to release the PDCP entity RLC SAP to use for RLC (TM/UM/AM). |
| CPDCP-SN-Req | RRC to PDCP | | Used in UTRAN to transfer PDCP SN to PDCP. |
| | | PDCP SN | The PDCP SN. |
| CPDCP-RELOC-Req | RRC to PDCP | | Initiates SRNS relocation for RBs supporting lossless SRNS relocation Next_Receive_SN at UE. |
| | | Next_Receive_SN | Two SNs (uplink and downlink) for next SN to be received. |
| CPDCP-RELOC-Cnf | PDCP to RRC | | Used to transfer the next send and receive SN to the upper layers. The Next_Send_SN is only used in UTRAN. |
| | | Next_Receive_SN | Two SNs (uplink and downlink) for next SN to be received. |
| | | Next_Send_SN | Two SNs (uplink and downlink) for next SN to be transmitted (UTRAN only). |

Figure 10.3 illustrates the operation of the PDP protocol in terms of the transfer of higher layer data from a sending entity (this could be the UE or the RNC) to the receiving entity (the RNC or UE). The PDCP-DATA-Req primitive is used by the user of the PDCP services to pass data to the PDCP layer for transfer to the peer entity. The functions provided by the PDCP layer depend on the configuration required, i.e. whether HC is applied and whether SN control is used for lossless SRNS relocation, for instance.

In this example, the PDCP is configured to use the TM or UM RLC. The data are passed to the RLC using the RLCxx-DATA-Req primitive (xx = UM or TM depending on RLC mode), and then across the radio interface to the peer RLC, using

## 10.1 PDCP architecture and operation

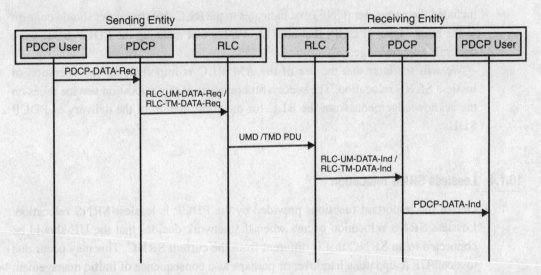

**Figure 10.3** Operation of PDCP entity.

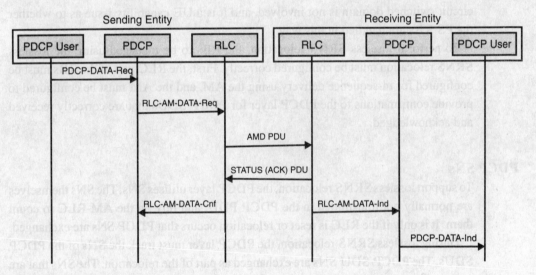

**Figure 10.4** PDCP operation via RLC AM.

either a UMD PDU or a TMD PDU. At the receiving side, the PDU is passed to the PDCP using the RLCxx-DATA-Ind primitive and to the PDCP service user using the PDCP-DATA-Ind primitive.

The next example, depicted in Figure 10.4, of the peer to peer protocol shows the use of the PDCP with an acknowledged mode RLC. In this instance, the AM RLC can confirm the successful transfer of a PDCP SDU (RLC-AM-DATA-Cnf primitive). This can occur if the RLC layer is correctly configured. The receipt of the RLC-AM-DATA-Cnf primitive by the PDCP only occurs if the RLC-AM-DATA-Req primitive included the parameter (CNF) that indicates to the RLC that the sender should confirm to the higher layer (PDCP) that all RLC PDUs that make up the PDCP SDU were correctly acknowledged.

We will see later that the use of the AM RLC is important for the operation of lossless SRNS relocation. The successful operation of this relocation service relies on the acknowledgements from the RLC for the confirmation of the delivery of PDCP SDUs.

### 10.1.4 Lossless SRNS relocation

One of the important functions provided by the PDCP is lossless SRNS relocation. Lossless SRNS relocation occurs when the network decides that the UE should be connected to an SRNC that is different from the current SRNC. This may occur due to a cell/URA update, a handover or perhaps as a consequence of traffic management within the network.

Note that not all RBs will need to be involved with lossless SRNS relocation. The circuit switched domain is not involved, and it is a UE capability issue as to whether the PS RBs are handled in this way.

To perform lossless SRNS relocation, the RBs to be managed using the lossless SRNS relocation must be configured correctly. First, the RLC part of the RBs must be configured for in-sequence delivery using the AM, and the AM must be configured to provide confirmations to the PDCP layer for PDCP SDUs that are correctly received and acknowledged.

**PDCP SNs**

To support lossless SRNS relocation, the PDCP layer utilises SNs. The SNs themselves are normally not transmitted in the PDCP PDU, and rely on the AM-RLC to count them. It is only if the RLC is reset or relocation occurs that PDCP SNs are exchanged.

During lossless SRNS relocation, the PDCP layer must track the SNs of the PDCP SDUs. The PDCP SDU SNs are exchanged as part of the relocation. The SNs that are exchanged define the PDCP SDU SNs that were transmitted but not acknowledged. After relocation, the transfer of PDCP SDUs starts at the first unconfirmed PDCP SDU. This occurs for both uplink and downlink.

The SN used by the PDCP is a 16 bit number (having a range 0–65535). In the UE this SN is managed as follows. A variable called the UL_Send PDCP SN is defined. This is initialised to zero when the PDCP entity is created. The SN of the first PDCP SDU that is sent is set to zero, and then the UL_Send PDCP SN is incremented by 1 for each PDCP SDU that is submitted to the lower layer for transmission.

The UE has a second variable called the DL_Receive PDCP SN. This quantity is initialised to zero when the first PDCP SDU is received from the lower layer. The variable increments by 1 for each subsequent PDCP SDU received by the PDCP from the lower layer.

The UTRAN numbers the PDCP SDUs in a similar manner to the UE. The SNs for the sent SDUs are DL_Send PDCP SN and for the received SDUs the UL_Receive PDCP SN.

## SN synchronisation

As part of the operation of the PDCP layer, if the SN synchronisation is lost, for instance due to an RLC reset procedure, then the PDCP SNs need to be resynchronised. The resynchronisation procedure occurs if the PDCP entity receives notification on a receive SN that is less than the variable xx_Send PDCP SN (xx is UL for UE and DL for UTRAN) or is greater than the first unsent PDCP SN.

In these circumstances the PDCP layers needs to resynchronise. To do this, the entity observing the out-of-synchronisation sends a PDCP SeqNum PDU to the lower layer. This PDU defines the SN as part of the header along with the data. At the receiver, the receive entity sets the variable xx_Receive PDCP SN (xx is DL in UE and UL in UTRAN) to the value of the SN received in the PDCP SeqNum PDU.

## Operation of lossless SRNS relocation

When lossless SRNS relocation is activated, the UE sends the next expected DL_Receive PDCP SN to the UTRAN and the UTRAN sends the next expected UL_Receive PDCP SN to the UE. Within the UTRAN, the source RNC sends to the target RNC the UL_Receive PDCP SN of the next expected PDCP SN, the DL_Send PDCP SN of the first transmitted but not acknowledged PDCP SDU and all of the transmitted but not acknowledged PDCP SDUs – including their PDCP sequence numbers, as well as the not yet transmitted PDCP SDUs. After relocation, the data transfer commences with the first unacknowledged PDCP SDU.

Figure 10.5 illustrates the message flows within the UE and between the UE and the network for the SRNS relocation procedure. In this example we are assuming that the SRNS relocation is triggered by a URA update within the UE. It could be a cell update or hard handover or some other cause that results in the network deciding to change the SRNC. In the following list, the paragraph number refers to the message SN in Figure 10.5.

1. The UE sends the URA UPDATE and the network decides to move the SRNC to a new RNC. Part of this procedure in the network requires the transfer of the Next_Send and Next_Receive PDCP SNs as well as any unacknowledged and unsent PDCP SDUs to the target RNC.
2. The target RNC responds with the URA UPDATE CONFIRM message, which contains the UL_Next_Receive PDCP SN for all RBs supporting lossless SRNS relocation.
3. The affected RLC entities are stopped by the RRC.

**344**   PDCP and BMC protocols

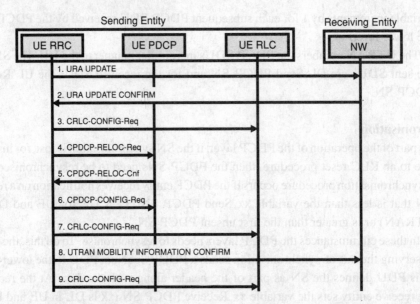

**Figure 10.5**   Message flows for lossless SRNS relocation.

4. The RRC sends the CPDCP-RELOC-Req primitive including the UL_Next_Receive PDCP SN from the UTRAN. This allows the PDCP layer to resynchronise the UL PDCP entity confirming any unacknowledged PDCP SDUs.
5. The PDCP layer responds with the CPDCP-RELOC-Cnf primitive which will include the DL_Next_Receive PDCP SN.
6. The RRC then requests the reinitialisation of the header compression algorithm if it is activated.
7. The RLC entity is reestablished.
8. The RRC in the UE responds to the network with the UTRAN MOBILITY INFORMATION CONFIRM message which will include the DL_Next_Receive PDCP SN for use in the network for synchronisation of the PDCP layer in the target RNC.
9. The RLC entities for the affected RBs are now reestablished and data transfer can recommence.

## 10.2   Broadcast/multicast control

Figure 10.6 illustrates the network architecture used for BMC functions. The network architecture includes a cell broadcast centre (CBC) which is connected to the UTRAN via the Iu–BC interface.

Figure 10.7 illustrates the protocol architecture for the BMC. The BMC messages are sent via the SCCPCH and the other channels as indicated in Table 10.3. The messages

## 10.2 Broadcast/multicast control

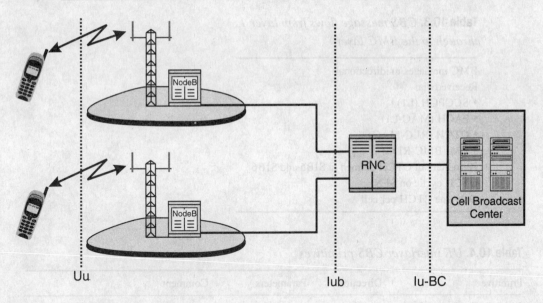

**Figure 10.6** BMC network architecture.

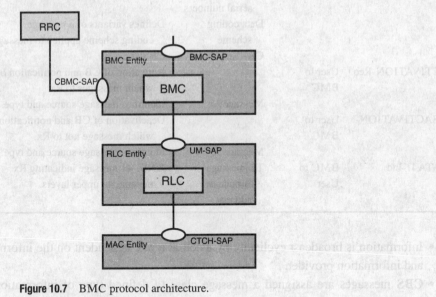

**Figure 10.7** BMC protocol architecture.

are periodic, with a periodicity defined by parameters that are broadcast to the UE in SIB5 and SIB6 messages.

The characteristics of the channel broadcast service (CBS) are basically those of GSM and are summarised below:
- The CBS page is 82 octets (equivalent to 93 characters).
- Up to 15 pages can be concatenated to form the CBS message.
- CBS messages include a message identifier and SN.

**Table 10.3.** *CBS message flows from layer 1 through to the BMC layer*

BMC messages unidirectional
Received via:
- SCCPCH (L1)
- FACH (MAC-L1)
- CTCH (RLC-MAC)
- UM (BMC-RLC)

Periodicity of CTCH defined in SIB5 and SIB6
CTCH cyclic on SFN
Only one CTCH per cell

**Table 10.4.** *UE interlayer CBS primitives*

| Primitive | Direction | Parameters | Comment |
|---|---|---|---|
| BMC-DATA-Ind | BMC to user | | Used to indicate received cell broadcast. |
| | | Message ID serial numbers | Identifies message source and type. |
| | | Data coding scheme | Defines variants of a message coding scheme applied to message. |
| | | CB-Data | CB data. |
| BMC-ACTIVATION-Req | User to BMC | | Activation of CB and notification of which messages to Rx. |
| | | Message Id | Identifies message source and type. |
| BMC-DEACTIVATION-Req | User to BMC | | Deactivation of CB and notification of which message not to Rx. |
| | | Message Id | Identifies message source and type. |
| BMC-DATA41-Ind | BMC to User | TL message-Broadcast address | ANSI 41 message indicating Rx message to upper layers. |

- Information is broadcast cyclically by a cell at a rate dependent on the information and information provider.
- CBS messages are assigned a message class (a defined type of information and language).
- Users can set the UE to ignore certain message types or languages.

Table 10.4 and Table 10.5 present the interlayer primitives used by the CBS service from the UE's perspective. Table 10.4 presents the primitives used between the BMC and the higher layers for passing CBS messages and configuration of the CBS service. Table 10.5 presents the primitives between the RRC and the BMC that are used to configure the CBS service, and to provide the RRC with the necessary scheduling information of broadcast messages of interest.

**Table 10.5.** *UE CBS configuration and control primitives*

| Primitive | Direction | Parameters | Comment |
| --- | --- | --- | --- |
| CBMC-RX-Ind | BMC to RRC | Action DRX selection | Used by BMC to indicate when CBS Rx should start and stop and when CB message of interest arrives in next period; start CBS Rx or stop CBS Rx; list of absolute CTCH BS of interest and which L1 should receive. |
| CBMC-CONFIG-Req | RRC to BMC | CTCH configuration | RRC to BMC to inform BMC on CTCH configuration; current CTCH BS index, FACH identification, TFS of FACH, reserved CTCH transmission rate. |

**Table 10.6.** *Summary of BMC PDUs*

| | |
| --- | --- |
| BMC CBS message | Carries cell broadcast data and address information for GSM based CBS. |
| CBS BMC schedule message | Defines CBS schedule including scheduling of next schedule message in next period. Allows support for DRX. |
| BMC CBS41 | Carries cell broadcast data and address information for ANSI-41. |

## 10.3 CBS PDU summary

Table 10.6 summarises the BMC PDUs, with Table 10.7 and Table 10.8 providing more details on the structure of the CBS message PDU and the Schedule PDU respectively.

### 10.3.1 CBS message and schedule message

The CBS message is the PDU that carries the CBS information, and includes message identification and version information that the UE uses to detect whether it needs to receive a specific message. The schedule message is the PDU that carries the scheduling information for future messages in what is referred to as a scheduling period. The scheduling information defines how many transport block sets there are in the scheduling period and also defines a bitmap that is used to indicate information on new CBS messages. The message description field is used to define the type of messages present in the bitmap.

**Table 10.7.** *CBS message PDU contents*

| Information element | Comment |
| --- | --- |
| Message type | Defines the PDU type (CBS message, schedule message or CBS41 message). |
| Message Id | Identifies the source and type of CBS message. |
| Serial number | Identifies variations of a CBS message. |
| Data coding scheme | Identifies alphabet/coding scheme and language in CBS. |
| CB data | CBS message. |

**Table 10.8.** *CBS schedule message PDU contents*

| Information element | Comment |
| --- | --- |
| Message type | Defines the PDU type (CBS message, schedule message or CBS41 message). |
| Offset to begin CTCH BS index | Points to first CTCH block set in the next CBS schedule period relative to current schedule message. |
| Length of CBS scheduling period | Number of CTCH block sets in the next CBS schedule period. |
| New message bitmap | Bitmap that identifies which CTCH BS includes all or part of a new CBS message. |
| Message description | Defines message type (repetition of new BMC message in period, new message, reading advised, reading optional, repetition of old BMC message in period, old message, schedule message, CBS41 message, no message); message ID – identifies source and type of CBS message (from upper layers); offset to CTCH BS index of first transmission – location of first transmission. |

## 10.4 Summary

In this chapter we have reviewed the structure and operation of the PDCP layer and the BMC layer. The PDCP layer provides support for two specific functions: lossless SRNS relocation and HC. Lossless SRNS relocation is used to ensure that packet data are not lost when the SRNC is moved from one RNC to another. HC is used to compress the data that are typically present in transmission control protocol (TCP)/IP headers. The BMC layer provides support for the CBS that was defined for the GSM system. The BMC allows the UE to select certain types of messages and notify the RRC of scheduling options for messages.

# 11 Layer 3 – RRC

## 11.1 Introduction

In this chapter we examine the structure and the operation of the RRC protocol. The RRC protocol is the main AS control protocol. It is responsible for the configuration and control of all of the different layers that create the radio connection between the UE and the UTRAN. It is a large and complex protocol and consequently, in this chapter, we consider only some key aspects of its operation, leaving the interested reader to consult the relevant specification [24] for a more thorough description.

We start this chapter with a review of the RRC protocol architecture before considering specific key elements of its operation.

### 11.1.1 Architecture and messages

The RRC protocol architecture is illustrated (from the perspective of the UE) in Figure 11.1. The key functions of the architecture are the dedicated control functional entity (DCFE), the paging and notification functional entity (PNFE) and the broadcast control functional entity (BCFE).

The RRC messages are passed between the UE and the UTRAN. They are used to configure and control the RRC connection between the UE and the UTRAN. The RRC messages can be loosely grouped into four categories: RRC connection management messages; RB control messages; RRC connection mobility messages and RRC measurement messages.

In Tables 11.1–11.4, we review the basic message types. It should be noted that in these tables we are not considering the individual messages, but rather a generic type of message. For instance, in Table 11.1 we look at an RRC CONNECTION message type. In fact, there are a number of such messages: RRC CONNECTION REQUEST; RRC CONNECTION SETUP; RRC CONNECTION SETUP COPMPLETE and RRC CONNECTION RELEASE. By considering just the message types, we can compactly represent the different messages in the tables.

The first category of messages is the RRC connection management messages, which are responsible for establishing and maintaining the RRC connection in whatever form

# Layer 3 – RRC

**Table 11.1.** *RRC connection management messages*

| Generic RRC message | Comment |
| --- | --- |
| RRC CONNECTION | Messages to establish, release and reject the creation of an RRC connection including the creation of SRBs. |
| SECURITY MODE | Messages to start, reconfigure confirm and indicate failure in the establishment of ciphering and integrity protection procedures. |
| COUNTER CHECK | Messages to request by UTRAN a check and provide a response to the current COUNT-C used for encryption and ciphering. |
| xx DIRECT TRANSFER | Messages to create a CN signalling connection (xx=INITIAL), send NAS PDUs on uplink (xx=UPLINK) and receive NAS PDUs on downlink (xx=DOWNLINK). |
| PAGING | Messages to send paging on common channels (Type 1) or using in-band dedicated channels (Type 2). |
| UE CAPABILITY | Messages to allow UTRAN to request and respond with the capabilities of the UE. |
| SYSTEM INFORMATION | Messages to carry from UTRAN system information and to indicate changes to it. |
| SIGNALLING CONNECTION | Messages to notify UE or UTRAN that signalling connection to CN is released. |

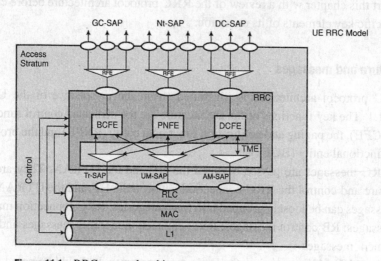

**Figure 11.1** RRC protocol architecture.

it may take. The messages include the RRC connection messages, security control messages, and system information broadcast messages, as well as messages for NAS data transfer. The DIRECT TRANSFER messages are considered in Section 11.4 when we examine the direct transfer procedure. The RRC CONNECTION establishment messages are considered in Section 11.5 when we look at the establishment of an RRC connection. The SECURITY MODE messages are considered in Sections 2.6 and the SYSTEM INFORMATION messages are considered in the next section.

## 11.1 Introduction

**Table 11.2.** *RB control messages*

| Generic RRC message | Comment |
| --- | --- |
| RADIO BEARER | Messages to establish, modify and release RBs and hence RABs. |
| PHYSICAL CHANNEL RECONFIGURATION | Messages used to assign, replace or release a set of physical channels. |
| TRANSPORT CHANNEL RECONFIGURATION | Messages to reconfigure a transport channel including the physical channels. |
| TRANSPORT FORMAT COMBINATION CONTROL | Messages to control the uplink TFC. |
| PUSCH CAPACITY REQUEST | [TDD] UE requesting uplink capacity on PUSCH. |
| UPLINK PHYSICAL CHANNEL CONTROL | [TDD] Message from UTRAN to transfer uplink physical channel information. |

**Table 11.3.** *RRC connection mobility messages*

| Generic RRC message | Comment |
| --- | --- |
| ACTIVE SET UPDATE | [FDD] To add, replace or delete radio links from the active set. |
| CELL CHANGE ORDER FROM UTRAN | Message from UTRAN to request cell change to another RAT cell. |
| CELL UPDATE | Messages to perform the cell update. |
| HANDOVER TO UTRAN | Message sent via another RAT to cause a handover to the UTRAN. |
| HANDOVER FROM UTRAN | Messages from UTRAN to cause handover to another RAT (e.g. GSM). |
| INTER-RAT HANDOVER INFO | Information from UE to UTRAN sent via another RAT prior to handover to UTRAN. |
| URA UPDATE | Messages to perform the URA update. |
| UTRAN MOBILITY INFORMATION | Messages used by UTRAN to allocate a new RNTI + other mobility information. |

The RB control messages shown in Table 11.2 are concerned with the establishment, modification and release of various aspects of the RBs and RABs created in the network. This set of messages can be used to configure all or individual layers of an RB. The RADIO BEARER control messages are considered in Section 11.6 when we review the RB establishment procedures, and the PHYSICAL CHANNEL RECONFIGURATION set of messages are covered in Section 11.7 when we consider some of the handover aspects.

The RRC connection mobility messages shown in Table 11.3 are concerned with the mobility aspects of the connection between the UE and the UTRAN. These messages include the soft-handover control messages (ACTIVE SET UPDATE) and handover

**Table 11.4.** *RRC measurement messages*

| Generic RRC message | Comment |
|---|---|
| ASSISTANCE DATA DELIVERY | Message from UTRAN to provide UE positioning assistance data. |
| MEASUREMENT CONTROL | Message from UTRAN to setup, modify or release a measurement. |
| MEASUREMENT REPORT | Message from UE to deliver measurement reports. |

messages to and from the UTRAN, as well as messages such as CELL and URA update. The procedures associated with handover are considered in more detail in Section 11.7 for both soft- and hard-handover.

The final categories of messages are the measurement control and measurement reporting messages that are shown in Table 11.4. The measurement messages are concerned with the controlling and reporting of the various measurements made by the UE and reported to the UTRAN. The subject of measurements is a large and complex issue and for these reasons it is addressed separately (Chapter 12).

## 11.2 System information broadcasting

We start looking at the design and operation of the RRC, beginning with system information broadcast messages. The system information broadcast messages are normally carried via the PCCPCH and the SCCPCH in the case of broadcast information used for DRAC. In this section we focus on the PCCPCH case.

### 11.2.1 Structure of broadcast system information

SIBs are system information that is transmitted from the UTRAN to the UE. The UE needs to locate and read the system information prior to establishing any radio connection to the UTRAN. One of the design problems associated with the SIBs is that the information comes in a variety of types. Some information is updated frequently (such as estimates of uplink interference levels as measured at the Node B) and some information does not need regular updating (cell and system IDs for instance). In addition, the messages can be long and also of varying lengths. The structure of the broadcast channels is designed, therefore, to cope with these differing constraints.

Before examining the structure and architecture of the broadcast channels, it is useful to examine the lower layers of the physical channels that carry the SIBs. The method of transporting the SIBs is via a common physical channel known as the PCCPCH. This physical channel is broadcast with a constant data rate and constant TF, so that it is easy for the UE to detect and decode the information that is carried by that specific channel.

## 11.2 System information broadcasting

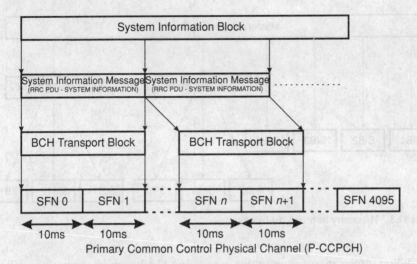

**Figure 11.2** Basic transmission structure of SIBs.

For the broadcast information, a specific transport channel is defined (the BCH). For the BCH, the TTI is fixed by the standard as being 20 ms, i.e. a transport block is delivered across the radio interface using two 10 ms radio frames. Figure 11.2 illustrates the basic transmission of the SIB messages and their relationship to the transport blocks and hence the PCCPCH. The radio interface has a frame structure that is based on a 10 ms frame with a cell SFN that counts the number of frames up to a total length of 4096. The SFN is used as the basis of the scheduling of the SIB information, as illustrated in the diagram.

### System information message

The SIBs are segmented and concatenated into system information messages with each system information message fitting into a BCH transport block. From Figure 11.2 it can be seen that an SIB is segmented into a number of system information messages, each of which becomes an RRC SYSTEM INFORMATION PDU. SIBs of different types can be concatenated into the same system information message. The scheduling of the system information messages is defined by information that is contained in a special broadcast message known as the master information block (MIB).

### 11.2.2 Example hierarchy of broadcast blocks

System information is organised with a tree-like hierarchy. Figure 11.3 illustrates an example of the hierarchy for the system information. At the top level, there is the MIB. The MIB, as the name implies, is the main controlling block that the UE needs to locate. The MIB contains either scheduling information for the SIBs directly, or

## Layer 3 – RRC

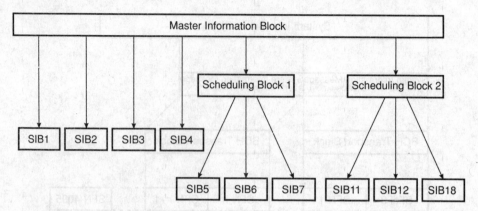

**Figure 11.3** Hierarchy of broadcast blocks.

scheduling information for up to two scheduling blocks which themselves define the scheduling for the SIBs. Only the MIB or the scheduling blocks can contain scheduling information.

The UE needs first to locate the MIB and from this it can locate the scheduling blocks (assuming any are being used). From the MIB and the scheduling blocks, the UE can identify the scheduling information for all of the SIBs. Each SIB is scheduled independently to allow different transmission rates for the SIBs.

### 11.2.3 Segmentation and concatenation of SIBs

In general, the SIBs are too large for the BCH transport blocks (the BCH transport block has a fixed size of 246 bits) and so segmentation and concatenation of the SIBs is required as shown in Figure 11.2. The SIBs are broken into system information messages, and each system information message is transported in an RRC SYSTEM INFORMATION PDU.

The segmentation and concatenation procedure is done at the RRC layer, and to facilitate the segmentation process a number of different segment types are defined:
- first segment,
- subsequent segment,
- last segment,
- complete.

In addition, the UTRAN can concatenate a number of segments from different SIBs if there is sufficient space within the system information message. For each segment type, there is header information as well as the data. The header information indicates the number of segments for a specific SIB and the subsequent and last segments contain a segment index to identify where they are within the segment for use when the reassembly of the segments is performed in the receiver.

## 11.2 System information broadcasting

**Table 11.5.** *Parameters used in the segmentation of SIBs*

| Parameter | Usage |
|---|---|
| SEG_COUNT | Defines the number of segments for a specific SIB. For no segmentation it equals 1. Values in the range 1, ..., 16. |
| SIB_REP | Defines the SIB repetition period, i.e. how many radio frames before the SIB is retransmitted. Only specific values allowed (4, 8, 16, 32, ..., 4096). |
| SIB_POS | Defines the SIB position within the SFN. Due to segmentation, this parameter can take multiple values for the same SIB. Must be a multiple of 2 (this is due to 20 ms transport block size and 10 ms frame size). |
| SIB_OFF | Defines the offset for subsequent segments of a segmented SIB. Must be a multiple of 2 for the same reason as above. SIB_OFF can be an array of elements, consequently the SIB offset can be variable. |

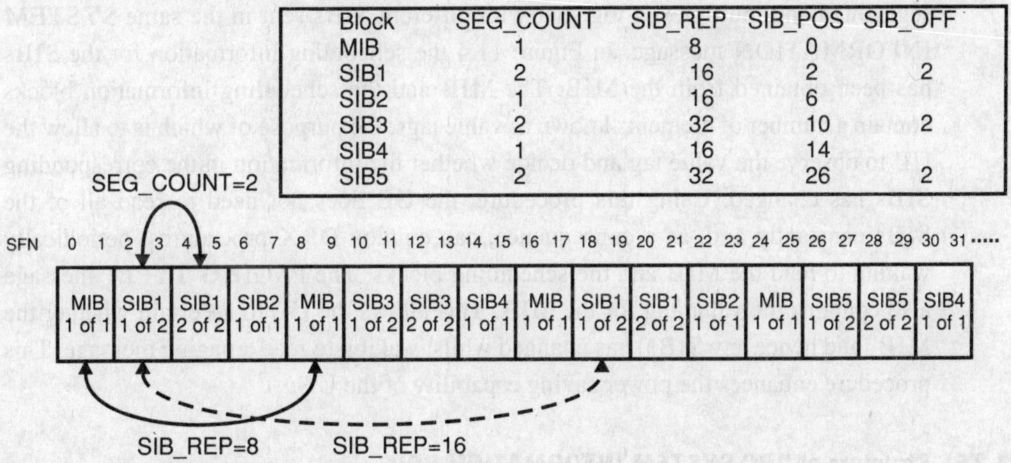

| Block | SEG_COUNT | SIB_REP | SIB_POS | SIB_OFF |
|---|---|---|---|---|
| MIB | 1 | 8 | 0 | |
| SIB1 | 2 | 16 | 2 | 2 |
| SIB2 | 1 | 16 | 6 | |
| SIB3 | 2 | 32 | 10 | 2 |
| SIB4 | 1 | 16 | 14 | |
| SIB5 | 2 | 32 | 26 | 2 |

**Figure 11.4** Example of SIB and MIB segmentation and scheduling.

### 11.2.4 Scheduling of system information

To define the scheduling of the SIBs a number of parameters are used as defined in Table 11.5, and illustrated in Figure 11.4. For the MIB, some of the parameters are fixed: $SIB\_POS = 0$ (i.e. it starts at the beginning of the SFN); $SIB\_OFF = 2$ frames (i.e. adjacent segments). The MIB repetition period ($SIB\_REP$) is 8 frames for the FDD mode and could be 8, 16 or 32 frames for the TDD mode. For the TDD mode the UE must attempt to determine the SIB_REP as no signalling information is used to indicate the value. In general, the segmented SIBs can be multiplexed together in the same SYSTEM INFORMATION message using the different message combinations considered in the following section.

The parameter $SIB\_POS$ defines the position of the first segment in the system frame, and $SIB\_OFF(i)$ is an array of offsets applied consecutively to define the location of subsequent segments (this is illustrated in the equation presented below). The number of segments is defined by the parameter $SEG\_COUNT$.

$$SIB\_POS(i) = SIB\_POS(i-1) + SIB\_OFF(i) \text{ for } i = 1, 2, \ldots, SEG\_COUNT\ 1$$

The MIB may be segmented and so the UE needs to read the contents of the first segment of the MIB to determine the parameter $SEG\_COUNT$ and consequently how many segments the MIB is in.

Figure 11.4 illustrates example parameters and scheduling of an MIB and some SIBs. In this example it is assumed that the MIB is not segmented ($SEG\_COUNT = 1$), and that the SIBs are transmitted individually, one per BCH transport block (per 20 ms). In reality, however, the MIB may be segmented (in which case the segments are sent in adjacent SYSTEM INFORMATION messages ($SIB\_OFF = 2$)). Also, SIBs are segmented and multiplexed together with different SIBs sent in the same SYSTEM INFORMATION message. In Figure 11.4 the scheduling information for the SIBs has been obtained from the MIB. The MIB and the scheduling information blocks contain a number of elements known as value tags, the purpose of which is to allow the UE to observe the value tag and decide whether the information in the corresponding SIBs has changed. Using this procedure, the UE does not need to read all of the SIBs constantly and, as a consequence, can employ DRX procedures, periodically waking to read the MIB and the scheduling blocks. The PAGING TYPE1 message also contains the value tag for the MIB. This allows the UE to ascertain whether the MIB (and hence any SIBs) has changed whilst waking to read a paging message. This procedure enhances the power saving capability of the UE.

### 11.2.5 Structure of RRC SYSTEM INFORMATION PDU

The UTRAN has the facility to segment and concatenate SIBs before transporting them via the BCH and PCCPCH to the UE. To achieve this a number of different combinations of segments have been defined. Table 11.6 identifies the different segments combinations, and their use. The structures for some of the different combinations of the RRC SYSTEM INFORMATION PDU are outlined in Figure 11.5. In all cases, the SYSTEM INFORMATION PDU must fit the BCH transport block size of 246 bits.

Figure 11.5 illustrates the structure of data that will become the transport block. The data include what is called the SFNPrime, which is the SFN with the least significant bit removed. The SFNPrime with 0 appended defines the SFN of the first radio frame used to carry the transport block, and SFNPrime with 1 appended defines the SFN of the second radio frame used to carry the transport block.

The UE initially does not know which of the two possible frames is the start of the 20-ms TTI that carries the BCH transport block. To locate the start of the 20-ms

## 11.2 System information broadcasting

**Table 11.6.** *List of possible combinations of SIBs in SYSTEM INFORMATION PDU*

| Combination | Description | Usage |
| --- | --- | --- |
| Combination 1 | No data | |
| Combination 2 | First segment | Used to carry the first part of a segmented SIB. |
| Combination 3 | Subsequent segment | Used to carry subsequent segments of a segmented SIB. |
| Combination 4 | Last segment | Used to carry the last (short) segment of a segmented SIB. |
| Combination 5 | Last segment<br>First segment | Used to carry the last (short) segment of a segmented SIB followed by the first segment of a following segmented SIB. |
| Combination 6 | Last segment<br>Complete SIBs | Used to carry the last (short) segment of a segmented SIB followed by a number of complete (unsegmented) SIBs. |
| Combination 7 | Last segment<br>Complete SIBs<br>First segment | Used to carry the last (short) segment of a segmented SIB followed by a number of complete (unsegmented) SIBs followed by the first segment of a segmented SIB. |
| Combination 8 | Complete SIBs | Used to carry a number of complete (unsegmented) SIBs. |
| Combination 9 | Complete SIBs<br>First segment | Used to carry a number of complete (unsegmented) SIBs followed by the first segment of a segmented SIB. |
| Combination 10 | Complete SIBs | Used to carry complete SIB of size 215–226 bits. |
| Combination 11 | Last segment | Used to carry the last segment of a segmented SIB of size 215–222 bits. |

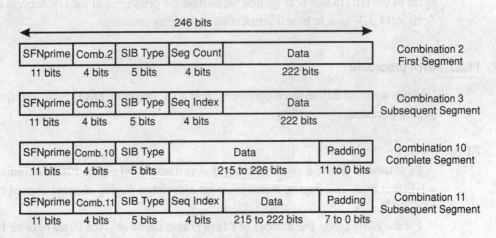

**Figure 11.5** Examples of SYSTEM INFORMATION PDU.

transport block, the UE uses the CRC bits that are attached to the transport block during the creation of the CCTrCH.

The UE attempts to decode the CRC bits for two possible start locations. One of the start locations is correct (CRC check passes), and the other is incorrect (CRC check fails). In this way, the UE can detect the start of the BCH transport channel and, from this, read the SFNPrime, which can be passed to the RRC via the MAC and RLC along with the contents of the transport block.

### 11.2.6 Purpose of SIBs

Table 11.7 defines the purpose of the different SIBs that are present in the system. All of the SIBs (except for SIB10) are sent via the BCH and the PCCPCH. SIB10 is used only in the FDD mode for dynamic resource control and is sent via the FACH and the SCCPCH.

## 11.3 Paging and DRX

### 11.3.1 DRX

DRX is used in systems such as UMTS to allow a UE to periodically move into a sleep mode. Whilst in this sleep mode, the UE is able to power down many of its normally operational functions, thus conserving battery power and prolonging the standby time.

One problem that is associated with the use of DRX is that the UE must be able to receive paging messages from either the UTRAN or the CN. If the UE is using DRX, this means that the paging messages need to be co-ordinated with the sleep cycle of the UE. In this next section we outline the principles of the DRX cycle and in Section 11.3.3, look at how it impacts on the paging process.

### 11.3.2 DRX procedure

Initially we must define three quantities: first a PI, second a DRX cycle and finally a paging occasion.

**PI**

A PI is used to define a short indicator that is transmitted on the PICH to indicate to a UE that there is a paging message on an associated paging channel carried by the SCCPCH.

For the FDD mode, the number of PIs per radio frame ($N_p$) (of 10 ms) can be 18, 36, 72 or 144. For the TDD mode the number of PIs per radio frame depends on a number of parameters (see [40]). The advantage in using the PI is that the detection in the UE is both easy and relatively fast. The following expression defines in the UE which PIs it should monitor (note there is also an additional layer 1 equation that defines which bits in the PICH the UE should monitor for a specific PI in a specific SFN – see Chapter 4):

$$PI = DRX\ Index\ mod\ N_p$$

## 11.3 Paging and DRX

**Table 11.7.** *Definitions for different SIB types*

| Name | Purpose |
| --- | --- |
| MIB | Main index for system information. Contains scheduling information on SIBs and up to two scheduling blocks. |
| Scheduling block 1 | Optional block used to provide scheduling information on SIBs. |
| Scheduling block 2 | Optional block used to provide scheduling information on SIBs. |
| SIB 1 | Contains NAS information (CN specific information) as well as information on timers for use in idle or connected mode. |
| SIB 2 | Contains information on the URAs that are available. There can be up to eight URAs in a cell. |
| SIB 3 | Contains information on the cell selection and reselection parameters that the UE should use whilst in idle mode. If SIB4 is not present it can also be used for UEs in connected mode. |
| SIB 4 | Contains information on the cell selection and reselection parameters that the UE should use whilst in connected mode. If SIB4 is not present the UE should use SIB3. |
| SIB 5 | Contains information on the common physical channels in the cell (PICH, AICH, P-CCPCH, PRACH, SCCPCH) for a UE in idle mode. If SIB6 is not present it can also be used for UEs in connected mode. |
| SIB 6 | Contains information on the common physical channels in the cell (PICH, AICH, P-CCPCH, PRACH, SCCPCH) for a UE in connected mode. If SIB6 is not present the UE should use SIB5. |
| SIB 7 | Contains information on fast changing cell parameters such as the uplink interference levels (used for open loop power control for the PRACH) and the dynamic persistence value (also used for PRACH). |
| SIB 8 | For FDD mode only. Contains static information for CPCH. |
| SIB 9 | For FDD mode only. Contains dynamic information for CPCH. |
| SIB 10 | For FDD mode only. Sent via FACH, contains information relevant to the DRAC procedures. |
| SIB 11 | Contains measurement control information for a UE in idle mode. If SIB 12 not present it can also be used for UEs in connected mode. |
| SIB 12 | Contains measurement control information for a UE in connected mode. If SIB 12 not present the UEs can use SIB 11. |
| SIB 13–13.4 | Contains information on ANSI-41 parameters used with ANSI-41 core networks. |
| SIB 14 | For TDD mode only. Contains outer loop power control information applied to dedicated and common physical channels. |
| SIB 15–SIB 15.4 | Contains information to be used for UE positioning methods such as GPS or OTDOA. |
| SIB 16 | Contains information on channel configuration (physical, transport and RB) to be stored in the UE for use during handover to UTRAN. |
| SIB 17 | For TDD mode only. Contains information on shared common channels to be used in connected mode. |
| SIB 18 | Contains PLMN identities for neighbouring cells to be considered for use by a UE that is in either idle or connected mode. |

where

DRX Index = IMSI div 8192

**DRX cycle**

The DRX cycle defines the periodicity of the DRX process (Table 11.8). The longer the DRX cycle, the longer the UE is in a sleep state, and the longer the delay before the UE can respond to a paging message. The DRX cycle length is defined by the DRX cycle length coefficient ($k$) thus:

DRX cycle length = $2^k$ frames for FDD mode

There can be a number of values for $k$ depending upon the current state of the UE. For the CN, each of the CN domains can have a different value for $k$. If the UE is attached to multiple CN domains, each with different DRX cycle lengths, then the UE selects the shortest cycle length. Similarly, there is also a DRX cycle length defined for the UTRAN.

**Paging occasion**

The paging occasion defines the SFN of the frame of which the UE must monitor the PICH to see whether a paging message is being sent to that UE. If the PI bits are set (i.e. equal to binary 1) in that paging occasion, the UE reads the paging message on the PCH transmitted on the associated SCCPCH. The paging occasion for the FDD mode is defined by:

$$\text{paging occasion (SFN)} = (\text{IMSI div } K) \bmod (\text{DRX cycle length}) + n^*\text{DRX cycle length} \quad (11.1)$$

where $n$ can take the values 0,1,2, ... up to a maximum such that the SFN is valid (i.e. < 4096) and $K$ is the number of SCCPCHs that carry a PCH. The paging occasion for the TDD mode is defined using a formula with slight modifications and defined in [40].

## 11.3.3 Example

Let us consider an example, in which we illustrate how a UE can estimate the paging occasion, the DRX cycle length and which PIs to look for. As we saw in Section 11.3.2, the information that the UE calculates is based in part on information received from broadcast messages, and also information that is calculated based on the IMSI. In this example we are assuming that there are four SCCPCHs ($K = 4$) that are carrying PCHs. First, the UE needs to ascertain which of the SCCPCHs it is using and to do this it uses the expression (IMSI mod $K$). The example shown in Figure 11.6 uses a specific IMSI and from this we find that the UE should be using SCCPCH1 from the four available (SCCPCH0–SCCPCH3).

## 11.3 Paging and DRX

**Table 11.8.** *Values for DRX cycle coefficient*

| Parameter | Values | Cycle length |
|---|---|---|
| 'UTRAN DRX cycle length coefficient' | 3–9 | 80 ms–5.12 s |
| 'CN domain specific DRX cycle length coefficient' | 6–9 | 640 ms–5.12 s |

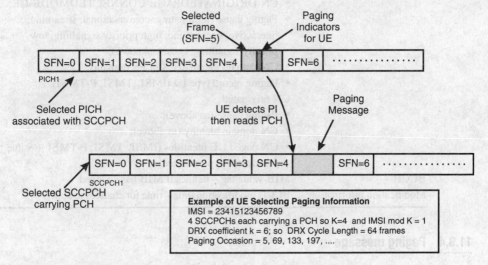

**Figure 11.6** Example of PI.

Next, the UE needs to calculate the DRX duty cycle and then the paging occasion. The DRX duty cycle is estimated using the expression presented earlier and in our example the DRX duty cycle coefficient $k$ is 6, which corresponds to a duty cycle of 64 frames. The paging occasion is calculated using the IMSI, $K$ and the DRX duty cycle using (11.1). In this example we find the paging occasion occurs on frame numbers 5, 69, 133 etc. and every 64th frame up to the maximum frame count of 4095.

The final thing that the UE needs to calculate is the PI from which it can calculate which part of the PICH to detect for a possible paging message. In this example, using the equation in Section 11.3.2, the PI is calculated as being 13.

Now, with all of this information, the UE can check whether the appropriate bits in SFN = 5 of the PICH are set. In this example (as shown in Figure 11.6) the PI bits are set and so the UE should read the paging message that will be transmitted on SCCPCH1 in the following frame.

The offset between the PICH and the SCCPCH is so defined to allow the UE to receive the PICH and then have time to read the paging message. For the FDD mode the TTI is 10 ms and for the TDD mode the TTI is 20 ms. This means that for the FDD mode the paging message is sent using a single radio frame.

**Table 11.9.** *Contents of paging message*

| Field | Comment |
|---|---|
| Paging record list | A list of paging records (1, . . . 8). |
| Paging record | Details of each paging record in the list includes: |
| | UTRAN originated: |
| | • u-RNTI; |
| | • CN-ORIGINATEDPAGE-CONNECTEDMODE-UE; |
| | • Paging cause (terminating: conversational; streaming; interactive; background; high priority signalling; low priority signalling; cause unknown); |
| | • CN domain identity (CS or PS domain); |
| | • Paging record type ID [IMSI, TMSI, P-TMSI] |
| | CN originated: |
| | • Paging cause (as above); |
| | • CN domain identity (as above); |
| | • CN paged UE identities (IMSI, TMSI, P-TMSI + value for identity type selected). |
| BCCH Modification information | MIB value tag – defines if MIB has changed. BCCH modification time – time for changes to apply. |

### 11.3.4 Paging message

The contents of the paging message are illustrated in Table 11.9. The paging message can include up to eight paging records. The paging message is from the CN, but could come via the UTRAN if the UE is in the CELL_PCH or URA_PCH states. In either case, the paging record defines the reason for the paging message (some type of mobile terminated transaction as shown in Table 11.9), and also the identity type and in the case of CN paging it also includes the identity itself.

## 11.4 RRC connection establishment

The next aspect of the RRC procedures that we consider is the establishment of an RRC connection. Figure 11.7 illustrates the basic RRC connection request procedure initiated at the request of higher layers and which is the first stage in establishing a signalling connection to the CN. At the start the UE is in the idle mode. In the following we examine the procedure outlined in Figure 11.7 in greater detail.

### 11.4.1 RRC CONNECTION REQUEST

This first message in the opening sequence is sent by the UE to the UTRAN using a CCCH logical channel, the RACH transport channel and the PRACH. The structure

## 11.4 RRC connection establishment

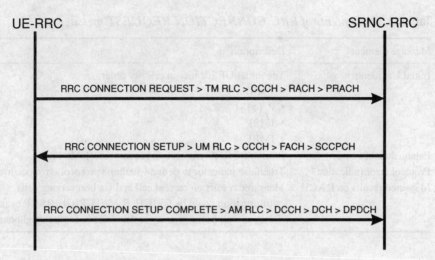

**Figure 11.7** RRC connection request procedure.

for these channels is part of SRB0 used by the UE on the uplink, and defined by the contents of the SIB5 broadcast message.

The UE sets the connection frame number (CFN) based on the SFN for the common channels as follows:

CFN = SFN mod 256

Next the UE maps the AC to ASC (this is considered in detail in Chapter 8) to facilitate PRACH parameter selection and performs the PRACH transmission procedure, sending the RRC CONNECTION REQUEST message.

The contents of the RRC CONNECTION REQUEST message are summarised in Table 11.10. One of the elements of the RRC CONNECTION REQUEST message is the establishment cause. The potential values for the establishment cause are presented in Table 11.11, it informs the UTRAN on the nature of the RRC connection required. Once the message is sent, the UE selects a SCCPCH carrying a FACH according to (initial UE identity mod $K$), where $K$ is the number of SCCPCHs that carry FACH, excluding those that only carry PCHs. Once the SCCPCH is selected, the UE monitors the SCCPCH and FACHs for the response from the UTRAN.

### 11.4.2 RRC CONNECTION SETUP

The action by the UTRAN to the RRC CONNECTION REQUEST is to return an RRC CONNECTION SETUP message to the UE and establish an RRC connection whose characteristics are defined within the setup message. As a minimum, three SRBs (SRB1, SRB2 and SRB3) are established with an optional fourth (SRB4) possible.

**Table 11.10.** *Contents of RRC CONNECTION REQUEST message*

| Message element | Description |
| --- | --- |
| Initial UE identity | The initial UE identity in priority order:<br>• TMSI + LAI;<br>• P-TMSI + RAI;<br>• IMSI;<br>• IMEI. |
| Establishment cause | Reason for RRC connection request (see Table 11.11). |
| Protocol error indication | True/false indicator to define whether a protocol error occurred. |
| Measured results on RACH | Measured results on current cell and six best serving cells. Information could be CPICH $E_c/N_o$, CPICH RSCP or path loss as well as the primary scramble code for the neighbour cells. |

**Table 11.11.** *Establishment causes*

| Signalling | MO call | MT call | Other |
| --- | --- | --- | --- |
| Originating HP | Conversational | Conversational | Emergency call |
| Originating LP | Streaming | Streaming | Inter-RAT cell reselection |
| Terminating HP | Interactive | Interactive | Inter-RAT cell change order |
| Terminating LP | Background | Background | Registration |
|  | Subscribed traffic |  | Detach |
|  |  |  | Terminating – unknown |
|  |  |  | Call Re-establishment |

Table 11.12 defines the basic contents of the RRC CONNECTION SETUP message and their purposes. Table 11.13 defines the processing steps that the UE follows, and the following section defines the contents of the RRC CONNECTION SETUP message in more detail.

## Selection of SCCPCH

Having transmitted the RRC CONNECTION REQUEST message, the UE needs to listen for the RRC CONNECTION SETUP message. First, the UE must identify the SCCPCH that is carrying the FACH that carries the CCCH that carries the SETUP message.

The UE will have listened to the SIB5 message that defines the structure of the common channels. Part of this message includes a list of the SCCPCHs present in the cell. The UE counts the number of SCCPCHs that carry a FACH (those SCCPCHs

## 11.4 RRC connection establishment

**Table 11.12.** *Contents of RRC CONNECTION SETUP message*

| Message element | Description |
| --- | --- |
| Initial UE identity | This should be the same as the one used by the UE. IE will look for this on selected SCCPCH/FACH. |
| RRC transaction identifier | An identifier (0–3) that is used to identify specific RRC messages. |
| Activation time | This defines the time (specified as the CFN) at which the parameters in the message will take effect. |
| New u-RNTI | UTRAN specific temporary identity allocated to the UE when entering connected mode. |
| New c-RNTI | Cell specific temporary identity allocated to the UE within a specific cell. |
| RRC state indicator | This defines which state the UE should enter. Only CELL_DCH and CELL_FACH are valid in the initial RRC connection setup message. |
| UTRAN DRX cycle length coefficient | The quantity '$k$' used to calculate the DRX cycle length. Value ranges from 3 to 9 for the UTRAN. |
| Capability update requirement | This defines whether the UE should supply FDD, TDD and other system (e.g. GSM) capability information. |
| Signalling RB information setup (multi 3–4) | This defines the three (optionally four) SRBs that need to be setup for the UE. The message also configures the RLC (uplink and downlink) and the mapping possible for the SRBs onto the different transport channels. |
| UL transport channel information common to all | This defines the transport channel information, TFSs and TFCs that are relevant to all transport channels on the uplink. |
| Added or reconfigured uplink transport channel information (multi nos. transport channels) | This defines information for transport channels that are being added (in this case). There is one message for each transport channel being defined. |
| DL transport channel information common to all | This defines common downlink transport channel information. |
| Added or reconfigured downlink transport channel information (multi nos. transport channels) | This defines downlink transport channel information for new/changed transport channels. |
| Frequency information | This defines UARFCN for uplink and downlink. |
| Maximum allowed uplink Tx power | This defines the maximum allowed uplink transmit power (–50 to 33 dBm). |
| Uplink DPCH information | This defines parameters for uplink DPCH such as scramble code type (short or long), scramble code number (0 to 16777215), (minimum) spreading factor. |
| CPCH set information | This defines the CPCH parameters if configured. |
| Downlink common for all RLs | This defines the physical channel parameters for the downlink RLs. Parameters include diversity type, compressed mode parameters, spreading factors. |
| Downlink information for each RL (multi) | This defines information specific to each RL in the active set. This also includes information for the PDSCH if that is also being allocated at the same time. |

**Table 11.13.** *Processing stages for RRC CONNECTION SETUP message*

| Step | Procedure |
|---|---|
| 1 | Process the 'activation time' information element. If it indicates that the time is 'now'; this means that the UE should activate the RRC connection on an appropriate TTI boundary. Alternatively, the 'activation time' could contain the CFN and hence the time that the UE should activate the connection. |
| 2 | Compute the DRX cycle length using the UTRAN DRX cycle length coefficient. This is used to define the paging occasions for the UE in connected mode. |
| 3 | UE selects state according to RRC state indicator. |
| 4 | Store the new c-RNTI received and use it for common channels (RACH, FACH and CPCH). |
| 5 | Store the new u-RNTI. |
| 6 | If requested compute the various capability information required for subsequent transmission to the UTRAN. |
| 7 | The UE establishes the different SRBs according to the information defined within the setup message. This information may also include multiplexing options that define which, and how, SRBs can be mapped onto the different possible transport channels (this can occur when moving from the CELL_DCH to CELL_FACH state and require different TFCs). In addition the MAC and RLC are configured, including the possibility that more than one logical channel is connected to the same transport channel (logical channel multiplexing). |
| 8 | The UE configures the TF set and TFCS according to the received messages. |
| 9 | The UE configures the physical layers according to the physical layer configuration messages. |

carrying only a PCH are ignored). If there are $K$ SCCPCHs carrying a FACH listed in the SIB5 message, the UE selects the SCCPCH according to:

Index of selected SCCPCH = Initial UE identity mod $K$

The index is a number in the range $0-(K-1)$ and identifies which of the SCCPCHs the UE should use. The first SCCPCH in the SIB5 list is index 0, the second index 1 and so forth.

The initial UE identity is the identity that the UE used in the RRC CONNECTION REQUEST message. The identity could be the IMSI, TMSI, P-TMSI, IMEI or DS-41 based identities. The UE converts the identity into an integer value prior to estimating the required index.

## Initial UE identity

Once the UE has identified the SCCPCH, it can listen for the RRC CONNECTION SETUP message that is intended for that UE. To do this, the UE has to decode each message on the selected SCCPCH and extract the initial UE identity contained within the message (note: the MAC UE identity field is not used for the CCCH carried by the FACH).

The initial UE identity is the identity that was used by the UE in the uplink and is used by the UTRAN in the downlink. This initial identity is only needed for the first exchange of information to allow the network to identify the UE prior to the allocation of the temporary UTRAN identity that will be used for subsequent messages.

**RRC transaction identifier**

The RRC transaction identifier is an integer in the range 0–3. The identifier is used to identify the different downlink procedures to allow multiple procedures. The UE uses the identifier for error trapping, such as the repeat transmission of the same message, or the transmission of a second RRC CONNECTION SETUP message.

**Activation time**

This is an integer between 0 and 255 that defines the CFN in which the changes specified in the remainder of the message should take effect. The activation time that is selected depends upon the CFN and the TTI boundary for all of the transport channels that are part of the CCTrCH. The CFN that is used for defining the activation time depends on whether the UE is being put into the CELL_FACH state or the CELL_DCH state.

In the CELL_FACH state the CFN is the same as that defined above. For the CELL_DCH state the CFN for calculating the activation time is given by:

$CFN = ((SFN*38400 - DOFF) \, div \, 38400) \, mod \, 256$     FDD mode

where DOFF is the default DPCH offset in steps of 512 chips for the FDD mode or

$CFN = (SFN - DOFF) \, mod \, 256$     TDD mode

where DOFF is the default offset in frames for the TDD mode. DOFF is defined in the part of the RRC CONNECTION SETUP 'downlink information common to all radio links'.

The activation time has a default value, and that default value is 'now'. A default of 'now' requires the UE to choose an activation time as soon as possible, which is short enough to allow the UE to respond to the RRC CONNECTION SETUP within a time in the region of 100 ms (the actual time is defined, but depends on a number of factors such as the Node B DPCH start time).

**New u-RNTI**

The u-RNTI is the UTRAN identifier for a UE. The u-RNTI is a 32 bit bit-string consisting of two parts: the SRNC identity (12 bits) and the s-RNTI (20 bits). The UE stores the u-RNTI and it is used when the UE is required to uniquely identify itself within the UTRAN.

**New c-RNTI**

The c-RNTI is a 16 bit bit-string used to uniquely identify a UE within a cell. The c-RNTI is an optional part of the RRC CONNECTION SETUP message and is only needed if the UE is being put into the CELL_FACH state.

**RRC state indicator**
>The RRC state indicator defines the state that the UE should move into after successfully completing the RRC CONNECTION REQUEST procedure. The UE is entering the connected state. There are only two valid states, i.e. the CELL_DCH state and the CELL_FACH state. The other two possible states (CELL_PCH and URA_PCH) are invalid states for a UE establishing an RRC connection and result in an error condition if received in the RRC CONNECTION SETUP message.
>
>A UE that is put into the CELL_DCH state is assigned a dedicated physical channel on both the uplink and the downlink. The UE can use the resources of the channel as required. A UE that is put into the CELL_FACH state is assigned a common physical channel (SCCPCH on downlink and PRACH or PCPCH on uplink). In this state, the UE must share the resources on the uplink with the other UEs in the cell that use these common channels.

**UTRAN DRX cycle length coefficient**
>The UTRAN DRX cycle length coefficient is an integer number in the range $(3, \ldots, 9)$ and is used by the UE to derive the length of the DRX period and the location of the paging occasions.

**Capability update requirement**
>This field defines whether UE capability information is required. The default value is false, indicating that capability information is not required. A value of true indicates that the UE should provide capability information.
>
>The capability update requirement can also request capability information on up to four other RATs, and for R99, GSM is defined as one of these RATs.

**SRB information to setup**
>The SRB information defines the SRBs that are being established as part of the RRC connection establishment procedure. Three SRBs and an optional fourth SRB are established as part of this procedure. SRBs are used as follows:
>>SRB1: UM RLC used for RRC signalling;
>>SRB2: AM RLC used for RRC signalling;
>>SRB3: AM RLC used for NAS signalling – high priority;
>>SRB4: AM RLC used for NAS signalling – low priority (optional).
>
>The contents of the SRB setup are as follows.

**RB identity**
>The RB identity defines the identity of the RB that is being established. The value of the first SRB is defined to be 1, and the value is incremented by 1 for each additional SRB. For the initial RRC connection establishment, we expect to see either three or four RBs and so the RB identity should be 1–3 or 1–4 respectively.
>
>For each RB, the information in the following subsections is defined.

## 11.4 RRC connection establishment

**Choice RLC info type**
The 'choice RLC info type' defines the RLC information for the SRBs. The choice keyword indicates that there may be more than one selection to choose from. The first selection allows the explicit definition of the RLC information (see below), or alternatively the RLC information can be the same as that for another RB ('Same as RB' option defined below).

**RLC info**
This first option for the RLC info type choice defines the RLC information explicitly. The details of this field of information define the configuration of the RLC layer for the SRB to which it applies. Both the uplink and the downlink fields can be defined, as can the three different RLC modes (TrM, UM and AM). For the mode selected, the subsequent fields define all of the parameters that should be configured for that mode. The parameters and configuration of the RLC layer are considered in detail in Chapter 9.

**Same as RB**
This is the second option for the RLC info type field. If this option is selected for the specific SRB being configured, then the RLC information is copied from an existing RB, and the value in the field defines the RB identity to copy from.

**RB mapping info**
The RB mapping information defines how the RB is mapped onto different possible transport channels. The basic objective of the RB mapping information is to define how the RBs (SRBs in this example) can be mapped onto different transport channels (possibly due to a UE being in either the CELL_DCH state or the CELL_FACH state).

The RB mapping information IE relates to a set of logical channels, and defines the logical channel identity and RLC configuration for these logical channels. Next, it associates these logical channels with those associated with a specific TFS IE either defined within the RRC CONNECTION SETUP message (as is the case for the DCH transport channels) or alternatively defined elsewhere (e.g. within the SIB5 or SIB6 message for the RACH). The mapping information can therefore be used to switch transport channels (for instance due to a change of state from CELL_FACH to CELL_DCH when instructed by the UTRAN) but still maintain the same logical channel, albeit with different TFC options in the transport channel, and different effective QoS for the transport channels.

At this stage, we are considering the SRBs, and these could be mapped onto either a DCH transport channel in the uplink and the downlink, or alternatively a RACH transport channel in the uplink and a FACH transport channel in the downlink. This situation is summarised in Figure 11.8.

**Number of uplink RLC logical channels**
The number of uplink RLC logical channels defines how many logical channels there are per RLC entity. This situation is discussed in detail in Chapter 9. Essentially, when

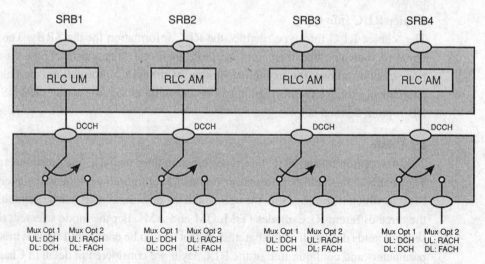

**Figure 11.8** SRB multiplexing options.

operating in the AM, the RLC requires a signalling path to transfer the RLC control information (such as STATUS PDUs). This signalling path may use the same logical channel as the data path (one logical channel option) or it may have its own logical channel (two logical channels option). In the example we are considering here, each RLC entity uses only one logical channel.

## Uplink transport channel type

For each multiplexing option and for each RLC logical channel the uplink transport channel type is defined. In our example, there is only one logical channel per RLC entity and two mapping options. For the first mapping option the uplink transport channel type is set to DCH, and for the second mapping option, the uplink transport channel type is set to RACH.

## Uplink transport channel identity

The uplink transport channel identity defines the identity of the transport channel that is being used if the transport channel is DCH or USCH (TDD mode only). If the transport channel is a RACH, this field is not present. In this situation, the UE has previously selected which of the available RACHs can be used, based on those available in the cell (defined in SIB6 or SIB5 if SIB6 does not contain the information) and the TTI usable with them. For these transport channels, the transport channel identity is defined in the appropriate SIB5/6 message for the RACH.

## Logical channel identity

The logical channel identity is used to distinguish the logical channels that are mapped to the same transport channel by the MAC. The logical channel identity is a number in

## 11.4 RRC connection establishment

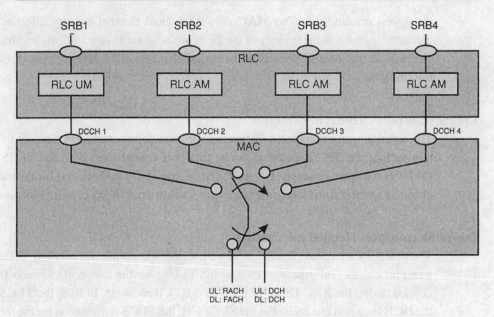

**Figure 11.9** Combined logical channel and transport channel multiplexing.

the range 1–15, so up to 15 logical channels can be multiplexed by the MAC onto the same transport channel. In this example, we are considering three or four SRBs being multiplexed onto the same transport channel. The logical channel identity, therefore, is 1 for SRB1, 2 for SRB2, 3 for SRB3 and 4 for SRB4 if SRB4 is present. The combination of transport channel switching and logical channel multiplexing is illustrated in Figure 11.9.

### Choice RLC size list

This field defines how the RLC sizes defined within the RB mapping field are related to the RLC sizes defined within the TFS for the specific transport channel that is being considered. If the transport channel is a DCH, the options are: 'all', 'configured', or 'explicit list'. 'All' means that all RLC sizes in the TFS defined for that specific transport channel are applicable for the logical channel. The option 'configured' means that the RLC sizes allowed are configured within the RRC CONNECTION SETUP message. The option 'explicit list' means that the RLC sizes are indexed within the RB mapping information IE, and the index relates to the TFS information defined for that transport channel. For the RACH transport channel, the only option is the 'explicit list'. The TFS parameters for the RACH transport channel are defined within the SIB6 message if available or else in the SIB5 message.

### MAC logical channel priority

The MAC logical channel priority is an integer number in the range 1–8 and defines the priority of the logical channel entering the MAC. The highest priority is 1, and the lowest priority is 8. The MAC uses the logical channel priority information to define things such as the priority of the TF combining in the case of dedicated transport channels, or the absolute priority of the logical channels in the case of common channel transmission.

**Uplink transport channel information**

The uplink transport channel information common defines transport channel information such as the TFCS for the different transport channels. In R99, the TFCS is for the DCH transport channels. For the PRACH, the TFCS information for the transport channels is not defined here, instead it is defined in the SIB 5/6 broadcast messages.

**Downlink transport channel information**

The downlink transport channel information, like the uplink information, defines the transport channel information (such as the TFCS) for the transport channels that are defined within the RRC CONNECTION SETUP message. In R99, the TFCS is for the DCH transport channels. For the SCCPCH, the TFCS information for the transport channels is not defined here, but is defined instead in the SIB 5/6 broadcast messages.

**Frequency information**

The frequency information IE defines the UARFCN for the carrier that the UE is tuned to. The UE needs to know which channel it is using so that it can correctly change to different frequencies for measurement purposes that may lead to a handover.

**Maximum allowed uplink Tx power**

This IE indicates the maximum allowed uplink $T_X$ power. The value is an integer in the range $(-50, \ldots, 33)$, where the integer value is defined in dBm.

**Uplink DPCH information**

This set of information defines aspects of the physical channel for the DPCH that the UE may have been assigned. The information within the IE relates to the physical channel and includes elements such as the minimum allowed spreading factor for the data part on the uplink, the scrambling code Id, the power control algorithm to use, the length of the power control preamble and the SRB delay.

The power control preamble and the SRB delay define the number of frames after the power control is activated for the DPCCH before the DPDCH and the SRBs, respectively, are transmitted.

**CPCH set information**

The CPCH set information contains the information required to establish the CPCH on the uplink (if required to do so by the UTRAN).

## 11.4 RRC connection establishment

**Table 11.14.** *RRC CONNECTION SETUP COMPLETE message contents*

| Message element | Description |
| --- | --- |
| RRC transaction identifier | Message identifier. |
| START multi CN | Multiple messages for each CN domain. |
| CN domain | This defines CN domains for which START will be sent to either the CS domain or PS domain. |
| START | Initialisation value of HFN used for each CN domain and used in security procedures. |
| UE radio access capability | This defines the UE radio access capability. |
| UE radio access capability extension | This defines extensions to radio access capability, e.g. for different frequency bands. |
| Inter-RAT UE radio access capability | This defines formats for inter-RAT capabilities. Currently cdma2000 and GSM are defined. |

**Downlink information**

The downlink information defines all of the information that is required to establish the downlink physical channels. This includes information common to all radio links including elements such as:
- power control information including offset between pilot and DPDCH;
- spreading factor information, TFCI information for dedicated channels;
- compressed mode information and transmit diversity information.

There is also information that is specific to each radio link such as:
- the primary CPICH scrambling code number;
- information to configure the PDSCH (if present);
- DPCH configuration information (frame offset, spreading factor and code number, scrambling code number (if different to primary scrambling code in cell), power control information and transmit diversity information).

### 11.4.3 RRC CONNECTION SETUP COMPLETE

Upon completion of the setup procedures, the UE sends an RRC CONNECTION SETUP COMPLETE message to the UTRAN. The contents of this message are defined in Table 11.14. Once the UE has created the RRC CONNECTION SETUP COMPLETE message, it is transmitted to the UTRAN using the appropriate logical, transport and physical channels for the current mode of operation of the UE (i.e. CELL_DCH or CELL_FACH states within the connected mode). The flow of the message through the layers was illustrated in Figure 11.7.

### 11.4.4 Summary of RRC connection setup

Figure 11.10 summarises the structure of the protocol architecture after the establishment of the SRBs for the uplink. Only the CELL_DCH architecture is shown. If the

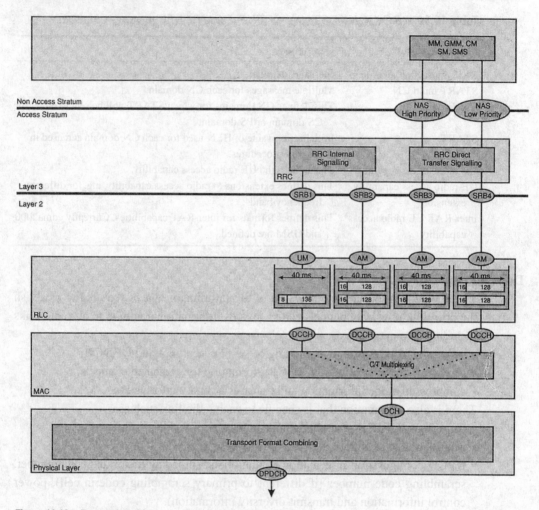

**Figure 11.10** Protocol architecture after configuration of SRBs.

UE was configured to operate also in the CELL_FACH state, then there would be a comparable structure but only using the common transport and physical channels. Two of the RBs are used for RRC signalling and two are used for NAS signalling. The RRC CONNECTION SETUP message contained all the information that was needed by the UE to configure all the layers shown in the figure.

## 11.5 Direct transfer procedure

The next procedure that we consider is that used to transfer the NAS messages from the UE to the appropriate CN domain and the reverse operation. This procedure is referred to as direct transfer.

Direct transfer is the mechanism that allows a UE to send a receive NAS messages from the CN. There are a number of direct transfer messages used across the radio

## 11.5 Direct transfer procedure

**Table 11.15.** *Direct transfer messages*

| Message | Direction | Description |
|---|---|---|
| INITIAL DIRECT TRANSFER | Uplink | Initial direct transfer message that also activates a signalling connection to a specific CN domain. |
| UPLINK DIRECT TRANSFER | Uplink | Subsequent direct transfer message on uplink. |
| DOWNLINK DIRECT TRANSFER | Downlink | Direct transfer on downlink using previously created signalling connection. |

interface and these are outlined in Table 11.15. The direct transfer messages shown in Table 11.15 are used to establish signalling connections (INITIAL DIRECT TRANSFER) and to exchange NAS messages between the UE and the CN. The NAS messages are carried within the direct transfer message.

We start by considering an example of an initial direct transfer procedure. This could be used to carry an 'ATTACH REQUEST' NAS message or some alternative NAS message such as a location update message.

### 11.5.1 Initial direct transfer

If we assume that the NAS in the UE wishes to send a message to the NAS in the CN (e.g. an ATTACH REQUEST message to the PS domain), first, the NAS must request the AS to create an RRC connection as described in the previous section. Next the NAS can create the NAS message and pass it to the AS, which can then send the message to the appropriate CN domain using the INITIAL DIRECT TRANSFER message. This procedure is outlined in Figure 11.11, and the contents of the INITIAL DIRECT TRANSFER message are defined in Table 11.16. In this example, we are assuming that the NAS has requested the establishment of a PS connection to the PS-domain using the ATTACH REQUEST message (the use and contents of this message are considered in more detail in Chapter 13).

Before we can establish a PS connection, an RRC connection must be established. Figure 11.11 illustrates the sequence of messages required using the RRC primitives shown in Figure 11.11, the NAS can request the establishment of a GMM context and in the process a PS signalling connection. The primitives include the NAS message (ATATCH REQUEST) as well as the establishment cause, signalling channel priority (high or low), CN identity, UE identity and RAIs and LAIs.

The RRC layer first needs to establish an RRC connection using the procedures outlined previously (the establishment cause in the RRC CONNECTION REQUEST message is that received from the NAS). Once the RRC connection is available, the INITIAL DIRECT TRANSFER message is sent (via SRB3 assuming that the signalling channel priority flag was set to high) and includes the NAS message.

## Table 11.16. *INITIAL DIRECT TRANSFER message*

| Message element | Description |
| --- | --- |
| Integrity check information | Used to check the message integrity has not been violated. |
| CN domain identity | Identifies the CN domain that is the intended recipient of the initial direct transfer message. |
| Intradomain NAS node selector | Defines a 10 bit routing parameter based on a TMSI or IMSI that can be used to identify a specific connection between the UE and the CN domain. |
| NAS message | Contains the NAS message to be transmitted transparently through the UTRAN to the CN domain. The length of the message is between 1 and 4095 octets. |
| Measured results on RACH | Set of measurements made by the UE on the current cell and up to seven monitored cells. |

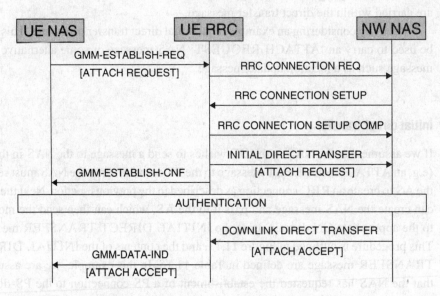

Figure 11.11  Example use of direct transfer to attach UE to network.

The RRC then confirms the creation of the signalling connection with the GMM-ESTABLISH-CNF primitive.

### 11.5.2 DOWNLINK DIRECT TRANSFER

If we assume that the UE sends an ATTACH REQUEST message, the response (ATTACH ACCEPT) is sent using the DOWNLINK DIRECT TRANSFER message. This part of the procedure is illustrated in the lower part of Figure 11.11. The contents of the DOWNLINK DIRECT TRANSFER message are illustrated in Table 11.17.

**Table 11.17.** *DOWNLINK DIRECT TRANSFER message*

| Message element | Description |
|---|---|
| RRC transaction identifier | Identifier to track the RRC messages. |
| Integrity check information | This is used to check the message integrity has not been violated. |
| CN domain identity | This identifies the CN domain that is the intended recipient of the initial direct transfer message. |
| NAS message | This contains the NAS message to be transmitted transparently through the UTRAN from the CN domain. The message is between 1 and 4095 octets long. |

## 11.6 RB setup

When the CN establishes a service to the UE, it has to create an RAB between the UE and the CN. The RAB in turn is composed of an RB and an Iu bearer. The relationship between the RAB and RB was illustrated in Chapter 2. The procedure that is performed to create a RAB is briefly outlined below:

- CN requests the RNC to create an RAB (RAB assignment request);
- RNC creates an Iu bearer between the RNC and the CN;
- RNC creates RLs between the Node B and the UE;
- RNC creates an RB between the RNC and the UE.

The last stage of this procedure creates the RB and involves the UE. The creation of the RB (and hence RAB) is achieved through two messages: RADIO BEARER SETUP from the UTRAN (SRNC) to the UE and in response the RADIO BEARER SETUP COMPLETE message from the UE to the UTRAN. To establish the RB the UTRAN sends the RADIO BEARER SETUP message to the UE, the contents of which are summarised in Table 11.18. Upon receipt of this message, the UE acts on the message appropriately and responds with the RADIO BEARER SETUP COMPLETE message.

The vast majority of the contents of the RADIO BEARER SETUP message are the same as the RRC CONNECTION REQUEST message and consequently we will not go through the details of these common areas here (please refer to Section 11.4). One specific area that differs (there are others such as the configuration of the PDCP layer and inclusion of DRAC information) is the RAB information that is included in the RADIO BEARER SETUP message.

The message RAB INFORMATION FOR SETUP is included below for reference.

### RAB INFORMATION FOR SETUP
The RAB INFORMATION FOR SETUP message defines the RAB specific information. This information includes elements such as:

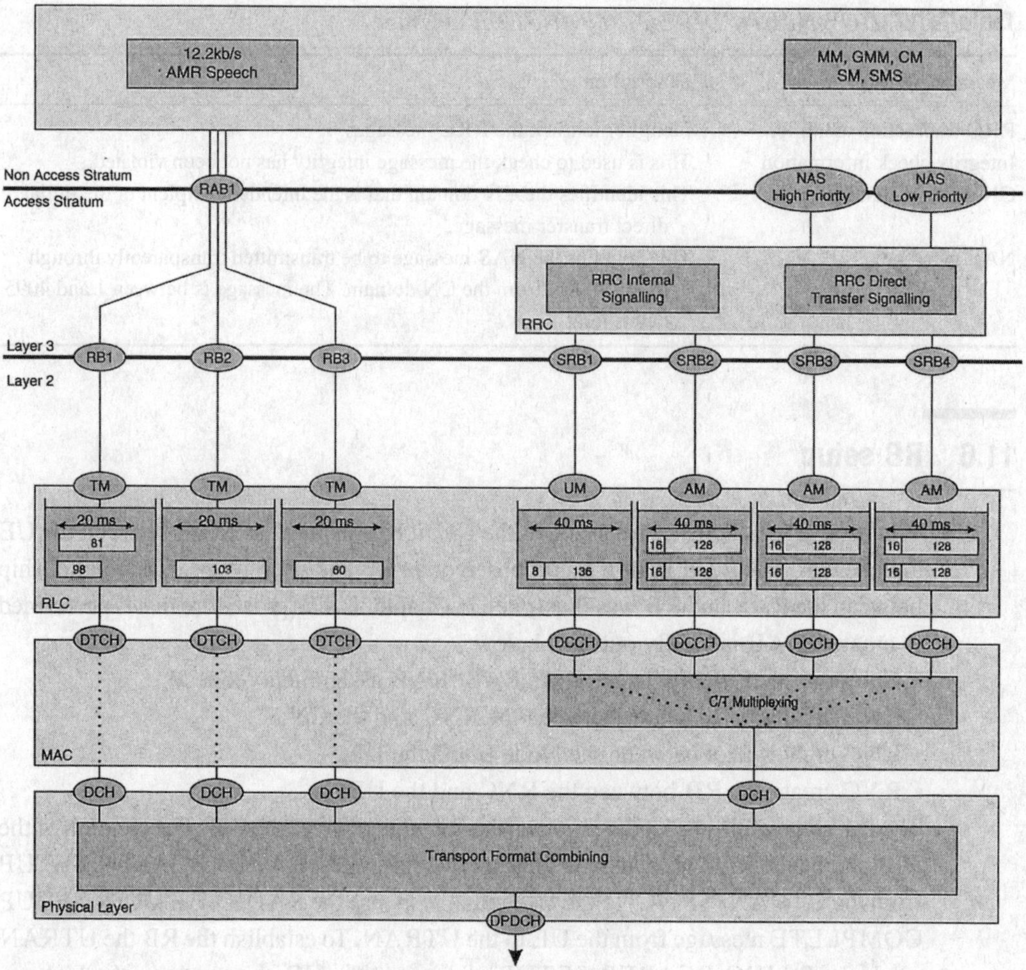

**Figure 11.12** UE configuration after CS connection establishment.

- RAB identity: an 8 bit string that links the CN domain, bearer and UE; it comprises of an SI for the CS domain, and a NSAPI (numbered 5–15) for the PS domain.
- CN domain identity: this defines whether it is the CS domain or the PS domain.
- NAS synchronisation information: this is used by NAS for synchronising the bearer.

### 11.6.1 RAB setup for CS connection

Figure 11.12 illustrates the configuration of the UE after the establishment of an RAB for a CS speech connection. The RAB establishment procedure may be triggered in the UE by the request to establish a CS connection such as a speech call. This very example is considered in detail in Chapter 13.

## 11.6 RB setup

**Table 11.18.** *RADIO BEARER SETUP message*

| Message element | Description |
| --- | --- |
| RRC transaction identifier | Identifies individual RRC transactions per message type. |
| Integrity check information | Used to check the message integrity not violated. |
| Integrity protection mode information | Activates and configures the integrity protection. |
| Ciphering mode information | Activates and configures the ciphering mode for the different RBs. |
| Activation time | Defines when the changes in the setup message should be applied. |
| New u-RNTI | Defines a new u-RNTI if required, replaces any old value. |
| New c-RNTI | Defines a new c-RNTI if required, replaces any old value. |
| New DSCH-RNTI | Defines a new DSCH-RNTI if required, replaces any old value. |
| RRC SI | Defines the RRC state the UE is to move into (CELL_DCH, CELL_FACH, CELL_PCH, URA_PCH). |
| UTRAN DRX cycle length coefficient | Defines the DRX cycle length coefficient. |
| URA identity | Defines the URA identity to be stored and used in the URA_PCH state to activate a URA UPDATE procedure if it differs from values broadcast in SIB2. |
| CN information | Contains: PLMN identity; GSM NAS system information; up to four CN domains NAS system information. |
| SRB information setup list | Contains the information that defines the SRBs: SRB identity, RLC information and mapping information, logical channel information. |
| RAB information setup list | Contains the information that defines the RABs: RAB identity, one or more RB identity, PDCP information, RLC information and logical channel mapping information. |
| RB information affected list | Modifies the RB mapping information. |
| Downlink counter synchronisation information | Used to synchronise the downlink counters used for security procedures. |
| Uplink common transport channel information | Configuration information for the uplink common transport channels. |
| Uplink deleted transport channel information | List of uplink transport channels that are being deleted. List only allows DCH and USCH (TDD). |
| Uplink add/reconfigured transport channel information | List of transport channels and transport channel information for new transport channels or reconfigured transport channels. |
| Mode specific transport channel information | FDD mode transport channel information defining CPCH set identity (if applicable) and DRAC parameters (if applicable). |
| Downlink common transport channel information | List defining downlink transport channel information for common transport channels. |
| Downlink deleted transport channel information | List of downlink transport channels that are being deleted. List only allows DCH and DSCH. |
| Downlink add/reconfigured transport channel information | List of transport channels and transport channel information for new transport channels or reconfigured transport channels. |
| Frequency information | Uplink UARFCN and downlink UARFCN. |
| Maximum uplink Tx power | Defines the maximum uplink transmit power that the UE can use. |
| Uplink channel requirement | Defines the uplink DPCH or CPCH set information. |
| Mode specific physical channel information | Defines DSCH information for FDD mode. |
| Downlink common RL information | Defines the downlink RL information common to all RLs. For FDD mode this includes Tx diversity information and compressed mode information. |
| Downlink information per RL | Defines the RL information for all downlink RLs, including all of those in the active set. Includes elements such as scrambling code numbers, spreading code numbers. |

Part of the RAB creation is the transmission of a RADIO BEARER SETUP message. The RADIO BEARER SETUP message includes all of the information that the UE requires for the AS part of the RAB creation process and is presented in Table 11.18; the contents of many of these fields were considered in Section 11.4. At the end of this procedure, the architecture of the UE is configured in a way similar to that presented in Figure 11.12. The details of the NAS part of the CS connection establishment (MO and MT) are considered in Chapter 13.

### 11.6.2 RAB setup for PS connection

We could assume that the next thing that the UE performs is the creation of a PS data connection (referred to as a PDP context). To do this, the UE performs a PDP CONTEXT ACTIVATION procedure, the details of which are considered in Chapter 13.

Part of the establishment of the PDP context is the creation of the RAB that supports the PDP context, and part of the establishment of the RAB is the creation of the RB using the RADIO BEARER SETUP message that we have just considered. The details of the NAS signalling and interlayer primitives between the NAS and AS are considered in Chapter 13 for this specific case.

On completion of this procedure, assuming that the CS call is still active, the NAS/AS architecture for the UE resembles that illustrated in Figure 11.13, which shows the simultaneous presence of the CS RAB and the PS RAB as well as the four SRBs used for the signalling messages between the UE and the network. The architecture shown is for the uplink and in the CELL_DCH case. There is an equivalent architecture for the downlink.

## 11.7 Handover

There are various forms of handover defined within the UMTS specifications. In this section we review the different forms of handover and then move on to explore some of the details associated with the different handover scenarios.

In general, before a handover can occur, the UE makes some signal measurements and reports them to the UTRAN. Based on the measurements, the UTRAN decides which type of handover to employ. The details of the various measurements that are made prior to a handover are considered in Chapter 12.

The types of handover considered in this section are soft-handover, hard-handover, handover to GSM and handover from GSM to UMTS. In addition we also consider cell change order, which is a cross between handover and cell reselection. Cell selection and reselection are considered in Chapter 12.

## 11.7 Handover

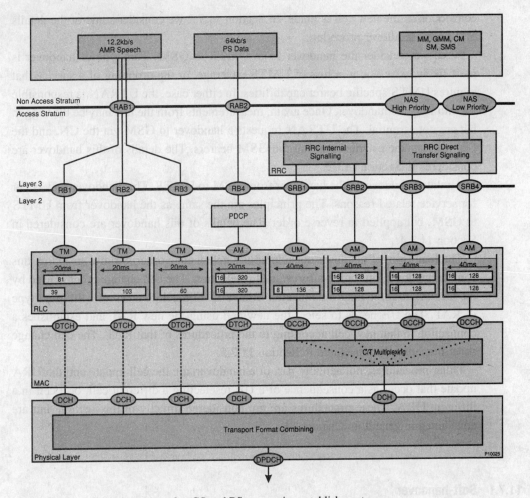

**Figure 11.13** UE configuration after CS and PS connection establishment.

Soft-handover is a type of intrafrequency handover whereby the UE can be simultaneously connected to more than one cell and it is only applicable to the FDD mode of operation. In soft-handover we define something called the active set as being the set of RLs via which the UE is actively transmitting and receiving. The soft-handover procedure is referred to as the ACTIVE SET UPDATE and only applies to UEs that are in the CELL_DCH state. We consider the details of soft-handover in Section 11.7.1.

Hard-handover can be to a different cell on the same or a different frequency, to a TDD mode cell, or maybe to the same cell but using a different spreading code. The hard-handover procedure, like the soft-handover procedure, is activated by the UTRAN after the receipt of measurement reports from the UE. A hard-handover is a break-before-make handover, where the connection to the old cell is lost before the connection to the new cell is made. In Section 11.7.2 we consider some of the details of the hard-handover procedure.

Next, we consider the handover from UMTS to GSM. This type of handover is most likely to be due to a loss of UMTS coverage, or the dropping of a service that requires UMTS specific bearer capabilities. In either case, the UTRAN is responsible for initiating the handover. Once again, measurements from the UE may be the trigger, but are not essential. The UTRAN requests a handover to GSM via the CN, and the CN requests the establishment of the GSM bearers. The details of this handover are considered in Section 11.7.3.

A related handover is the handover from GSM to UMTS. This handover may occur for service related reasons. The principles are the same as the handover from UMTS to GSM, but applied in reverse order. The details of this handover are considered in Section 11.7.4.

The final form of handover considered is a cell change order. Strictly speaking, this is not a handover, but rather a forced cell reselection. The cell change order is used by the network to force the PS connection from a UMTS cell to a cell of a different type of RAT. The UE needs to select the new cell using the new RAT and establishes a connection to that new cell according to the procedures of that RAT. The cell change order procedure is considered in Section 11.7.5.

Other procedures, not actually part of a handover, are the cell update and the URA update that occur as a consequence of a UE reselecting a different cell, or a cell in a different URA. These procedures are not considered directly in this section, but are considered in general in Chapter 14.

### 11.7.1 Soft-handover

An active set is defined in [41] as a 'set of radio links that are simultaneously involved in a specific communication service between an UE and a UTRAN access point'. Figure 11.14 illustrates this basic concept of an active set. The UE has a logical connection to the CN via the SRNS, but is in addition receiving signals from the DRNS. It is the responsibility of the UTRAN to ensure that the transmissions from each of the Nodes B arrive at the UE within the same nominal time window. The rake receiver in the UE assigns a 'finger' to each of the transmissions. The number of fingers that a UE contains is an implementation issue, but as [24] stipulates the maximum number of RLs in the active set is eight, this implies that up to eight fingers are required in the rake receiver. Each finger in the rake receiver may be set to collect energy from the specific Node B it is monitoring with the UE configured with the channelisation and scrambling codes used. On the uplink, the SRNS is responsible for the combining of the information flows received by the different cells. Selection diversity combining is the technique most likely to be used for this purpose.

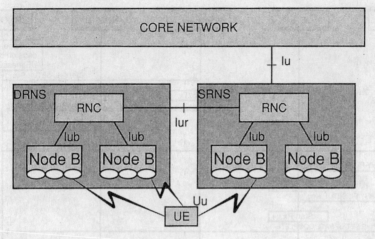

**Figure 11.14** Active set operation.

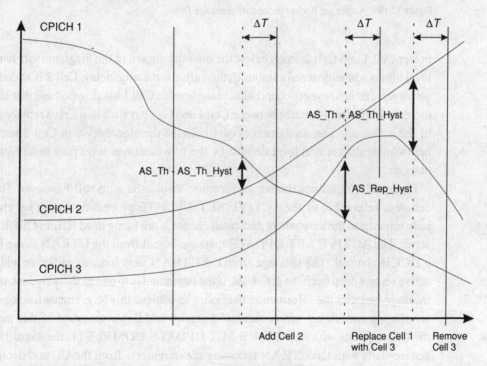

**Figure 11.15** Typical active set measurements made by the UE.

Figure 11.15 shows an example of the measurement processes that need to be implemented to add and replace cells in the active set. The diagram represents the signal power that the UE is measuring from the primary pilot channels from a number of neighbouring cells. At the start, the UE is only connected to Cell 1 (whose power is indicated by CPICH 1). The power in Cell 2 increases and the difference between the

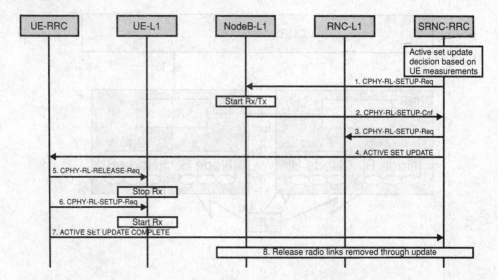

**Figure 11.16** Active set update procedure message flows.

power Cell 1 and Cell 2 drops below the quantity shown in the diagram (soft-handover level minus some hysteresis amount), then after some time delay, Cell 2 is added to the active set. The next event occurs because the level in Cell 1 has dropped and that in Cell 3 increased, and the difference between Cell 1 and Cell 3 is such that Cell 3 replaces Cell 1 in the active set. The final event occurs because the signal level in Cell 3 has fallen below some differential level defined by the soft-handover level plus some hysteresis margin.

Figure 11.16 outlines the basic procedure associated with soft-handover. The procedure is referred to as the ACTIVE SET UPDATE procedure. Active set update is only relevant to the case where dedicated channels are being used (UE in CELL_DCH state). The ACTIVE SET UPDATE message is sent from the UTRAN to the UE via a DCCH channel. The message from the UTRAN may indicate either an add to the active set or a drop from the active set. If the command is to add to the active set, then the message includes the information necessary to achieve this (e.g. channelisation codes, scrambling codes etc.). Upon successful receipt and implementation of this message, the UE responds with an ACTIVE SET UPDATE COMPLETE message. The procedure starts with the UTRAN receiving measurements from the UE and deciding to change the active set.

1–3. The UTRAN configures the target cell resources assuming that RLs are being added to the active set.

4. The UTRAN sends the ACTIVE SET UPDATE message to the UE. The update message contains all of the necessary information on the RLs that are being added and removed.

5–6. The UE modifies the physical layer removing RLs as required and configuring the new RLs that are to be added.

**Table 11.19.** *Message contents for ACTIVE SET UPDATE message*

| Message element | Description |
| --- | --- |
| General elements | Transaction identifier, activation time, integrity protection information, ciphering information. |
| RNTI and CN information | New u-RNTI, CN information, PDCP SN information. |
| RL addition list | Physical channel parameters for RL being added. |
| RL removal list | Scrambling codes of RL to be removed. |
| Diversity information | Type of Tx diversity used if any. |

7. The UE sends the ACTIVE SET UPDATE COMPLETE message.
8. The UTRAN frees any resources that are no longer used for the UE.

Table 11.19 outlines the basic contents of the ACTIVE SET UPDATE message. The message includes the general elements such as the activation time and ciphering and integrity protection information. The message also includes the physical channel information for the RLs that are to be added (if any) and the scrambling codes for the radio links to be deleted (if any). Diversity information is also included to indicate what transmit diversity is used, if any.

### 11.7.2 Hard-handover

Figure 11.17 outlines the hard-handover procedure. Hard-handover is implemented using the PHYSICAL CHANNEL RECONFIGURATION message (step 4, Figure 11.17) that changes some elements of the physical channel. This requires the UE to modify the physical channel (steps 5–8, Figure 11.17) before responding with the PHYSICAL CHANNEL RECONFIGURATION COMPLETE message (step 9, Figure 11.17).

Table 11.20 outlines the basic elements of the PHYSICAL CHANNEL RECONFIGURATION message. Many of the elements in Table 11.20 have been defined in Section 11.4 and Section 11.6 and consequently they are not considered in detail here.

On receipt of the PHYSICAL CHANNEL RECONFIGURATION message, the UE responds with the PHYSICAL CHANNEL RECONFIGURATION COMPLETE message; the elements of which are presented in Table 11.21. The PHYSICAL CHANNEL RECONFIGURATION COMPLETE defines mainly counters and timing elements. The elements include the START time used by ciphering and ciphering activation time, the PDCP sequence numbers used by lossless relocation, and the integrity protection configuration parameters.

### 11.7.3 Handover from UMTS to GSM

Handover from UMTS to GSM is possible even if no measurements are made on the target cell. The UE must be in the CELL_DCH state. Handover is possible with no

**Table 11.20.** *Contents of PHYSICAL CHANNEL RECONFIGURATION message used for hard-handover*

| Message contents | Description |
|---|---|
| General information | Integrity protection, ciphering info, activation time, RNTI, RRC state, CN information, URA identity, DRX information. |
| Uplink channel requirement | Defines the uplink physical channels (FDD/TDD). |
| Downlink common information | Defines downlink physical channels (FDD/TDD). |
| Downlink information per RL | Defines physical channel information. |

**Table 11.21.** *Contents of PHYSICAL CHANNEL RECONFIGURATION COMPLETE message used for hard-handover*

| Message contents | Description |
|---|---|
| General information | Integrity protection, transaction identifier. |
| Integrity protection activation information | Defines the time when integrity protection should be activated. |
| Uplink counter synchronisation | START value for CN domains and PDCP sequence numbers. |
| Uplink RB cipher activation time | Defines timing of ciphering activation in terms of RLC sequence numbers. |

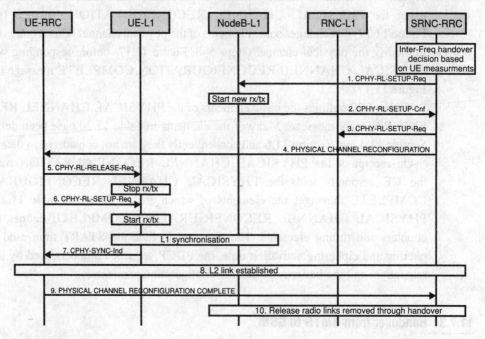

**Figure 11.17** Hard-handover procedure.

## 11.7 Handover

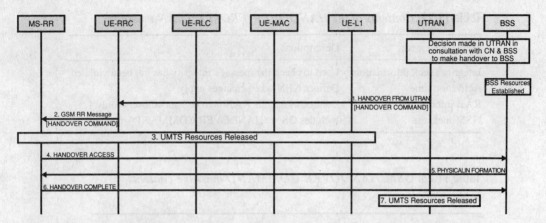

**Figure 11.18** Handover from UMTS to GSM procedure.

RABs, CS RABs, CS and PS RABs. For R99, however, only one CS RAB can be handed over from UMTS to GSM. The network indicates the RAB identity (RAB Id) for the RAB that is to be handed over to GSM.

Figure 11.18 outlines the basic steps taken in the change from UMTS to GSM. This starts with a decision in the UTRAN that a change in RAT is required, based on measurements made by the UE, although the UE can be requested to handover without having made measurements. The numbered messages in Figure 11.18 are explained in the numbered clauses below.

1. The HANDOVER FROM UTRAN command is sent to the UE. This message also includes the GSM HANDOVER COMMAND message. The HANDOVER COMMAND includes all of the information that the MS needs to continue the connection in the GSM system.
2. The RRC in the UE sends the contents of the message to the GSM RR layer in the MS (GSM mode handset).
3. The UMTS radio resources are released by the UE.
4. The MS sends the GSM HANDOVER ACCESS access messages to the BSS as it would in GSM.
5. The BSS sends the PHYSICAL INFORMATION message including timing advance information.
6. The MS sends the HANDOVER COMPLETE message.
7. The UTRAN resources are released.

Table 11.22 presents the outline contents of the HANDOVER FROM UTRAN message. Moving from UMTS to GSM, multicall is not currently supported in GSM, and so only one of the CS RABs can be handed over to GSM. The RAB information defines which RAB should continue.

The details of the GSM part of the handover are contained in the GSM HANDOVER COMMAND, which is considered next. Table 11.23 presents the outline

**Table 11.22.** *Contents of the HANDOVER FROM UTRAN message*

| Message element | Description |
| --- | --- |
| Integrity check information | Used to check the message integrity has not been violated. |
| Activation time | Defines CFN when changes apply. |
| RAB information | Defines which CS RAB is to be handed over – only 1. |
| GSM message | Includes GSM 'HANDOVER COMMAND' message. |

**Table 11.23.** *GSM HANDOVER COMMAND message contents (abbreviated)*

| Information | Description |
| --- | --- |
| Cell description | PLMN colour code, BS colour code, BCCH ARFCN. |
| Channel description | Defines the channel type, hopping information. |
| Handover reference | Defines a reference value used to identify the HO access. |
| Multislot configuration | Multislot configuration information. |
| Time information | Information on cell time offset and timing advance. |
| Codec information | Information on multirate codec. |

contents of the GSM HANDOVER COMMAND message, which is included within the HANDOVER FROM UTRAN message. The HANDOVER COMMAND message provides information on the frequency and frequency hopping patterns, the channel configuration information and the bearer service configuration information that allows the UE to continue the call in the GSM system.

As part of the transfer of the radio connection to the GSM frequency, the MS (the name for the UE in the GSM system) continues by transmitting the GSM HANDOVER ACCESS message using the GSM RACH. The HANDOVER ACCESS message is sent a number of times for reliability and it also includes the handover reference value that the MS received in the HANDOVER COMMAND to identify the handover attempt.

The network responds with the PHYSICAL INFORMATION message, which is used to pass information on the timing advance to the MS and to stop the MS sending more HANDOVER ACCESS messages. The MS completes the procedure by sending the HANDOVER COMPLETE message, which includes an optional mobile observed time difference.

### 11.7.4 Handover from GSM to UMTS

Figure 11.19 outlines the basic steps taken in the change from GSM to UMTS; the meaning of the different messages is identified by the numbered clauses below. Similarly to the previous case, the handover is triggered by the source radio network (BSS) in

## 11.7 Handover

**Table 11.24.** *Contents of the INTER SYSTEM TO UTRAN HANDOVER COMMAND message*

| Message element | Description |
|---|---|
| RR protocol disc | RR protocol discriminator. |
| Skip indicator | Skip indicator. |
| Message type | Defines INTERSYSTEM TO UTRAN HANDOVER command. |
| Handover to UTRAN | UMTS HANDOVER TO UTRAN message. |

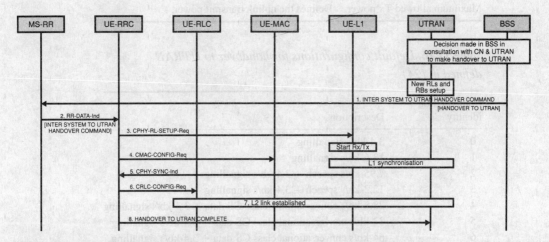

**Figure 11.19** Handover from GSM to UMTS procedure.

consultation with the target radio network (UTRAN) (although not directly, rather via their respective CN elements).

1. The procedure starts with the MS receiving an INTER SYSTEM TO UTRAN HANDOVER COMMAND, which includes the UMTS HANDOVER TO UTRAN message.
2. The contents of this message are passed internally in the handset from the MS-RR to the UE-RRC layers.
3–7. The UE configures the L1 and L2 entities and synchronises with the L1 transmissions from the Node B.
8. The UE sends the HANDOVER TO UTRAN COMPLETE message when it has successfully moved to the target UMTS cell.

Table 11.24 outlines the basic contents of the INTER SYSTEM TO UTRAN HANDOVER COMMAND that the MS receives, including the UMTS RRC message: HANDOVER TO UTRAN.

The MS extracts the HANDOVER TO UTRAN part of the message, it changes to the UMTS mode of operation defined in the message and synchronises with the UMTS network. The response is sent and the connection continues via the UMTS network.

**Table 11.25.** *Contents of the HANDOVER TO UTRAN message*

| Message element | Description |
|---|---|
| New U-RNTI | Allocates a new u-RNTI. |
| Ciphering algorithm | Defines which ciphering algorithm to use. |
| Specification mode | Complete specification – L2 explicitly defined. |
| | Predefined configuration – most of L2 defined by BSS. |
| | Default configuration – configuration Id defined in [24] for all L1 parameters defined. |
| Maximum allowed Tx power | Defines the uplink transmit power. |

**Table 11.26.** *Default configurations for handover to UTRAN defined in [24]*

| Configuration identity | Description |
|---|---|
| 0 | 3.4-kb/s signalling |
| 1 | 13.6-kb/s signalling |
| 2 | 7.95-kb/s speech + 3.4-kb/s signalling |
| 3 | 12.2-kb/s speech + 3.4-kb/s signalling |
| 4 | 28.8-kb/s conversational class CS data + 3.4-kb/s signalling |
| 5 | 32-kb/s conversational class CS data + 3.4-kb/s signalling |
| 6 | 64-kb/s conversational class CS data + 3.4-kb/s signalling |
| 7 | 14.4-kb/s streaming class CS data + 3.4-kb/s signalling |
| 8 | 28.8-kb/s streaming class CS data + 3.4-kb/s signalling |
| 9 | 57.6-kb/s streaming class CS data + 3.4-kb/s signalling |
| 10 | Multimode speech + 3.4-kb/s signalling |

Table 11.25 outlines the basic contents of the HANDOVER TO UTRAN message. In addition to the typical message contents, the message includes the specification mode elements. The specification mode defines how the UE is told which of the UTRAN cell parameters it should use. Currently there are three options. This first is called complete specification, in which all of the details of L1 and L2 are specified within the message. The second option is a predefined configuration that the UE received from the original RAT (e.g. from the BSS in GSM) and which is identified using a tag value. The UE can be told to use one of these preconfigured configurations. The third option is to use a default configuration that is defined in [24] and also in Table 11.25. The default configurations list all of the necessary L2 and some L1 parameters that the UE needs to activate certain services. The default configurations are defined using a configuration identifier. Currently there are 11 default configurations, which are listed in Table 11.26.

**Table 11.27.** *Contents of the CELL CHANGE ORDER FROM UTRAN message*

| Message contents | Description |
| --- | --- |
| General elements | Transaction identifier, activation time, |
| RAB information list | RAB Ids, CN domain, NAS synch indicator. |
| Inter RAT target cell desc | Defines target cell characteristics such as GSM, BSIC, frequency band, ARFCN. |

### 11.7.5 Cell change order

The CELL CHANGE ORDER FROM UTRAN is used to change from a UTRAN cell to a cell in another RAT. It is a little like a handover, but it is intended for non-real-time services, and so is analogous to a forced cell reselection. It is applicable to the CELL_FACH state and the CELL_DCH state (PS connections only), and is used to allow UTRAN to force UEs not using the CS domain to use other RAT cells.

The contents of the CELL CHANGE ORDER FROM UTRAN message can be seen in the Table 11.27. In the reverse direction, it is possible that the other RAT may also cause a CELL CHANGE ORDER TO UTRAN. In this case the other RAT, through some means, encourages the UE to change to a UTRAN cell using information that is provided on that cell. When camped on the UTRAN cell, the UE performs an RRC connection establishment procedure with the establishment cause defined as 'Inter-RAT Cell Change Order'.

## 11.8 Miscellaneous RRC procedures

In this section we consider a number of associated RRC procedures, starting with the definition and calculation of the CTFC. Then we consider the dynamic resource allocation control (DRAC), which is used to dynamically change uplink data rates for the UEs in a cell in the CELL_DCH state.

### 11.8.1 CTFC calculation

The CTFC is used as an efficient method of signalling the TFC from the RNC to the UE. The definition and use of the TFC is considered in Chapter 8.

The TFCs are signalled to the UE using the CTFC. Here we examine the structure of the CTFC and how it is interpreted by the UE to form the TFC. The use of and operation of the CTFC are best considered via an example.

In the first instance, the quantity $P_i$ is calculated for each of the transport channels in the TFCS from:

$$P_i = \prod_{j=0}^{i-1} L_j$$

Assume 3 transport channels $TFI_1 \in \{0, 1\}$, $TFI_2 \in \{0, 1, 2\}$, $TFI_3 \in \{0, 1, 2, 3\}$

$$P_1 = L_0 = 1$$
$$P_2 = L_0 \times L_1 = 1 \times 2 = 2$$
$$P_3 = L_0 \times L_1 \times L_2 = 1 \times 2 \times 3 = 6$$

| $TFI_1$ | $TFI_2$ | $TFI_3$ | CTFC | TFCI |
|---|---|---|---|---|
| 0 | 0 | 0 | 0x1 + 0x2 + 0x6 = 0 | 0 |
| 0 | 1 | 0 | 0x1 + 1x2 + 0x6 = 2 | 1 |
| 0 | 2 | 0 | 0x1 + 2x2 + 0x6 = 4 | 2 |
| 1 | 0 | 1 | 1x1 + 0x2 + 1x6 = 7 | 3 |
| 1 | 2 | 2 | 1x1 + 2x2 + 2x6 = 17 | 4 |
| 1 | 1 | 3 | 1x1 + 1x2 + 3x6 = 21 | 5 |
| 1 | 2 | 3 | 1x1 + 2x2 + 3x6 = 23 | 6 |
| 1 | 0 | 3 | 1x1 + 0x2 + 3x6 = 19 | 7 |

**Figure 11.20** Calculation of CTFC.

where $j = 1, 2, \ldots, L$ and $L_0 = 1$. Next the CTFC for a specified transport channel format $TFI_i$ is given by:

$$CTFC(TFI_1, TFI_2, TFI_3, TFI_4, \ldots TFI_I) = \sum_{i=1}^{I} TFI_i \cdot P_i$$

The example shown in Figure 11.20 illustrates the basic procedure associated with calculating the CTFC. Assume that there are three transport channels. The first has two TFs, the second has three TFs, and the third has four TFs. First we calculate the values for $L_i$, which is the number of TFs for transport channel $i$; $L_0$ is equal to 1. The values for $P_i$ are calculated from the equations above and are illustrated in Figure 11.20, for different TFs. By using the equation presented earlier, it is possible to derive the CTFC for each combination.

The CTFC uniquely defines the specific TFC combination. It is the CTFC numbers that are signalled to the UE and which the UE uses when creating the TFCS.

### 11.8.2 DRAC procedure

DRAC of uplink DCH is used as a means to dynamically control the uplink load from one or several UEs. The use of the DRAC procedure is indicated to the UE by the presence of a static DRAC information element in the messages that establish or modify the dedicated channels. Part of the DRAC static information is the DRAC class that is assigned to the UE. The DRAC class is used as part of the selection of the TF in the operational phase of the DRAC procedure. The DRAC procedure is activated by the reception of a DRAC static information IE. This occurs during RB establishment, RB reconfiguration, RB release or transport channel reconfiguration.

## 11.8 Miscellaneous RRC procedures

**Table 11.28.** *Example TFCS table and the data rate calculation*

| TFCI | DCH1 | DCH2 | DCH3 | TOTAL | BIT RATE |
|------|------|------|------|-------|----------|
| 0 | 0 | 0 | 0 | 0 | 0 |
| 1 | 0 | 100 | 0 | 100 | 10 kb/s |
| 2 | 0 | 200 | 0 | 200 | 20 kb/s |
| 3 | 200 | 0 | 200 | 400 | 40 kb/s |
| 4 | 200 | 200 | 400 | 800 | 80 kb/s |
| 5 | 200 | 100 | 800 | 1100 | 110 kb/s |
| 6 | 200 | 200 | 800 | 1200 | 120 kb/s |
| 7 | 200 | 0 | 800 | 1000 | 100 kb/s |

DRAC is achieved by the UTRAN broadcasting a number of resource control parameters on the downlink SIB10 message transported using a SCCPCH. These parameters control how the uplink resources are allocated. The DRAC procedure is only valid for DCH transport channels configured to follow the DRAC procedures. In addition, it is only available to UEs that can support both a DPCH (physical channel that carries the DCH transport channels) and a SCCPCH (physical channel that transports the FACH transport channel).

The SIB10 message includes a TRANSMISSION_PROBABILITY parameter and a MAXIMUM_BIT_RATE parameter. TRANSMISSION_PROBABILITY has the range of 0.125–1 in steps of 0.125. MAXIMUM_BIT_RATE has a range of 0–512 kb/s with a step of 16 kb/s.

The DRAC procedure works as follows. The UE obtains a pair of parameters (transmission probability, maximum bit rate) from the SIB10 messages broadcast in each of the cells in the active set and selects the pair that has the lowest product. Having established the lowest set of quantities, this defines the transmission probability and the maximum bit rate to use for the remainder of the procedure.

In the next stage, the UE reduces the set of available TFCs. Only the TFCs whose data rate (summed over all DCHs contributing to the peak data rate) is less than the selected maximum bit rate will be included in the new set. Consider an example of the three DCHs defined in the previous section where we considered the CTFC, each with TTI of 10 ms:

- DCH1: {0, 200} bits per TTI
- DCH2: {0, 100, 200} bits per TTI
- DCH3: {0, 200, 400, 800} bits per TTI

Assume the same TFCS as per the CTFC example previously, and assume that the maximum bit rate is set at 96 kb/s. The peak bit rate is calculated for each TFC as shown in Table 11.28. As the maximum bit rate is set at 96 kb/s, TFCI1–TFCI4 are allowed in the new TFCS, but TFCI5–TFCI7 are not.

The dynamic part of the DRAC procedure proceeds as follows. It commences as soon as the SIB10 message is received by the UE. At the start of the next TTI, the UE randomly selects a parameter $p$ from the range $\{0,1\}$. If $p$ is less than the TRANSMISSION_PROBABILITY, then the UE can transmit with a TFC obtained from the new set of TFs. The transmission will occur for $T_{validity}$ frames, after which the process is repeated – starting with the selection of the random number. If the random number is greater than the transmission probability, the UE waits for a period of $T_{retry}$ frames before starting the process again.

## 11.9 Summary

The RRC protocol is a large and complex protocol that provides for the configuration and control of the radio connections between the UE and the UTRAN. In this chapter we have reviewed some of the key aspects of the RRC protocol starting with the receipt of the initial system broadcast information, the establishment of RRC connections and RABS, and then moving on to look at issues such as handover.

# 12 Measurements

## 12.1 Introduction

The UTRAN contains many terminals spread across the coverage area. WCDMA is an RAT that is dependent upon the interference levels within the network. To maintain an acceptable operating point, therefore, the UTRAN needs to have precise knowledge of the system status at any given point in time. To provide this understanding, measurements are made throughout the radio network, and the measurement information delivered to the UTRAN for processing.

The measurements themselves are made by the UE and the UTRAN. The types of measurements made by the UE range from cell signal quality measurements to traffic volume measurements. In general, the type of measurement that a specific UE makes at a given point in time is defined by the UTRAN.

One overriding concept that we meet through the course of this chapter is the relationship between the RRC states and the cell signal measurements that are made by the UE. We will see that there is a distinct difference between the measurement activities of a UE in the CELL_DCH state compared with its activities in all other states (CELL_FACH, CELL_PCH, URA_PCH and idle mode). In CELL_DCH the UE makes measurements and reports them to the UTRAN for action by the UTRAN, in all other states the measurements are not reported to the UTRAN and the UE takes action based on the measurements. The key difference, therefore, is that the UTRAN directly controls which cell a UE uses in the CELL_DCH state, but in all other states the UE makes the decisions on which cell to use based on parameters received from the UTRAN via broadcast messages.

There are two main methods for controlling the measurements of a UE within the UTRAN, a direct method using a measurement control command and a method that relies on broadcast information to configure some of the measurement control information. We next explore the details for both of these methods.

Following this, we actually look at some of the different measurement scenarios that are likely in the system. The types of measurements that are made depend upon the state of the UE and the transitions between the states. We explore what types of measurements are possible for different scenarios and in addition we examine some of the performance issues that impact on these measurements in terms of speed and accuracy.

Once the measurements have been made, they must be filtered. The filters can be used to smooth out rapid and undesirable fluctuations in the measurement quantities, as well as to reduce the rate of flow of measurement information to the UTRAN. We examine what these filters are, how they work and how they are configured by the UTRAN.

The final stage of the measurement process is the reporting of the measurements. Again, there are different possibilities in terms of the reporting. The measurements could be reported continuously, they could be reported periodically, or they could be reported as the consequence of some event. We examine all of these different possibilities, including the configuration and the control of the reporting procedures.

In the final section we consider the specific issues associated with GSM measurements. We look at the measurements that are required prior to handover to GSM and also the measurement reports that are required for these measurements.

To start this chapter, however, we begin with an introduction to the basic principles of measurements, what the measurements are, what is being measured, why it is being measured and how the information is collected together.

### 12.1.1 Measurement types

In this section, we start by considering the different measurement types that a UE should be capable of making – although some of them depend upon the capability of the UE.

The first type of measurement is generically referred to as signal measurements. We see in this section that there are a number of different types of signal measurements, and that these signal measurements are made on a number of different cell types. The signal measurements are to provide the UTRAN with information on the signal levels in the current cell and the surrounding cells for a UE in the CELL_DCH state. For a UE that is in any of the other states the cell signal measurement information is used within the UE for the cell reselection procedure.

Next we consider the traffic volume measurements (TVMs). These measurements are important for the UTRAN to ensure that the flow of data across the radio interface is maintained at some controlled level.

The quality measurements are reported to the UTRAN by the UE to provide the UTRAN with an indication of the RB quality at a specific point in time. The RB quality is defined in terms of the quantity called the BLER.

UE internal measurements are measurements on internal parameters of the UE such as transmit power and time-offsets.

Finally, positioning measurements are measurements made by the UE to enable the UTRAN to report the position of the UE to any external nodes that require the UE position information.

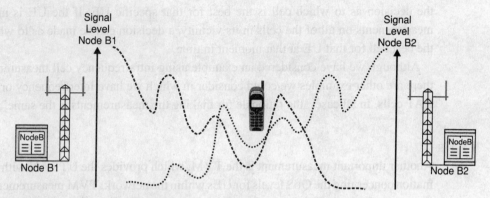

**Figure 12.1** Typical signal measurement scenario.

## Signal measurements

Signal measurements are measurements of items such as path loss, RSCP, $E_c/N_o$ and received signal strength indicator (RSSI) and some of the more important measurements that need to be made within the UTRAN by both the UE and the Node B.

There are three specific types of signal measurements summarised as:

- Intrafrequency measurements – the measurement of cells on the same frequency.
- Interfrequency measurements – the measurement of cells on a different frequency.
- Inter-RAT measurements – the measurement of cells using a different technology such as GSM or IS2000.

The potential outcome for each of these three types of measurement is the same – either a handover to the measured cell (CELL_DCH state), or a cell reselection to the measured cell (all other states).

It is useful to explore why we need to consider the signal measurements. To do this, we start by looking at Figure 12.1, which considers a realistic scenario in which a UE is located roughly equidistant from two cells (Node B1 and Node B2). The signal level from these cells slowly decays with distance (this phenomenon is well known and applies to most propagating signals). Besides this slow decay, there is also a variation in the signal level, often referred to as 'slow fading', which comes from variations in the terrain surrounding the UE as well as the presence of obstacles such as buildings and hills that will cast a shadow from the signal. The consequence of this signal fading is that the best signal for the UE to choose is not necessarily the one that comes from the closest cell (best in this context is the one with the greatest amplitude, but we see later that the best signal is not necessarily the one with the greatest amplitude). In this example, we see that the best signal changes a number of times as we move from one cell to the other.

If we now imagine that there are a number of other cells (typically there may be six relatively close cells), we can see that the selection of the best cell is not simply based on some measure of the location of the UE. Introducing measurements simplifies the decision as to which cell is the best for that specific UE. If the UE is making measurements on all of the cells in its vicinity, a decision can be made as to which is the best cell for that UE at that moment in time.

Although we have considered an example using intrafrequency cell measurements, there are other examples we could consider in which we have interfrequency or inter-RAT cells. In all cases, the rationale for making the measurements is the same.

## TVMs

Another important measurement is the TVM, which provides the UTRAN with information concerning the QoS levels for UEs within the network. TVM measurements are important to ensure that the RB for each UE is kept at some optimum configuration. The UTRAN has the freedom to manage the links to the UE in a variety of ways, examples of which include the choice of channels for the UE (common or dedicated) as well as the configuration of the channels.

The UTRAN objective is the allocation of the minimum required resources to meet the QoS requirements for the UE. If these resources are insufficient for the traffic that the UE is offering to the network, there is a build-up of traffic in buffers in the UE.

In using TVM, the UTRAN can monitor the flow of data from UEs and in doing so can apply corrective action before an overload condition occurs. TVMs are applicable to a UE that is in either the CELL_DCH state or the CELL_FACH state. In all other states no transmission is allowed and so TVMs are not required.

## Quality measurements

Quality measurements are used to assess the quality of the RB to a UE. The quality measurements are an assessment of the number of errors that occur across the RB at a specific point in time. Quality measurements are needed to ensure that the configuration of the RL to a UE at a specific point in time is acceptable. If the quality of the bearer is poor, the UTRAN needs to take remedial action (usually this may simply consist of an increase in the average transmit power). If the quality is very high, the UTRAN may make the decision to reduce the link quality, thereby conserving cell resources that may be allocated to other users in the network. Quality measurements are applied to UEs in the CELL_DCH state.

## UE internal measurements

UE internal measurements report a mixture of parameters that are measured by the UE. These parameters are the UE transmit power, the received RSSI value and the UE Tx–Rx time difference. The use of these measured quantities is different for the different measurements:

- The UE transmit power indicates to the UTRAN how close the UE is to reaching its maximum transmit power.

- The received RSSI value indicates the total measured power at the UE, which is useful for downlink load estimations within the UTRAN.
- The Tx–Rx time difference indicates the time offset between transmit and receive.

**UE positioning measurements**

The final category of measurements is UE positioning measurements, which are made to allow an estimate of the UE position. The measurements that are made by the UE include GPS based measurements and observed time difference of arrival (OTDOA) measurements. In both cases, the objective is for the UE to make measurements that are then used by the UTRAN to estimate the position of the UE within the coverage area.

### 12.1.2 Measurement sets

The signal measurements that were considered previously are based on measurements of cells in areas surrounding the UE. A number of different sets have been defined to characterise the different types of cells. In this section, we consider what the different sets are and how they are used as part of the measurement procedure.

**Active set**

The active set is the collection of cells to which the UE is currently in soft-handover. This means that each of the cells has an RL to the UE.

**Monitoring set**

The monitoring set is the collection of cells (excluding the cells in the active set) that the UE should be measuring, and which could become members of the active set assuming that the signal levels and quality reach the required level. The monitoring set is equivalent to the neighbour list in GSM.

**Detected set**

The detected set is the group of cells that the UE has found, which have an acceptable signal level and signal quality, but are not currently part of the monitoring set, probably due to a configuration error for that part of the network.

**Virtual active set**

The virtual active set is a collection of cells that are on a different frequency and which could become the active set cells assuming a handover to that frequency. It is important to be able to handover to a group of cells, because the UE may be at the edge of coverage and need to handover to a group of cells to remain within the coverage area.

## 12.2 Measurement control

In this section we examine the mechanisms that are defined to control the measurements that the UE makes. We see that there are two basic approaches. First, for a UE that is in any state except CELL_DCH, the measurement control information is received via the SIB messages (SIB11 and SIB12 to be specific). Second, for a UE that is in the CELL_DCH state, the measurements are controlled via a MEASUREMENT CONTROL message received directly from the serving RNC.

### 12.2.1 Measurement control via SIB messages

Whilst in the idle mode, CELL_FACH, CELL_PCH or URA_PCH states, the UE needs to perform cell signal measurements that are used within the UE for cell reselection. The cells being measured can be using the same frequency (intrafrequency measurements), a different frequency (interfrequency measurements) or a different system (inter-RAT measurements). The UE receives the details of the measurements to make via the SIB messages within the cell that it is currently camped on. There are two basic types of SIB message called SIB11 and SIB12. SIB11 is the broadcast message that the UE detects and uses whilst in the idle mode and SIB12 is the broadcast message that the UE uses whilst in the connected mode. If SIB12 is not present in a specific cell, then the contents of SIB11 are normally used for a UE in connected mode. The outline contents of the SIB11 message are illustrated in Table 12.1. The contents of the SIB12 message are virtually the same except for the first field, i.e. the SIB12 indicator is not present.

SIB11 and SIB12 have two main elements as seen in the table. First there is the FACH measurement occasion information and secondly the measurement control information. The FACH measurement occasion information defines the times when the UTRAN halts downlink transmissions to UEs in the CELL_FACH state to allow them to make measurements on other cells on other frequencies. This is considered in more detail next. The measurement control information defines the measurements that the UE needs to make, it defines the cells that the UE needs to measure (intrafrequency, interfrequency and inter-RAT cells) and it defines the configuration for the measurement reporting.

**FACH measurement occasion information**

Part of the SIB11 or SIB12 message is the FACH measurement occasion information element. This information element provides information on when the UE can make measurements whilst in the CELL_FACH state. The reason this is an issue is because whilst in the CELL_FACH state, the UE receives information using the SCCPCH.

## 12.2 Measurement control

**Table 12.1.** *Outline of SIB11 message contents*

SIB12 indicator
 This defines whether the UE should read SIB12 in connected mode or use SIB11.
FACH measurement occasion information
 This defines the parameters for the FACH measurement occasion. It includes the cycle length coefficient described in Section 12.2 and a flag that defines whether interfrequency (separately for FD and TDD) and/or inter-RAT measurements are required during the measurement occasions.
Measurement control system information
 This defines:
 - HCS use in the cell (used or not)
 - Cell selection and reselection criteria ($E_c/N_o$ or RSCP)
 - 32 intrafrequency cells, cell signal measurements and measurement reporting criteria
 - 32 interfrequency cells, cell signal measurements and measurement reporting criteria
 - 32 inter-RAT cells, cell signal measurements and measurement reporting criteria
 - Traffic volume measurements (for use in CELL_FACH state only)

The information is not necessarily continuous, but the UE does not know when the information is being transmitted as there is no scheduling information available for the UE. The solution is to define certain radio frames when the UE can make the measurements that it needs to make.

To identify the frames that the UE uses there is a formula that defines the *SFN* of the frame(s) in which it can make a measurement. The formula is as follows:

$$SFN \text{ div } N = C\_RNTI \text{ mod } M\_REP + n \times M\_REP \tag{12.1}$$

where

$$M\_REP = 2^k. \tag{12.2}$$

Here $k$ is the FACH measurement occasion cycle length coefficient, $N$ is the largest TTI (in frames) in use on the FACH, $C\_RNTI$ is the c-RNTI of the UE in that cell and $n = 0, 1, 2, \ldots$.

The measurement frames (given by the SFN) are selected whenever (12.1) is satisfied for any value of $n$. If $N$ is greater than 1 (i.e. a TTI of more than 10 ms in duration), then there may be multiple consecutive frames that can be used for measurements. The measurement occasion is repetitive. The FACH measurement occasion cycle length coefficient ($k$) defines the periodicity of these measurements given by $M\_REP$ in (12.2).

**Table 12.2.** *The SFN, and left and right hand sides of Equation (11.1)*

| SFN | SFN div N | C_RNTI mod M_REP + n*M_REP |
|-----|-----------|----------------------------|
| 0   | 0         | 2                          |
| 1   | 0         | 10                         |
| 2   | 0         | 18                         |
| 3   | 0         | 26                         |
| 4   | 1         | 34                         |
| 5   | 1         | 42                         |
| 6   | 1         | 50                         |
| 7   | 1         | 58                         |
| 8   | 2         | 66                         |
| 9   | 2         | 74                         |
| 10  | 2         | 82                         |
| 11  | 2         | 90                         |
| 12  | 3         | 98                         |
| 13  | 3         | 106                        |
| 14  | 3         | 114                        |
| 15  | 3         | 122                        |
| 16  | 4         | 130                        |
| 17  | 4         | 138                        |
| 18  | 4         | 146                        |
| 19  | 4         | 154                        |

Consider an example with $N = 4$, $C\_RNTI = 1210$, $k = 3$. Table 12.2 illustrates the values for the SFN and the left and right hand sides of the equation. Examining Table 12.2, we can see that (12.1) is satisfied for an SFN of 8, 9, 10 and 11 and so in these frames the UE is free to make measurements. The next measurement opportunity does not occur until $SFN$ div $N$ equals 10, which happens in SFN 40, 41, 42, and 43. Both the UE and the UTRAN use this equation to define when the UE is making measurements.

### 12.2.2 Measurement control via the MEASUREMENT CONTROL message

The MEASUREMENT CONTROL message is used by the UTRAN whenever it decides to control the measurement procedure in a specific UE with configuration settings different from those used in the SIB11 and SIB12 messages that we have just examined. The basic structural elements of the MEASUREMENT CONTROL message are laid out in Table 12.3.

Upon receipt of this message, the UE reads and stores the information in the MEASUREMENT_IDENTITY variable. The message can either establish new measurements, or reconfigure existing measurements. The message can also activate the use

## 12.2 Measurement control

**Table 12.3.** *Outline of the elements of the MEASUREMENT CONTROL message*

Standard RRC message headers
  The MEASUREMENT CONTROL message includes all of the standard RRC message headers such as the message type, the transaction identifier and the integrity check information.
Measurement identity
  An integer number that defines the measurement identity; up to 16 values allowed.
Measurement command
  Measurement control commands: setup, modify, release.
Measurement reporting mode
  This defines which RLC mode to use (AM or UM) and the reporting type (periodic or event triggered reports).
Additional measurements list
  This defines the additional measurements to be reported with this measurement.
Intrafrequency measurement
  This element configures intrafrequency measurements. It contains the cell information list (up to 32 cells), the measurement and reporting quantities (RSCP, $E_c/N_o$ (FDD), pathloss, ISCP (TDD)). It also contains the reporting cell status that defines the number and type of cells to report, the measurement validity indicator that defines the valid states for the measurement (CELL_DCH, all states, all except CELL_DCH), and the report criteria ( periodic reporting criteria, or intrafrequency measurement reporting criteria).
Interfrequency measurement
  This element configures interfrequency measurements. It includes the cell info list (up to 32 cells) and the same basic elements as the intrafrequency measurement with additions. It also includes the interfrequency set update information that allows for the update of the virtual active set. The report criteria define periodic reporting criteria, or intrafrequency measurement reporting criteria, or interfrequency measurement reporting criteria.
Inter-RAT measurement
  This element configures inter-RAT measurements. It includes the cell information list (up to 32 cells), the measurement quantity (UTRAN quality, GSM RSSI, BSIC verification, IS2000 parameters), the reporting quantity (UTRAN quality, GSM cell time, GSM RSSI), the reporting cell status (defines number and type of cells to report) and the report criteria (periodic reporting criteria, or inter-RAT measurement reporting criteria).
UE positioning measurement
  This element configures UE positioning measurements. It includes the positioning configuration information, the reporting criteria (event triggered or periodic reporting) and the assistance data for OTDOA and GPS measurements.
TVM
  This element configures the TVMs. It includes the measurement object (transport channel to be measured), the measurement and reporting quantity (current, average and variance of RLC buffer payload), the measurement validity state (CELL_DCH, all states, all except CELL_DCH) and the reporting criteria (event triggered or periodic reporting).
Quality measurement
  This element configures the quality measurement. It includes the reporting quantity (BLER of transport channel) and the reporting criteria (event triggered or periodic reporting).
UE internal measurement
  This element configures UE internal measurements. It includes the measurement and reporting quantity (UE Tx power, UTRA RSSI, Tx–Rx time difference) and the reporting criteria (event triggered or periodic reporting).
DPCH compressed mode status information
  This defines DPCH compressed mode status information.

**Table 12.4.** *Information stored in CELL_INFO_LIST*

Intrafrequency measurement
> This contains up to 32 possible cell entries, including the serving cell. It also includes the cell individual offset, which is a measurement bias used to include/exclude certain cells, the reference time difference between the current cell PCCPCH and the neighbour cell PCCPCH. It defines whether the UE should read the measured cell SFN. Cell information for the FDD mode (primary CPICH scrambling code and Tx power, Tx diversity indicator) and the TDD mode (PCCPCH information, PCCPCH Tx power, timeslot information). Cell selection and reselection information (*Qoffsets*, maximum uplink Tx power, HCS neighbour cell information, *Qqualmin* (FDD) and *Qrxlevmin* (FDD/TDD/GSM)) is used to define the cell selection and reselection procedures.

Interfrequency measurement
> This contains the same information as intrafrequency cell information, plus frequency information (uplink UARFCN and downlink UARFCN (FDD), UARFCN (TDD)).

Inter-RAT measurement
> This element contains information for the inter-RAT measurement cells such as the GSM cell selection and reselection information (BSIC; BCCH ARFCN) and the IS2000 information (frequency; timeslot; colour code; output power; PN offset).

of compressed mode for the interfrequency and the inter-RAT measurements. In this case, the message indicates what connection frame number should be used for the start of the compressed mode pattern.

## 12.3 Measurement variables

CELL_INFO_LIST and MEASUREMENT_IDENTITY are variables that the UE uses to store the information used as part of the measurement procedures. In this section, we consider what type of information is stored in the CELL_INFO_LIST and the MEASUREMENT_IDENTITY variables, and in the following section we will see how these variables are used by the UE in the different states.

### 12.3.1 CELL_INFO_LIST

CELL_INFO_LIST contains a list used to define the cells employed for the three basic types of cell signal measurement: intrafrequency cell; interfrequency cell and inter-RAT cell measurements. For each of the three cell types, the variable CELL_INFO_LIST can store information on up to 32 cells.

Table 12.4 summarises the information that is stored for the three types of cells. Not all of the information that is defined in [24] is included, but the table provides an idea of the type of information that is being stored.

## 12.3 Measurement variables

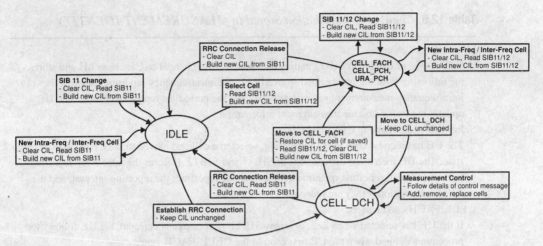

**Figure 12.2** Changes to apply to CELL_INFO_LIST as the UE state changes.

As the UE changes state as a consequence of changes in activity levels in the UE or due to directions from the UTRAN, the contents of the CELL_INFO_LIST variable also need to change to reflect these state changes. Figure 12.2 summarises the changes that the UE follows as it selects new cells or moves to new states.

### 12.3.2 MEASUREMENT_IDENTITY

The variable MEASUREMENT_IDENTITY stores additional information that defines, for instance, the quantities to be measured, the quantities to be reported, and when and how they should be reported. At the highest level, the MEASUREMENT_IDENTITY variable stores some or all of the IEs in the ZMEASUREMENT CONTROL message, some or all of the IEs in the SIB11 message, and some or all of the IEs in the SIB12 message.

The information that is stored, when it is stored, and when it is cleared is defined by the RRC protocol and is dependent on the specific state that the UE is in at a given time. The contents of the MEASUREMENT_IDENTITY variable are the measurement configuration information that the UE received in the SIB11/SIB12 message or in the MEASUREMENT CONTROL message. The essential contents of the MEASUREMENT CONTROL message are presented in Table 12.3 and from this we can see the elements that are stored in the MEASUREMENT_IDENTITY variable (except for the normal RRC message header information). As the UE moves through the coverage area it can change both its state and its cell. Table 12.5 summarises the possible changes and their effects on the MEASUREMENT_IDENTITY variable.

**Table 12.5.** *Changes to Information Stored in MEASUREMENT_IDENTITY*

Idle mode cell change
: Cell change or change to SIB11 causes the changes listed here. The UE clears MI and stores the following information obtained from SIB11: the intrafrequency reporting quantity, the intrafrequency measurement reporting criteria and the periodical reporting criteria. The UE updates the traffic volume measurement information.

Select cell – idle mode to CELL_FACH
: The UE has requested an RRC connection, selected a cell and moved to the CELL_FACH state. The UE then reads the contents of SIB11 and SIB12. It stores the inter-RAT reporting quantity, the cell reporting quantities, the amount of reporting, the reporting interval, and the TVM information if they are received.

CELL_FACH cell change
: If the UE has selected a new cell, or a new SIB11/12 message is detected, the UE follows the procedures defined when the UE first entered the CELL_FACH state.

CELL_FACH measurement control
: If the UE receives a MEASUREMENT_CONTROL message whilst in the CELL_FACH state, it overwrites any TVMs obtained from the message.

Move to CELL_DCH
: There is no direct change to the MEASUREMENT_IDENTITY variable, but there may be some changes to the measurements.

CELL_DCH measurement control
: When the UE receives a measurement control message whilst in the CELL_DCH state it adds, modifies or removes the measurements from the MEASUREMENT_IDENTITY according to the message.

Establish RRC connection – idle mode to CELL_DCH
: When moving from the idle mode to the CELL_DCH state, there is no change in the MEASUREMENT_IDENTITY variable.

Move to CELL_FACH
: When moving from the CELL_DCH state to the CELL_FACH state, the UE deletes the inter-RAT, quality and UE internal measurement control information from the MEASUREMENT_IDENTITY variable.

RRC connection release to idle mode
: When returning to the idle mode from the CELL_FACH or the CELL_DCH states, the UE clears the MEASUREMENT_IDENTITY variable, and then follows the procedures defined for an idle mode cell change as laid out above.

## 12.4 Cell signal measurement procedures

We have seen the different types of measurements that the UE needs to make. In this section we focus on the measurements of the cell signal levels and quality. As we saw previously, these measurements are made for intrafrequency cells, interfrequency cells and inter-RAT cells.

## 12.4 Cell signal measurement procedures

**Table 12.6.** *Number and type of cells to be monitored in the CELL_DCH state*

| Cell type | Number of cells |
|---|---|
| Intrafrequency | 32 cells, which includes the active set. |
| Interfrequency | 32 cells comprising: |
| | 1. FDD cells on up to 2 FDD carriers; |
| | 2. TDD cells on up to 3 TDD carriers. |
| Inter-RAT | 32 GSM cells on up to 32 GSM carriers. |

First, let us consider the basic measurement capabilities of the UE, which define how many cells must be measured in the different cell states. Next, we consider the timing requirements for the measurements. Finally, we consider the measurements themselves, what is being measured and how the measurements change as the UE moves between different states.

This section concentrates on the intrafrequency and interfrequency cases. The special case of inter-RAT measurements for the GSM system is considered in a later section.

### 12.4.1 UE measurement capabilities

**CELL_DCH state monitoring capabilities**

In the CELL_DCH state the UE has to be able to monitor the following number of cells in different cell configurations. The term monitor means that the UE is making measurements on these cells and reporting the measured quantities back to the UTRAN if appropriate.

Table 12.6 illustrates how many cells the UE should be capable of monitoring for the different cell types. For the intrafrequency cell measurement, the cells that the UE is monitoring include cells in the active set.

**CELL_FACH state monitoring capabilities**

Similarly to the CELL_DCH state, the UE in the CELL_FACH state needs to monitor the cells in the intrafrequency, interfrequency and inter-RAT cases. Table 12.7 identifies the requirements for these measurements.

### 12.4.2 Measurement period

The time that a UE is given to make the different types of measurements depends upon the capabilities of the UE and the number of sets of measurements that need to be made at a given point in time. This section outlines some of the time constraints for the cell measurements. A more detailed consideration of the cell measurement constraints can be found in [23].

**Table 12.7.** *Number and type of cells to be monitored in the CELL_FACH state*

| Cell type | Number of cells |
|---|---|
| Intrafrequency | 32 cells. |
| Interfrequency | 32 cells comprising: |
| | 1. FDD cells on up to 2 FDD carriers; |
| | 2. TDD cells on up to 3 TDD carriers. |
| Inter-RAT | 32 GSM cells on up to 32 GSM carriers. |

### CELL_DCH FDD intrafrequency cells

The measurement period for the intrafrequency cell measurements defines how often the physical layer needs to make measurements and pass the measurement results to the higher layers. For intrafrequency cells, the measurement period is 200 ms. In this time period, the UE should make the required measurements (measurement quantities such as RSCP, pathloss and $E_cN_o$) on the primary CPICH of eight intrafrequency cells. If, however, the UE is also making measurements on other cells such as interfrequency or inter-RAT cells, then there is a *pro rata* reduction in the number of cells that the UE is expected to measure, based on the time that is available from the 200 ms for the intrafrequency measurements.

### CELL_DCH FDD interfrequency cells

If the compressed mode is not required for interfrequency measurements, the UE should be capable of monitoring six interfrequency cells on the same FDD carrier in a time period of 480 ms. If the compressed mode is required (the compressed mode is described in greater detail in Chapter 2) the measurement period is greater, depending upon the number of FDD carriers to measure and the measurement time taken.

### CELL_DCH inter-RAT GSM measurements

To permit handover to a GSM system, or to facilitate cell reselection to a GSM system, the UE needs to make measurements on the GSM system. The measurements associated with GSM and some of the timing issues are covered in more detail in Section 12.6, which examines the measurements required for handover to the GSM system in greater depth.

### CELL_FACH intrafrequency cells

As for the CELL_DCH state, a UE in the CELL_FACH state must be able to measure eight intrafrequency cells in the monitored set in 480 ms, assuming that the UE is not making any other measurements. If the UE is also making measurements on interfrequency and inter-RAT cells, the measurement time is reduced in proportion to the amount of time spent on the other measurements. The only additional point is that these other measurements are only made when a FACH measurement occasion arises. The

## 12.4 Cell signal measurement procedures

**Table 12.8.** *UE measurement quantities excluding positioning measurements*

| Measurement quantity | Measurement type |
|---|---|
| CPICH $E_c/N_o$ (FDD) | Intrafrequency, interfrequency |
| CPICH RSCP (FDD) | Intrafrequency, interfrequency |
| Pathloss | Intrafrequency |
| PCCPCH RSCP (TDD) | Intrafrequency, interfrequency |
| Timeslot ISCP (TDD) | Intrafrequency |
| GSM Carrier RSSI | Inter-RAT |
| Base station identity code (BSIC) verification | Inter-RAT |

frequency of this depends upon the FACH measurement occasion length coefficient, which was considered in Section 12.2.

### CELL_FACH interfrequency cells

If a UE does not require a FACH measurement occasion to make measurements on the interfrequency cells, it should be able to measure six interfrequency cells in 480 ms. If the UE does require a measurement occasion, the measurement period increases and is dependent on the number of frequencies that are to be measured and the number of other measurements being made.

### CELL_FACH inter-RAT GSM cells

The measurements associated with GSM and some of the timing issues are covered in more detail in Section 12.6, which examines the measurements required for cell reselection to the GSM system.

### 12.4.3 Cell signal measurements

The cell signal measurements that can be configured for intrafrequency, interfrequency and inter-RAT cell signal measurements are summarised in Table 12.8 and presented in more detail in Table 12.9.

The measurements that the UE makes will depend upon the state that the UE is in and the changes that occur to the UE. Here we will summarise these measurements in the different RRC states and the changes to these measurements as the UE moves between the states.

### Idle mode

The cell signal measurements that are made while the UE is in the idle mode are used for the purposes of cell selection and reselection. As we have seen the cells that could be selected/reselected may be on different frequencies and even on different systems. The measurements and cells to measure are controlled by the information that the UE

**Table 12.9.** *Explanation of the measurement quantities for the cell signal level measurements*

CPICH $E_c/N_o$ measurement quantity (FDD mode only)

    Primary CPICH $E_c/N_o$ is a measure of 'RF quality' of a cell. The information is used internally in the UE for cell reselection or passed to the UTRAN for use by the UTRAN.

    The measurement quantity is the energy per chip ($E_c$) divided by the noise and interference power spectral density ($N_o$). The measurement of $E_c$ requires the UE to measure the code power for just the primary CPICH transmissions in the presence of all of the other downlink code signals.

    $E_c/N_o$ is a relative measurement and is similar to the signal to interference ratio measurements in GSM.

CPICH RSCP measurement quantity (FDD mode only)

    The primary CPICH RSCP is a measure of the absolute received primary CPICH signal. The UE receives a number of physical channels simultaneously and so it has to estimate what part of these received signals comes from the primary CPICH.

    The RSCP is a useful quantity because it can be used to define how close the UE is to the edge of the cell.

Pathloss

    The pathloss is a measure of the transmission loss between the UE and the UTRAN. The pathloss for FDD mode (in dB) is defined as:

        Primary CPICH Tx power – Primary CPICH RSCP

    The pathloss for the TDD mode (also in dB) is defined as:

        Primary CCPCH Tx power – Primary CCPCH RSCP

    It is basically the difference between the transmitted power and the measured received power.

Primary CCPCH RSCP measurement quantity (TDD mode only)

    In TDD mode the primary CCPCH is used to measure the RSCP. The measurement concept is the same as for the FDD mode; however, the UE needs to measure the fraction of Rx power that is from the PCCPCH. The RSCP provides the UE with some estimate of how close it is to the edge of a cell, and can be used as part of the cell reselection procedures.

Timeslot ISCP measurement quantity (TDD mode only)

    The interference signal code power (ISCP) is the interference on the received signal in a timeslot. The measurement is made by the UE using the midamble that is present within a TDD slot.

GSM carrier RSSI

    The GSM carrier RSSI is the received power measured within the relevant channel bandwidth measured on a GSM BCCH carrier. Further details on this are in Section 12.6.

BSIC verification

    As part of the GSM measurements, the UTRAN can request that the UE verifies the BSIC of the cell that it is measuring. This is an additional stage to the measurement process, and requires the UE to synchronise to the GSM cell. Further details on this are in Section 12.6.

has received in the SIB11 message. Any change in state (or even cell) that returns the UE to the idle mode requires the UE to reread the contents of the SIB11 message and apply them to the measurement control.

### CELL_FACH/CELL_PCH and URA_PCH states

A UE is in the connected mode if it is in any of these states. In this case all of the measurement control information is obtained from the cell's SIB12 message if present else the SIB11 message. The main differences between the measurements here and those in the idle mode relate to the CELL_FACH state. In the CELL_FACH state the UE could be expected to make TVM measurements. Like the idle mode, a change to the cell or the state putting the UE back into one of these states requires the UE to reread the SIB11/12 messages for the cell the UE is in and apply the contents of these messages to the measurements that the UE is making.

## CELL_DCH state

We mentioned in the introduction to this chapter that the main difference between the CELL_DCH and all the other states is the use of the measurements and the actions taken with those measurements. In the CELL_DCH state the UE performs measurements and reports the measurements to the UTRAN via the reporting criteria that are considered in the following section. It is now the UTRAN's responsibility to decide whether to make any changes to the UEs connections to adjacent cells using the various handover strategies considered in Chapter 11.

The information that the UE obtains for controlling the measurements in the CELL_DCH state comes initially from the SIB11/12 broadcast messages. In the CELL_DCH state the UE cannot monitor the broadcast channels and so any changes that the UTRAN wishes to make to the measurement configurations to a specific UE need to come via the MEASUREMENT CONTROL message via a dedicated control channel (SRB2 in most cases).

### 12.4.4 Measurements

In this section we consider the measurements that the UE needs to make and how these measurements change as the UE moves between the different states. Some of the measurements are only for internal use by the UE. Some are reported back to, and used by, the UTRAN.

## Measurements in the idle mode

When a UE enters the idle mode, the measurements it has made are used for the purposes of cell reselections whilst in the idle mode. We assume that the UE is camping on a cell in the idle mode and receiving the SIBs. For idle mode measurement control, this information is carried in SIB11. The UE populates the CELL_INFO_LIST and MEASUREMENT_CONTROL variables from the SIB11 message as defined in Section 12.2. Using the information on cells obtained from SIB11 and the information obtained from the SIB3 broadcast message the UE can perform cell selections and reselections.

To facilitate the decision as to which cell a specific UE should be camped on, the UE must make measurements on the serving cell and the neighbour cells, which could be on the same frequency, a different frequency or even a different RAT. If the UE crosses a cell boundary into a new cell, the UE performs a cell reselection to the new cell based on the criteria that are outlined in Chapter 14.

## UE moving from idle mode to CELL_FACH state

To move from the idle mode to the CELL_FACH state, the UE has to establish an RRC connection (described in greater detail in Chapter 11). When this connection is established, the UE moves into the connected mode and, in this case, the CELL_FACH state. Whilst in the CELL_FACH state, the measurements are mainly for the internal use of the UE for cell reselections.

Let us consider an example. In this example, we assume that the SIB12 message is available (if it were not, the UE would rely upon the SIB11 message), and the UE reads and responds to this message appropriately. After the UE has read the SIB11 and SIB12 messages, it stores the data in the CELL_INFO_LIST and the MEASUREMENT_IDENTITY variables as discussed in Section 12.2. The UE makes the appropriate measurements on intrafrequency, interfrequency and inter-RAT cells as defined in CELL_INFO-LIST, according to the timings outlined in Section 12.2 and the timing and measurement controls that apply to the UE whilst in the CELL_FACH state as part of the cell reselection procedure that is described in Chapter 14. In addition to the internal signal level measurements, the UE also starts making and reporting the traffic volume measurements.

## UE moving from CELL_FACH to CELL_PCH or URA_PCH state

In moving from CELL_FACH to CELL_PCH or URA_PCH there is little change in the activity of the UE regarding measurements. The UE still needs to perform the required cell based measurements for the purpose of cell reselections.

## UE moving from CELL_FACH to CELL_DCH

In the CELL_DCH state, the UE is allocated dedicated physical resources from one or more cells. The UE may be in soft-handover to multiple cells, or it may be connected only to a single cell. Whilst in the CELL_DCH state, the UE is making measurements. These measurements are transmitted back to the UTRAN, which is responsible for processing them.

In the CELL_DCH state, the UE does not make cell reselections as it does in all of the other states; instead, the UE moves from cell to cell using handovers. The handovers could be intrafrequency handovers (also called soft-handovers), interfrequency handovers (often called hard-handovers) or inter-RAT handovers (to GSM for instance), which are also called hard-handovers.

To facilitate the handovers, the UE must make measurements on the current cells, the intrafrequency cells, the interfrequency cells or the inter-RAT cells on a different system. The measurements that the UE needs to make are either predefined, based on the measurement system information that the UE received in the SIB11 or SIB12 messages, or alternatively could come via the MEASURMENT CONTROL message that is used to setup, modify, or release a specific measurement.

## 12.4 Cell signal measurement procedures

In moving from CELL_FACH to CELL_DCH, the set of measurements that the UE needs to make in the CELL_DCH state are stored in the MEASUREMENT_IDENTITY variable. The measurements stored in this variable are either from the SIB11 and SIB12 messages that the UE received whilst in the CELL_FACH state, or possibly from a previous time that the UE was in the CELL_DCH state.

Assuming that the UE is in the same cell as it was whilst in the CELL_FACH state, the CELL_INFO_LIST variable is still applicable to the UE. At any time whilst in the CELL_DCH state, the UE can receive a MEASURMENT CONTROL message that can be used to modify the contents of both the MEASURMENT_IDENTITY and the CELL_INFO_LIST variables.

**Moving from idle mode to CELL_DCH**

A UE in idle mode moving to the CELL_DCH state does so as a consequence of an RRC connection being established. Whilst in the idle mode, the UE is receiving the SIB11 messages. It stores the relevant contents of the SIB11 message in the MEASURMENT_IDENTITY and the CELL_INFO_LIST variables. Not all of the contents of the SIB11 message are applicable to a UE in idle mode. Some of the contents may be intended for use by the UE when it eventually moves into the CELL_DCH state. The UE knows which measurements to make and report after the transition to the CELL_DCH state through the use of the measurement validity IE that defines the state in which the measurement is applicable.

Once the UE is in the CELL_DCH state it can start to make the measurements and report them according to the reporting criteria that were established with the measurements. The UTRAN can change the measurements and reporting criteria using the MEASURMENT CONTROL message at any time.

**Moving from CELL_DCH to CELL_FACH / CELL_PCH / URA_PCH**

Moving from CELL_DCH to CELL_FACH / CELL_PCH / URA_PCH may occur as a result of activity from a specific UE dropping to a low level. On moving to these new states, the UE ceases reporting measurements for the interfrequency, intrafrequency and inter-RAT cell measurements. It stops reporting these measurements because they are only reported whilst in the CELL_DCH state. The UE still needs to monitor the intrafrequency, interfrequency and inter-RAT cells to allow the cell reselection procedures to continue. The reporting of TVMs depends on whether the TVMs have been configured for these states. If the measurements are configured for these states, they continue.

If, as a consequence of the state change, the UE changes its cell, the UE needs to monitor the intrafrequency, interfrequency and inter-RAT cell lists received in the SIB12 or SIB11 messages for that cell.

## Moving from CELL_DCH to Idle_Mode

In moving from CELL_DCH back to idle mode, the UE stops reporting all measurements that it was previously performing. It clears the variable MEASUREMENT_IDENTITY. The UE then monitors the intrafrequency, interfrequency and inter-RAT cells that are defined by the SIB11 message for use by UEs in the idle mode, and uses this information as part of a cell reselection procedure.

## 12.5 Reporting the measurement results

So far, we have considered the basic principles of the measurements and how the measurements are configured. In this section we consider the issue of measurement reporting. Measurement reporting concerns the process of sending measurement results back to the UTRAN. The reporting of results may be periodic, where the results are sent over a predefined period, or they may be event-triggered reporting, whereby some event triggers the activation of the measurement reporting. To start, we consider some of the terminology that is applied to the measurement reporting activity. After that, we consider some of the specific details of the measurement reports for the intrafrequency, interfrequency and inter-RAT cell measurements, and then we consider the other measurement types and the reports that are defined for them.

### 12.5.1 Reporting terminology

**Measurement reports**

The measurement reports define the measurements that the UE should report, assuming that either an event-triggered report or a periodic report is activated. The measurement reporting information is configurable. The measurement report information for individual measurements is presented below.

**Measurement report criteria**

For the event-triggered reports, the measurement report criteria define the conditions that need to be met before a measurement report is created. They define cell measurements such as pathloss and RSCP, as well as other conditions that need to be achieved for the other measurement types before a measurement report is triggered.

**Measurement filtering**

Figure 12.3 illustrates the measurement model that is defined for the measurements considered here. The measurement model consists of a Layer 1 filter that samples and filters the measurement data. Next there is a Layer 3 filter which is responsible for filtering the measurement information that was provided by Layer 1 to Layer 3. The details of the Layer 3 filter are considered below. The reporting criteria evaluation stage

## 12.5 Reporting the measurement results

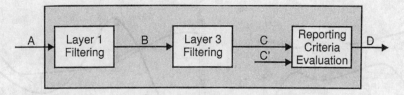

**Figure 12.3** Measurement model and measurement filtering.

is where the criteria for the reports are assessed against the measurement information to decide when to generate a report. The reason for the Layer 3 measurement filter is to smooth the measurements before they are either acted upon or reported to the UTRAN. The Layer 3 measurement filter is controlled using a filter coefficient that can be configured independently for each measurement. The equation that defines the Layer 3 measurement filtering is

$$F_n = (1 - a)F_{n-1} + aM_n$$

where

$$a = 1/2^{(k/2)} \text{ and } F_0 = M_0$$

$M_n$ is the measurement quantity, $F_n$ is the filtered measurement and $F_0$ is the first filtered measurement, which is set equal to the first measurement $M_0$. The quantity $a$ is derived from the filter coefficient $k$, which the UE receives when the measurement is configured.

The UE should be able to support at least two different measurement filters per measurement type (e.g. two for the intrafrequency measurement, two for the interfrequency measurement and two for the inter-RAT measurement).

### Additional measured results

The measurement reports can be configured to provide additional measurement information as part of the measurement report. If reporting of additional measured results is configured when a measurement report is triggered, the UE can also include other measurements as defined in the additional measured results configuration information.

### Reporting range

The reporting range is defined for events 1A and 1B (defined in Section 12.5.2). The reporting range is based on an offset from the best primary CPICH. Figure 12.4(a) represents the basic principles behind the reporting range. Only when a signal enters the reporting range can it cause an event. In this example, it causes periodic reports to be triggered.

### Event-triggered periodic reporting

Figure 12.4(a) illustrates the principles of event-triggered periodic reporting. An event may occur, such as a primary CPICH entering the reporting range or a non-active

# 416 Measurements

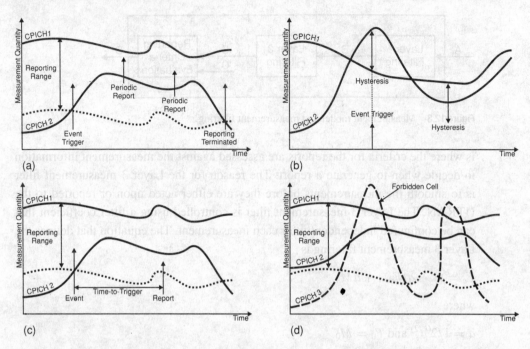

**Figure 12.4** Measurement reporting concepts: (a) event-triggered periodic reporting; (b) use of hysteresis to limit reporting events; (c) use of time-to-trigger; (d) use of forbidden cells.

primary CPICH becoming better than an active primary CPICH. Once the event has occurred, the periodic reporting that is configured as part of the event reporting is activated. The UE then performs periodic reporting until either the amount of reporting is exceeded, or the primary CPICH that triggered the event falls below some threshold.

## Hysteresis

The measurement quantities are continuously varying with time. As a consequence, a quantity that is varying around the trigger threshold induces a number of reports from the same basic event. To prevent this undesirable situation, hysteresis is applied to the measurements to ensure that a large number of measurements are not triggered from a single event. Figure 12.4(b) illustrates the basic principles behind the use of hysteresis. The event only triggers if it is greater than (or less than) the measured quantity by an amount equal to the hysteresis value defined for that event trigger. The figure shows two cases: one where the condition is met and the event is generated and one where the measurement fails to fall below the hysteresis value and so the event is not triggered.

## Time-to-trigger

Time-to-trigger is another type of hysteresis. If a measurement report uses time-to-trigger before the report is generated and sent, then it means that the trigger needs to have been active for a period equal to the time-to-trigger before the measurement report is sent. This process also reduces the number of reports occurring, and only allows ones that show a genuine trend in the quantity that is being measured. Figure 12.4(c) illustrates the use of the time-to-trigger concept. The event occurs only if the conditions are maintained for the entire period 'time-to-trigger'.

**Cell individual offsets**

Cell individual offsets are quantities added to cell measurements to control whether or not a report is generated. The cell offset is applied on a per cell basis, and so is a good method of controlling the measurements from specific reports in either a positive or a negative way.

One specific use of the cell offset is for cells that suffer from 'street corner' effects, such as micro-cells, where it is important that cells are measured even though the cell signal strength may be quite low, because it will suddenly rise in value when the street corner is turned.

**Forbid cell to affect reporting range**

Certain cells can be prevented from affecting the reporting range (in events 1A and 1B for instance). These cells are defined in the intrafrequency measurement reporting criteria message. The objective is that the forbidden cells are excluded from the evaluation and report generation procedures for those specific events. The reasons for doing this are varied, but one reason is knowledge that a specific CPICH is very unstable due to rapid shadowing in a specific cell. Figure 12.4(d) illustrates the idea behind the forbidden cell.

**Periodic reporting**

Periodic reporting is the simplest form of reporting available for the UE. If the UE is configured to perform periodic reporting in place of event-triggered reporting, it is told how much reporting it can do (amount of reporting) and also the reporting period (ranging from 0.25 s to 64 s). Once activated, the UE reports the measurement results at a rate that is defined by the reporting rate until the amount of reporting is achieved. Figure 12.4(a) is an example of event-triggered periodic reporting. Message-activated periodic reporting (via the MEASUREMENT CONTROL message) can also be used.

### 12.5.2 Intrafrequency reporting for the FDD mode

In this section, we consider the basic measurement reporting modes that are applicable to the intrafrequency reporting, and in particular the FDD mode. Before reviewing that in detail, however, there are a couple of general points on the event reporting to be considered.

## Intrafrequency reporting quantities

The following are the quantities that can be reported as part of the measurement reports that are configured for the intrafrequency cell measurements:
- SFN–SFN observed time difference;
- cell synchronisation information;
- cell identity;
- downlink $E_c/N_o$ (FDD);
- downlink pathloss;
- downlink RSCP (primary CPICH FDD, PCCPCH TDD);
- downlink timeslot ISCP (TDD);
- proposed TGSN (TDD).

## Intrafrequency measurement event triggers

The following can be the cause of an event that will trigger a measurement report:
- downlink $E_c/N_o$;
- downlink pathloss;
- downlink RSCP.

There are a number of different events defined and they are listed below:
- 1A – a primary CPICH enters the reporting range;
- 1B – a primary CPICH leaves the reporting range;
- 1C – a non-active primary CPICH becomes better than an active primary CPICH;
- 1D – change of best cell;
- 1E – a primary CPICH becomes better than an absolute threshold;
- 1F – a primary CPICH becomes worse than an absolute threshold.

In the following we outline the details for each of the different event reporting modes.

## Event 1A – a primary CPICH enters the reporting range

The trigger occurs when the pathloss or the $E_c/N_o$ or RSCP appears to be better than a weighted measure of the current measured cells. The measured cells can be configured as one or more of the monitored set and detected set. For R99, active set cells are not included in the triggering criteria.

In the estimation of the trigger condition, UTRAN defined weighting factors and hysteresis values are included in the equation that defines whether or not an event trigger is activated. Additionally, the event trigger has to be active for a period of time (time-to-trigger) before the event is registered.

The final parameter to consider is the reporting range, also set by the UTRAN, which controls the range over which events can be reported. If the current measurement satisfies the required conditions, a periodic reporting event is activated according to the configuration for that periodic event.

The event remains active until the cells that triggered it fall below some defined threshold based on other cell measurements and parameters defined by the UTRAN.

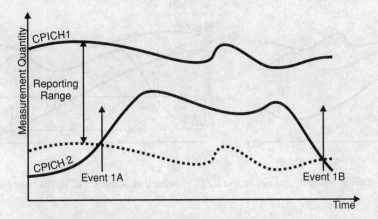

**Figure 12.5** Event 1A, a primary CPICH entering the reporting range, and event 1B, a primary CPICH leaving the reporting range.

The reporting range is applied relative to the defined best cell performance. The use of the reporting range present in event 1A is illustrated in Figure 12.5.

### Event 1B – a primary CPICH leaves the reporting range

This second event is defined to send a measurement report when a specific primary CPICH drops below certain predefined levels. For R99 the trigger is only activated for the active set cells. Although the monitored set cells are included in the protocol, the UE behaviour is unspecified if the monitored set cells are used to trigger this event. The measurement report includes a range of measurement quantities. Finally, the cell that has dropped out of the reporting range is removed from the list that can trigger the generation of a measurement report. Event 1B is illustrated in Figure 12.5.

### Event 1C – a non-active primary CPICH becomes better than an active primary CPICH

This measurement report event relates specifically to the activities of the active set. The objective of this procedure is to ensure that the best cells not in the active set are being measured. The measurements can be pathloss, $E_c/N_o$ or RSCP, with the term best being applied appropriately for each measurement (i.e. for pathloss the best cell is the lowest, but for $E_c/N_o$ or RSCP the best cell is the highest).

Figure 12.6(a) illustrates the measurements for two cells in the active set, and a cell from the monitoring set. When the measurements from a cell that is not in the active set are better than those from a cell that is in the active set (including offsets and hysteresis margins) for a defined time period (time-to-trigger), that cell causes event 1C to trigger. Once the event is triggered, periodic reports of these new 'better' cells as well as the active set cells that are 'worse' will be sent to the UTRAN. A second set of conditions

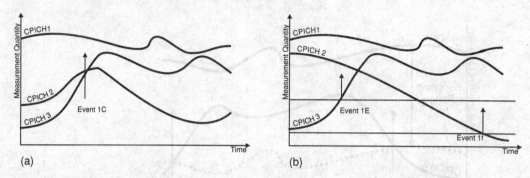

**Figure 12.6** Examples of event reporting: (a) non-active CPICH better than active CPICH; (b) examples of event 1E and event 1F.

is used to terminate the periodic reporting. If the new cells drop in performance with respect to the active set cells, or the new cell is added to the active set, the reporting process will halt.

### Event 1D – change of best cell

This event is used to indicate which cell is currently the best cell (defined in terms of pathloss, $E_c/N_o$ or RSCP). Initially, the best cell is set to the best in the active set (as defined by the measurement quantity). The trigger can occur from any other cell. The measurement report is a 'single shot' measurement with a report sent at the transition.

### Event 1E – a primary CPICH becomes better than an absolute threshold

The objective for this measurement report is to inform the UTRAN when a specific cell rises above some absolute threshold. For this, the trigger is that a primary CPICH becomes better than some absolute threshold for the quantity being measured. The measured cells can be configured as one or more of the monitored set and detected set. For R99, active set cells are not included in the triggering criteria. Figure 12.6(b) illustrates the basic principles of this specific measurement.

### Event 1F – a primary CPICH becomes worse than an absolute threshold

This event is the reverse of event 1E. The event is triggered when the primary CPICH for a cell falls below some absolute threshold for a defined time period (time-to-trigger). For R99, the trigger is only activated for the active set cells. Although the monitored set cells are included in the protocol, the UE behaviour is unspecified if the monitored set cells are used to trigger this event. If the conditions are met, the event trigger causes the measurement report to be sent to the UTRAN. This scenario is illustrated in Figure 12.6(b).

## 12.5.3 Intrafrequency reporting for the TDD mode

In this section we consider the TDD mode event reporting criteria and reports. The TDD mode event report triggers are summarised below:
- 1G – change of best cell;
- 1H – timeslot ISCP below a certain threshold;
- 1I – timeslot ISCP above a certain threshold.

**Event 1G – change of best cell**

This event is specific to the TDD mode system, but is similar to the FDD mode equivalent event. For the TDD mode there is no active set, and so the events are based on all of the cells included in the reporting criteria. When a new cell becomes better than a previous best cell for the time-to-trigger, then an event 1G is generated and a report is sent back to the UTRAN. The triggers for this event includes cell offsets and hysteresis values as described previously.

**Event 1H – timeslot ISCP below a certain threshold**

The timeslot ISCP is measured and filtered and compared with a threshold value defined by the UTRAN. If the measured ISCP falls below the threshold for a time-to-trigger, then an event 1H is generated and a measurement report is sent to the UTRAN. The contents of the measurement report are defined as part of the measurement report information that is received from the UTRAN.

**Event 1I – timeslot ISCP above a certain threshold**

This event is the inverse of the previous event. The measurements of the timeslot ISCP are made, filtered and if they remain above a threshold value defined by the UTRAN for a time-to-trigger, an event is created and a measurement report is sent to the UTRAN including measurements of cells that generated the event as well as any additional measurement reports commanded to do so by the UTRAN.

## 12.5.4 Interfrequency reporting

**Interfrequency measurement quantity**

The interfrequency measurement quantity is defined when an interfrequency measurement is configured for a specific UE. The measurement quantity is used to determine whether a measurement event has occurred. The defined interfrequency measurement quantities are:
- downlink $E_c/N_o$;
- downlink RSCP after despreading.

## Virtual active set

The virtual active set is a set of cells on a carrier frequency that is different from the current frequency. The idea behind the virtual active set is that the UE can track the measurements of the virtual active set, and if an interfrequency (hard-)handover is required, it can be done in a way that immediately puts the UE into soft-handover using an active set that may comprise the elements from the virtual active set.

One method of creating the virtual active set is for the UE to perform what is called an autonomous update of the virtual active set. In this the UE sets up an initial virtual active set and this is then updated at each interfrequency measurement.

The initial virtual active set is defined by the UE based on the measurements that the UE is making on the non-used frequency. The configuration of the virtual active set is controlled either by information obtained as part of the interfrequency measurement control, or using information that was obtained from the intrafrequency measurement control.

The basic idea for the initial virtual active set is that the UE selects the best cells from those available on the non-used frequency up to a maximum that is defined by the configuration information. Once the virtual active set is selected, it can be updated based on the ongoing measurements that the UE is making. The control of the virtual active set depends upon the specific configuration, but the basic objective is to have the best cells present in the virtual active set at a given point in time.

## Interfrequency reporting quantity

The interfrequency measurement reports can contain the following elements:
- cell identity;
- SFN–SFN observed time difference;
- cell synchronisation information;
- downlink $E_c/N_o$ (FDD);
- downlink pathloss;
- downlink RSCP (primary CPICH FDD, PCCPCH TDD);
- downlink timeslot ISCP (TDD);
- proposed TGSN (TDD);
- UTRA carrier RSSI.

## Frequency quality estimate (FDD)

For interfrequency event reporting, a frequency quality estimate is defined which can be used as part of the frequency quality estimation procedures. The frequency quality estimate is based on the sum of the weighted measurements from the best cell and the weighted sum from the other cells that are in the virtual active set on a specific frequency.

### Event 2A – change of best frequency

This interfrequency measurement event occurs when the frequency quality on one frequency becomes better than the frequency quality of the current best frequency. For the FDD mode, this is based on the contents of the virtual active set for the non-used frequency, and the active set for the used frequency.

### Event 2B – used frequency quality below threshold non-used frequency above threshold

This interfrequency measurement event is intended to indicate to the UTRAN that the used frequency has fallen below some threshold and that a non-used frequency is currently above the threshold. For the FDD mode the frequency quality is based on the contents of the virtual active set, and for the TDD mode, it is based on measured cells on the different frequencies. If the criteria for the event are satisfied for a period (defined by the time-to-trigger), a measurement report is sent with the configured contents based on the measurements as well as any additional measurement information that was configured as part of the measurement.

### Event 2C – estimated quality of non-used frequency is above a certain threshold

This event is similar to the previous case, except that only the non-used frequency is considered. The objective for this event is to send a measurement report when the non-used frequency quality exceeds some threshold (also including a hysteresis value) for a time period (time-to-trigger). Additional measurement reports can also be sent depending upon the configuration of the measurement.

### Event 2D – estimated quality of used frequency is below a certain threshold

This event is also similar to event 2B, except that it only considers the used frequency. If the used frequency falls below some threshold for a time period (time-to-trigger) then the event is triggered and a measurement report sent to the UTRAN. Similarly to the other cases, additional measurement results can also be sent as part of the measurement report.

### Event 2E – estimated quality of non-used frequency is below a certain threshold

This is similar to event 2C, but the measured frequency has now fallen below some defined threshold.

### Event 2F – estimated quality of used frequency is above a certain threshold

This is similar to event 2D but the estimated quality of the used frequency has gone above some defined threshold.

**Table 12.10.** *Traffic volume measurement quantities*

| TVM | Definition |
| --- | --- |
| Buffer occupancy | Defines the RLC buffer payload (0k–1024k) bytes. |
| Average buffer occupancy | Defines average RLC buffer payload (0k–1024k) bytes. |
| Variance of buffer occupancy | Defines variance of RLC buffer payload (0k–16k) bytes. |

### 12.5.5 Traffic volume reporting

Three quantities are defined for TVM reporting and are illustrated in Table 12.10. The quantities relate to the RLC buffers that are used for the RBs across the radio interface. The data are calculated per radio bearer. There are two types of TVMs, depending on the trigger method used: periodic measurement, and event-triggered measurement. Periodic TVM uses the same basic concept as cell quality measurements. It is a report that is repeated on a certain number of occasions, with the period between reports configurable by the UTRAN.

**Transport channel traffic volume (TCTV)**

The TCTV is a quantity that is used as a basis for event triggers. It is the sum of the buffer occupancy for all logical channels that are mapped onto a transport channel and it is calculated at least every TTI. The TCTV is used as the basis for event triggers for the TVMs. There are two basic events that can be configured for the TVMs and these are:
- Event 4A – TCTV becomes larger than an absolute threshold;
- Event 4B – TCTV becomes smaller than an absolute threshold.

With no other parameters set, these events are triggered when the TCTV exceeds the threshold or falls below the threshold. To limit the amount of reporting and provide some hysteresis, however, two additional elements are added to the event trigger decision process.

In the first, a pending-after-trigger timer is used. The purpose of this timer is to restrict the rate at which consecutive event reports are sent. Once a trigger is activated, further reports cannot be sent until the timer has expired and even then only if the event condition is still present.

The second event conditioning procedure is called the time-to-trigger. In this case, a timer is started when the event occurs (in the case of event 4A going above threshold or event 4B going below threshold). If the event is still valid when the timer expires, the event report is sent. If the event condition disappears before the timer expires, the timer is stopped. The time-to-trigger process provides a degree of hysteresis that can be used to prevent rapid changes in TCTV from generating events. Once an event report is triggered, the measurement report that was configured is sent to the UTRAN.

## 12.6 Measurements for interoperation with GSM

**Table 12.11.** *UE internal measurement quantities*

| UE internal measurement | Definition |
|---|---|
| UE Tx power | The UE transmit power (over a timeslot for TDD). |
| UE received signal strength | The received RSSI value. |
| UE Tx–Rx time difference | The difference in time between transmit and receive. |

**Table 12.12.** *The different events defined for the UE internal measurements*

| UE internal measurement event | Definition |
|---|---|
| Event 6A | UE Tx power becomes larger than an absolute threshold. |
| Event 6B | UE Tx power becomes less than an absolute threshold. |
| Event 6C | UE Tx power reaches its minimum value. |
| Event 6D | UE Tx power reaches its maximum value. |
| Event 6E | UE RSSI reaches the UE's dynamic receiver range. |
| Event 6F | The UE Rx–Tx time difference for an RL in the active set becomes greater than an absolute threshold. |
| Event 6G | The UE Rx–Tx time difference for an RL in the active set becomes less than an absolute threshold. |

### 12.5.6 Quality measurement reporting

The quality measurement report describes the quality of the received data on the downlink. To do this it measures the CRC failures for data blocks on the downlink. A report is sent if the number of bad CRCs exceeds a quantity (bad CRC count) when counted over a number of CRCs called the total CRC. Once again, to reduce the amount of reporting, a timer 'pending-after-trigger' is used to prevent any additional reports until the expiry of that timer.

### 12.5.7 UE internal measurement reporting

The UE internal measurement reports are defined in Table 12.11. There are also a number of events associated with the UE internal measurement reporting and these are illustrated in Table 12.12. The meaning of the measurements is self-evident from the table. The measurements are configurable, and also include the time-to-trigger parameter that can be used to reduce the effects of measurement bounce.

## 12.6 Measurements for interoperation with GSM

In this section we explore the measurements and the reports that the UE performs on GSM cells whilst attached to UTRAN FDD mode cells. The measurements are used prior to decisions (by UE) on cell reselections while in the idle mode, CELL_FACH, CELL_PCH and URA_PCH states and prior to decisions (by UTRAN) on handovers while in the CELL_DCH state.

We focus on two specific examples: the measurements made while in the CELL_FACH state and the measurements made while in the CELL_DCH state. These two sets of measurements are complicated by the fact that the UE is potentially transmitting and receiving and does not readily have an opportunity to make the measurements. For the CELL_FACH state we use the FACH measurement occasions described in Section 12.2 and for the CELL_DCH state we use the compressed mode that was described in Chapter 2. For the other states the measurements are the same as for the CELL_FACH state, but the inactive nature of these states means that the UE has more opportunities for these measurements.

### 12.6.1 CELL_FACH inter-RAT GSM cell measurements

A UE that is in the CELL_FACH state may have the ability to make measurements independently of the downlink activity (it may have two separate receivers, for instance), or alternatively, the UE may need to use the FACH measurement occasions that were described in Section 12.2.4.

**Measurement occasions not required by UE**

If the UE has the ability to make the measurements without measurement occasions, the UE makes the GSM BCCH signal level measurements (also referred to as RSSI measurements) with a measurement period of 480 ms.

**Measurement occasions required by UE**

If the UE is using the FACH measurement occasions, the time is equally divided between the different types of measurements that use the measurement occasions (interfrequency FDD, interfrequency TDD and inter-RAT GSM). The time separation between the measurements is given by:

$$T_{meas} = [(N_{FDD} + N_{TDD} + N_{GSM}) \, TTI_{Long} K]$$

where $N_{FDD}$, $N_{TDD}$, $N_{GSM}$ are all equal to 1 if there is a measurement of that type, and 0 if there is no measurement of that type. $TTI_{Long}$ is the longest TTI in the SCCPCH on the downlink and $K$ is the FACH measurement occasion length coefficient that is received in a system information broadcast message. The values for $K$ are defined in [23, 42] and depend upon the value for $TTI_{Long}$.

Let us consider an example. Assume that $N_{FDD}$ and $N_{GSM}$ are equal to 1 and $N_{TDD} = 0$; $TTI_{Long} = 40$ ms and $K = 3$, and in this case, $T_{meas} = 240$ ms. With a measurement occasion cycle length of 240 ms, the UE must make one measurement on all of the defined systems within this period.

## 12.6 Measurements for interoperation with GSM

For the inter-RAT GSM cell measurements, the use of these measurement occasions is further subdivided. The GSM measurement occasions are used as follows. Every second measurement occasion is used for initial BSIC identification (in the example this means an initial BSIC identification every 480 ms). Taking the remaining measurement occasions in groups of four, they three out of the four can be used for GSM BCCH signal level measurements and the fourth for BSIC reconfirmation.

The diagram below illustrates the possible scheduling for the different activities. An X in the top line indicates that the measurement occasion is intended for use by GSM measurements (this is every 240 ms in our example). In the bottom line, an I indicates that the measurement occasion is used for initial BSIC identification (every 480 ms), an S indicates a BCCH signal measurement and an R a BSIC reconfirmation. It should be noted that the specifications do not define how the available occasions are scheduled for BCCH signal measurement or BSIC reconfirmation other than giving their relative frequency.

```
X X X X X X X X X X X X X X      X      X      X
I S I R I S I S I S I R I S I    S      I      S
```

## BSIC reconfirmation

BSIC reconfirmations are required to be performed periodically by the UE. If the UE needs to use the measurement occasions to perform the reconfirmation, the frequency of the BSIC reconfirmations depends upon the measurement periodicity, but typically ranges from just over 1 for small measurement periodicities around 80 ms, to over 1 min when the measurement occasion periodicities are greater than 3 seconds. We need to be careful to differentiate between the time for the measurements to take place and the total time required for the successful completion of the measurement.

If the UE does not need measurement occasions, it attempts to check the six strongest carriers at least every 10 seconds. If the measured BSIC agrees with the expected BSIC, it is reconfirmed or verified. If the measured BSIC does not agree with the expected BSIC it is referred to as non-verified.

## Initial BSIC identification

The task of initial BSIC identification is one to which a large amount of the measurement time is devoted. The UE performs BSIC identification in a cell order dictated by the RSSI of the cell whose BSIC is required, starting with the strongest cell.

The BSIC identification procedures concentrate on one cell, until it is identified, and then move to the next cell in the list. The specification defines the time taken for the BSIC identification, but typically it ranges from just over a second to some tens of seconds, depending upon the lengths of the measurement occasions and the number of frames per TTI.

## Use of measurement information

The measurement information that is collected by the UE is used for the cell reselection procedures detailed in Chapter 14 which considers the procedures for a UE in idle mode and certain substates of the connected mode.

### 12.6.2 CELL_DCH inter-RAT GSM cell measurements

In this section we examine the measurements and the procedures that a UE in the CELL_DCH state needs to follow prior to making a handover from a UTRAN FDD mode cell to cell in the GSM system. Firstly, we consider some of the general issues of handover to GSM before moving on to consider some of the specific issues that need to be addressed.

## Reasons for GSM handover

There are a few reasons why a UE may need to consider a handover to the GSM system. The main ones are:
- loss of coverage in UMTS cell;
- operator preference.

Whatever the reason, the UE needs to make some measurements on the GSM system and report these measurements to the UTRAN. Based on the measurements that the UE is making in the serving cells, the intrafrequency cells, the interfrequency cells and the inter-RAT GSM cells, the UTRAN may make the decision to handover to GSM.

Here we focus on the measurement aspects of this procedure, although we touch briefly on the messaging that is associated with a handover to GSM. The UTRAN FDD mode is a full duplex radio technology; consequently the UE is continuously transmitting and receiving (unlike the TDD mode where there are some measurement opportunities). To facilitate the measurements, the UE may use the compressed mode technique, which was introduced in Chapter 2. In this section we examine how compressed mode is used to make measurements on the GSM system.

## Measurements for inter-RAT GSM handover

In the discussion of the CELL_FACH state in Section 12.6.1, we saw that the measurements the UE needs to make are the BCCH RSSI measurements, BSIC identification and BSIC reconfirmation. The reason for the BSIC measurements is that the reporting of the BCCH RSSI can be done with the BSIC either verified or non-verified.

It is an option as to whether the UE needs to use compressed mode for these measurements or whether it can make the measurement without compressed mode. A further refinement to this point is also whether the UE needs compressed mode in both the uplink and the downlink or in just one of the links.

We next consider both the case where the UE needs to use compressed mode to make the measurements, and the general case of requiring both the uplink compressed mode

**Table 12.13.** *The different TGMPs for GSM measurements*

GSM carrier RSSI measurement
GSM initial BSIC identification
GSM BSIC reconfirmation

as well as the downlink compressed mode. Following this, we consider the alternative scenario where the UE does not need compressed mode.

**GSM measurements using compressed mode**
The UE can be requested to measure and report the GSM BCCH RSSI carrier level either with the BSIC verified or the BSIC non-verified. There are, therefore, two parts to this measurement: first, there is the requirement to make the RSSI measurements; and second, there is the requirement to decode the BSIC (or not, depending on the measurement type).

*BCCH RSSI measurement*
First, we will consider the BCCH carrier RSSI measurement, and following that the issues related to the BSIC verification. To make BCCH RSSI measurements in the compressed mode, the UE needs to receive a set of compressed mode parameters. As well as the parameters defining the compressed mode gap locations, the UE also receives a transmission gap sequence measurement purpose (TGMP) information element. This information element defines the purpose of the specific compressed mode. Table 12.13 shows those TGMPs defined for GSM measurements.

The UE should only use the compressed mode patterns whose purpose matches the measurements to be made. In total, there can be up to six different compressed mode sequences active at the same time for different purposes.

The compressed mode pattern defines both the width and the repetition of the gap. For the GSM RSSI measurements, the UE can use gap lengths of 3, 4, 5, 7, 10 and 14 in numbers of slots. There are also patterns defined where both the first gap (TGL1) and the second gap (TGL2) are defined within the pattern. The number of measurement samples that the UE makes per gap depends upon the gap length. Table 12.14 defines the minimum number of GSM carrier RSSI samples per gap that is required for a UE.

In the CELL_DCH state, a measurement period of 480 ms is defined for the BCCH RSSI measurement. In this time period, the UE needs to measure all cells in the monitoring set for the GSM BCCH RSSI measurements (up to 32 cells on up to 32 carriers), taking at least three BCCH RSSI samples per carrier, assuming that the compressed mode patterns have been allocated in such a way as to permit this behaviour.

If there is insufficient time for these measurements, the UE should measure as many carriers as possible with the same measurement constraints (as outlined above) within

**Table 12.14.** *The number of RSSI samples per gap period*

| TGL | Number of RSSI samples per gap |
|---|---|
| 3 | 1 |
| 4 | 2 |
| 5 | 3 |
| 7 | 6 |
| 10 | 10 |
| 14 | 15 |

the measurement period, and measure the remaining carriers in the following measurement period.

*Initial BSIC identification*

The next stage of measurement is the BSIC identification. For the purposes of initial BSIC identification, the UE needs to use the compressed mode pattern of the same name. To achieve this initial BSIC identification, the gaps in the compressed mode pattern need to be relatively large (gaps of 7 slots, 10 slots and 14 slots are defined for this purpose). The UE also needs to identify the BSIC of the eight strongest GSM carriers defined by the BCCH RSSI. The UE performs the identification task serially. Starting with the strongest carrier, the UE attempts to decode the BSIC using all of the measurement gaps allocated for this purpose before moving to the next carrier.

In the initial BSIC identification procedure, the UE does not know what the timing of the GSM cell is, and so part of the procedure requires the UE to locate the GSM synchronisation channel (SCH) prior to decoding the BSIC. The specifications allow the UE a reasonably large amount of time to achieve this identification. In the worst cases, the UE has between 1.5 s and 5.5 s to identify the BSIC before aborting the attempt on that carrier and coming back and trying again after finding all the others of the eight that the UE needs to identify.

*BSIC reconfirmation*

Once the UE has identified the BSIC of the carrier, it needs to periodically confirm that BSIC. This process is referred to as BSIC reconfirmation, and a compressed mode pattern has been defined specifically for that purpose. After the initial BSIC identification procedure, the UE will have identified the timing of the GSM cell whose BSIC needs reconfirming. It is the responsibility of the UE to track this cell time internally, updating it whenever the BSIC is reconfirmed. To facilitate the BSIC reconfirmation, transmission gaps of 7 slots, 10 slots or 14 slots may be used. The UTRAN defines which pattern sequence should be used, and this information is passed to the UE via the MEASUREMENT CONTROL message.

## 12.6 Measurements for interoperation with GSM

**Table 12.15.** *The different event triggers for GSM cell events*

| Event | Trigger |
|---|---|
| 3A | Current cell quality below threshold and GSM cell RSSI above threshold. |
| 3B | Current cell quality below threshold. |
| 3C | GSM cell RSSI above threshold. |
| 3D | Change of best GSM cell. |

**GSM measurements not requiring the use of compressed mode**

*BCCH RSSI measurement*

If the UE does not require the compressed mode, it makes the measurements in a period of 480 ms, taking three samples per carrier (uniformly spread across the measurement period) for all of the carriers in the monitoring set.

*Initial BSIC identification and BSIC reconfirmation*

The UE needs to perform the initial BSIC identification and BSIC reconfirmation according to the relevant GSM specification [GSM 05.08].

### 12.6.3 Measurement reporting modes and criteria for GSM measurements

As with the intrafrequency and interfrequency cases, there are a number of different methods of generating a measurement report. The first big difference is whether periodic reporting or an event driven report is activated. Periodic reports are similar to those considered previously for the intrafrequency and interfrequency measurement cases and so we do not consider them in detail here. Event driven reports, however, are quite different to those considered previously. For the GSM event reporting, there are several different event report trigger types, which are summarised in Table 12.15.

**Event 3A**

This event is designed to notify the UTRAN when the current cell falls below a certain threshold, at the same time as one or more GSM cells rise above a defined threshold. There are two other basic options that depend upon whether the GSM cell measurements were requested to be for BSIC verified or BSIC non-verified. If the measurements are for the BSIC verified case, the measurement report is only sent if the BSIC has been verified. If the measurements are for the BSIC non-verified case, then the measurement reports are sent with only the BCCH ARFCN to define the cell identity. The triggering of the measurement reports is illustrated in Figure 12.7, which shows the three possible cases that can occur. In Figure 12.7(a), the measurement report is sent if the current cell is below threshold and the new cells are above threshold and remain so for a period of time referred to as the time-to-trigger, then the measurement report is sent. In Figure 12.7(b), if the wanted signal rises above the upper threshold, the trigger is cleared, and no measurement report needs to be sent. In Figure 12.7(c), if one of the target GSM cells

**432** Measurements

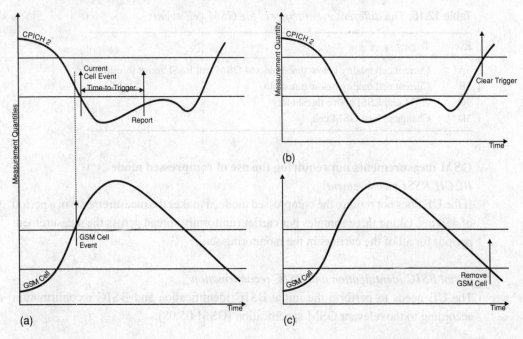

**Figure 12.7** GSM reporting event 3A: (a) trigger and report; (b) current cell trigger removal; (c) GSM cell removed.

falls below the lower threshold set for them, these cells are removed from the variable that controls the trigger.

### Event 3B

The second set of event triggers is a subset of the previous case. In this case, it is only the current cell that causes the trigger. The event is triggered if the current cell falls below the threshold for a period of time (the time-to-trigger) after which a measurement report is sent. The measurement report is configured to include all of the necessary measurements such as current cell and GSM cell quantities. If the current cell rises above the upper threshold, the trigger is cleared and no measurement report is sent.

### Event 3C

Event 3C is similar to event 3A, but only relates to the set of GSM cells. When the GSM cell quality rises above threshold and remains so for the time-to-trigger, the measurement report is sent. If the GSM cell falls below the lower threshold, the trigger is cleared and no measurement report is sent.

**Table 12.16.** *Summary of the different measurement techniques employed for location services in R99*

| Location services type | Description |
|---|---|
| Cell ID based positioning | The network uses the cell identity and Node B measured round trip time for positioning. |
| OTDOA positioning | OTDOA uses measurements in the UE and Node B. |
| Network assisted GPS positioning | UE based GPS measurements assisted by the network. |

**Event 3D**

The final case starts with the UE sending a measurement report to the UTRAN to indicate what the current best GSM cell is. If another GSM cell rises above this and remains so for a period time-to-trigger, a measurement report is sent.

## 12.7 Location services measurements

In this section, we consider some of the basic techniques that are applied to the location services measurements. Much of the detail for the location services measurement is avoided due to the high degree of complexity, which could legitimately fill a book in its own right.

### 12.7.1 Types of location services measurement

For R99 there are three basic types of location services measurement that can be performed within the network. These are presented in Table 12.16.

### 12.7.2 Cell ID based positioning

This technique is the simplest – and the least accurate – method of identifying the current location of the UE within the cell. The cell ID method is based on a decision within the UTRAN as to which cell the UE is currently attached to. For a UE that is in an inactive state (such as idle mode or CELL_PCH), it is necessary to first page the UE. In response, the UE contacts the cell that it is camped upon and from that information and the round trip time from the Node B a location estimate can be made. If the UE is in the CELL_DCH state and in particular in soft-handover, the network uses the measurements obtained from all of the cells in the active set to determine the current cell location.

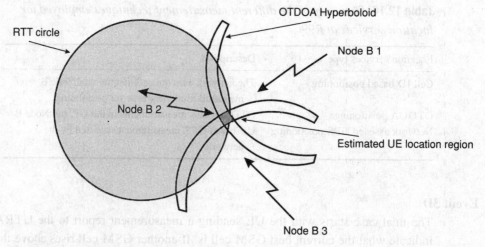

Figure 12.8 OTDOA measurement scenario.

### 12.7.3 OTDOA positioning

The next method of position location is OTDOA. This technique relies upon measurements in the UE and the Node B for a specific UE. The UE measures the arrival time of signals from different Nodes B (SFN–SFN time offsets), and the Node B measures the RTT for that particular UE. The basic situation is depicted in Figure 12.8, where the UE is measuring two Nodes B, and a third Node B is measuring the RTT.

There are two modes for the OTDOA measurement: UE assisted OTDOA and UE based OTDOA. In each, there is an additional consideration as to whether the cell site uses an idle period on the downlink (IPDL). If it does, it means that that cell is periodically blanked to allow the UE to make the measurements on distant cells that are blocked by the transmissions from the close cells.

**UE assisted OTDOA**

With UE assisted OTDOA, the UE makes the measurements of the arrival time from a number of Nodes B and delivers these measurements to the UTRAN, which is then responsible for converting the measurements into position information. In the case of the UE assisted OTDOA, the UE receives information about neighbour cells to measure in either a MEASUREMENT CONTROL message, or via SIB messages (SIB 15.4 and SIB 15.5 for UE based OTDOA). Using this information, the UE can make a better measurement due to the more accurate timing information available in SIB 15.4.

**UE based OTDOA**

With UE based OTDOA, the UE is responsible not only for making the measurements, but also for computing the location based on additional measurement information received from the UTRAN. Using the MEASUREMENT CONTROL message, the

## 12.7 Location services measurements

**Table 12.17.** *Information passed to UE for different OTDOA modes*

| Information sent to UE | UE assisted | UE based |
| --- | --- | --- |
| Intrafrequency cell information | Yes | Yes |
| Ciphering information for UE positioning | No | Yes |
| Measurement control information (idle period locations) | Yes | Yes |
| Sectorisation of the neighbouring cells | No | Yes |
| RTD measurements intrafrequency cells | No | Yes |
| RTD accuracy | No | Yes |
| Measured RTT for primary cell | No | Yes |
| Geographical position of primary cell | No | Yes |
| Relative neighbour cell positions | No | Yes |
| Accuracy range for geographic positions | No | Yes |
| IPDL parameters | Yes | Yes |

UE receives information about the neighbour cells. This neighbour cell information includes the exact geographic co-ordinates of the cells as well as cell based measurements for that UE such as the RTT. Table 12.17 summarises the information that is required for UE based measurement and also UE assisted measurement.

### 12.7.4 GPS based positioning methods

**Network assisted GPS positioning**

The final type of measurement is referred to as network assisted GPS positioning. In this case, the UE makes GPS measurements and the network assists these measurements to different degrees. The network provides information that can be used by the UE in making the GPS measurements. The data include information such as the reference time for GPS, which speeds up the acquisition process, clock correction information and almanac data that provide the UE with a coarse estimate of the location of the satellite. This information either is signalled directly to the UE via the MEASUREMENT CONTROL message, or the UE receives it via the block of SIB15 messages defined for use with GPS positioning.

**UE assisted GPS positioning**

With UE assisted GPS positioning, the UE uses a reduced complexity GPS receiver. The receiver performs the code phase measurements which are passed to the network, which further processes the data to provide an improved accuracy position estimate.

**UE based GPS positioning**

In this alternative technique, the UE includes a complete GPS receiver and the UE performs all of the measurements that are required to estimate a position.

## 12.8 Summary

The importance of both accurate measurement capabilities and the ability to process that information accurately cannot be underestimated in the WCDMA system, in which levels of interference have such an influence on the integrity of the system. In this chapter we have reviewed the measurements that the UE needs to make, the effects that the RRC modes and states have on these measurements and the reporting that needs to be made of the measurement results. For a UE in the CELL_DCH state, the measurements are made and reported to the UTRAN for use in handovers for the cell signal measurements or perhaps to optimise the RBs for the other measurements. For the other states the cell signal measurements are used internally within the UE for the cell reselections. The cell signal measurements are made on the same frequency cells, different frequency cells and same or different mode cells (FDD or TDD) and also on different systems such as GSM. The UE can be configured for a range of different measurement reporting options. The objective behind the reporting configuration is to limit the reporting data to events that are significant as far a specific UE is concerned. The final set of measurements, a huge subject in its own right, are those relating to UE positioning. The different techniques are cell-based measurements; OTDOA measurements or GPS based measurements.

# 13 NAS

## 13.1 Introduction

In this chapter, we consider some of the details of the NAS protocols. As we have seen in the earlier chapters, the NAS is the upper layer family of protocols between the UE and the CN and is defined in [32]. Currently, there are two main sets of NAS protocols: those between the UE and the CS domain, which are based on the GSM upper layer protocols; and those between the UE and the PS domain, which are based on the GPRS upper layer protocols.

In this chapter we start by looking at the basic NAS protocol architecture before moving on to consider some of the specific details of the NAS protocols.

## 13.2 NAS architecture

Figure 13.1 illustrates the basic architecture of the NAS for the R99 phase of the UMTS specifications for a UE that is capable of supporting both CS and PS services. There is also a UMTS NAS protocol architecture for a PS-only UE, but we will not consider that here and the interested reader is referred to [33] for the details.

The protocol architecture for the UMTS NAS protocol is different from that used in the GSM/GPRS modes of operation. In terms of the common elements between the modes of operation of GSM/GPRS and UMTS, we have the same basic elements in the CM sublayer and the MM sublayer. The differences are the removal of the logical link control (LLC) sublayer for the support of the PS services, and the introduction of a radio access bearer manager (RABM) entity.

The functions of the LLC have, to some extent, been absorbed into the functions in the AS. The RABM entity is there to provide a link between the RABs provided by the AS and the NAS control functions. Within the other elements of the architecture, the UMTS CM sublayer and the UMTS MM sublayer are very similar to the equivalent GSM/GPRS protocol elements.

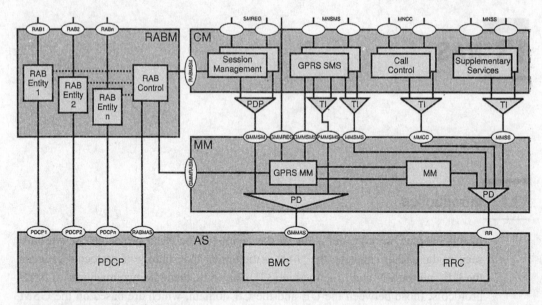

**Figure 13.1** NAS protocol architecture.

### 13.2.1 MM sublayer

The MM sublayer comprises an MM entity that supports the MM functions for the CS domain and a GPRS MM (GMM) entity that supports the MM functions for the PS domain.

The MM sublayer is responsible for:

- Supporting the mobility of the UE (initial registration and location/routing area updates).
- Providing authentication services (in UMTS this is bi-directional authentication with the network validating the UE as well as the UE validating the network).
- Providing user identity protection (through the allocation of temporary identities).
- Providing connection management services for use by the upper layers (CM sublayer). The upper layers request an MM connection or a GMM context that they use to transfer the upper layer messages. The CS MM connection is used for the CS NAS protocols (CC, SMS via CS and SS) and the GMM context for PS NAS protocols (SM or SMS via PS). There can be multiple MM connections active simultaneously and/or a single GMM context active.

The MM sublayer uses the services of the AS through either the radio resource service access point (RR-SAP) or the GPRS mobility management access stratum service access point (GMMAS-SAP). The AS is responsible for creating an RR connection on request from the MM sublayer. We will see that there can be multiple MM connections activated by the MM sublayer using a single RR connection.

In general, the CS domain NAS entities use the RR-SAP for the transfer of upper layer messages, and the PS domain NAS entities use the GMMAS-SAP for the transfer of the upper layer messages. There are some circumstances, which are considered in Section 13.7, in which the GMM entity and the MM entity co-operate to provide what is known as combined services; in this situation, the selection of RR-SAP or GMMAS-SAP depends upon the specifics of what is required.

Within the AS, the RR-SAP and the GMMAS-SAP map onto the RRC dedicated control SAP (DC-SAP) (this was introduced in Chapter 11. In turn, the DC-SAP utilises either the high priority NAS AM service (SRB3) or the low priority NAS AM service (SRB4). The choice of the service priority is either 'hard defined' within the NAS protocol, or indicated to the AS using a priority parameter in the service request primitive. The high priority AM service is equivalent to the GSM SAPI-0; the low priority AM service is equivalent to the GSM SAPI-3.

The MM entity, its structure and the operation of the MM protocol are considered in greater detail in Section 13.4. The GMM entity, its structure and operation are considered in Section 13.7.

### 13.2.2 CM sublayer

Above the MM sublayer, there is the CM sublayer. The CM sublayer uses the communication transfer services provided by the MM sublayer to transfer the CM protocol messages between the peer entities in the UE and the CN.

The CM sublayer comprises a number of different entities. Some of the entities are specific to the CS domain, some are specific to the PS domain, and some apply to both the CS domain and the PS domain. The entities that we can see as part of the CM sublayer in Figure 13.1 are the SM entity, the GPRS short message service (GSMS) entity, the CC entity and the supplementary service (SS) entity.

**SM entity**

The SM entity is responsible for the PS domain data connections that are referred to as PDP contexts. There can be multiple SM entities to support multiple PDP contexts. For each active PDP context, there is an SM entity. The individual SM entities are identified through the transaction identifier (TI) related to a specific PDP context. The PDP contexts themselves are identified via the network service access point identifier (NSAPI). The TI is used to identify the SM controlling entity (this is in the control plane), and the NSAPI is used to identify the PDP context data flow (this is in the user plane).

The SM entity is responsible for requesting the creation of a PDP context on request from higher layers, modifying that PDP context and releasing it. Additionally, the SM entity interacts with the RABM responsible for monitoring and controlling the RAB that is used by a specific PDP context to transfer the user data.

The RAB that is created for a specific PDP context has an AS identity (RAB-Id) which is associated with the NAS identity (NSAPI) for that specific PDP context (often referred to as a binding between the two identifiers).

The SM entity uses the services of the GMM sublayer (in particular the GMM context) for the transfer of the SM messages between the UE and its peer in the CN.

The SM entity, the SM protocols and the creation and release of PDP contexts are considered in more detail in Section 13.8.

## GSMS entity

The GSMS entity extends the capabilities of the GSM SMS service entity that was defined for SMS in the CS domain. The GSMS entity supports the transmission and reception of short messages via both the PS domain and the CS domain.

There can be multiple instances of the GSMS entity. As a minimum, the UE supports two GSMS entities for CS domain operation and two GSMS entities for PS domain operation [34]. The two entities for the CS domain and PS domain allow the UE to both send and receive short messages simultaneously via whichever domain is currently active. The architecture and the operation of the GSMS entity are considered in greater detail in Section 13.9.

## CC entity

The CC entity is the CS domain entity responsible for the establishment, release and modification of CS services between the UE and the CS domain. Within the UE, there can be multiple CC entities, each responsible for a specific CS connection. For each CC entity, there is a separate MM connection, and the messages from the different CC entities are identified through the use of the TI that is included within the protocol header.

The CC entity uses the GSM call control protocol to establish, release and modify the CS connections. To transfer the CC protocol messages between the UE and its peer entity in the CN requires an active MM connection to be available.

For UMTS, a CS domain service known as multicall has been defined. With multicall, it is possible to establish multiple simultaneous CS connections. Each CS connection has an RAB associated with it. An identifier called the stream identifier (SI) is used to identify a specific CS connection and it is associated with the RAB-Id that defines the user plane flow across the AS.

The architecture and operation of the CC entity is considered in greater detail in Section 13.7.

## SS entity

The SS entity is responsible for the support of SSs, which include elements such as call barring, call hold and call waiting services. The SS entity also uses the MM connection facilities. Further details on the support of the SS entities can be found in [35].

### 13.2.3 RABM entity

The RABM entity is responsible for providing an interface between the RAB management in the AS (in particular the RRC protocol) and the PDP context control protocols in the NAS (the SM entities). It is likely that, in the course of a PDP context, some aspects of the RAB used by the PDP context may change. If, for instance, an RAB is released during an active PDP context, the RABM can request the reestablishment of that service. Alternatively, the QoS of a PDP context may be modified, in which case the RABM is told that a modification is underway by the SM entity. The RABM notifies the SM entity when the QoS of the RAB has been modified and is available for use. Some aspects of the use of the RABM entity are considered in Section 13.8.

## 13.3 MS classes and network modes

Before studying the details of the operation of the different layers of the NAS protocol, we should consider one final important factor, namely the interrelationship between the PS domain protocols and the CS domain protocols.

In GSM/GPRS, a number of different classes of terminal are defined to allow for different functional offerings for the combination of CS and PS modes of operation. UMTS also specifies an equivalent means of defining equipment capability. The CN can also have different architecture options that have an impact on the performance and operation of the network. These architecture options are referred to as network modes. There are some network modes defined for GSM/GPRS interoperation and some equivalent modes defined for the UMTS CS domain/PS domain interoperation.

In this section, we review the GSM/GPRS definitions for these quantities before considering the equivalent quantities used in UMTS. The reason for considering these issues is that the type of equipment and the operational mode of the network have a major bearing on the operation of some of the protocols, in particular the MM sublayer protocols.

### 13.3.1 MS classes

GSM/GPRS define a number of classes of mobile that determine whether the MS can simultaneously access GSM for CS services whilst accessing GPRS for PS services. There are three classes, which are summarised below:

Class A: GSM and GPRS services accessed simultaneously.
Class B: Can monitor control channels for both GSM and GPRS, but only access services from one at a time.
Class C: GPRS services or GSM services (assuming the capability) but not both at the same time.

UMTS defines three modes of operation, but these modes are slightly different to the classes defined for GSM/GPRS:

Mode PS/CS: The UE can simultaneously support CS and PS services (similar to GSM/GPRS class A).

Mode PS: The UE can only support PS services (similar to GSM/GPRS class C).

Mode CS: The UE can only support CS services (not a defined GPRS class).

### 13.3.2 Network modes

GSM/GPRS define three network modes, mode I, mode II and mode III, and these are defined in more detail below:

Mode I: The UE can receive PS or CS paging via packet connection (either packet paging or packet data channel).

Mode II: The UE can receive paging for CS and PS operation via the same CS paging channel. If there is a packet data channel active, however, the UE still needs to monitor the CS paging channel.

Mode III: The UE needs to monitor both the CS paging channel and the PS paging channel to receive paging for both CS and PS operation.

For UMTS, only two network modes are defined, mode I and mode II. Mode I indicates that the optional Gs interface between the MSC and SGSN is present. For mode II, this interface is not present. The use of the different modes is outlined below:

Mode I: Combined procedures between the CS and PS domains are supported (these also include attach and area updates).

Mode II: Combined procedures between CS and PS domains are not supported (separate attach and area update procedures are required).

## 13.4 MM protocol entity

In this section we consider the architecture and the operation of the MM protocol entity. The MM protocol is used in particular for the support of services for the CS domain. The MM entity provides functions such as registration and authentication, location management and communication management.

We start by considering the architecture and MM entity state machine, the layer to layer protocol for the MM entity and then some of the specifics of the peer to peer protocol between the UE and the CS domain. In Section 13.5.3 we will see how the MM entity interacts with the other layer entities in the creation of specific types of service.

### 13.4.1 MM entity states

Figure 13.2 illustrates the state diagram for the MM sublayer. As UMTS is not currently supporting the group call services, the states that are specifically for the support of the

## 13.4 MM protocol entity

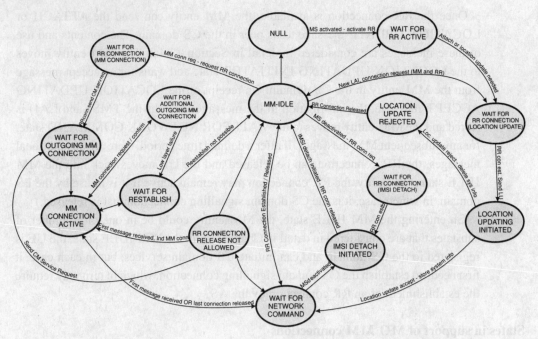

**Figure 13.2** MM entity state diagram.

voice group call service (VGCS) have been excluded. The MM IDLE state in the state diagram includes a number of substates. These substates are not shown in the state diagram or considered in detail in this chapter, but are covered in Chapter 14, where the details of the operation of the UE whilst in MM IDLE and equivalent states are considered.

We can see that the state diagram has been defined to support three basic functions: location updating, IMSI detach and the establishment of one or more MM connections to be used either by the MM sublayer or by the CM sublayer and higher layers.

### States in support of location updating

The UE starts in the NULL state; this indicates that the UE needs to perform a registration (or a location update procedure) and potentially an authentication procedure. The registration procedure is often referred to as IMSI attach, and indicates that the MM entity is going to register with the CS domain. As an alternative to using the IMSI attach procedure, the MM entity could perform a location update procedure. Which approach the MM entity takes for this initial registration step is a network-defined option. In either case, the MM entity first needs to activate the RR entity (the MM entity moves to WAIT FOR RR ACTIVE state). Once active, the MM entity requests the establishment of an RR connection from the AS (the MM entity moves to the WAIT FOR RR CONNECTION (LOCATION UPDATE) state).

Once the RR connection is available, the MM entity can send the ATTACH or LOCATION UPDATE message to its peer in the CS domain. The contents and use of these messages are considered in detail in Section 13.4.3. The MM entity moves to the LOCATION UPDATING INITIATED state and waits for an accept message from the MM entity in the CS domain. On receipt of the LOCATION UPDATING ACCEPT message, the information in the message (such as the TMSI and LAI) is stored and the MM entity moves to the WAIT FOR NETWORK COMMAND state, for any subsequent MM messages. If after a defined time period, there are no additional messages, the RR connection can be released and the UE moves back into the MM IDLE state. (In reality, the RR connection may remain active if it is in use by the PS domain, in which case, it is the CS domain signalling connection that is released.)

On entering the MM IDLE state, the MM entity could be in one of a number of substates that are considered in detail in Chapter 14. In the MM IDLE state, the UE is registered to the CS domain and can initiate CS domain services, but in each case, it first needs to establish the CS domain signalling connection, which in turn may require the establishment of an RR connection by the AS.

## States in support of MO MM connections

The states towards the left hand side of Figure 13.2 are present to support the establishment of MM connections that can be used by the CM sublayer entities. We assume that we are now starting from the MM IDLE state as the UE will have performed the registration procedure outlined above.

The first stage is the request for the establishment of a CS domain signalling connection and an RR connection if the RR connection does not already exist or just the establishment of a CS domain signalling connection if the RR connection does exist (to support the PS domain activities for instance). This takes the MM entity to the WAIT FOR RR CONNECTION (MM CONNECTION) state.

When the RR connection/CS domain signalling connection is available, the MM entity can request the establishment of an MM connection by sending the CM SERVICE REQUEST message to the peer entity in the CS domain. The MM entity moves to the WAIT FOR OUTGOING MM CONNECTION state.

The receipt of notification that encryption and/or integrity protection is activated or receipt of the CM SERVICE ACCEPT message confirms that the MM connection has been established and the MM entity moves into the MM CONNECTION ACTIVE state. From this state, the MM entity can request the establishment of additional MM connections (for additional services for instance) in which case the MM entity moves to the WAIT FOR ADDITIONAL OUTGOING MM CONNECTION state until the MM connection is confirmed with the CM SERVICE ACCEPT message.

In the situation where there are multiple MM connections, the protocol discriminator (PD) and the TI define the CM entity and the specific instance of the CM entity respectively; this assumes that we have multiple instances of the specific CM entity.

Each of the CM entities has a unique PD, and for each instance of a specific CM entity, a new TI is used. Together, therefore, {PD,TI} define which CM entity and which instance within that entity a message originates from or is destined for. The CM entity includes the PD and TI elements within the message header that allow the MM entity to route messages to the correct destination entity in the peer system.

**States in support of MT MM connections**

If the network requests the establishment of an MM connection to a specific UE, it starts by using a paging message. We assume that the UE is in the MM IDLE state. On receipt of the paging message, the UE establishes an RR connection and responds with the PAGING RESPONSE message (this is a GSM/GPRS RR message and so the PD should indicate this for reasons of backwards compatibility). The MM entity then moves to the WAIT FOR NETWORK COMMAND state.

Next, the MM common procedures may occur (e.g. authentication). Finally, the CM entity message is received (such as the CC SETUP message). On receipt of this message, the MM entity moves to the MM CONNECTION ACTIVE state, and the MM connection for that specific CM entity is available for use.

**States in support of IMSI detach**

The final sequence of state transitions involves the IMSI detach procedure. IMSI detach is the process that the MM entity follows when the registration for the specific UE is to be deactivated; usually, this would occur as part of the UE being powered down.

The MM entity starts in the MM IDLE state. To perform the detach, first an RR connection and then a CS signalling connection must be established. The MM entity moves to the WAIT FOR RR CONNECTION (IMSI DETACH) whilst waiting for the establishment of the RR connection. When the RR connection is available, the MM entity sends the IMSI DETACH message and then moves to the IMSI DETACH INITIATED state. If successful, the RR connection is released and the MM entity moves to the MM IDLE state.

### 13.4.2 MM entity service primitives

In this section, we review the MM service primitives used by the MM service users to obtain services from the MM entity. For the MM entity the service users from the CM sublayer are: the CC entity; the SS entity; the SMS entity; and the LCS entity. There are four SAPs (MMCC-SAP, MMSS-SAP, MMSMS-SAP and MMLCS-SAP) used for the exchange of service primitives between the MM entity and the CM entities, and these are illustrated in Figure 13.1.

**Table 13.1.** *List of MM entity service primitives*

| Primitive | Direct | Parameters | Comment |
|---|---|---|---|
| MMxx-EST-REQ | CM to MM | CM SERVICE REQUEST parameters | Request from CM entity to establish an MM connection. |
| MMxx-EST-IND | MM to CM | First CM message | MT MM connection established and message delivered to CM entity. |
| MMxx-EST-CNF | MM to CM | | Confirmation of MM connection established. |
| MMxx-REL-REQ | CM to MM | Cause | CM request to release MM connection and PD/TI. |
| MMxx-REL-IND | MM to CM | Cause | Indication that MM connection released includes cause. |
| MMxx-DATA-REQ | CM to MM | CM message | CM request for acknowledged message transfer. |
| MMxx-DATA-IND | MM to CM | CM message | Receipt via acknowledged transfer of CM message. |
| MMxx-UNIT-DATA-REQ | CM to MM | CM message | CM request for unacknowledged message transfer. |
| MMxx-UNIT-DATA-IND | MM to CM | CM message | Receipt via unacknowledged transfer of CM message. |
| MMCC-SYNC-IND | MM to CM | Cause, Res.Ass. list{RAB-Id, NAS synch Ind} | CC entity only. Indicates channel assigned or modified. Includes RAB-Id list and NAS synchronisation indicators. |
| MMxx-REEST-REQ | CM to MM | | Request to reestablish interrupted MM connection. |
| MMxx-REEST-CNF | MM to CM | | Confirmation of successful reestablishment. |
| MMxx-ERR-IND | MM to CM | Cause | Indication of lower layer failure of MM connection. CM to decide whether to reestablish or release. |
| MMxx-PROMPT-IND | MM to CM | | Indication that MM connection completed for network initiated MO CM connection request. |
| MMxx-PROMPT-REJ | CM to MM | | Rejection of request to establish MM connection for network initiated MO CM connection request. |

Table 13.1 lists the service primitives that are defined for the MM entity. These primitives apply to all of the CM sublayer entities. In the table, the primitive MMxx should be read as MMCC for the CC service entity, MMSS for the SS service entity, MMSMS for the SMS service entity and MMLCS for the LCS service entity.

In the examples that follow in this chapter, we will see the use of some of these service primitives between the CM entities and the MM entity.

**Table 13.2.** *Summary of common, specific and MM connection management procedures*

| MM procedure | Description |
| --- | --- |
| Common | Common procedures are initiated when an RR connection exists. They include:<br>• TMSI allocation;<br>• UE identification and authentication;<br>• IMSI detach procedure. |
| Specific | Specific procedures are only initiated if no other MM procedures are active or no MM connection is active. They include:<br>• normal and periodic location updating;<br>• IMSI attach procedure. |
| MM connection management | MM connection management procedures are used to establish and release an MM connection for use by the CM layer. They can only be performed if no MM specific procedure (itemised above) is running and there may be more than one MM connection active at the same time. |

### 13.4.3 MM procedures – peer to peer protocol

Having examined the architecture, states and service primitives for the MM entity within the MM sublayer, we now turn to examine the MM procedures themselves.

The MM procedures are divided into three distinct types: common, specific and MM connection management. The common and specific procedures are concerned with the transfer of MM messages to perform an MM function such as registration or authentication. The difference between the common and specific procedures is that the specific procedures can only commence if there is no other MM procedure or MM connection active. MM common procedures, however, can be started during an existing MM specific procedure except for the IMSI detach procedure. The MM connection management procedures are concerned with the establishment and control of an MM connection to be used by the entities in the CM sublayer.

The different procedures are summarised in Table 13.2, and each is considered in more detail in the sections that follow.

**MM common procedures**

The MM common procedures are procedures that include TMSI reallocation, authentication, identification and IMSI detach. We will review each of these common procedures in turn.

**TMSI reallocation**

The TMSI is a four-octet number (32 bits), which is used as a temporary identity for the UE. It is used to identify the user in place of the full UE identity (IMSI) in order to

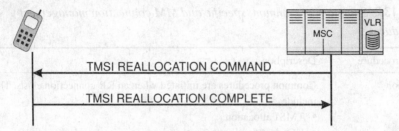

**Figure 13.3** TMSI reallocation message exchanges.

provide identity confidentiality for the user. The TMSI is unique only within the LA in which it was allocated. To provide a unique identity, the TMSI is combined with the location area identifier (LAI) and this uniquely identifies the UE anywhere.

Although not mandated, it is expected that the TMSI is reallocated at least every change in LA as part of the LA update procedure. In cases where the TMSI is not recognised or cannot be traced back to the network element assigning it, the serving network needs to obtain the IMSI from the UE (using the identification procedure described below) before a new TMSI can be allocated.

As with all of the common procedures, the TMSI can be allocated at any time as long as an RR signalling connection exists. The MM sublayer uses the existing RR signalling connection to the CN to exchange the TMSI reallocation messages. Most probably, the TMSI reallocation will occur as part of another set of procedures such as location area update, or a CS call setup procedure.

Figure 13.3 illustrates the message exchanges between the CN and the UE for the TMSI reallocation procedure. The TMSI REALLOCATION COMMAND message contains the message type, the new LAI (the LAI comprises of MCC + MNC + location area code (LAC)) and the mobile identity (IMSI, IMEI (international mobile equipment identity) or TMSI). The IMSI and IMEI are possible options for this message, and these indicate that the UE should delete the old TMSI and use the IMSI or IMEI. Normally, the TMSI REALLOCATION COMMAND message is sent across an encrypted link to prevent eavesdropping.

The TMSI REALLOCATION COMPLETE message contains the message type to indicate that the TMSI reallocation has been completed. There is a timer associated with the procedure (it is called T3250), which is used as part of the error trapping. The timer is started when the TMSI REALLOCATION COMMAND is sent, and stopped when the TMSI REALLOCATION COMPLETE message is received.

**Authentication**

The authentication procedures used in UMTS are slightly different to those used in GSM. The details of authentication and encryption are considered more fully in Chapters 2 and 11, but outlined here.

## 13.4 MM protocol entity

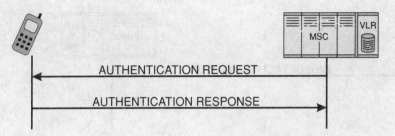

**Figure 13.4** Authentication procedure.

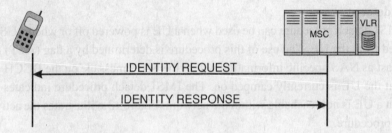

**Figure 13.5** Identity check message exchanges.

Authentication in UMTS has four main purposes:
1. To allow the network to validate the UE identity (same concept as in GSM).
2. To provide new information that the UE can use to generate a new UMTS encryption key (same concept as in GSM).
3. To provide new information that the UE can use to generate a new UMTS integrity key (not currently present in GSM).
4. To allow the UE to validate the network (not currently present in GSM).

The basic message flows (excluding the exceptions cases) are illustrated in Figure 13.4.

**Identification**

The MM sublayer in the CN can use the identification procedure to request a UE to provide specific identification information such as the IMSI or IMEI. As in the TMSI reallocation and authentication case, the MM sublayer can use an active RR signalling connection to exchange the identification messages. The identification procedure can be used as part of the authentication procedure if the network cannot identify which CN element allocated a specific TMSI.

The basic message exchanges for the identification procedure are illustrated in Figure 13.5. The IDENTITY REQUEST message contains information on the message type and the type of identity requested (e.g. IMSI, IMEI or TMSI). The IDENTITY RESPONSE message contains the message type and the mobile identity that was requested.

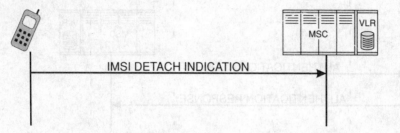

**Figure 13.6**  IMSI detach procedure.

### IMSI detach

The IMSI detach procedure can be used when a UE is powered off or when a USIM is removed from the UE. The use of this procedure is determined by a flag (ATT) that is broadcast as NAS specific information within the SIB1 message on the BCCH of the cell that the UE is currently camped on. The IMSI detach procedure indicates to the CN that a UE is not reachable within the network. Figure 13.6 illustrates the activation of this procedure.

From the state diagrams presented in Figure 13.2, the IMSI detach procedure can be activated whilst a UE is in the MM IDLE state or the WAIT FOR NETWORK COMMAND state. In the first case, an RR signalling connection is required before the UE transmits the IMSI DETACH INDICATION message and moves to the IMSI DETACH INITIATED state.

There are a number of conditions that can affect the procedure, not least of which are the unavailability of a cell and the presence of a 'specific MM connection' as defined earlier. In these cases, the UE may need to wait before performing the IMSI detach procedure.

The IMSI DETACH INDICATION contains the message type, the MS classmark 1 information and the mobile identity as defined previously. MS classmark 1 is defined in Table 13.3.

### MM specific procedures

The MM specific procedures are procedures that require special consideration in the allocation and control of the MM connections. MM specific procedures cannot be started if there is an existing MM connection, or if a separate MM specific procedure is active. Subsequent MM specific procedures cannot be started until the MM connection has been dropped. Whilst an MM specific procedure is active, MM connections for use by the CM sublayer cannot be started until the MM specific procedure is completed and the MM connection released.

There are three main types of MM specific procedure, which are considered in greater detail below. These are: normal location updating, periodic location updating and IMSI attach. The LOCATION UPDATING REQUEST message is used for

## 13.4 MM protocol entity

**Table 13.3.** *Contents of LOCATION UPDATING REQUEST message*

| | |
|---|---|
| LU type | Normal location updating, periodic updating or IMSI attach. In this section we are considering normal location updating. |
| CKSN | The number that identifies which ciphering and integrity protection keys the UE has. |
| LAI | The number stored in the USIM: LAI = MCC + MNC + LAC. |
| MS classmark 1 | Information on UE capabilities e.g. GSM encryption algorithm support, UE GSM (mode) Tx power. |
| Mobile identity | IMSI, IMEI or TMSI. |
| MS classmark 2 | Additional UE capability information, mainly for compatibility with GSM. |

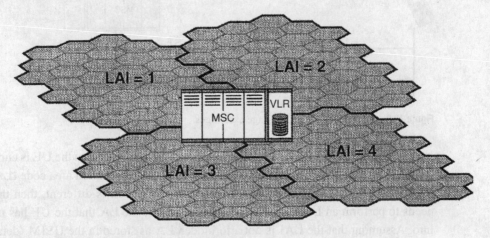

**Figure 13.7** Relationship between LAs and MSCs.

three MM specific procedures: normal updating (this will be considered first), periodic updating and IMSI attach.

**Location updating**

Location updating is a procedure used to ensure that the CN knows of the location of the UE whilst the UE is in the MM IDLE state. In this case, the UE does not have an active signalling connection to the CS CN and consequently the CS CN is responsible for tracking the location of the UE. If, however, the UE does have an active signalling connection to the CS CN and also an active MM connection, then it is the UTRAN that is responsible for tracking the location of the UE. Here, we consider the former case; the latter case was discussed in Chapter 11. Figure 13.7 illustrates the relationships between a collection of cells called the LA, the cells that comprise the LA, and the CN entities that are responsible for the location areas.

In general, location updating requires the UE whilst in the MM IDLE state to decode the LAI of its current cell. The LAI information is broadcast as NAS specific

**Table 13.4.** *Contents of LOCATION UPDATING ACCEPT message*

| | |
|---|---|
| LAI | LAI for the new LA. |
| Follow on proceed | Indicates if a follow on MM connection is to use the same RR signalling connection. |
| CTS permission | Defines whether the UE can use the GSM cordless telephony system (CTS) in the LA. |
| Equivalent PLMNs | List of equivalent PLMNs that can be used as part of the idle mode procedures for PLMN selection and reselection. |

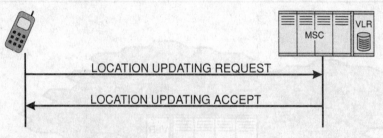

**Figure 13.8** Location updating message exchanges.

information within the SIB1 message on the BCCH of the cell that the UE is currently camped on. The LAI comprises the MCC, MNC and the location area code (LAC).

If the UE moves to a new cell and the LAI of that cell is different, then the UE needs to perform an LAU so that the CN is aware of the LA that the UE has moved into. Assuming that the LAI is not a forbidden LA as stored in the USIM (details of the USIM and data stored in it are covered in more detail in Chapter 14) in elementary file $EF_{PLMN}$, the UE can proceed with the location updating procedure. The purpose of this location awareness in the CN is to control the amount of paging required by the CN to reach specific UEs. By using LAs, the CN needs to page only the specific cells that are part of an LA.

Figure 13.8 illustrates the basic message exchanges that are required for location updating by the UE. As mentioned previously, the UE is in the MM IDLE state and so requires the establishment of an RR connection and a CS signalling connection. Once the signalling connection is active, the UE can transmit the LOCATION UPDATING REQUEST message to the CN. The network proceeds with any network internal procedures required for the updating, and responds with the LOCATION UPDATING ACCEPT message. The contents of the LOCATION UPDATING REQUEST message are presented in Table 13.3 and those of the LOCATION UPDATING ACCEPT message in Table 13.4.

Once the LAU has been successfully completed, the USIM can set the location update status as 'Updated' in the USIM elementary file $EF_{LOCI}$ ($EF_{LOCI}$ defines the LA and the location update status of the LA that the UE has registered on).

### Periodic updating

The periodic updating procedure uses the same messages as the normal updating procedure (except that the location updating type indicates a periodic update). The periodic update procedure for a UE that is in the normal service substate of the MM IDLE state (this is considered in Chapter 14) is triggered by the expiry of a timer (T3212). The timeout value for the timer is broadcast as NAS specific information within the SIB1 message on the BCCH of the cell that the UE is currently camped on. It has a value defined in decihours in a range of 1–255 decihours (6 minutes to 25.5 hours).

The periodic update procedure provides the CN with information on a specific UE that has not crossed any LAI boundaries. If a UE had lost coverage or the battery was exhausted before an IMSI detach could take place, the CN would think the UE was still attached. Through the use of periodic updates, any mobiles that do not update at the correct time are designated as detached by the CN.

If the UE has left the coverage area when the periodic update timer expires, the periodic update procedure will be delayed until the UE returns to the coverage area.

### IMSI attach

The IMSI attach procedure indicates to the CN that a specific IMSI is active within the network. The IMSI attach procedure uses the same messages as the normal location update, except that the location updating type indicates that it is an IMSI attach.

In UMTS, a flag (ATT) is broadcast as NAS specific information within the SIB1 message on the BCCH of the cell that the UE is currently camped on. The ATT flag indicates whether the UE needs to perform an IMSI attach/detach procedure in that cell ($ATT = 0$: UE shall not perform attach/detach in cell; $ATT = 1$: UE shall perform attach/detach in cell). There are some additional constraints that define whether the UE should use the IMSI attach, or alternatively the location update procedure.

### MM connection establishment

An MM connection can be established when the MM sublayer is in one of the following states (see the state diagram in Figure 13.2): MM IDLE, RR CONNECTION RELEASE NOT ALLOWED, or MM CONNECTION ACTIVE.

### MO MM connection establishment

The MO MM connection establishment procedure is activated through the transmission of the CM SERVICE REQUEST message by the MM entity. If an MM specific procedure (e.g. IMSI attach or location updating) is active at the time, the MM connection establishment procedure may be delayed or even cancelled until the MM specific procedure is completed.

An MM connection requires an RR signalling connection, and that RR signalling connection can support a number of simultaneous MM connections. If there is no active RR signalling connection, the MM entity requests the establishment of one

**Table 13.5.** *Contents of CM SERVICE REQUEST message*

| | |
|---|---|
| CM service type | MO CS call, PS connection, emergency call, SMS, SS, location service. |
| Cipher key sequence number (CKSN) | Number that identifies which ciphering and integrity protection keys are used by UE without requiring an authentication. |
| MS classmark 2 | UE capability information, mainly for compatibility with GSM. |
| Mobile identity | IMSI, IMEI or TMSI. |
| Priority level | Defines the priority of the CM SERVICE REQUEST message (for enhanced multilevel precedence and pre-emption (eMLPP) priority services). |

(using the RR-EST-REQ primitive, which contains the layer 3 MM message CM SERVICE REQUEST). To transfer the CM SERVICE REQUEST message to the CN, the AS in the UE uses the RRC protocol message INITIAL DIRECT TRANSFER (see Chapter 11 for details), which establishes the RR signalling connection as well as transferring the encapsulated NAS message to the CS CN. If an MM connection (and hence RR signalling connection) exists, the MM entity can request the establishment of an additional MM connection again using the CM SERVICE REQUEST message. In this instance, within the AS, the NAS message is encapsulated within the RRC protocol message UPLINK DIRECT TRANSFER and transferred using the existing signalling connection to the CS CN.

The MM entity can support multiple MM connections from the same or different CM entities. The PD differentiates between the different CM entities, and a TI discriminates between different instances for MM connections from the same CM entity.

In response to the CM SERVICE REQUEST, the network has a number of options. If this is the first MM connection being established to the CN domain, then the AS activates the security mode control procedure. If MM connections exist, the network sends a CM SERVICE ACCEPT message that indicates the acceptance of this new MM connection and the MM entity moves into the MM CONNECTION ACTIVE state.

The CM SERVICE REQUEST message contains the elements presented in Table 13.5. The CM SERVICE ACCEPT message simply contains the message type: CM SERVICE ACCEPT.

### MT MM connection establishment

The network can request the establishment of an MM connection if an existing MM connection is not available. To do this, the network first pages the UE using the AS paging procedure. The UE then establishes an RR connection and responds with a PAGING RESPONSE message that acts like the CM SERVICE REQUEST message.

After authentication procedures and the security mode control procedure are completed, the network sends the first CM entity specific message. The receipt of this message indicates to the MM entity that the MM connection has been activated and

## 13.4 MM protocol entity

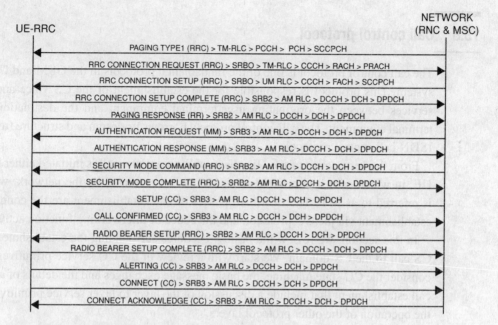

**Figure 13.9** Example MM connection establishment.

the MM entity can move into the MM CONNECTION ACTIVE state. Figure 13.9 illustrates part of this procedure. In this instance, it is a MT CS call establishment that is being performed and so, in this instance, the SETUP message from the network is the first CM entity message to be received and it causes the MM entity to move into the MM CONNECTION ACTIVE state.

**MM connection information transfer**

Once an MM connection has been established, the CM sublayer can use the connection for the transfer of information. There can be a number of MM connections (one or more for each CM entity), each MM connection being differentiated by a PD and a TI.

The PD and TI are inserted into the CM sublayer message headers by the CM entity responsible for the creation of the CM layer message at the transmitter. The MM entity in the receiver reads the contents of the CM header and routes the CM sublayer messages to the appropriate CM layer entity in the receiver.

**MM connection release**

The MM connection is released locally without the need to signal its release to its peer entity. The associated RR signalling connection can also be released when there are no other active MM connections (the AS needs to consider whether the RR connection is used by the GMM entity for PS connections) and also uses a timer to control how quickly the release of the RR connection can occur.

## 13.5 Call control protocol

The call control protocol is one of the fundamental protocols in the GSM and UMTS system. This protocol is responsible for the establishment of the CS voice and data services between the originating user terminal equipment and the destination user terminal equipment. The call control protocol borrows its design and structure from the ISDN call control protocol.

From the perspective of the UE, the establishment of a call is initiated either by the UE, in which case it is referred to as MO or it is initiated by the network, when it is referred to as MT. The two main phases of call establishment are the connection establishment via the CC protocol, and the codec or interworking function activation.

In this section, we review the basic procedures surrounding the establishment of a CS call to the CS domain. We start with a review of the CC service primitives, then consider the CC state machine, the basic message exchanges and the details of an MO call establishment that include a discussion of the layer to layer service primitives and the operation of the other protocol layers.

### 13.5.1 CC service primitives

Table 13.6 outlines the service primitives that are defined at the MNCC-SAP (illustrated in Figure 13.1) in the UE. There is a similar set of primitives defined for the network, and the interested reader is referred to [33]. These primitives are between the mobile network (MN) layer and the CC entity in the CM sublayer.

### 13.5.2 Call control state machine

Figure 13.10 illustrates the CC states as defined in [32]. It shows the different call states and also the service primitives used to define the transitions between the different states. Figure 13.10 presents three sets of procedures: MO call establishment, MT call establishment and call release and disconnection. We will consider the states that are used for each of these procedures.

**MO call establishment**

The MO call establishment procedure passes through the call states on the left hand side of Figure 13.10. The exchange of CC messages between the CC entity in the UE and the CC entity in the network is summarised in Figure 13.11. Here, we consider the basic relationship between the CC state diagram in Figure 13.10, the primitives in Table 13.6 and the message sequence flows in Figure 13.11. In Section 13.5.3 we will examine in more detail the contents of the various messages, as well as the interaction with other layers of the NAS and AS protocols.

## 13.5 Call control protocol

**Table 13.6.** *Call control service primitives*

| Primitive | Direct | Parameters | Comment |
|---|---|---|---|
| MNCC-SETUP-REQ | MN to CC | SETUP or EMERGENCY SETUP | Request to establish an MO CS call (SETUP) or emergency call (EMERGENCY SETUP). |
| MNCC-SETUP-IND | CC to MN | SETUP | Start of MT CS call establishment. |
| MNCC-SETUP-RES | MN to CC | CONNECT | MT user – call accepted. Send CONNECT to acknowledge. |
| MNCC-SETUP-CNF | CC to MN | CONNECT | MO call accepted by remote user. |
| MNCC-SETUP-COMPLETE-REQ | MN to CC |  | Request to send MO CONNECT ACK: call accepted by MN. |
| MNCC-SETUP-COMPLETE-IND | CC to MN |  | MT CONNECT ACK RX. Call setup completed user plane activated. |
| MNCC-REJ-REQ | MN to CC | RELEASE COMPLETE | MN request to reject MT call. Call refused or cannot be setup. |
| MNCC-REJ-IND | CC to MN | Cause | MO call rejected reason: 'Cause' |
| MNCC-CALL-CONF-REQ | MN to CC | CALL CONFIRMED | MN confirm MT call request. Could offer different bearer capability. |
| MNCC-CALL-PROC-IND | CC to MN | CALL PROCEEDING | MO call setup started in network. No more setup info will be accepted. |
| MNCC-PROGRESS-IND | CC to MN | PROGRESS | Indicates call is progressing. Receipt of PROGRESS message/IE. |
| MNCC-ALERT-REQ | MN to CC | ALERTING | Local user alerting initiated. |
| MNCC-ALERT-IND | CC to MN | ALERTING | Remote user alerting initiated. |
| MNCC-NOTIFY-REQ | MN to CC | NOTIFY | Request to send user call state information, e.g. user suspended to NW. |
| MNCC-NOTIFY-IND | CC to MN | NOTIFY | Indication of remote user call state information, e.g. user suspended. |
| MNCC-DISC-REQ | MN to CC | DISCONNECT | Send request to disconnect call. |
| MNCC-DISC-IND | CC to MN | DISCONNECT | Receive disconnect call notification. Call cleared. |
| MNCC-REL-REQ | MN to CC | RELEASE | Request to send RELEASE. To clear call and MM connection. |
| MNCC-REL-IND | CC to MN | RELEASE | Receipt of RELEASE. Notifies that call and MM connection be released. |
| MNCC-REL-CNF | CC to MN | RELEASE or RELEASE COMPLETE | Confirms MN release request. The release of call and MM connection may proceed. |
| MNCC-FACILITY-REQ | MN to CC | Facility | Request to transmit facility IE in support of SS. |
| MNCC-FACILITY-IND | CC to MN | Facility | Indication of the receipt of facility IE in support of SS. |
| MNCC-START-DTMF-REQ | MN to CC | START DTMF | Request to transmit START DTMF – starts DTMF control. |

(*cont.*)

**Table 13.6.** (*cont.*)

| Primitive | Direct | Parameters | Comment |
|---|---|---|---|
| MNCC-START-DTMF-CNF | CC to MN | START DTMF ACK or START DTMF REJ | Confirmation that DTMF control is acknowledged or it is rejected. |
| MNCC-STOP-DTMF-REQ | MN to CC | STOP DTMF | Request to transmit START DTMF – stops DTMF control. |
| MNCC-STOP-DTMF-CNF | CC to MN | STOP DTMF ACK | Confirmation that DTMF control is stopped. |
| MNCC-MODIFY-REQ | MN to CC | MODIFY | Request to start MO in-call modification. |
| MNCC-MODIFY-IND | CC to MN | MODIFY | Notification of start MT in-call modification. |
| MNCC-MODIFY-RES | MN to CC | MODIFY COMPLETE | Response by MN indicating MT in-call modification complete. |
| MNCC-MODIFY-CNF | CC to MN | MODIFY COMPLETE | Notification by NW indicating MO in-call modification complete. |
| MNCC-SYNC-IND | CC to MN | Cause (resources assigned, RAB-Id, Chan Mode modify) | Indicates dedicated resources assigned and/or channel mode is changed. |

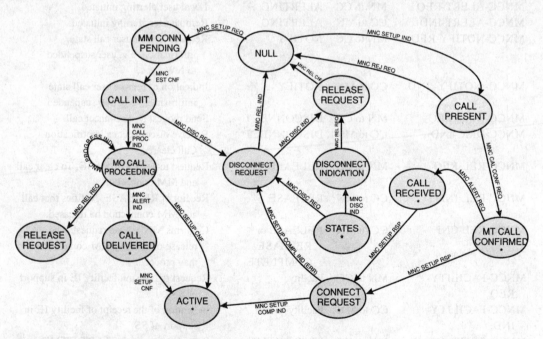

**Figure 13.10** Call control entity state diagram.

## 13.5 Call control protocol

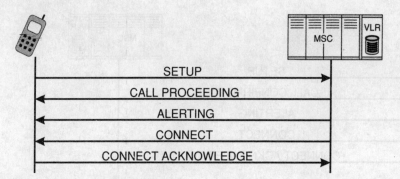

**Figure 13.11** MO call establishment.

The UE starts in the NULL state. We assume that the MN layer entity has requested the establishment of a CS call by sending the MNCC-SETUP-REQ primitive to the CC entity. The call could be a normal CS call, or alternatively it could be an emergency call. Here we will assume that it is a normal call.

On receipt of a MNCC-SETUP-REQ primitive from the MN layer, the CC entity is requested to establish a CS call. The contents of the primitive include the parameters that will become the contents of the SETUP message to be sent to the network. The CC entity requests the establishment of an MM connection by sending an MM-CC-EST-REQ primitive to the MM entity and then moves into the MM CONNECTION PENDING state. When the MM connection is available the MM layer sends on MM-EST-CNF primitive to the CC entity. The CC entity sends the SETUP message to the MM entity using the MMCC-DATA-REQ primitive and moves into the CALL INIT state. If the UE supports the multicall supplementary service, the parameter SI identifies this particular call (stream). If it is the only call, SI is set to 1.

If the CC entity receives a CALL PROCEEDING message from the network via the lower layers, it passes the contents of this message to the MN layer using the MNCC-CALL-PROC-IND primitive and moves into the MO CALL PROCEEDING state. While in the CALL PROCEEDING state, the network may need to update the CC entity on aspects of the call progress (such as the involvement of interworking functions). The CC entity may receive a PROGRESS message, which can be passed to the MN layer via the MNCC-PROGRESS-IND primitive.

At this stage in the call flow, the network requests the establishment of an RAB to carry the user plane information. The RAB is assigned a RAB-Id that is the same as the SI used in the SETUP message. In Section 13.5.3 we will see a more detailed call flow that indicates this activity.

When the CC entity receives an ALERTING message from the network, it sends an MNCC-ALERT-IND primitive containing the contents of the ALERTING message to the MN layer and moves into the CALL DELIVERED state. The receipt of the

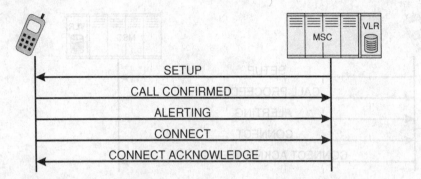

**Figure 13.12**  MT call establishment.

ALERTING message indicates that the user of the destination terminal is being alerted. If the user plane is active, the network generates the alerting signal; if it is not, the alerting signal is generated within the terminal.

When the destination user accepts the call, the network delivers the CONNECT message to the CC entity in the MO terminal. The CC entity notifies the MN layer via the MNCC-SETUP-CNF primitive (which includes the CONNECT message contents), attaches the user plane connection, sends a CONNECT ACKNOWLEDGE message, terminates the alerting indication (if generated locally) and moves into the ACTIVE state.

**MT call establishment**

The MT call establishment procedure passes through the call states on the right hand side of Figure 13.10. The message sequence flows for MT call establishment are summarised in Figure 13.12. The UE starts in the NULL state. For the MT call establishment that we will examine, the network needs to establish an MM connection for use by the CC entity in the network.

Once the MM connection is active, the CC entity in the network sends the SETUP message to the CC entity in the UE. The contents of the SETUP message include network capabilities surrounding the support of multicall services. After checking that the compatibility information in the SETUP message is acceptable, the CC entity passes the SETUP message to the MN layer via the MNCC-SETUP-IND primitive and moves to the CALL PRESENT state.

The MN entity responds to the CC entity via the MNCC-CALL-CONF-REQ primitive, which contains the contents of the CALL CONFIRMED message. If the UE supports the multicall SS and the network supports the multicall service, the SI should be set to show a new bearer if no existing CS calls are in progress. If there is an existing bearer in use, the CC entity should indicate a new value (not currently used) for the SI for CS calls. If the UE supports multicall and the network does not support multicall then the SI is set equal to 1. After passing the CALL CONFIRMED message to the

MM layer for transmission to the CC entity in the network, the CC entity in the UE moves to the MT CALL CONFIRMED state.

In the next step, the CC entity can move to accept the call or to start the user alerting procedure. In the example illustrated in Figure 13.12, we are assuming that the alerting function is activated as the next stage in the procedure.

Next, the network requests the establishment of an RAB for the CS call. The RAB-Id that is returned to the UE in the AS RADIO BEARER SETUP message is the SI that was included in the CALL CONFIRMED message. The activation of the radio resource is indicated to the CC entity by the MM entity and this is indicated to the MN layer using the MNCC-SYNCH-IND primitive that contains the details of the assigned resources (e.g. RAB-Id).

The MN layer can then start the user alerting procedure and request the transmission of the ALERTING message using the MNCC-ALERT-REQ, which includes the contents of the ALERTING message. The CC entity in the UE transmits the ALERTING message to the CC entity in the network and then moves to the CALL RECEIVED state.

When the user accepts the call, the MN layer indicates this by passing the MNCC-SETUP-RES primitive to the CC entity which includes the contents of the CONNECT message. The CC entity sends the CONNECT message to the CC entity in the network and moves to the CONNECT REQUEST state. If it is a speech call that is being established, the user plane for the speech call is activated.

When the CC entity receives the CONNECT ACKNOWLEDGE from the CC entity in the network, the contents of the message are passed to the MN layer via the MNCC-SETUP-COMPLETE-IND primitive, the CC entity moves to the ACTIVE state and for data calls, the user plane is activated.

### 13.5.3 MO basic call establishment

In this section, we now want to return to the scenario of an MO call establishment. We will go over some of the steps introduced in Section 13.5.2, but we also want to introduce the interactions that the CC entity has with the MM entity, the interactions that the MM entity has with the AS in the UE, and the relationship between the AS in the UE and the AS in the network.

Figure 13.13 illustrates the basic call establishment procedure, which includes the MM layer, the CC layer and the RRC layer. We now consider each stage of the call establishment process in turn. We are assuming that the UE is requesting the establishment of a voice call using the AMR voice codec.

In Figure 13.13, the different states of the CC, MM and RRC protocols are shown. The CC entity starts in the NULL state, the MM entity is in the MM IDLE state (we assume the UE has registered previously) and the RR protocol is in the idle mode. The following list steps through each of the call stages numbered in Figure 13.13.

# NAS

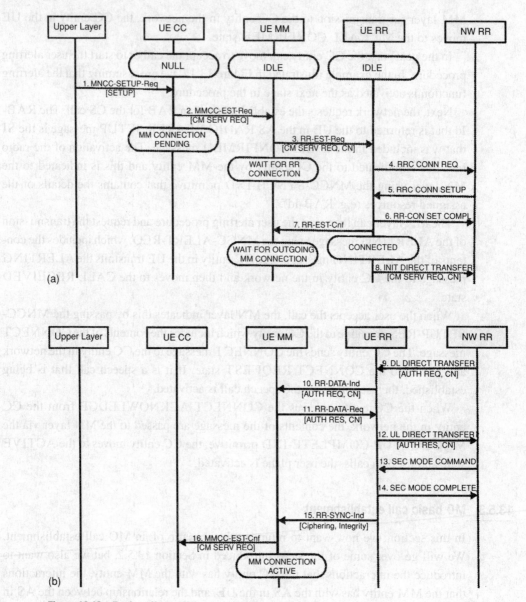

**Figure 13.13** Basic call establishment procedure: parts (a)–(e) fit together to show the whole procedure.

1. The MN layer requests the establishment of a CS call by sending the MNCC-SETUP-REQ primitive that includes the parameters that will form the SETUP message to the CC entity.
2. We are assuming that there is no existing MM connection and consequently, the CC entity requests an MM connection from the MM entity using the MMCC-EST-REQ primitive. The primitive includes the MM message CM SERVICE REQUEST.

## 13.5 Call control protocol

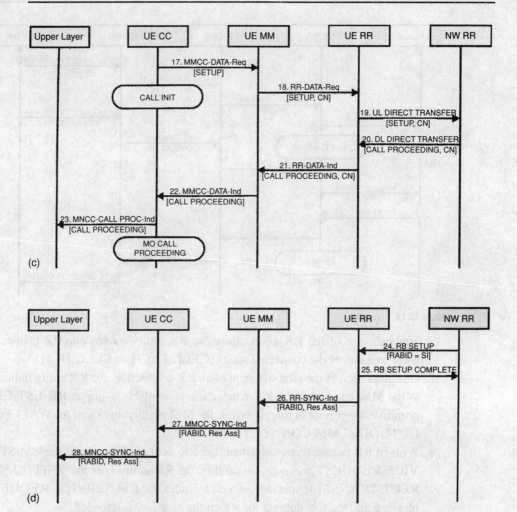

**Figure 13.13** (cont.)

3. To establish an MM connection, the MM entity first needs to establish an RR connection. Currently the MM entity is in the MM IDLE state. To request an RR connection, the MM entity sends the RR-EST-REQ message to the RRC including the CM SERVICE REQUEST message as a parameter and the required CN domain (CS domain in this example). The MM entity moves to the WAIT FOR RR CONNECTION (MM CONNECTION) state as shown in Figure 13.2.

4–6. After receiving the RR-EST-REQ primitive, which includes the establishment cause, the RRC entity requests the establishment of an RRC connection to the NW. The NW establishes up to four SRBs (SRB1, SRB2 SRB3 and optionally SRB4). SRB3 and SRB4 are used for the NAS messages based on the priority of the message received with the message from the NAS. On completion of the

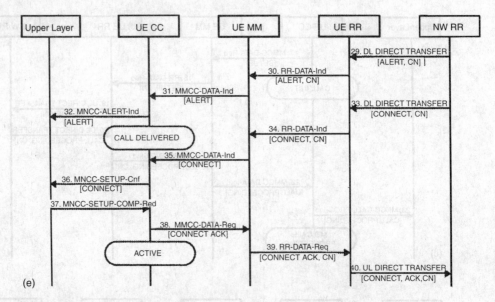

**Figure 13.13**  (*cont.*)

establishment of the RR connection, the RR entity moves into the connected mode and one of the connected states (CELL_FACH or CELL_DCH).

7. On completion of the establishment of an RR connection, the RR entity indicates to the MM entity that the RR connection is available using an RR-EST-CNF primitive. On receipt of this primitive, the MM entity moves into the WAIT FOR OUTGOING MM CONNECTION state.

8. With an RR connection established, the RR entity can now send the CM SERVICE REQUEST message. To do this, the RR entity uses the INITIAL DIRECT TRANSFER message, which includes the CM SERVICE REQUEST message and the CN domain for which the message is intended.

9. The MM entity in the network receives the CM SERVICE REQUEST message and may respond with an AUTHENTICATION REQUEST message (containing the parameters RAND and AUTN as defined previously). The AUTHENTICATION REQUEST message is sent as a parameter in the message from the RR entity – DOWNLINK DIRECT TRANSFER.

10. The RR entity in the UE forwards the AUTHENTICATION REQUEST message to the MM entity using an RR-DATA-IND primitive.

11. The MM entity in the UE processes the AUTHENTICATION REQUEST message as defined in Section 13.4.3; this includes the generation of the parameter RES to be returned to the MM entity in the network. The response from the MM entity in the UE is the AUTHENTICATION RESPONSE message. The MM entity requests the transmission of this message by passing it to the RR entity via the RR-DATA-REQ primitive, which contains the AUTHENTICATION RESPONSE message as a parameter.

12. Using the UPLINK DIRECT TRANSFER message, the RR entity passes the AUTHENTICATION RESPONSE to the network as a parameter.
13. On command from the higher layers in the network, the RR entity in the network requests the activation of the ciphering and integrity protection functions by sending the SECURITY MODE COMMAND.
14. The receipt of the SECURITY MODE COMMAND causes the RR entity in the UE to configure the ciphering and integrity protection functions according to the instructions in the message. When complete, SECURTIY MODE COMPLETE is passed to the RR entity in the network.
15. The RR entity in the UE indicates to the MM entity that the integrity protection and ciphering are activated. This is achieved through the passing of the RR-SYNC-IND primitive. The receipt of this message by the MM layer is the indication that the MM connection establishment is successful.
16. The MM entity indicates to the CC entity that the MM connection is active by passing the MMCC-EST-CNF primitive to the CC entity. The MM entity now moves into the MM CONNECTION ACTIVE state shown in Figure 13.2.
17. With the MM connection active, the CC entity can now request the transfer of the SETUP message to the CC entity in the network. To achieve this, the CC entity uses the MMCC-DATA-REQ primitive that includes the SETUP message as a parameter.

    The contents of the SETUP message are illustrated in Table 13.7.
18. The MM entity passes the SETUP message to the RR layer for transmission via the recently created MM connection using the RR-DATA-REQ primitive.
19. The RR entity sends the SETUP message to the CS domain using the UPLINK DIRECT TRANSFER message.
20. The network responds with the CALL PROCEEDING message sent to the RRC in the UE using the DOWNLINK DIRECT TRANSFER message. The contents of the CALL PROCEEDING message are illustrated in Table 13.8.
21. The RR entity sends the CALL PROCEEDING message to the MM layer via the RR-DATA-IND primitive.
22. The MM entity sends the CALL PROCEEDING to the CC entity via the MMCC-DATA-IND primitive.
23. The CC entity sends the CALL PROCEEDING message to the upper layer using the MNCC-CALL PROC-IND primitive. The CC entity moves into the MO CALL PROCEEDING state.
24. The AS in the network is requested by the CS CN to establish an RAB for the speech call, and so the RADIO BEARER SETUP message is sent. This includes the RAB-Id, which is the same as the SI used in the SETUP message to associate the CS call with the RAB.
25. The RRC acknowledges the RADIO BEARER SETUP with the RADIO BEARER SETUP COMPLETE message when the RB is established.

**Table 13.7.** *Contents of MO CC SETUP message*

| Information element | Comment |
|---|---|
| CC protocol discriminator | Identifies CC messages. |
| TI | Identifies CC instance. |
| Message type | Identifies SETUP message. |
| Bearer capability (RI, 1, 2) | Defines the type of bearer requested and configuration information. |
| Facility | Used to transfer supplementary service information. |
| Calling party subaddress | Used to carry the calling party subaddress. |
| Called party binary coded decimal (BCD) number. | Called party phone number. |
| Called party subaddress | Used to carry the called party subaddress. |
| Low layer compatibility (RI, 1, 2) | Used to allow the destination equipment to perform compatibility check, for modems for instance. |
| High layer compatibility (RI, 1, 2) | Used to allow the destination equipment to perform high layer compatibility check. |
| User to user | Mechanism to transfer data between the users prior to the establishment of a call. |
| SS version | Defines the version of the SS protocol to be used to interpret the facility IE. |
| CLIR suppression | Used to suppress the calling line identification restriction (CLIR) SS. |
| CLIR invocation | Used to invoke the CLIR SS. |
| CC capabilities | Identifies the maximum number of bearers and speech bearers supported. For R99, the maximum number of speech bearers restricted to 1. |
| SI | Number sent in SETUP message and returned as RAB-Id. |

**Table 13.8.** *Contents of the CALL PROCEEDING message*

| Information element | Comment |
|---|---|
| CC protocol disc | Identifies CC messages. |
| TI | Identifies CC instance. |
| Message type | Identifies CALL PROCEEDING message. |
| Bearer capability (RI, 1, 2) | Defines the network selection of bearer capability information. |
| Facility | Used to transfer SS information. |
| Progress indicator | Call progress indication such as interworking information. |
| Priority granted | Priority granted by network if e-MLPP supported. |
| Network call control capabilities | Indication on whether the network supports multicall. |

**Table 13.9.** *Contents of the ALERTING message*

| Information element | Comment |
| --- | --- |
| CC protocol disc | Identifies CC messages. |
| TI | Identifies CC instance. |
| Message type | Identifies ALERTING message. |
| Facility | Used to transfer supplementary service information. |
| Progress indicator | Call progress indication such as interworking information. |
| User–user | User to user information. |

26. The RR entity notifies the MM entity that the resources are assigned using the RRC-SYNC-Ind primitive, which includes the RAB-Id.
27. The MM notifies the CC entity that the resources are assigned using the MMCC-SYNC-Ind primitive, which includes the RAB-Id.
28. The CC entity notifies the MN layer that the resources are assigned using the MNCC-SYNC-Ind primitive which includes the RAB-Id.

29–32. The RR entity receives the ALERTING message from the network and passes it through to the upper layer via the MM and CC entities as shown. The ALERTING message indicates that the end user's phone is ringing. The CC entity moves into the CALL DELIVERED state. The contents of the ALERTING message are illustrated in Table 13.9.

33–36. When the destination user accepts the call, the CC entity in the network generates the CONNECT message that is sent to the CC entity in the UE. The CONNECT message is received and sent to the upper layers in the manner shown in the diagram. The user plane is activated and the voice data can flow via the RAB to the end user when approved by the network. The contents of the CONNECT message are illustrated in Table 13.10.

37–40. The MN layer sends the CONNECT ACKNOWLEDGE message to the network indicating that the call is accepted. The CC entity moves into the ACTIVE state. The voice data can now flow between the two users. The contents of the CONNECT ACKNOWLEDGE message are illustrated in Table 13.11.

## 13.6 GMM protocol states

The GMM entity is within the MM sublayer alongside the MM entity that we have reviewed. The GMM entity and the GMM protocol are responsible for the MM control of the PS domain, in a similar manner to the MM protocol being in control of the MM for the CS domain.

In this section, we start by reviewing the GMM state machine, and in the following sections we will consider the GMM protocol operation, and review the main PS domain

**Table 13.10.** *Contents of CONNECT message*

| Information element | Comment |
| --- | --- |
| CC protocol disc | Identifies CC messages. |
| TI | Identifies CC instance. |
| Message type | Identifies CONNECT message. |
| Facility | Used to transfer SS information. |
| Progress indicator | Call progress indication such as interworking information. |
| Connected number | Identifies the connected party of the call. |
| Connected subaddress | Identifies the subaddress for the connected party. |
| User–user | User to user information. |

**Table 13.11.** *Contents of CONNECT ACKNOWLEDGE message*

| Information element | Comment |
| --- | --- |
| CC protocol disc | Identifies CC messages. |
| TI | Identifies CC instance. |
| Message type | Identifies CONNECT ACKNOWLEDGE message. |

protocol that lies in the CM sublayer, which is called the SM protocol. In that section, we will also review some examples of the interoperation of the GMM and the SM protocols.

### 13.6.1 GPRS update status

Before examining the details of the GPRS states, it is worth considering the GPRS update status, which defines the current position regarding the registration of the user. There are three update statuses defined: GU1, GU2 and GU3. The update status is stored in non-volatile memory in the USIM. In this section, we outline the meaning of each of these update status values. To understand the definitions in this section we refer to some identifiers such as the P-TMSI and the RAIs and some GPRS MM procedures such as GPRS attach and RAU. This terminology is similar to the CS domain equivalents (TMSI, LAI, IMSI attach and location area updating) that we saw in Section 13.4. Later in this section we will formally define the meanings of these different terms.

GU1 (UPDATED) means that the last GPRS attach or RAU was successful. The USIM stores the P-TMSI and RAI assigned as part of the last attach or RAU. The USIM may also contain some of the various security keys and information assigned by the network.

GU2 (NOT UPDATED) means that the last attach or RAU failed for some 'technical' reason rather than reasons of service restrictions. The USIM may contain information on P-TMSI and RAI and security information, but the UE may be expected to delete some of this under certain conditions.

## 13.6 GMM protocol states

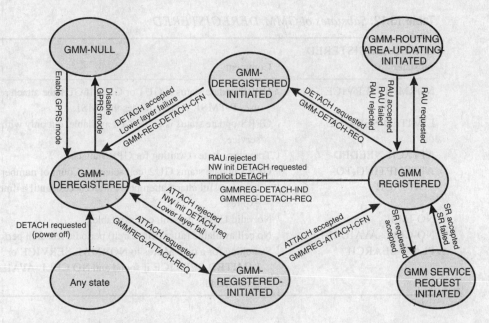

**Figure 13.14** GPRS MM entity state diagram.

GU3 (ROAMING NOT ALLOWED) means that the last attach or RAU was performed correctly but was rejected by the network for service registration reasons such as roaming restrictions. The same comments apply regarding P-TMSI, RAI and security information as applied to GU2.

### 13.6.2 GPRS MM states

Figure 13.14 illustrates the basic state diagram for the GPRS MM state model in the UE. As in the previous section, we focus on the state machine from the perspective of the UE. Readers interested in the state machine from the network perspective are referred to [32]. In this section, we will review the purpose and rationale for each of the states, and the transitions between the states.

**GMM-NULL state**
This is the state that the GMM entity starts from and ends in as the GPRS service is activated and deactivated. In the GMM-NULL state there are no GPRS MM functions available.

**GMM-DEREGISTERED state**
In this state the GPRS functionality is active, but the GMM entity is not registered and needs to perform either a GPRS ATTACH or a combined GPRS/IMSI ATTACH. The GMM-DEREGISTERED state has a number of substates. The

**Table 13.12.** *Substates of GMM-DEREGISTERED*

| GMM-DEREGISTERED substate | Comment |
| --- | --- |
| NORMAL-SERVICE | GPRS update status is GU1 or GU2. If GU1 the attach request uses P-TMSI. If GU2 it uses the IMSI. |
| LIMITED-SERVICE | GPRS update status is GU3. Cell available, but only with limited service. |
| ATTACH-NEEDED | Transitory state – waiting for GPRS attach. |
| ATTEMPTING-TO-ATTACH | GPRS update status GU2, cell selected. Count of number of unsuccessful attach attempts restricts access until a timer expires. |
| NO-IMSI | No valid USIM present, cell available. |
| NO-CELL-AVAILABLE | No cell available (after cell search) periodic research performed. |
| PLMN-SEARCH | Searching for a cell. Moves to NORMAL SERVICE or LIMITED SERVICE if found and NO CELL AVAILABLE if not found. |

substates are identified as GMM-DEREGISTERED-SUBSTATE, where the different values for SUBSTATE are defined in Table 13.12. The details of the substate selection are considered in greater detail in Chapter 14.

There are many reasons for the transitions between the GMM-DEREGISTERED state (and its substates) and the other GMM states. Table 13.13 presents the main transitions and some of the reasons for these transitions. For further details on this, the interested reader is referred to [32].

## GMM-REGISTERED-INITIATED state

In this state, the GMM entity has initiated either a GPRS attach or a combined GPRS and IMSI attach. The GMM entity is waiting for a response from the GMM entity in the network.

The transitions between the GMM-REGISTERED-INITIATED state and other GMM states are shown in Table 13.13 and Table 13.15.

## GMM-REGISTERED state

In the GMM-REGISTERED state, the GMM entity has successfully completed either a GPRS attach, or a combined GPRS/IMSI attach, and as a consequence a GMM context exists.

The GMM-REGISTERED state has a number of substates. Any specific substate defines the functions available in that substate. In general, the state allows the UE to establish PDP contexts, transfer data, respond to paging requests and perform cell and RAUs.

## 13.6 GMM protocol states

**Table 13.13.** *Transitions to and from GMM-DEREGISTERED state*

| From GMM | To GMM | Comment |
|---|---|---|
| NULL | DEREGISTERED | UE switched on or GPRS capability activated. |
| REGISTERED-INITIATED | DEREGISTERED | Authentication failed with the AUTHENTICATION AND CIPHERING REJECT message received. |
| REGISTERED-INITIATED | DEREGISTERED | Attach rejected (ATTACH REJECT received) due to a range of possible reasons such as GPRS services not allowed, roaming not allowed, PLMN not allowed, no suitable cells. |
| DEREGISTERED-INITIATED | DEREGISTERED | GMM entity in UE is performing a GPRS DETACH (without switch-off) and has received a GPRS DETACH ACCEPT. Low layer failure and time-out will also cause the transition prior to DETACH ACCEPT being received. |
| REGISTERED | DEREGISTERED | UE and network initiated detach via DETACH REQUEST. |
| REGISTERED | DEREGISTERED | Network initiated via paging using IMSI to recover from possible network failure. |
| REGISTERED | DEREGISTERED | Network rejects RAU using ROUTING AREA UPDATE REJECT. |
| REGISTERED | DEREGISTERED | Network rejects service request using SERVICE REJECT indicating that the service request has failed for some reason. |
| DEREGISTERED | REGISTERED-INITIATED | GPRS attach or combined GPRS/IMSI attach via ATTACH REQUEST initiated. |
| REGISTERED-INITIATED | DEREGISTERED | MO or network originated DETACH REQUEST causes the GMM entity to abandon the GPRS attach procedure. |

Table 13.14 defines the substates of the GMM-REGISTERED state with some comments on the functions available in the different substates. The transitions between the substates of the GMM-REGISTERED state are caused by cell selection/reselection, change of routing area and loss/regain of coverage.

Table 13.15 shows the transitions between the GMM-REGISTERED state and the other GMM states and includes some of the rationale for these transitions.

### GMM-DEREGISTERED-INITIATED state

A request to release GMM context via GPRS detach or combined GPRS/IMSI detach has been initiated. GMM moves into this state if UE is not being 'powered-off'. The reasons for the transitions between this state and the other GMM states are provided in Tables 13.13 and 13.15.

**Table 13.14.** *Substates of GMM-REGISTERED*

| GMM-REGISTERED substate | Comment |
|---|---|
| NORMAL-SERVICE | Data and signalling exchanged normally. Cell selection and reselection performed, normal and periodic RAU, respond to paging requests. |
| UPDATE-NEEDED | No data or signalling allowed. RAU currently not allowed due to access class restrictions. GMM must wait until cell allows RAU or UE changes cell. |
| LIMITED-SERVICE | Selected cell does not support normal service. GMM moves to NORMAL-SERVICE substate when appropriate cell selected. |
| ATTEMPTING-TO-UPDATE | RAU update failed due to possible network error. When permitted (limits due to timers and counters; cell change) GMM will attempt RAU. No data exchange allowed. |
| ATTEMPTING-TO-UPDATE-MM | MM part of combined update failed (GMM part success). Timers and counters control new attempt. PS data and signalling exchanges allowed. Cell reselection allowed. |
| IMSI-DETACH-INITIATED | GMM entity awaiting response to combined GPRS detach for non-GPRS services only. PS data and signalling exchanges allowed. |
| NO-CELL-AVAILABLE | GPRS coverage lost. UE can only perform cell and/or PLMN reselection. |
| PLMN-SEARCH | Searching for PLMN. No service available. Will move to NORMAL-SERVICE or LIMITED-SERVICE if cell found and NO-CELL-AVAILABLE if cell not found. |

### GMM-SERVICE-REQUEST-INITIATED state

The service request procedure is activated. The GMM entity is awaiting response from network. The transitions between this state and the GMM-REGISTERED state are presented in Table 13.15.

### GMM-ROUTING-AREA-UPDATING-INITIATED state

The GMM entity has initiated RAU and is awaiting response from the network. The transitions between this state and the GMM-REGISTERED state are presented in Table 13.15.

### 13.6.3 Links between GMM states and PS mobility management (PMM) states

There are a number of state machines surrounding the GMM entity. The GMM state machine defined below is a detailed low level state machine. There is also a higher level state machine (PMM states) defined in [36]. In this section we consider the definition of these PMM states and their relationships to the GMM states presented in the previous section. The PMM states are used to define the state of the PS signalling connection between the UE and the PS CN. There are three states to this PS signalling connection and they are: PMM-DETACHED, PMM-IDLE and PMM-CONNECTED. In the

## 13.6 GMM protocol states

**Table 13.15.** *Transitions to and from GMM-REGISTERED state*

| From GMM- | To GMM- | Comment |
|---|---|---|
| REGISTERED | DEREGISTERED | Network initiated detach via DETACH REQUEST. |
| REGISTERED | DEREGISTERED | Network initiated paging using IMSI to recover from possible network failure. |
| REGISTERED | DEREGISTERED INITIATED | MO detach initiated by the transmission of a DETACH REQUEST (not power off). |
| REGISTERED | ROUTING-AREA-UPDATING-INITIATED | GMM entity performing an RAU or combined RAU/LAU by sending a ROUTING AREA UPDATE REQUEST message. |
| REGISTERED | SERVICE-REQUEST-INITIATED | GMM entity requests a GMM connection to move into PMM connected state for possible use by CM sublayer or for signalling transfer. GMM sends SERVICE REQUEST message. |
| REGISTERED-INITIATED | REGISTERED | Network accepts GPRS attach or combined GPRS/IMSI attach (GMM entity received ATTACH ACCEPT). |
| ROUTING-AREA-UPDATING-INITIATED | REGISTERED | Network may respond to RAU with ROUTING AREA UPDATE REJECT and the GMM entity moves to the LIMITED-SERVICE substate. |
| ROUTING-AREA-UPDATING-INITIATED | REGISTERED | Network may respond to RAU or combined RAU/LAU with ROUTING AREA UPDATE ACCEPT and the GMM entity moves to the NORMAL-SERVICE substate. |
| SERVICE-REQUEST-INITIATED | REGISTERED | If network accepts service request in PMM idle mode a response with security mode control indicates acceptance. If in PMM connected mode SERVICE ACCEPT indicates acceptance. |
| SERVICE-REQUEST-INITIATED | REGISTERED | If network rejects (SERVICE REJECT) due to roaming not allowed or no suitable cells moves to LIMITED-SERVICE substate. |

following we define the characteristics of these PMM states and then finish by relating them to the GMM states that we have seen.

## PMM-DETACHED

In this state there is no communication between the UE and the SGSN, there is no valid RA information and the SGSN has no means of reaching the UE. To leave this state, the UE must perform a GPRS attach and establish a GMM context, in which case it moves into the PMM-CONNECTED state.

## PMM-IDLE

In this state the location of the UE is known in the SGSN to an RA, the UE has previously established a GMM context but there is no active PS signalling connection between the UE and the PS CN. It should be noted that there may be an active RRC connection for the UE (an active CS signalling connection to the CS CN for instance), but that does not imply an active PS signalling connection.

To contact the UE, the SGSN needs to page that UE within the RA. The UE is responsible for performing RAUs whenever it notices that the RA identity has changed (the RA identity is broadcast in SIB1 of the system information broadcast messages of the cell that the UE is camped on). The UE needs to perform periodic RAUs whilst in this state based on a timer that is set by the network.

**PMM-CONNECTED**

In this state the location of the UE is known in the SGSN to an RNC, the UE has previously established a GMM context (or is about to do so) and there is an active PS signalling connection between the UE and the PS CN. The location of the UE is tracked by the UTRAN, but the UE still needs to perform RAUs whenever the UE identifies that the RA has changed. With an active PS signalling connection, the UE has an active RRC connection, and so the UE is notified of changes in the RAI either through broadcast messages (CELL_FACH, CELL_PCH or URA_PCH states) or through the RRC mobility information signalling messages (CELL_DCH state). The UE does not need to perform periodic RAUs whilst in the PMM-CONNECTED state.

The PMM-CONNECTED state is entered from the PMM-DETACHED state as soon as a PS signalling connection is established, and this is the first stage towards performing a GPRS attach. The PMM-CONNECTED stage can also be entered from the PMM-IDLE state with the re-establishment on the PS signalling connection. From the PMM-CONNECTED state the UE can move to the PMM-DETACHED state as a consequence of a GPRS detach procedure or a routing update request failure in which case the PS signalling connection and the GMM context will no longer exist. The UE can move to the PMM-IDLE state if the PS signalling connection is released.

Table 13.16 summarises the relationship between the PMM states and the GMM states. It shows the relationship between the GMM entity states and the PMM signalling connection states. A yes in the table indicates a coincidence in the states, i.e. in the GMM-NULL state the PS connection is in the PMM-DETACHED state. With the exception of the GMM-REGISTERED state there is a single connection between the GMM states and the PMM states. For the GMM-REGISTERED state, however, the PS signalling connection can be either active or not, but the GMM entity can still be registered, hence there are two entries for the GMM-REGISTERED state.

### 13.6.4 Combined GPRS and non-GPRS procedures

The network protocols and procedures have been defined to allow some of the procedures in the MM sublayer to be combined. This means that the MM entity and

## 13.6 GMM protocol states

**Table 13.16.** *Relationship between the GMM entity state and the PMM PS signalling connection states*

| GMM states | PMM states | | |
|---|---|---|---|
| | PMM-DETACHED | PMM-CONNECTED | PMM-IDLE |
| GMM-NULL | Yes | No | No |
| GMM-DEREGISTERED | Yes | No | Yes |
| GMM-REGISTERED-INITIATED | No | Yes | No |
| GMM-REGISTERED | No | Yes | Yes |
| GMM-ROUTING-AREA-UPDATING-INITIATED | No | Yes | No |
| GMM-SERVICE-REQUEST-INITIATED | No | Yes | No |
| GMM-DEREGISTERED-INITIATED | No | Yes | No |

the GMM entity within the MM sublayer cooperate in some of the procedures being performed.

The conditions that are required to permit this activity are that the network is operating in Mode I as defined in Section 13.3.2 (the optional Gs interface is available between the CS MM entity and the PS GMM entity), and that the UE is capable of simultaneous CS and PS activity. If we assume that these conditions are satisfied, then the following additional points can be made about this combined activity.

### Combined function network entity

The combined activity surrounds the attach (GPRS attach and IMSI attach) and area update functions (RAU and LAU). The peer entity in the CN with which the UE establishes a dialogue is the PS domain entity (serving GPRS support node (SGSN)). The attach and the area updates are directed towards the SGSN, and the SGSN relays the CS domain elements towards the MSC responsible for the CS domain functions.

### Combined function constraints

If the network is operating in Mode I, and the UE has the capability (mode CS/PS), the UE should use the combined RAU/LAU procedures when it is simultaneously GPRS and IMSI attached. The UE knows what network mode the network operates in from the NAS specific broadcast messages sent via SIB1 in the cell the mobile is currently operating in.

For the detach procedure (described in Section 13.6.1), IMSI detach is performed using the GMM combined GPRS detach procedure. If an MM authentication procedure (CS domain) is rejected by the network, the GMM procedures are halted and the GMM registration cancelled.

**Table 13.17.** *Summary of GPRS MM procedures*

| GMM procedure | Description |
|---|---|
| Common | GPRS P-TMSI reallocation |
| | GPRS authentication and ciphering procedure |
| | GPRS identification procedure |
| | GPRS information procedure |
| | GPRS service request procedure |
| Specific | GPRS attach procedure |
| | GPRS detach procedure |
| | GPRS routing area update procedure |

## 13.7 GMM procedures

The GMM procedures are outlined in Table 13.17. The procedures are divided into common GMM procedures and specific GMM procedures. In this section, we review aspects of these different procedures, starting with the GPRS attach procedure.

### 13.7.1 GPRS attach

GPRS attach is the procedure that the GMM entity needs to undertake to register with the PS domain (SGSN) prior to any services being activated. In doing so, the UE establishes a GMM context. There are two variants of the GPRS attach procedure. The first is as a normal attach *rephrase* for PS services. The second is as a combined attach for both PS and CS services.

As part of the operation of the GPRS attach procedure, a GPRS attach attempt counter is defined to count up the number of attach attempts. Too many attempts can cause a temporary cessation of subsequent attempts. The counter can be reset by a number of events such as UE powered off, GPRS attach successfully completed, new RA entered.

The GPRS attach procedure, if successful, starts with the GMM entity in the GMM-DEREGISTERED state, moves to the GMM-REGISTERED-INITIATED state after the transmission of an ATTACH REQUEST message, and results in the GMM entity in the GMM-REGISTERED state if the ATTACH ACCEPT was received.

If the GPRS attach was unsuccessful, for instance as a result of a network rejection (ATTACH REJECT) (there are other possible reasons for rejection some of which are identified in Table 13.13), then the GMM entity returns to the GMM-DEREGISTERED state.

The GMM entity could also be performing a combined attach (GPRS attach and IMSI attach). This is only possible with the Network Type I, outlined in Section 13.3.2.

## 13.7 GMM procedures

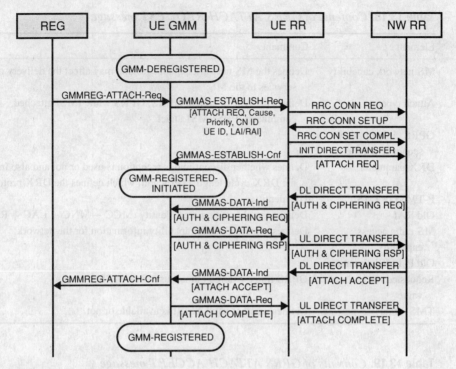

**Figure 13.15** GPRS attach procedure with interlayer primitives.

In this case, the same procedures are followed except that the attach type indicates that it is a joint attach that is being requested. The ATTACH REQUEST is still directed towards the PS domain (SGSN).

Figure 13.15 illustrates the basic procedures associated with a successful GPRS attach function. In Figure 13.15 we can see the primitives between the layers in the UE used to transfer the messages and parameters, as well as the changes in state of the GMM entity in the UE.

### GPRS attach not accepted and abnormal cases

As mentioned above, there are several reasons why the network may reject the GPRS attach request (both normal and combined). Some of these reasons include illegal UE, GPRS services not allowed and roaming not allowed in this LA. In these and other cases, the GMM entity follows a sequence of events that would normally put the GMM entity into the GMM-DEREGISTERED state or substate and change the USIM GPRS update status to GU3. The precise details for these and other failure conditions can be found in [32].

### ATTACH REQUEST

Table 13.18 defines the parameters that are contained in the ATTACH REQUEST message from the UE to the SGSN.

**Table 13.18.** *Contents of GPRS ATTACH REQUEST message*

| Element | Comment |
|---|---|
| MS network capability | Defines the MS network capabilities that may affect the delivery of services to the MS. |
| Attach type | Defines the type of attach: GPRS, GPRS while IMSI attached, combined GPRS/IMSI attach. |
| GPRS ciphering key sequence number | |
| DRX parameter | Defines whether discontinuous reception is used or not and also includes the DRX cycle length coefficient, which defines the DRX protocol. |
| P-TMSI or IMSI | UE identity. |
| Old RAI | Defines the old routing area identity (MCC + MNC + LAC + RAC). |
| MS radio access capability | Defines some UE radio capability information for the network. |
| Old P-TMSI signature | |
| Requested ready timer value | GPRS timer – stored but not used in UMTS. |
| TMSI status | Defines whether valid TMSI is available or not. |

**Table 13.19.** *Contents of GPRS ATTACH ACCEPT message*

| Element | Comment |
|---|---|
| Attach result | Specifies the result of the attach. GPRS only attach or combined GPRS/IMSI attach. |
| Force to standby | Value stored, not used in UMTS. |
| Periodic RA update timer | Defines timer value for periodic RAUs. |
| Radio priority for SMS | Defines the radio priority for SMS messages (1–4 with 1 highest). |
| RAI | RAI that the UE is in. |
| P-TMSI | Defines P-TMSI to be used by UE. |
| GMM cause | Defines reasons why IMSI attach not successful during a combined procedure. |

## ATTACH ACCEPT

Table 13.19 defines the contents of the ATTACH ACCEPT message that is returned by the SGSN in reply to the ATTACH REQUEST.

### 13.7.2 GPRS authentication

Part of the attach procedure shown in Figure 13.15 is the authentication procedure, which is started when the network (SGSN in this case) receives the ATTACH REQUEST message from the UE. If the UE is unknown to the SGSN, it can request the authentication information from the previously used SGSN (this is indicated by the contents of the ATTACH REQUEST message). If the previously used SGSN cannot be traced, the SGSN can request identification information from the UE and then authentication information from the HLR/AuC. In our example, we will assume that the SGSN does know of the UE, and that the authentication information (authentication vector defined in Chapter 2) is available.

The SGSN sends an authentication vector with 'RAND, AUTN, CKSN' parameters to the GMM entity in the UE using the AUTHENTICATION AND CIPHERING REQUEST message. Within the message, RAND is a random number, AUTN is the authentication token and CKSN is the cipher key sequence number.

Upon receipt, the USIM verifies the AUTN and, if it is accepted, computes 'RES' which is returned to the SGSN in the AUTHENTICATION AND CIPHERING RESPONSE for comparison with 'XRES' obtained from the AuC. If all signatures agree, the USIM uses computed values for 'CK' and 'IK' and the SGSN selects 'CK' and 'IK' (CK is the cipher key and IK the integrity key).

### 13.7.3 MS and network initiated GPRS / IMSI detach

The GPRS detach procedure is used to detach the UE from PS domain services, either singly or as part of a combined detach which also includes the IMSI detach. The detach can also be used by the network in cases where the network wants to force a subsequent reattachment, possibly as remedial action to some network failure. The remedial action of GPRS attach also includes deactivation of PDP contexts present prior to the detach.

The detach procedure is activated for a range of reasons such as the power off of the UE, USIM removed from the UE, PS domain service deactivated in the UE or network initiated detach for network reasons. At the completion of the detach, the GMM context is released.

The UE initiated detach occurs whilst the GMM entity is in the GMM-REGISTERED state and is activated by the transmission of the DETACH REQUEST message. There are various types of detach definable, depending upon whether it is a normal or combined detach and whether it is due to 'power off' or due to service deactivation. The network needs to know the specific type, partly to define what signalling exchanges to expect and partly to define what other network elements may need to be informed.

The completion of the UE initiated detach occurs when the UE receives the DETACH ACCEPT (if the detach is not due to 'power off') and the GMM entity ends in the GMM-DEREGISTERED state. If the detach is due to 'power off', it is completed by the transmission of the DETACH REQUEST by the GMM entity.

In a similar manner to the attach procedure, there are a number of abnormal and failure cases that the GMM entity is designed to cope with.

Figure 13.16 illustrates the procedures that occur when a mobile imitates a combined GPRS IMSI detach due to PS services being deactivated.

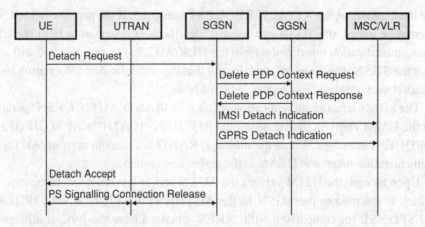

**Figure 13.16** Example of MS initiated detach including network actions.

The GMM entity sends DETACH REQUEST to SGSN including detach type and P-TMSI signature. The P-TMSI signature is used to test the authenticity of the DETACH REQUEST.

For GPRS detach, the SGSN requests the GGSN to delete any active PDP contexts. The procedure includes IMSI detach, and so the SGSN sends an 'IMSI detach indication' message to the visitor location register (VLR) responsible for that UE. If the procedure does not include IMSI detach, the SGSN still sends the 'GPRS detach indication' message to MSC/VLR to allow the VLR to remove the association with SGSN that is used for paging to the CS domain. As the detach is not due to switch off, the SGSN sends the DETACH ACCEPT message. When the mobile is detached and in the GMM-DEREGISTERED state, the SGSN requests the release of the PS signalling connection and the PMM state moves to PMM-DETACHED.

### 13.7.4 RAU

The RAU is used in a manner similar to the LAU used in the CS domain. The objective is to update the current location of the registration for the GMM entity in terms of the RAs defined in the system. Additionally, the RAU procedure can be used in place of the attach procedures to perform GPRS attach. It is also used as a mechanism to perform an intersystem change between UMTS based PS services and GSM based PS services. The RAU procedure is only ever initiated by a UE, and it must be one that is in the GMM-REGISTERED state.

In a manner similar to the attach procedure, there is a counter that counts the number of rejected or unsuccessful RAU attempts with a view to limiting the number made. The counter for rejected attempts causes a delay in subsequent attempts when a certain number is exceeded, and is cleared under a variety of conditions. Example events that clear the counter are a successful RAU procedure and when a new RA is entered. The RAU may be a normal procedure, a combined procedure with an LAU, a periodic RAU, or a combined periodic RAU and LAU.

As in the CS domain, certain RAs can be defined as forbidden areas which the UE must track. Forbidden areas can be defined through broadcasts or as a result of RAU rejects from the network. The UE should not perform RAU in forbidden areas.

The normal and periodic single or combined RAU is initiated by a GMM entity in the GMM-REGISTERED state by the transmission of the ROUTING AREA UPDATE REQUEST message. The message indicates the type of update and after transmission the GMM entity moves to the GMM-ROUTING-AREA-UPDATING-INITIATED state shown in Figure 13.14.

A successful update results in the ROUTING AREA UPDATE ACCEPT being sent by the network to the GMM entity, which extracts parameters from the message such as new P-TMSI (if present) and RAI. If the ROUTING AREA UPDATE ACCEPT message includes a P-TMSI, the GMM entity needs to respond with ROUTING AREA UPDATE COMPLETE to acknowledge receipt of the new P-TMSI. The GMM entity can then return to the GMM-REGISTERED state.

The GMM entity knows that an unsuccessful RAU has occurred if it receives a ROUTING AREA UPDATE REJECT. This message includes the cause information, which is used by the GMM entity to define the subsequent actions.

Figure 13.17 illustrates an example of an RAU procedure, which includes an optional location update procedure. The ROUTING AREA UPDATE REQUEST includes information on the RAU type (e.g. RA update, periodic update, combined RA/LA update). If the RAU is to a new SGSN, the new SGSN will request GMM context information from the old one.

After the transfer of the GMM context information and any data to be transmitted to the UE, the SGSN initiates security functions. In the example shown in Figure 11.17, the UE has an active PDP context (the details of PDP contexts are considered in the next section). The old SGSN forwards any stored data packets to the new SGSN, the GGSN is notified of the new SGSN (PDP context is updated) and any new data packets for the PDP context are sent directly from the GGSN to the new SGSN. Once this is completed, the SGSN can start the LAU procedure if requested by the GMM entity. In this procedure, the MSC responds by performing an update location procedure to old MSC and transfer of data from HLR. On completion, the MSC sends LOCATION UPDATE ACCEPT to SGSN.

The SGSN sends a ROUTING AREA UPDATE ACCEPT message to the GMM entity. It is a new SGSN and MSC and so we expect a new TMSI and P-TMSI. In this case, the GMM entity finishes by sending the ROUTING AREA UPDATE COMPLETE message to SGSN confirming the new TMSI and P-TMSI. The SGSN confirms the new TMSI reallocation to MSC/VLR.

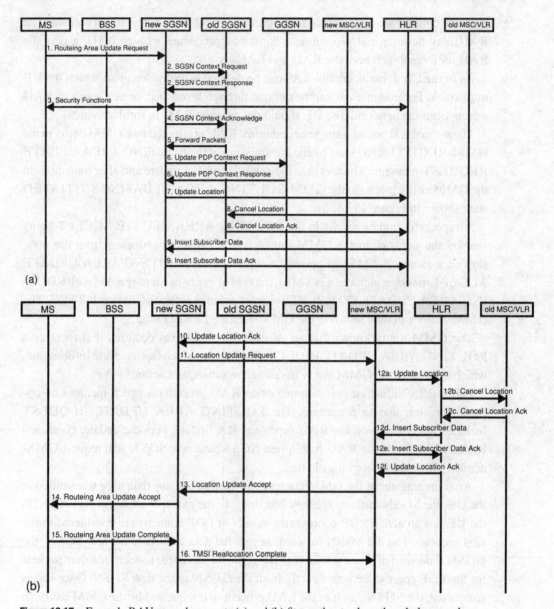

**Figure 13.17** Example RAU procedure: parts (a) and (b) fit together to show the whole procedure.

### 13.7.5 Service request procedure

The GPRS service request procedure is similar to the CM service request procedure used in the CS domain considered in Section 13.4. The service request procedure is used to establish a PS connection that can be employed for either signalling or data. There are two states in which we use the service request procedure: the PMM-IDLE state and the PMM-CONNECTED state. We consider each of these in turn.

A UE in the PMM-IDLE state has an active GMM context, but does not have an active PS signalling connection. The occasions when the UE may need an active signalling connection are:
- to transfer user data via an existing PDP context in either the uplink or the downlink;
- to transfer signalling messages from CM sublayer entities such as the SMS or SM entity.

In the PMM-IDLE state, the UE in the uplink sends the SERVICE REQUEST message and moves into the GMM-SERVICE-REQUEST-INITIATED state. The indication that the procedure is finished is the successful completion of the security mode control procedure in which case the GMM entity moves into the GMM-REGISTERED state and the UE connection into the PMM-CONNECTED state.

For the downlink, the network pages the UE with an indication of the cause such as 'high priority signalling'. The UE responds with the SERVICE REQUEST message indicating a service type 'paging response', and then follows the remainder of the procedure as outlined above.

A UE in the PMM-CONNECTED state has both an active GMM context and an active PS signalling connection. There are occasions, however, when the UE may need to send a service request. If the UE has an active PDP context but no active RAB for that context and the UE wishes to send data on the uplink, then the SERVICE REQUEST message is used to request the re-establishment of the RAB for the PDP context. In this case the SERVICE REQUEST message indicates the request type was 'data'. The response indicating success on the downlink is the SERVICE ACCEPT message (this is different to the previous case when the UE was in the PMM-IDLE state as the UE already has an active security connection).

## 13.8 SM protocol and PDP contexts

In this section, we will consider the SM protocol and PDP contexts. PDP contexts are activated and deactivated by the SM entity within the CM sublayer, but this can only be done if a GMM context exists. If there is no active GMM context, then the SM entity needs to request the establishment of a GMM context. The CM sublayer uses the services of the GMM entity to establish and maintain GMM contexts.

The PDP contexts are packet data connections that are created for the flow of user data from the UE to some destination terminal or network. The PDP contexts are established, modified and terminated through the SM protocol.

In this section, we start with a simple review of the PDP context. Next, we examine the architecture and operation of the SM protocol layer. Finally, we look at some examples of the establishment and modification of PDP contexts, drawing in the use of the other protocol layers as we do.

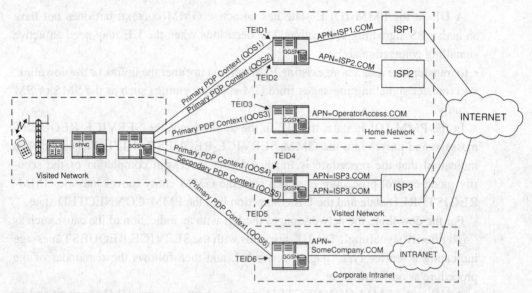

**Figure 13.18** Basic concepts for PDP contexts.

### 13.8.1 Introduction to PDP contexts

Figure 13.18 illustrates the basic concepts associated with PDP contexts (in this case multiple simultaneous PDP contexts). A PDP context is a packet data session, which is defined by a PDP address, a defined QoS, a quantity referred to as an access point name (APN) and the NSAPI.

The PDP context is a user-plane connection that extends between the UE and a GGSN selected from the APN. Figure 13.18 shows that the GGSN could be in the home network or the visited network. The GGSN acts as the access point to the external network (such as ISPs or corporate intranets as shown). The external networks are identified through the APN.

The APN is an identifier used to define the external access point that the UE is attempting to access. The user portion of the APN is a uniform resource locator (URL) (such as isp2.com shown in the diagram). The network adds to this a network supplied identifier that allows the APN to be routable within the mobile network. Typically the full APN (created by the SGSN) has the form isp2.com.mnc.mcc.gprs.

The UE has the ability to support multiple PDP contexts. The PDP contexts are identified in the UE by the NSAPI with each PDP context being allocated an NSAPI out of 11 available ones. Each PDP context can have the different QoS requirements (the QoS definition is considered in more detail in Section 13.8.4) and can share the same PDP address such as an IP address. A PDP context that shares an IP address is referred to as a secondary PDP context, the primary PDP context being the one that was initially associated with the IP address. Primary PDP contexts and secondary PDP contexts use the same PDP address and APN, but can have different QoS settings.

## 13.8 SM protocol and PDP contexts

**Table 13.20.** *SMREG SAP service primitives*

| Primitive | Direction | Parameters | Comment |
|---|---|---|---|
| SMREG-PDP-ACTIVATE-REQ | HL to SM | Address, QoS, NSAPI, APN | PDP context activation request |
| SMREG-PDP-ACTIVATE-CNF | SM to HL | Address, QoS, NSAPI, | PDP context active and RAB active |
| SMREG-PDP-ACTIVATE-REJ | SM to HL | Cause, NSAPI | PDP context rejected by NW |
| SMREG-PDP-ACTIVATE-IND | SM to HL | Address, APN | NW requests PDP context activation |
| SMREG-PDP-ACTIVATE-REJ-RSP | HL to SM | Cause, address, APN | NW requested PDP context active failed |
| SMREG-PDP-DEACTIVATE-REQ | HL to SM | NSAPI(s), tear ind, cause | Request to deactivate all coupled contexts |
| SMREG-PDP-DEACTIVATE-CNF | SM to HL | NSAPI(s) | Requested contexts deactivated |
| SMREG-PDP-DEACTIVATE-IND | SM to HL | NSAPI(s), tear ind, cause | NW initiated context deactivation done |
| SMREG-PDP-MODIFY-IND | SM to HL | QoS, NSAPI | NW initiated context modify finished |
| SMREG-PDP-MODIFY-REQ | HL to SM | QoS, NSAPI, TFT | Request to modify context |
| SMREG-PDP-MODIFY-CNF | SM to HL | QoS, NSAPI | Requested context modification complete |
| SMREG-PDP-MODIFY-REJ | SM to HL | Cause, NSAPI | Request to modify context rejected by NW |
| SMREG-PDP-ACTIVATE-SEC-REQ | HL to SM | QoS, NSAPI, TFT, P NSAPI | Request to activate secondary context |
| SMREG-PDP-ACTIVATE-SEC-CNF | SM to HL | QoS, NSAPI | Secondary context activation success |
| SMREG-PDP-ACTIVATE-SEC-REJ | SM to HL | Cause, NSAPI | Secondary context activation failed |

The classic example is a user accessing e-mail and an audio streaming service from the same provider, but requiring different QoS for each.

### 13.8.2 SMREG SAP service primitives

The SMREG SAP shown in Figure 13.1 illustrates the access point used by the higher layers (HLs) to access SM services.

Table 13.20 illustrates the service primitives that are used by the HL entity to access SM protocol services such as the establishment of primary and secondary PDP contexts, the deactivation of PDP contexts and the modification of PDP contexts. The primitive names and descriptions in the table are self explanatory. The parameters of the primitives

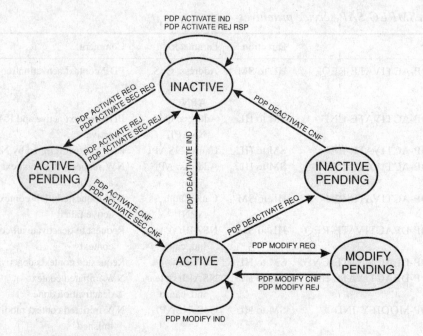

**Figure 13.19** SM entity state diagram.

include QoS, traffic flow templates (TFTs), NSAPI, PDP address, APN, teardown indicators and cause.

In the sections that follow, we will consider further details and examples that define the use of the different parameters in the control of the PDP contexts.

### 13.8.3 SM protocol state machine

The SM protocol is a CM layer entity responsible for the establishment, modification and release of PDP contexts. The SM entity operates a state machine that defines the current state of the SM entity. Figure 13.19 illustrates the UE SM state machine. There is a similar state machine for the network and the interested reader is referred to [32].

Let us now review the different states, the use of the different states and the transitions between the different states in terms of both the primitive exchanges between the SM entity and the HL entity and also the SM PDUs that are passed by the SM entity in the UE to the SM entity in the peer system. To keep Figure 13.19 simple, the primitive names exclude the prefix 'SMREG-' that is used in Table 13.20 and below.

**INACTIVE state**

The INACTIVE state is used to define the state of an SM entity in which there is no PDP context active. The INACTIVE state is the starting state from which the SM entity moves to create the PDP contexts.

There are a number of transitions to and from the INACTIVE state. First we start with the transitions from the INACTIVE state back to the INACTIVE state. There are two primitives associated with this transition. The first is the SMREG-PDP-ACTIVATE-IND, which is the result of the network sending a REQUEST PDP CONTEXT ACTIVATION message indicating that the network wants to establish a PDP context. The SM entity responds by starting the procedure to establish a PDP context. The next transition between an INACTIVE state and an INACTIVE state is triggered by the SMREG-PDP-ACTIVATE-REJ-RSP primitive. This primitive is used to carry the REQUEST PDP CONTEXT ACTIVATION REJECT message that is employed by the HL entity to reject the network requested PDP context activation. This rejection may be due to insufficient resources in the UE.

The next transition from the INACTIVE state is to the ACTIVE PENDING state. The SMREG-PDP-ACTIVATE-REQ primitive is associated with this transition and is used by the HL entity to request the SM entity establish a primary PDP context. The primitive carries the parameters that are required for the ACTIVATE PDP CONTEXT REQUEST message that is sent by the SM entity in the UE to its peer entity in the network. These parameters are the PDP address (e.g. IP address), QoS parameters (defining data rate, delays and error criteria for the PDP context), the NSAPI (an endpoint identifier used in the UE and network to define a specific PDP context) and the APN (used to identify the network access point and GGSN).

For the other transition from INACTIVE to ACTIVE PENDING, the SMREG-PDP-ACTIVATE-SEC-REQ primitive is used by the HL entity to request the establishment of a secondary PDP context. The parameters present are the QoS and NSAPI for the secondary PDP context, the NSAPI for the primary PDP context with which it is associated, the IP address which it will share and the TFT parameters, which are used by the GGSN for traffic flow filtering in the downlink at the GGSN (more on this in Section 13.8.4).

**ACTIVE PENDING state**

The ACTIVE PENDING state is the state that the UE moves to when the PDP context is activated. The transitions from this state are to the ACTIVE state or back to the INACTIVE state. We now consider these transitions.

SMREG-PDP-ACTIVATE-REJ and SMREG-PDP-ACTIVATE-SEC-REJ are the primitives associated with the transition from the ACTIVE PENDING to the INACTIVE state. These two primitives are from the SM entity to the HL entity. They indicate that the SM entity received an ACTIVATE PDP CONTEXT REJECT or ACTIVATE SECONDARY PDP CONTEXT REJECT message from the network, showing that the requested PDP context activation requests were rejected. The primitive contains the parameter 'cause', indicating the reason for the rejection (such as insufficient resources, missing/unknown APN, activation rejected by GGSN) and also the NSAPI that would have been used by the PDP context.

SMREG-PDP-ACTIVATE-CNF and SMREG-PDP-ACTIVATE-SEC-CNF are the primitives associated with the transition from the ACTIVE PENDING to the ACTIVE states. The first primitive applies to a primary PDP context activation procedure, the second to a secondary PDP context activation procedure. These primitives are sent from the SM entity to the HL entity on receipt of an ACTIVATE PDP CONTEXT ACCEPT message for the primary context and an ACTIVATE SECONDARY PDP CONTEXT ACCEPT message for the secondary context. The parameters of the primitive for the primary PDP context include the assigned PDP address, the negotiated QoS and the NSAPI to be used by the context. The parameters of the primitive for the secondary PDP context include the QoS negotiated and the NSAPI assigned for the secondary PDP context.

These transitions indicate the activation of the PDP contexts. A primary PDP context is activated first, and it is only if a secondary PDP context is required (i.e. a packet data connection to the same APN but with a different QoS requirement) that we see one or more secondary PDP contexts being activated.

The QoS that was requested by the HL entity in the initiation procedure is referred to as the requested QoS. The QoS that is allocated by the network is the negotiated QoS and this may be different from the requested QoS.

## ACTIVE state

The ACTIVE state is the state in which a specific PDP context is available for use by the HL entity for the transfer of user data. There can be multiple PDP contexts active at the same time, and consequently, there will be multiple state machines (one per SM entity) controlling these PDP contexts.

There are four possible transitions from the ACTIVE state: one back to the ACTIVE state, one to the INACTIVE state, one to the MODIFY PENDING state and one to the INACTIVE PENDING state. We now examine each of these transitions in turn, as well as the primitives and messages associated with these transitions.

First, the transition from ACTIVE state back to ACTIVE state occurs when the SM entity has received a MODIFY PDP CONTEXT REQUEST message and passed it to the HL entity using the SMREG-PDP-MODIFY-IND. The message is used by the network to modify the PDP context. The SM entity controlling the PDP context is defined by the NSAPI. The QoS information received in the MODIFY PDP CONTEXT REQUEST contains the modified QoS defined by the network. This new QoS information is put into the SMREG-PDP-MODIFY-IND primitive before being passed to the HL entity.

The next transition is from the ACTIVE state to the MODIFY PENDING state. This transition occurs when the HL entity requests the modification of a PDP context. The HL entity uses an SMREG-PDP-MODIFY-REQ primitive to request the SM entity to send a MODIFY PDP CONTEXT REQUEST message to its peer entity in the network. The contents of the primitive and the message include the NSAPI used to identify the PDP context to modify, a TFT if one is associated with the PDP context and the QoS that is being requested.

The transition from the ACTIVE state to the INACTIVE PENDING state is initiated by the HL entity that is requesting the deactivation of a specific PDP context. The deactivation is triggered by the HL entity sending a SMREG-PDP-DEACTIVATE-REQ primitive to the SM entity and the SM entity constructing and sending a DEACTIVATE PDP CONTEXT REQUEST message to its peer entity in the network. The contents of the primitive, and hence the message, include the NSAPI(s) of the PDP context(s) to deactivate, a cause value (such as regular PDP context deactivation or network failure) and an optional teardown indicator that indicates whether any associated secondary PDP contexts (sharing the same PDP address) should also be deactivated.

The final transition is from the ACTIVE state to the INACTIVE state. This transition occurs when the SM entity notifies the HL entity via the SMREG-PDP-DEACTIVATE-IND that a network initiated PDP context deactivation has occurred (i.e. a DEACTIVATE PDP CONTEXT REQUEST message was received from the peer entity in the network). The contents of this message include the NSAPI(s) affected and the teardown indicator, which is used for the same purpose as described previously.

**INACTIVE PENDING state**

The transition from the INACTIVE PENDING state is to the INACTIVE state. This occurs when the SM entity notifies the HL entity of the receipt of a DEACTIVATE PDP CONTEXT ACCEPT message. The notification is made using a SMREG-PDP-DEACTIVATE-CNF primitive. The message and primitive are the final stages of a UE initiated PDP context deactivation procedure. The parameters of the message and hence primitive include the NSAPI(s) that are being deactivated.

**MODIFY PENDING state**

The MODIFY PENDING state is a temporary state that the SM entity passes through as a PDP context is being modified. There is one transition from this state and it is to the ACTIVE state. There are two possible cases for this transition. In the first, the SM entity notifies the HL entity that the PDP modification procedure is completed. The SM entity has received a MODIFY PDP CONTEXT ACCEPT that acknowledged the UE initiated modification procedure. The SM entity notifies the HL entity using the SMREG-PDP-MODIFY-CNF primitive, which contains the negotiated QoS for the PDP context and the identifier for the context (NSAPI) that has been modified.

The second transition from the MODIFY PENDING to the ACTIVE state is from the SM entity notifying the HL entity that the UE initiated PDP context modification is being rejected. The SM entity has received a MODIFY PDP CONTEXT REJECT message and has indicated to the HL entity using the SMREG-PDP-MODIFY-REJ that the modification is rejected. The parameters of the rejection message include the

NSAPI identifying the PDP context and the cause value that defines the reason for the rejection (such as insufficient resources, syntactic error in TFT operation, semantic error in packet filters).

### 13.8.4 SM protocol

In this section, we examine some examples of the SM protocol that we have partially discussed through the consideration of the SM state machine. We consider some of the operational scenarios for the PDP contexts. First we examine from a system perspective the message flows between the UE and the different elements of the network for the establishment of a PDP context and a secondary PDP context. Following on from that we look in slightly greater detail at the interlayer procedures within the UE associated with the establishment of a PDP context.

**QoS definition**

The establishment of a PDP context to be used for packet data transfer consists of a request from the UE and a response from the network. The request from UE includes the QoS parameters that define the service characteristics of the PDP context. The UE defines these QoS parameters and sends them via the request to the network (referred to as requested QoS), the network evaluates what network resources are available and what the UE has subscribed to and responds with the negotiated QoS.

Table 13.21 defines the parameters that are used to characterise the requested and the negotiated QoS. These QoS parameters define aspects such as peak and average data rate, delay characteristics and error characteristics for the PDP context.

**Primary PDP context establishment**

Figure 13.20 illustrates the procedure that the UE uses to activate a PDP context. We are assuming that the UE is to establish a PDP context for Internet access and that a dynamic IP address will be requested from the network. The procedure, in this example, commences at the UE with the UE sending an ACTIVATE PDP CONTEXT REQUEST message to the SGSN in the core network. The SGSN through negotiation with the UTRAN about QoS requirements requests the establishment of an RAB that is configured to transport the PDP context data across the radio interface to the SGSN.

Next, the SGSN identifies which GGSN should be used as the CN end-point for the PDP context. To identify which GGSN should be used, the APN is used. The SGSN sends the CREATE PDP CONTEXT REQUEST message to the GGSN. This procedure requests the establishment of a PDP context in the GGSN and the creation of a CN bearer between the SGSN and the GGSN to transport data between them. Part of this procedure includes QoS negotiation, and the assignment of an IP address by the GGSN, assuming that dynamic IP addresses are being supported. The

## 13.8 SM protocol and PDP contexts

**Table 13.21.** *QoS definition parameters*

| Quantity | Description | Range |
|---|---|---|
| Traffic class | Defines four types of traffic class that characterise the service delay characteristic:<br>• conversational: low delay and constant delay – typical usage is speech conversation;<br>• streaming: constant delay – typical usage is streaming audio and video;<br>• interactive: high priority – typical usage is interactive games;<br>• background: best efforts delivery – typical usage is file download. | See description |
| Delivery order | Defines whether the SDUs are required to be delivered in order. | Yes/No |
| Delivery of erroneous SDU | Defines whether known erroneous SDUs are to be delivered. | Yes/No |
| Max SDU size | Defines the maximum SDU size that can be transmitted in either the uplink or downlink direction. | 10–1520 octets |
| Max bit rate | Defines the maximum bit rate available for a specific PDP context. There are two fields: one that defines the uplink requirement and one that defines the downlink requirement. | 1–8640 kb/s |
| Residual BER | Defines the BER when erroneous SDUs are delivered to higher layers, and the undetected BER when erroneous SDUs are not delivered to higher layers. | $6 \times 10^{-8} - 5 \times 10^{-2}$ |
| SDU error rate | Defines the fraction of SDUs lost or detected as in error. | $1 \times 10^{-6} - 1 \times 10^{-1}$ |
| Transfer delay | Is ignored for the interactive and background classes. It defines the 95th percentile of the delay distribution for the service. | 10–4100 ms |
| Traffic handling priority | Defines the relative importance of SDUs from the defined bearer compared with other bearers. Used only for the interactive class. | Priority 1 (highest) to 3 (lowest) |
| Guaranteed bit rate | Defines the guaranteed bit rate for the service. | 1–8640 kb/s |

CREATE PDP CONTEXT RESPONSE from the GGSN to the SGSN includes the IP address allocated by the GGSN or an entity attached to the GGSN.

The SGSN then delivers the ACTIVATE PDP CONTEXT ACCEPT message to the UE. This message includes the allocated IP address and the QoS negotiated by the SGSN, the UTRAN and the GGSN based on network resource criteria as well as service subscription criteria.

The SM entity in the UE indicates to the HLs that the context has been accepted and the QoS negotiated. If the QoS is acceptable, the context can be used by the UE; if the QoS is not acceptable, the UE needs to deactivate the PDP context.

**Table 13.22.** *ACTIVATE PDP CONTEXT REQUEST message contents*

| Element | Comment |
| --- | --- |
| Protocol discriminator | Defines the message as being from the SM entity. |
| TI | Used to identify SM entity in CM sublayer. |
| Activate PDP context request message type | Identifies the message type. |
| Requested NSAPI | A number between 5 and 15 allocated by the SM entity to identify the PDP context. |
| Requested LLC SAPI | Used in GSM/GPRS to identify logical link. |
| Requested QoS | QoS requested by HL – see Table 13.21. |
| Requested PDP address | Defines requested PDP address type (PPP, IPV4, IPV6) and address if static address. |
| APN | Defined as URL routable via DNS maximum length 102 characters. |
| PDP options | Protocol options such as PPP protocol options. |

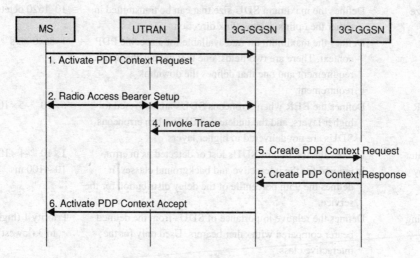

**Figure 13.20** PDP context activation procedure.

## ACTIVATE PDP CONTEXT REQUEST

As we have seen, this message is sent from the UE through the UTRAN to the SGSN. The contents of the message are presented in Table 13.22.

## ACTIVATE PDP CONTEXT ACCEPT

If the ACTIVATE PDP CONTEXT REQUEST is successful, the SGSN replies with the ACTIVATE PDP CONTEXT ACCEPT as shown in Figure 13.20. The contents of the ACTIVATE PDP CONTEXT ACCEPT message are considered in Table 13.23.

## 13.8 SM protocol and PDP contexts

**Table 13.23.** *ACTIVATE PDP CONTEXT ACCEPT message contents*

| Element | Comment |
| --- | --- |
| Protocol discriminator | Defines the message as being from the SM entity. |
| TI | Used to identify SM entity in CM sublayer. |
| Activate PDP context accept message type | Identifies the message type. |
| Requested LLC SAPI | Used in GSM/GPRS to identify logical link, not needed in UMTS. |
| Negotiated QoS | QoS offered by network. |
| Radio priority | Used in GSM/GPRS, not needed in UMTS. |
| PDP Address | Defines assigned PDP address type (PPP, IPV4, IPV6) and address if dynamic address. |
| PDP options | Protocol options for the PDP context. |
| Packet flow identifier | Identifier for packet flows associated with the PDP context. |

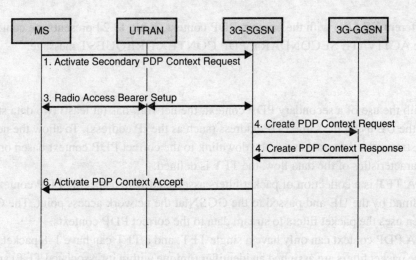

**Figure 13.21** Secondary PDP context activation procedure.

### Secondary PDP context establishment

Figure 13.21 illustrates the procedure required to activate a secondary PDP context. The secondary PDP context activation procedure is used to activate a new PDP context (e.g. with different QoS) whilst reusing the same PDP address and APN. The procedure allocates a new TI and NSAPI for the secondary PDP context. For multiple PDP contexts using the same PDP address and APN, a TFT is required for all except one PDP context.

The ACTIVATE SECONDARY PDP CONTEXT REQUEST contains similar parameters to the ACTIVATE PDP CONTEXT REQUEST message, but also includes linked TI which is used to associate the secondary PDP context (which has a

**Table 13.24.** *ACTIVATE SECONDARY PDP CONTEXT REQUEST message contents*

| Element | Comment |
| --- | --- |
| Protocol discriminator | Defines the message as being from the SM entity. |
| TI | Used to identify SM entity in CM sublayer. |
| Activate secondary PDP context request message type | Identifies the message type. |
| Requested NSAPI | A number between 5 and 15 allocated by the SM entity to identify the PDP context – will be different to primary. |
| Requested LLC SAPI | Used in GSM/GPRS to identify logical link. |
| Requested QoS | QoS requested by HL – see Table 13.21. |
| Linked TI | TI of primary PDP context associated with this secondary. |
| TFT | See below. |

different NSAPI) with the primary PDP context. Table 13.24 presents the contents of the ACTIVATE SECONDARY PDP CONTEXT REQUEST message.

## TFT

With the use of a secondary PDP context, the network has (at least) two data streams to the UE using the same PDP address (such as the IP address). To allow the network to stream the data packets on the downlink to the correct PDP context based on some characteristics of the data flow, the TFT is defined.

A TFT is a collection of packet filters associated with a specific PDP context. It is defined by the UE and passed to the GGSN at the network access point. The GGSN then uses the packet filters to stream data to the correct PDP contexts.

A PDP context can only have a single TFT, and a TFT can have 1–8 packet filters. The packet filters are assigned an identifier (unique within its associated TFT) and also evaluation precedence. The purpose of the TFT and packet filters is to classify and stream incoming PDP PDUs to the correct PDP context based on elements that would typically be found in the headers of IPv4 or IPv6 datagrams or TCP/UDP datagrams such as source IP address, source and destination port ranges, type of service (TOS) field etc., and flow identifiers. Table 13.25 outlines the fields used to define the TFT when the PDP context is being created or modified.

The GGSN receives data from external networks with an IP address that could correspond to a number of PDP contexts (at least one primary and a number of secondary contexts could be established). For each incoming data packet, the GGSN evaluates the packet headers based on information specified in the packet filters within the TFT for each of the PDP contexts. The GGSN then forwards the incoming packet to the PDP context whose TFT matches the incoming data packet. Through this procedure,

## 13.8 SM protocol and PDP contexts

**Table 13.25.** *TFT IE contents to configure TFT packet filters*

| Element | Comment |
| --- | --- |
| TFT IEI | Defines the information element (IE) identifier for the TFT IE. |
| Length of TFT IE | Defines the length of the TFT IE. |
| TFT operation code | Defines the TFT operation: create new TFT; delete existing TFT; add packet filter to existing TFT; replace packet filter in existing TFT; delete packet filter from existing TFT. |
| Number of packet filters | Defines the number of packet filters being added, deleted or replaced. Must be 8 or less and equal 0 if delete existing TFT is selected as the operation code. |
| Packet filter list | A list of packet filters, packet filter identifier, precedence and packet filter contents. Contents: IPv4 or IPv6 address; protocol/next header identifier; destination port (single or range); source port (single or range); security parameter; ToS; flow label type. |

the network can ensure that the QoS required for the data packet is matched to the QoS of the PDP context established for that packet within the UMTS network.

### 13.8.5 SM service primitives

Previously, whilst discussing the SM state machine, we considered the service primitives between the SM entity and the HL entity (SMREG-SAP). In this section we want to consider some additional service primitives that are associated with the SM protocol layer and which we will see in operation in a more detailed example of PDP context activation.

As we saw in Figure 13.1 there are a number of SAPs surrounding the use of the SM protocol. These SAPs include: the RABMSM-SAP between the RAB manager and the SM entity; the RABMAS-SAP between the RAB manager and the AS; the GMMAS-SAP between the GMM entity and the AS; the GMMSM-SAP between the GMM entity and the SM entity. All of these SAPs have a set of service primitives associated with them. In this section we review the use of the SAPs and the service primitives, and in the following section we will conduct a more detailed examination of the activate PDP context request procedure.

**RABMSM-SAP primitives**

Table 13.26 lists the primitives that are defined between the SM entity and the RABM within the NAS. The RABM is responsible for the interactions between the SM entity in the NAS and the RAB creation functions in the AS via the RABMAS-SAP. The RABM provides the SM entity with notification on the establishment of RABs and modification to RABs.

**Table 13.26.** *List of service primitives at the RABMSM-SAP*

| Primitive | Direction | Parameters | Comment |
|---|---|---|---|
| RABMSM-ACTIVATE-IND | SM to RABM | NSAPI, QoS | Part of a PDP context activation procedure notifying RABM of NSAPI and QoS. |
| RABMSM-ACTIVATE-RSP | RABM to SM | NSAPI | Response from RABM indicating that the NSAPI is in use and the associated RAB established. |
| RABMSM-DEACTIVATE-IND | SM to RABM | NSAPI | Part of a PDP context deactivation indicating that an NSAPI is deallocated and cannot be used. |
| RABMSM-DEACTIVATE-RSP | RABM to SM | NSAPI | Sent to confirm that the NSAPI is no longer in use and that the associated RAB is released. |
| RABMSM-DEACTIVATE-REQ | RABM to SM | NSAPI | Only used for RT bearers to indicate to SM that the RAB is released. |
| RABMSM-MODIFY-IND | SM to RABM | NSAPI, QoS | Part of PDP context modification procedure indicating a change to QoS for a specific NSAPI. |
| RABMSM-MODIFY-RSP | RABM to SM |  | Indicates that modification to NSAPI is complete and RAB changes have been completed. |
| RABMSM-STATUS-Req | RABM to SM | Cause | Indicates a lower layer failure (defined by cause) and that the RABM cannot continue. |

**Table 13.27.** *List of service primitives at the RABMAS-SAP*

| Primitive | Direction | Parameters | Comment |
|---|---|---|---|
| RABMAS-RAB-ESTABLISH-IND | AS to RABM | RAB-Id list | AS notifying the RABM that the RABs defined by list are in the process of being created. |
| RABMAS-RAB-ESTABLISH-RES | RABM to AS |  | RABM informing AS that NSAPI (in RAB-Id) is in the process of being activated by SM. |
| RABMAS-RAB-ESTABLISH-REJ | RABM to AS |  | RABM notifying AS that requested NSAPIs (in RAB-Id) are not being activated by SM. The RAB setup is to be rejected. |
| RABMAS-RAB-RELEASE-IND | AS to RABM | RAB-Id list | Indication from the AS that the RABs associated with the NSAPI have been released. |
| RABMAS-RAB-RELEASE-RES | RABM to AS |  | Response from RABM indicating that the RAB-Id has been released. |
| RABMAS-STATUS-IND | AS to RABM | Cause | Method of transferring failure information to the RABM. |

## RABMAS-SAP primitives

Table 13.27 presents the primitives that are defined between the RABM in the NAS and the AS. The primitives are used to exchange information on the status of RABs as they are established and released.

Between the two entities there is a one-to-one relationship between the NSAPI and the RAB-Id. In fact, the RAB-Id used by the AS is the same binary representation of

**Table 13.28.** *List of service primitives at the GMMAS-SAP*

| Primitive | Direction | Parameters | Comment |
|---|---|---|---|
| GMMAS-SECURITY-IND | AS to GMM | | Indication ciphering and integrity protection to be started – indicates the service request is complete. |
| GMMAS-SECURITY-RES | GMM to AS | CK, IK | Ciphering and integrity keys assigned to AS. |
| GMMAS-ESTABLISH-REQ | GMM to AS | L3 PDU, cause, priority, CN, MS-Id, LAI/RAI | Request to establish NAS signalling connection and transfer L3 PDU. |
| GMMAS-ESTABLISH-CNF | AS to GMM | | NAS signalling connection ready. |
| GMMAS-ESTABLISH-REJ | AS to GMM | | Request for signalling connection rejected by network. |
| GMMAS-RELEASE-REQ | GMM to AS | CN | Request to release signalling connection. |
| GMMAS-RELEASE-IND | AS to GMM | Cause | Indication that signalling connection is released. |
| GMMAS-DATA-REQ | GMM to AS | L3 PDU, priority, CN | Request to transfer NAS data. |
| GMMAS-DATA-IND | AS to GMM | L3 PDU | Indication of receipt of NAS data. |
| GMMAS-PAGE-IND | AS to GMM | MS-Id type, paging cause | Indication of receipt of paging message. |
| GMMAS-STATUS-IND | AS to GMM | Cause | Transfer of failure information from AS. |

the NSAPI IE used in the establishment of the PDP context that will be carried using the RAB.

### GMMSM-SAP primitives

Table 13.28 presents the service primitives that are defined between the GMM entity and the AS. These primitives are for general use by the GMM entity and are not specifically for the SM entity, although we will see some specific examples of their use. The table identifies the different primitives and their use. In general, the primitives are there to establish GMM contexts, to transfer CM entity data and to control specific functions such as security and paging.

### 13.8.6 PDP context activation

Figure 13.22 illustrates the stages of a PDP context activation looking at the interlayer messages. In this example, we assume that a GMM context exists and that the UE is in the PMM CONNECTED state.

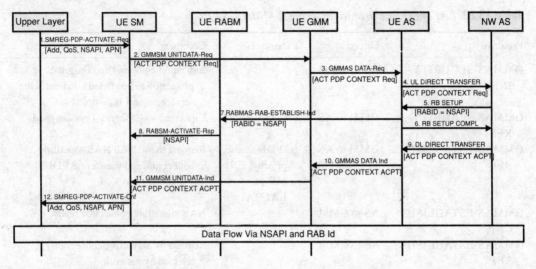

**Figure 13.22** PDP context activation in detail.

1. HL entity requests SM entity for a PDP context defining QoS, APN, NSAPI and address (if static address allocated).
2. SM entity requests GMM to send ACTIVATE PDP CONTEXT REQ.
3. GMM entity requests AS to send ACTIVATE PDP CONTEXT REQ.
4. AS uses UL DIRECT TRANSFER to send NAS message.
5–6. RB (and hence RAB) setup and RAB-Id assigned.
7. AS indicates to RABM that RAB is active and NSAPI available for use. NSAPI is returned as RAB-Id in RAB setup messages.
8. RABM entity indicates to SM that RAB is active and NSAPI available for use.
9. The network (NW) allocates PDP context and notifies AS with ACTIVATE PDP CONTEXT ACCEPT. QoS can be different, address assigned if required.
10. AS sends ACTIVATE PDP CONTEXT ACCEPT to GMM entity.
11. GMM sends ACTIVATE PDP CONTEXT ACCEPT to SM entity.
12. SM entity indicates to HL entity that the PDP context is active indicating address, QoS, NSAPI and APN using SMREG-PDP-ACTIVATE-Cnf.

## 13.9 SMS protocol

### 13.9.1 CM sublayer SMS protocol entities

Figure 13.23 illustrates the protocol architecture for the SMS entity in the CM layer. The architecture consists of the SMS MM entity (SMSMM), the short message control – GPRS protocol (SMC-GP) entity and the short message control – circuit switched (SMC-CS) entity.

## 13.9 SMS protocol

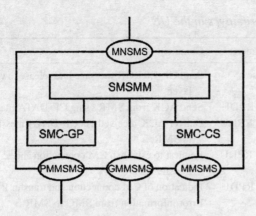

**Figure 13.23** SMS protocol entities.

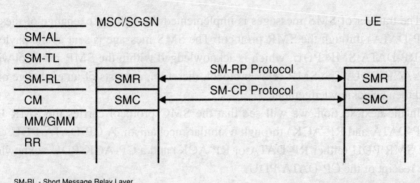

SM-RL - Short Message Relay Layer
SMR - Short Message Relay (entity)
SMC - Short Message Control (entity)
SM-RP - Short Message Relay Protocol
SM-CP - Short Message Control Protocol

**Figure 13.24** SMS protocol architecture.

The SMSMM is responsible for deciding how to deliver the SMS. It can use either CS or PS domain resources depending upon what is currently available. The SMSMM can use the GMMSMS-SAP to enquire as to the current registration state for the GMM entity.

The SMS messages can be sent or received via the PS domain through the SMC-GP entity or via the CS domain through the SMC-CS entity. It is the choice of the SMSMM as to whether the SMC-GP or the SMC-CS is used for MO SMS messages.

### 13.9.2 SMS protocol architecture

Figure 13.24 illustrates the SMS protocol architecture. The protocol architecture consists of an SMS relay layer controlled by the SMR protocol, which is responsible for the relay of SMS messages across the radio interface. To facilitate this, the SMR entity uses the services of the SMC entity for the transfer of the SMS messages.

**Table 13.29.** *SM control protocol service primitives in the UE*

| Primitive | Direction | Parameter | Comment |
|---|---|---|---|
| MNSMS-ABORT-Req | SMR to SMC | Cause | Send CP-ERROR message and release lower layer. |
| MNSMS-DATA-Req | SMR to SMC | MT RPDU | Send ACK from SMR using CP-DATA message. |
| MNSMS-DATA-Ind | SMC to SMR | MO RPDU | Deliver ACK received via CP-DATA message to SMR. |
| MNSMS-EST-Req | SMR to SMC | MO RPDU | Request to establish CM connection and transfer RPDU. |
| MNSMS-EST-Ind | SMC to SMR | MT RPDU | Indication of CM connection and transfer RPDU. |
| MNSMS-ERROR-Ind | SMC to SMR | Cause | Error information from SMC to SMR. |
| MNSMS-REL-Req | SMR to SMC | Cause | Request to release CM connection. |

The transfer of SMS messages is implemented through the exchange of messages (RP-DATA) through the SMR protocol. The SMS message is sent as the payload of an RP-DATA SMR PDU, which is acknowledged within the SMR protocol with an RP-ACK PDU. The SMR relay protocol, therefore, consists of an exchange of data and then an acknowledgement.

In the next section, we will see that the SMC protocol carries the SMR PDUs (RP-DATA and RP-ACK) through a similar mechanism. A CP-DATA PDU carries the SMR PDU (either RP-DATA or RP-ACK) and a CP-ACK PDU acknowledges the receipt of the CP-DATA PDU.

### 13.9.3 SM-CP primitives

Table 13.29 presents the primitives defined for the SM control protocol (CP) entity in the UE. The primitives are used by the SMR entity to request the establishment of a CP connection, to send and receive data via the CM connection, to report errors in SMC to SMR, to release the CM connection and to abort specific on-going transactions. For MO SMS a CM connection is required and so the MNSMS-EST-REQ is used to establish the connection and transfer the RP-DATA PDU. The response is an RP-ACK which is received using an MNSMS-DATA-IND primitive. For MT SMS the RP-DATA PDU are received via an MNSMS-EST-IND primitive that transfers the RP DATA PDU and establishes the CM connection. The RP-ACK in acknowledgement is sent using the MNSMS-DATA-REQ primitive using the established CM connection.

### 13.9.4 SM-RL primitives

Table 13.30 presents the primitives defined for the SMR entity in the UE. The primitives are used by the short message transfer layer (SM-TL) entity to request the transfer of SM-TL PDUs, and by the SMR to deliver the received SM-TL PDUs.

## 13.9 SMS protocol

**Table 13.30.** *SM-RL service primitives in the UE*

| Primitive | Direction | Parameter | Comment |
|---|---|---|---|
| SM-RL-DATA-Req | SM-TL to SMR | MO SMS_TPDU | Request from SM-TL to relay SMS-TPDU to peer. |
| SM-RL-DATA-Ind | SMR to SM-TL | MT SMS_TPDU | Delivery to SM-TL of Rx SMS-TPDU from peer. |
| SM-RL-MEMORY AVAILABLE-Req | SM-TL to SMR | | Request for information on memory availability. |
| SM-RL-REPORT-Req | SM-TL to SMR | | Used by SMR to relay RP-ACK and RP-ERROR messages. |
| SM-RL-REPORT-Ind | SMR to SM-TL | | Delivery of RP-ACK or RP-ERROR messages received. |

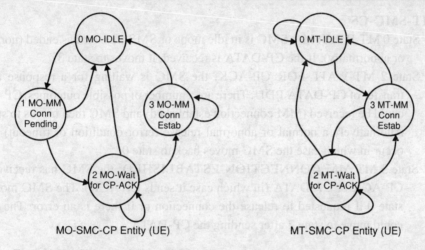

**Figure 13.25** SM control protocol CS entity state diagram.

SM-RL-DATA-Req and SM-RL-DATA-Ind are used to respectively send and receive the SM-TL PDUs. The SM-RL-MEMORY AVAILABLE-Req primitive is used to transfer to the network the information that the UE has enough memory to receive one or more SMS. The SM-RL-REPORT primitives are used to exchange acknowledgements and error information between the SM-TL and the lower layers.

### 13.9.5 SMC-CS states

Figure 13.25 presents the SMC-CS state machines for the MO and MT SMS cases.

**MO-SMC-CS**

State 0 MO-IDLE: MO-SMC is in idle mode or SMS transfer has ended (normally or abnormally).

State 1 MO-MM-connection pending: the SMC has requested the establishment of an MM connection. If the connection is confirmed it can transfer CP DATA

PDU and move to state 2, or if the connection is released normally or abnormally it can return to state 0.

State 2 MO WAIT FOR CP-ACK: the SMC is waiting for a response to the transfer of CP-DATA PDU. There are a number of possible outcomes. CP-ACK could be received (MM connection established) and SMC then moves to state 3. Alternatively a normal or abnormal release (error condition or timeout) could occur in which case the SMC moves back to state 0.

State 3 MO MM connection established: the SMC has received the CP-ACK or CP-DATA (in which case it sends CP-ACK). The SMC moves to state 0 if requested to release the connection or if there is an error. The SMC could stay in state 3 after sending the CP-ACK, waiting for further commands.

**MT-SMC-CS**

State 0 MT IDLE: MO SMC is in idle mode or SMS transfer has ended (normally or abnormally). If the CP-DATA is received, it moves to state 3.

State 2 MT-WAIT FOR CP-ACK: the SMC is waiting for a response to the transfer of CP-DATA PDU. There are a number of possible outcomes. CP-ACK could be received (MM connection established) and SMC then moves to state 3. Alternatively a normal or abnormal release (error condition or timeout) could occur in which case the SMC moves back to state 0.

State 3 MT MM CONNECTION ESTABLISHED: the SMC has received the CP-ACK or CP-DATA (in which case it sends CP ACK). The SMC moves to state 0 if requested to release the connection or if there is an error. The SMC could move to state 2 after sending the CP-DATA.

### 13.9.6 SMC-GP states

Figure 13.26 presents the SMC-GP state machines for the MO and MT SMS cases.

**MO-SMC-GP**

State 0 MO IDLE: MO SMC is in idle mode or SMS transfer has ended (normally or abnormally). SMC moves to state 1 while waiting for a PS signalling connection.

State 1 MO GMM CONNECTION PENDING: the SMC has requested a PS signalling connection. If the SMC sends a CP-DATA PDU is moves to state 2. If an error occurs it will move back to state 0.

State 2 MO WAIT FOR CP-ACK: the SMC has initiated the transfer of CP-DATA PDU. If the SMC receives a CP-ACK it moves to state 3. If an error occurs, or an abort is requested, the SMC moves back to state 1. A timer can cause the resending of CP-DATA 'max' times; in this case it returns to state 2 until a CP-ACK is received.

## 13.9 SMS protocol

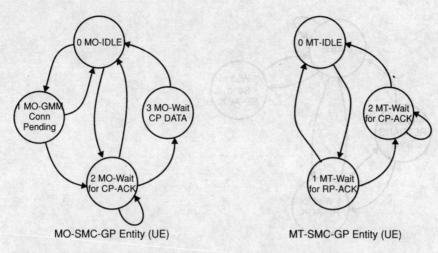

**Figure 13.26** SM control protocol – PS entity state diagram.

State 3 MO WAIT FOR CP-DATA: the SMC has received the CP-ACK PDU. The SMC is waiting for the CP-DATA and will send a CP-ACK in response, in which case it moves to state 0. If an error condition, an abort or a request to release the CM connection is received, the SMC moves to state 0.

**MT-SMC-GP**

State 0 MT IDLE: MO SMC is in idle mode or SMS transfer has ended (normally or abnormally). The SMC on receipt of CP-DATA sends CP-ACK, and moves to state 1.

State 1 MT WAIT FOR RP-ACK: the SMC has received CP-DATA and sent a CP-ACK in response. The SMC sends CP-DATA and moves to state 2, or if an error occurs or an abort is received it moves back to state 0.

State 2 MT WAIT FOR CP-ACK: the SMC has initiated the transfer of CP-DATA PDU. If it receives a CP-ACK it moves to state 0. A timer may cause the retransmission of CP-DATA ('max' times) before returning to state 2. An abort, request to release the CM connection or an error may cause the SMC to move to state 0.

### 13.9.7 SMR states

Figure 13.27 presents the SMR state machine for both the MO and MT SMS cases.

State 0 IDLE: SMR is in idle mode, or SMS transfer has ended (normally or abnormally). The SMR moves to state 1 after sending an RP DATA or RP SMMA PDU and awaiting an RP ACK. It moves to state 3 after receiving an RP DATA PDU.

State 1 WAIT FOR RP-ACK: this is used for MO SMS or notification transfer. RP-DATA or RP-SMMA is sent to SMC. After receiving an RP-ACK PDU

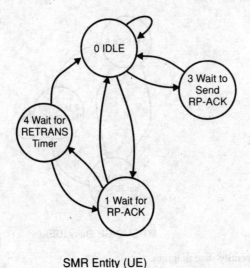

SMR Entity (UE)

**Figure 13.27** SM relay protocol entity state diagram.

and passing it to SM-TL the SM relay protocol entity state machine moves to state 0. If an error occurred or a timer has expired the SMR may request the release of the connection and move to state 0. A temporary network failure may cause the SMR to wait for the expiry of the retransmission timer in which case it moves to state 4.

State 3 WAIT TO SEND RP-ACK: this is used for MT SMS after the SMR receives RP-DATA and passes it to SM-TL. When the SMR receives the RP-ACK from the SM-TL it transfers it to its peer, requests connection release and moves to state 0. An error event in this state causes a move to state 0 and the release of the connection.

State 4 WAIT FOR RETRANS timer: SMR is waiting to retransmit RP-SMMA PDU. When the timer has expired the RP-SMMA PDU is sent and the SMR moves to state 1. If an abort is received from SM-RL it moves back to state 0.

### 13.9.8 SMC PDUs

Table 13.31 presents the SMC PDUs. The SMC is an acknowledged protocol. The HL messages (CP user data) that are sent across the link are acknowledged by the peer entity using the CP-ACK. The CP error PDU is used to pass error information relating the SMS service or the underlying network.

### 13.9.9 SMR PDUs

Table 13.32 presents the SMR PDUs. The SMR is an acknowledged protocol. The high layer messages (RP user data) that are sent across the link are acknowledged by the peer entity using the RP-ACK. The RP-DATA, RP-ACK, RP-SMMA and

## Table 13.31. SMC PDUs

| PDU | Comment |
|---|---|
| CP-DATA | Used to relay CP user data. Protocol disc, TI, message type, CP-User data. |
| CP-ACK | Acknowledges receipt of CP-DATA. Protocol disc, TI, message type. |
| CP-ERROR | Error message with cause. Protocol disc, TI, message type, CP-Cause. |

## Table 13.32. SMR PDUs

| PDU | Comment |
|---|---|
| RP-DATA | Used to relay TPDUs. Originator address (1 octet), destination address (12 octets), RP user data ($\leq 233$ octets). |
| RP-SMMA | Notification from UE to network. UE has memory for one or more messages. |
| RP-ACK | Acknowledgement of RP-DATA or RP-SMMA. |
| RP-ERROR | Error message with cause. |

RP-ERROR messages are all carried using the CP-DATA PDU by the SMC. The RP-ACK also acknowledges the RP-SMMA PDU, which is used by the UE to notify the network of its memory availability for one or more messages. RP-ERROR is used to report SMR errors.

### 13.9.10 MO SMS via SMC-CS

Figure 13.28 illustrates message flows for the MO SMS using the SMC-CS. The message uses the MM layer for the transfer of the SMS. In the diagram CM represents the SMC entity.

1. The SM-TL creates the message SMS SUBMIT and passes to the short message relay layer (SM-RL) for delivery using the SM-RL-DATA-Req primitive.
2. The SM-RL receives the SM-TL message and sends it to the CM sublayer using the MNSMS-Est-Req. This requests the establishment of a CM connection. The SM-RL PDU is encapsulated in the RP-DATA parameter. The SMR moves from the IDLE state (state 0) to 'Wait for RP-ACK' (state 1).
3–4. The SMC sublayer requests the establishment of an MM connection. The SMC moves from MO IDLE (state 0) to 'MO-MM-CONNECTION PENDING' (state 1).
5. On receipt of the MMSMS-EST-Cnf indicating the MM connection is ready, the CM layer sends the RP-DATA message in the CP-DATA PDU across the

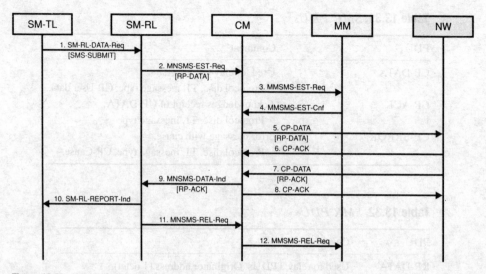

**Figure 13.28** MO SMS CS domain message flows.

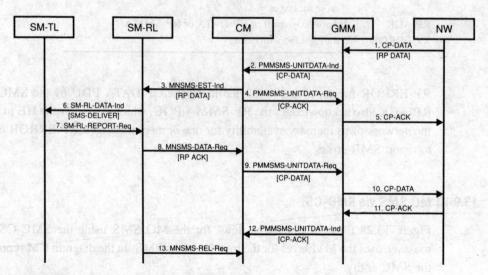

**Figure 13.29** MT SMS PS domain message flows.

radio interface (via UPLINK DIRECT TRANSFER). The SMC moves to 'MO WAIT FOR CP-ACK' (state 2).

6–9. SMC receives the CP-ACK acknowledging the CP-DATA message and the SMC moves to 'MO-MM CONNECTION ESTABLISHED' (state 3). Next the CM receives the CP-DATA which contains the RP-ACK. The SMC sends the RP-ACK to the SM-RL using MNSMS-DATA-Ind, sends the CP-ACK to the network and moves back to 'MO MM CONNECTION ESTABLISHED' (state 3).

10–11. The SMR receives the RP ACK, sends the SM-RL-REPORT-Ind to SM-TL request the release of the CM connection (by sending MNSMS-REL-Req) and moves to IDLE (state 0).

12. On receipt of the MNSMS-REL-Req, the SMC requests the release of the MM connection (MMSMS-REL-Req) and moves to 'MO IDLE' (state 0).

### 13.9.11 MT SMS via SMC-GP

Figure 13.29 illustrates message flows for the MT SMS using the SMC-GP. The message uses the GMM layer for the transfer of the SMS. In the diagram, CM represents the SMC entity. There is an assumption that a GMM context exists for the SMS transfers.

1. In the network, the SM-TL creates the message SMS DELIVER, which is sent as an RP-DATA PDU within a CP-DATA PDU. The message is transferred to the GMM using DOWNLINK DIRECT TRANSFER.

2–4. The SMC (in 'IDLE' (state 0)) receives the CP-DATA PDU via the PMMSMS-UNITDATA-Ind primitive from the GMM. It transfers it to the SM-RL with the MNSMS-DATA-Ind primitive, sends the CP-ACK acknowledging the CP-DATA via the GMM using the PMMSMS-UNITDATA-Req and moves to 'WAIT FOR RP-ACK' (state 1).

5. The GMM sends the CP-ACK to the network using the UPLINK DIRECT TRANSFER.

6. SMR in 'IDLE' (state 0) receives the MNSMS-EST-Ind with the RP-DATA PDU, extracts the TL layer PDU (SMS DELIVER) and transfers it to the SM-TL using the SM-RL-DATA-Ind primitive. The SMR moves to 'WAIT TO SEND RP-ACK' (state 3).

7–8, 13. The SM-TL acknowledges the RP-DATA with an RP ACK sent in the SM-RL-REPORT-Req to SMR. The RP-ACK is sent to SMC using the MNSMS-DATA-Req. Finally, the SMR requests the release of the CM connection sending the MNSMS-REL-Req primitive and then moves to 'IDLE' (state 0).

9–10. The SMC receives the MNSMS-DATA-Req carrying the RP-ACK and passes it to the GMM using the PMMSMS-UNITDATA-Req with RP-ACK carried across the radio interface in the CP-DATA PDU using the UPLINK DIRECT TRANSFER. The SMC moves to 'WAIT FOR CP-ACK' (state 2).

11–12. On receipt of the CP-ACK via DOWNLINK DIRECT TRANSFER and PMMSMS-UNITDATA-Ind, the SMC moves to 'IDLE' (state 0).

# 14 Idle mode functions

In this chapter we consider the actions of the UE when it does not have an active radio connection (RRC idle mode state) and also when it is in certain RRC connected mode states (CELL_PCH, URA_PCH and CELL_FACH). For historical reasons, these actions are generally referred to as the idle mode procedures; here we follow the same terminology but add comments where required for the applications of the procedures to the other states.

In the idle mode there are two closely related topics that we need to consider together and which interrelate with one another. These topics are the idle mode substate machine and the idle mode procedures. The idle mode substate machine defines the different states that the UE may be in, the potential transitions between these states and the reasons for being in the states and making the transitions between states. The idle mode procedures define the set of actions that are undertaken by the UE, the outcome of which drives the transitions between the different states in the state machine.

Now, the idle mode procedures and the idle mode substates span the NAS and the AS. Aspects of the procedures and substates apply to both the AS and NAS and so we will consider these in turn. The NAS procedures and substates relate specifically to the activities of the MM sublayer, which is considered in Chapter 13. These NAS issues are further separated into the CS domain activities (MM entity) and the PS domain activities (GMM entity).

To simplify the considerations, in this chapter we focus on the MM entity and state machines relevant to the MM entity. There are similar deliberations that apply to the GMM entity. The GMM entity was defined after the MM entity and consequently many of the MM procedures and states have been reused partly to have commonality but also to help support combined procedures that involve both the MM and, the GMM entities. The GMM equivalent substates are considered in Chapter 13 and the details of the procedures as they apply specifically to the GMM entity can be found in [32, 37].

In this chapter, we start by considering the twin issues of the idle mode substates and the idle mode procedures, before moving on to the AS issues. When considering the AS issues, we will see that many of the procedures that apply to a UE in the idle mode also apply in the connected mode, in particular whilst in the CELL_FACH, CELL_PCH

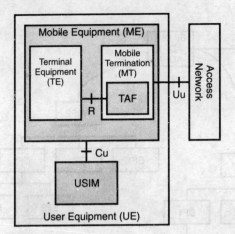

**Figure 14.1** Basic architecture for user equipment.

and URA_PCH states. We will, therefore, also consider the impact of the procedures considered here on the UE whilst in the connected mode.

To start, let us examine the USIM, as it is, to a large extent, the starting point for what happens in the idle mode procedures. The USIM is the semi-permanent store of information that pertains to the user's subscription, as well as information that controls the access procedures to the network. We will explore the basic structure and operation of the USIM, which we will need when we go on to consider the NAS and AS idle mode procedures.

## 14.1 USIM architecture and operation

### 14.1.1 UE architecture

Figure 14.1 illustrates the basic architecture of the UE. The architecture of the UE is slightly different to that of the GSM MS, in that there are fewer interfaces and MT types in UMTS than in GSM. The UE comprises the mobile equipment (ME) and the USIM. They are connected via a Cu interface. The ME, on the other hand, comprises terminal equipment (TE) and the mobile termination (MT), and they are linked via an R interface. The MT terminates the radio interface (Uu interface for UMTS) and provides bearer services for the TE via the terminal adaptation function (TAF).

### 14.1.2 UICC file structure

The USIM is an application that operates on the universal integrated circuit card (UICC) [38]. Figure 14.2 illustrates the logical architecture of the UICC, which comprises a collection of data referred to as 'records', which are stored in what is called an

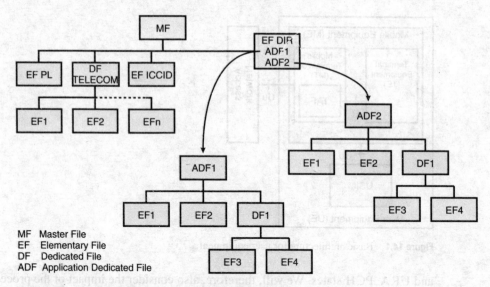

**Figure 14.2** Logical architecture of UICC including EFs.

elementary file (EF). In addition to the EFs, there are also dedicated files (DFs), which allow for grouping of EFs.

At the top level there is the master file (MF), which contains the identifier for the EFs defined in the UICC. There are four mandatory EFs: $EF_{PL}$, $EF_{ICCID}$, $EF_{DIR}$ and $EF_{ARR}$. These provide the preferred language, a unique ID, the directory of applications and the access rule reference respectively. Also within the UICC there are a number of application dedicated files (ADFs). The ADFs contain all EFs and DFs required by an application. The application identifiers (AIDs) define the ADFs and are stored in the $EF_{DIR}$. The USIM is an example of an ADF, which provides the USIM functionality.

### 14.1.3 USIM ADF

Figure 14.3 illustrates elements of the data structure of the USIM ADF. The USIM ADF comprises DFs for GSM access and MExE amongst others, and a number of EFs that are required for the operation of the USIM application, some of which are considered in the following sections.

### 14.1.4 UE application architecture

Figure 14.4 illustrates the protocol architecture between the USIM (UICC) and the ME. At the top is the application layer, where the USIM application resides. Below this is the USIM application toolkit (USAT). The transport layer defines the transmission protocol between the USIM and the ME. The data link layer furnishes a layer 2 type data link, providing block transfer and sequence error detection amongst other functions.

## 14.1 USIM architecture and operation

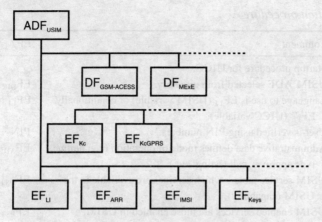

**Figure 14.3** Basic structure of USIM ADF.

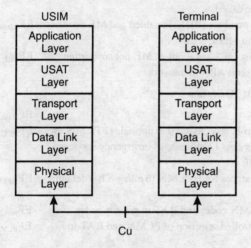

**Figure 14.4** Communication protocols between UE and UICC.

The final layer is the physical layer that is responsible for the transportation of bits between the UICC and the ME.

### 14.1.5 ME–USIM activation procedure

In this section we consider the activation procedure between the ME and the USIM. The activation procedure defines the order in which events occur and data are read from the USIM into the ME. Table 14.1 defines the basic activation procedure performed by the ME and the USIM, and the EF that is read from the USIM by the ME where appropriate.

The basic philosophy is that the ME reads data stored on the USIM. We can interpret the data stored in Table 14.1 as follows. At the start, the UICC powers on and performs

**Table 14.1.** *ME-USIM activation procedure*

| Procedure | Comment | EF |
|---|---|---|
| UICC activation | Startup procedure for UICC. | |
| USIM app selected | USIM ADF selected from directory. | $EF_{DIR}$ |
| Language indication | Language to use is $EF_{LI}$ (USIM variable) or conditionally $EF_{PL}$ (UICC variable). | $EF_{LI}$ or $EF_{PL}$ |
| User verification | User is verified using PIN numbers. | PIN |
| Administrative info req. | Administrative data defines mode of operation (e.g. normal, type approval, cell testing, etc.). | $EF_{AD}$ |
| USIM service table request | USIM service table – which services are available in the USIM (capability). | $EF_{UST}$ |
| Enabled services table request | USIM enabled services are those enabled in USIM (activation). | $EF_{EST}$ |
| Check if FDN enabled | If fixed dialling numbers (FDNs) enabled only defined numbers can be called. | $EF_{FDN}$ |
| Check if BDN enabled | If barred dialling numbers (BDNs) enabled and ME does not support CC – no MO-CS calls. | $EF_{BDN}$ |
| Check if ACL is enabled | If APN control list (ACL) is enabled, ME not supporting ACL may not send APN to network. | $EF_{ACL}$ |
| | Start 3G Session | |
| IMSI request | Read IMSI | $EF_{IMSI}$ |
| Read ACC | Read access control class (ACC) 0–9 normal, 11 PLMN use, 12 security services, 13 utilities, 14 emergency services, 15 PLMN staff. | $EF_{ACC}$ |
| HPLMN search period | Time between searches for HPLMN (6 min – 8 h – default 1 h). | $EF_{HPLMN}$ |
| HPLMN selector | Defines the HPLMN codes and RAT in priority order. | $EF_{HPLMNwAcT}$ |
| User controlled PLMN selector | Defines user controlled selection of PLMN and RAT in priority order. | $EF_{PLMNwAcT}$ |
| Operator controlled selector | Defines operator controlled selection of PLMN and RAT in priority order. | $EF_{OPLMNwAcT}$ |
| Read RPLMN | Reads PLMN ID and access technology previously used. | $EF_{RPLMNAcT}$ |
| GSM initialisation req | If ME supports GSM compact access, scan information for CPBCCH requested. | $EF_{InvScan}$ $EF_{CPBCCH}$ |
| Location information request | Read LAI/RAI, TMSI/P-TMSI and location status (updated or not updated). | $EF_{LOCI}$ |
| Read keys | ME reads CK, IK and key set identifier (KSI) for CS and PS domain. | $EF_{Keys}$ $EF_{KeysPS}$ |
| Read forbidden PLMNs | PLMNs ME cannot access. | $EF_{FPLMN}$ |
| Initialise HFN | ME reads values of $START_{CS}$ and $START_{PS}$ used as initialisation values for the HFN. | $EF_{START-HFN}$ |
| Read START max | Defines the maximum values for $START_{CS}$ and $START_{PS}$. | $EF_{THRESHOLD}$ |
| Read CBMID | Defines what type of cell broadcast the user wishes the UE to accept. | $EF_{CBMI}$ |
| Other | Other EFs to be read next. | |
| Ready for 3G service | USIM indicates ready for service | STATUS |

any necessary UICC activation procedures before calling the USIM application based on the identifier contained in the $EF_{DIR}$. Next, the language is selected. Normally this is defined by the $EF_{LI}$ (language indicator) defined in the USIM application, but this may be overridden by the $EF_{PL}$ (preferred language) field in the UICC under certain conditions.

After user verification (using a personal identification number (PIN)), the ME starts to read data from the USIM, beginning with the administrative information request that is used to define the mode of operation of the USIM (e.g. it could be for normal use or for cell testing). Next, the USIM service table and the enabled service tables are read. The USIM service table defines the services available in the USIM, and the enabled services define those services that are enabled. For a service to be usable, it must be both available and enabled. The next batch of data read by the ME defines things such as fixed dialling numbers (FDNs) and barred dialling numbers (BDNs).

Once the ME has read this preliminary data, the 3G session is started. This involves the transfer of data that is specific to the operation of the 3G application, starting with the reading of the IMSI and the ACC.

The HPLMN search period defines how often the ME should search for the home network. The range is from 6 minutes to 8 hours, with a default of 1 hour. There is also an option to halt the ME searching for the home network.

The HPLMN selector is a sequence of records that defines the home network (defined as the PLMN) and the access technology associated with that network. The access technology associated with the HPLMN is listed in priority order. There can be one or more entries in the table.

The next two entries read by the ME are the user controlled and the operator controlled selections on PLMN and RAT. The RPLMN is the PLMN that the ME was previously registered to including the RAT that was used.

If the GSM access is enabled and the ME has the capability to support the compact packet BCCH (CPBCCH) then the ME reads the investigation scan EF and the CPBCCH EF. These are used when the UE needs to support the CPBCCH.

The ME reads the location information. This includes the TMSI and P-TMSI and the LAI and RAI, as well as the update status of the location information. The security keys are read (CKs, IKs and key set identifiers (KSIs)) for both the CS and the PS domain.

Next the forbidden PLMN list is read. This defines the PLMNs that the ME cannot automatically access. The ME may later add to this list if a location update attempt is rejected for specific reasons.

The HFNs (there is one for the CS domain and one for the PS domain) are initialised to a value $START_{CS}$ for the CS domain and a value $START_{PS}$ for the PS domain. The parameter that defines the type of cell broadcasts that the user wishes to accept is read, followed by the other EFs required for additional services if supported by the USIM.

Finally, the USIM has completed this initialisation and is ready to support the 3G services, and this is indicated to the ME in a STATUS message.

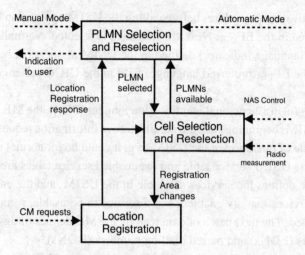

**Figure 14.5** Summary of idle mode procedures.

## 14.2 Idle mode overview

Having considered the USIM, some of the operation issues related to the USIM and the data stored in the USIM, we now need to consider the two main topics for this chapter, namely the idle mode substate machine and the idle mode procedures.

The idle mode substate machine is mainly an NAS topic, and so we consider it in its entirety in Section 14.3. The idle mode procedures, however, are divided between the AS and the NAS. Both of these are significant topics and consequently we consider them separately, but with the appropriate links between the two of them.

When considering the idle mode procedures, we will see that some of these procedures are also relevant to what occurs in the UE whilst in the connected mode. For this reason, we also consider the connected mode aspects of these idle mode issues.

In the next section, we start with the idle mode substate machine and some of the related issues. First, though, we introduce the basic concepts of the idle mode procedures that we will return to in Section 14.4 for the NAS aspects and Section 14.5 for the AS aspects.

Figure 14.5 summarises the idle mode procedures and the relationships between them. There are three main elements to the idle mode procedures: the PLMN selection and reselection, the cell selection and reselection and the location registration. All of these procedures are interrelated, as shown in the figure. A change incurred in one procedure will most probably trigger a change in one of the others. The main objective for these procedures can be summarised as follows: to ensure that a UE is registered on the most appropriate cell in the most appropriate PLMN at any given point in time. To make sure that this occurs, the NAS and the AS have certain

## 14.3 Idle mode substate machine

**Table 14.2.** *Summary of NAS and AS aspects of idle mode procedures*

| Idle mode process | UE NAS | UE AS |
|---|---|---|
| PLMN selection and reselection | Maintain list of PLMNs in priority. Ask AS to select cell from PLMN (with highest priority) or belonging to manually selected PLMN. Automatic mode: if higher priority PLMN found ask AS to select cell. | Report available PLMNs to NAS on request from NAS or autonomously. |
| Cell selection | Control cell selection, e.g. by maintaining list of forbidden registration areas and a list of NAS defined service areas in priority. | Perform measurements needed for cell selection. Detect and synchronise to broadcast channel. Receive and handle broadcast information. Forward NAS information to NAS. Search for suitable cell in requested PLMN. Select suitable cell to camp on. |
| Cell reselection | Control cell reselection, e.g. by maintaining list of forbidden registration areas and a list of NAS defined service areas in priority. | Perform measurements for cell reselection. Detect and synchronise to broadcast channel. Receive and handle broadcast information. Forward NAS information to NAS. Change cell if more suitable cell found. |
| Location registration | Register UE as active after power on. Register UE in registration area (periodically or on entry). Deregister UE when shutting down. | Report registration area information to NAS. |

responsibilities for each of these different functions. These responsibilities are outlined in Table 14.2.

## 14.3 Idle mode substate machine

In this section, we consider the idle mode substate machine, its functions and the reasons and rationale for the transitions between the substates. The transitions between the substates are driven by events within the idle mode procedures that were introduced in the previous section and which are considered in more detail in Section 14.4 for the NAS aspects and Section 14.5 for the AS aspects.

The idle mode substate machine is an expansion of the idle state presented in the MM state machine introduced in Chapter 13. The idle mode state machine defines the activities of the UE when it does not have an active signalling connection or an active call in progress. In the idle mode state, the UE is responsible for many of its own actions. These actions include the selection of both the correct PLMN and the most appropriate cell within that PLMN, and registering with the network within that cell within a specific area known as an LA.

The existence of an active registration for a specific USIM is defined by its location update status. We will encounter a number of forbidden networks and regions with barred cells (or they may be forbidden LAs). To proceed, let us first define the update status (US) that is stored in the USIM, and then consider the concepts of forbidden LAs and forbidden PLMNs before proceeding with the main topic of idle mode substates.

### 14.3.1 US

The US is used to define the status of the subscriber (USIM) with regard to the location updating procedure. In Chapter 13 we considered the US for the PS domain, here we consider it for the CS domain. There are three US states: U1, U2 and U3. The information is used as part of the control of the MM state machine to define the specific actions and services that a UE is able to perform. The meaning of each of these states is defined below.

**U1**

The location updating attempt last performed was successful. The USIM contains a valid LAI, TMSI and possibly valid security keys (CKs and integrity protection keys). The USIM defines the location update status as 'updated'.

**U2**

In this state, the last location updating attempt failed for some reason out of the control of the network, such as a signalling failure. The USIM does not contain a valid LAI, TMSI or keys. The USIM defines the location update status as 'not updated'.

**U3**

In this state, the network rejected the last location updating attempt, maybe due to roaming or service restrictions. The USIM does not contain a valid LAI, TMSI or keys. The USIM defines the location update status as 'location area not allowed'.

## 14.3 Idle mode substate machine

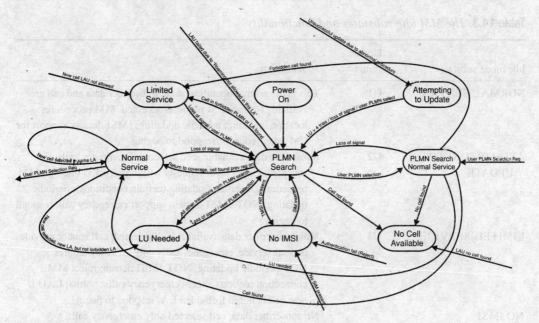

**Figure 14.6** Idle mode substates.

### 14.3.2 Forbidden PLMNs and forbidden LAs

The network can define a list of forbidden PLMNs and forbidden LAs. The UE should use the knowledge of these when performing the location updating procedure. In certain circumstances, if no other PLMN is available, the UE may select a cell from a forbidden PLMN or forbidden LA and provide a limited service in which only emergency calls are available.

The list of forbidden LAs is deleted when the UE is switched off and is recreated when it switched back on. A specific cell in the LA can be added to the list if a location update fails for the reason 'no suitable cells in location area'.

A PLMN is added to the forbidden PLMNs list if an LAU is rejected for the reason 'PLMN not allowed'. In this case the forbidden PLMNs remain in semi-permanent memory in the USIM, and is only removed if a manual initiated LAU attempt is successful.

### 14.3.3 MM idle state – substate machine

Figure 14.6 presents the substates for the MM idle state machine. The substates revolve around the PLMN SEARCH substate, and the outcome of this specific procedure defines which of the other substates are used. The PLMN SEARCH substate is the state that the MM entity is in when the UE is performing the PLMN selection and

**Table 14.3.** *The MM idle substates and functionality*

| Idle mode substate | Update status | Functionality |
|---|---|---|
| NORMAL SERVICE | U1 | This is the normal operating state. Subscriber data and cell are available for use. CM requests accepted. MM procedures: location updating; periodic updating; IMSI detach; support for CM layer requests; respond to paging. |
| ATTEMPTING TO UPDATE | U2 | Subscriber data available, cell selected. Emergency calls accepted. Upper layer requests will trigger a LU attempt. MM procedures: location updating (certain conditions); periodic updating; NOT IMSI detach; support emergency call; respond to paging. |
| LIMITED SERVICE | U3 | Valid subscriber data available, cell selected, cell cannot provide normal service, emergency calls only. MM procedures: NOT periodic updating; NOT IMSI detach; reject MM connection requests except emergency calls; normal LAU if new cell is not in forbidden LA; respond to paging. |
| NO IMSI | | No subscriber data, cell selected only emergency calls. NOT location updating; NOT periodic updating; NOT IMSI detach; reject MM connection requests except emergency calls; NOT respond to paging; perform cell selection. |
| NO CELL AVAILABLE | | No cell can be selected (coverage reasons) after thorough search, periodic (low periodicity) search, no MM procedures available. |
| LOCATION UPDATE NEEDED | U1 | Subscriber data available, location update required, waiting to perform update – delay may be due to update access restrictions. |
| PLMN SEARCH | | UE searching for PLMN. Leaves state when cell found. Next state depends on outcome and shown in Figure 14.6. No MM procedures available. |
| PLMN SEARCH, NORMAL SERVICE | U1 | Subscriber data available. PLMN search and cell selection. Next state depends on cell that is found (if any). MM procedures: location updating triggered in this state but performed in NORMAL SERVICE state; periodic updating triggered in this state but performed in NORMAL SERVICE state; IMSI detach; support requests for MM connections; respond to paging. |

reselection procedures that are illustrated in Figure 14.5. The outcome of this PLMN search defines the transition to the other substates.

The ideal objective is for the UE to end in the NORMAL SERVICE substate, which would result in the UE being able to perform all of the functions defined for the MM entity. Unfortunately, however, things occur that mean that the UE does not always end in this ideal state. The UE could leave radio coverage for instance; it may only be able to find a cell in a PLMN to which it cannot register. To cater for these many eventualities, the other substates shown in Figure 14.6 have been defined.

In this section, we consider the basic structure for the substate machine, and in Section 14.4.1, we will explore some of the background to the state changes that occur.

We start by considering Table 14.3, which shows the definition and use for each of the substates illustrated in Figure 14.6. The substate NORMAL SERVICE is the one the UE should be in; this represents a UE that is functioning correctly. The other substates are present due to some change in circumstance (out of coverage, update required, update failed) that means the UE can no longer remain in that NORMAL SERVICE substate. The contents of the table are there to represent what levels of MM functionality are available from the MM entity whilst the UE is in that specific substate.

In the following sections we consider the relationship between Figure 14.5 and the functions of the UE whilst in the idle mode.

## 14.4 NAS idle mode functions and interrelationship

Having considered the idle mode state machine, we can now address the NAS functions that occur whilst a UE is in idle mode, and from this we will see how the state machine and the idle mode procedures interact.

In Section 14.5, we will consider the AS idle mode functions and procedures, and Section 14.6 we will consider an example which combines all of the elements we have considered: the USIM EFs, the idle mode substates, the NAS idle mode procedures and the AS idle mode procedures.

We start by considering the basic idle mode procedures outlined previously in Figure 14.5 and separated into NAS and AS procedures in Table 14.2. The NAS functions are divided into three main parts: the PLMN selection and reselection, cell selection and reselection and location registration. These functions are linked. The PLMN selection and reselection procedure needs information from the cell selection and reselection process as to which PLMNs are available. The PLMN selection process can select a cell for a specific PLMN. The selection of a new cell in a new PLMN requires a location registration, and hence the linkage with the location registration functions.

The inputs to the NAS process are decisions as to the type of PLMN selection, whether it is manual or automatic (this decision normally comes from the user). The other inputs come from the CM sublayer requesting an MM connection or other NAS control inputs that can and do affect the interoperation between these different functions.

We start with the PLMN selection and reselection procedure and that leads us into the other functions. In considering these procedures, we also make reference to the state machine described previously. As mentioned in the introduction to this chapter, we consider the specific case of the MM entity idle mode functions; similar comments can be made concerning the GMM entity.

# 520 Idle mode functions

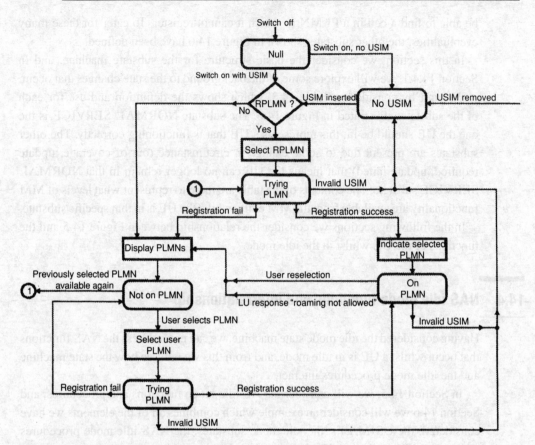

**Figure 14.7** Manual PLMN selection procedure state diagram.

## 14.4.1 PLMN selection and reselection

To start, we consider the NAS PLMN selection procedure and in the course of this we review where the UE obtains the information that is necessary in order for it to perform the PLMN selection procedure.

There are two methods by which the UE can select the PLMN, referred to as the manual mode and the automatic mode. It is a user decision, normally via the user interface, whether the manual mode or the automatic mode is selected. The objective of these two modes is to define a procedure to select the PLMN, in this case, immediately after the UE has been activated. In the manual mode, the user selects the PLMN based on a list of available PLMNs. With the automatic mode, the PLMNs are selected automatically, based on some priority information that is available to the UE from the USIM.

Figure 14.7 presents the PLMN selection state diagram for the manual mode; Figure 14.8 presents the PLMN selection state diagram for the automatic mode. First

## 14.4 NAS idle mode functions and interrelationship

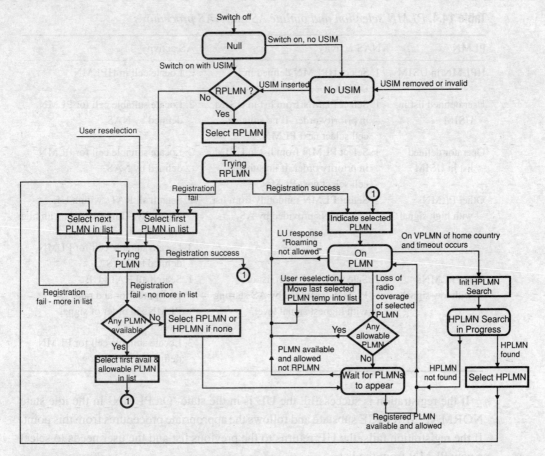

**Figure 14.8** Automatic PLMN selection procedure state diagram.

we consider the manual PLMN selection procedure and following that we consider the automatic mode.

**PLMN manual selection**

After switching on and checking the presence of the USIM (which we assume is present), the UE checks to see if the PLMN that it was previously registered on (RPLMN) is available. If the RPLMN is available, the UE attempts to register on that PLMN (this includes the cell selection procedure (see Section 14.5) followed by the location registration (LR) procedure (see Section 14.4.3). Once registered to the selected PLMN, the UE is 'On PLMN' and in the NORMAL SERVICE state within the idle mode substate diagram.

If the RPLMN is not available, the UE requests the AS to search for the available PLMNs (the AS PLMN search procedure is considered in detail in Section 14.5). The NAS then displays all of the PLMNs available to the user. The user selects a PLMN and the UE attempts to register to that PLMN ('Trying PLMN').

**Table 14.4.** *PLMN selection and outline AS and NAS procedures*

| PLMN | NAS activity | AS activity |
| --- | --- | --- |
| HPLMN in USIM | 1. Select HPLMN defined in USIM. | 2. Locate cell in HPLMN. |
| User defined list in USIM | 1. Select PLMN from list in USIM in priority order. If no suitable cell select next PLMN. | 2. Locate suitable cell for PLMN defined by NAS. |
| Operator defined list in USIM | 1. Select PLMN from list in USIM in priority order. If no suitable cell select next PLMN. | 2. Locate suitable cell for PLMN defined by NAS. |
| Other PLMNs with high signal | 2. Select PLMN randomly from list of PLMNs provided by AS. | 1. Search all RATs within UE capability – find PLMNs with high signal level. 3. Locate suitable cell for PLMN defined by NAS. |
| Other PLMNs with low signal | 2. Select PLMN from list of PLMNs provided by AS starting with highest signal level. | 1. Search all RATs UE configured for and rank PLMNs in order of signal level. 3. Locate suitable cell for PLMN defined by NAS. |

If the registration is successful, the UE is in the state 'On PLMN' in the idle state NORMAL SERVICE substate and follows the appropriate procedures from this point. If the registration fails, the UE returns to the previous list and the user needs to select a new PLMN from the list.

**Automatic PLMN selection**

The automatic PLMN selection procedure is presented in Figure 14.7. Here we review some of the decisions and procedures that are followed as the UE goes through the selection of the PLMN.

The UE starts with the same procedure as used for the manual mode, namely the decision as to whether the PLMN is a RPLMN or whether the UE needs to consider other PLMNS within the coverage area. Here we assume that the RPLMN is not being used, and that the UE needs to locate other PLMNs available within the coverage area.

There are a number of different options for the PLMN selection that the UE can follow. The USIM defines a list of PLMNs in priority order and Figure 14.7 illustrates the procedures to decide which PLMN to select if one is not available from the prioritised list.

Table 14.4 indicates the PLMN selection process in priority order and the role that the AS has in either generating PLMNs in the list or working with a selected PLMN.

**Table 14.5.** *Summary of transitions from PLMN SEARCH state*

| To substate | Comment |
|---|---|
| NORMAL SERVICE | The MM entity returns to this state after loss of coverage if location registration not needed. |
| LIMITED SERVICE | Selected cell is in forbidden PLMN or forbidden LA. Emergency calls only. |
| NO CELL AVAILABLE | The MM entity is in this substate until a cell is found. |
| NO IMSI | No USIM present. |
| LOCATION UPDATE NEEDED | If the MM entity has selected a cell in an LA in a PLMN in which the UE is not registered, then the MM entity moves to this substate until an LR has occurred. |

The details for the AS part of this activity, in particular the definition of a suitable cell, are considered in Section 14.5.

Table 14.4 provides two specific pieces of information. First it defines the prioritised order in which the UE should look for a PLMN to use (the table excludes the RPLMN, which it will try first). Second, for each type of PLMN selected, the table outlines the basic procedure split between the NAS and the AS used to find a suitable cell in the most appropriate PLMN.

In the table, the numbers indicate the rough ordering of the different processes. In some cases, the underlying procedures are quite involved, and the table is only intended as a rough guide to the PLMN selection procedure. The concept behind the table, however, is that the UE starts by looking for a PLMN with the highest priority (this is the HPLMN). If after a cell search, the HPLMN cannot be found, the NAS requests the AS in the UE to look for a suitable cell in a specific PLMN defined by the prioritised list in the USIM, starting with the highest priority PLMN. If a suitable cell cannot be found, then the next highest priority PLMN is selected, and the procedure is repeated. If at the end of this process, no suitable cell is found in any of the PLMNs in the user defined or operator defined list, the NAS then moves down to the 'Other PLMNs' selection procedure shown in Table 14.4 and follows a similar procedure, as defined there.

Returning to the idle mode substate diagram presented in Figure 14.6, we can see that the MM entity can move both to and from the PLMN SEARCH substate to a number of other substates. The state transitions from and to the PLMN SEARCH state and the reasons for those transitions are given in Tables 14.5 and 14.6 respectively.

### 14.4.2 Cell selection and reselection procedure

A key part of the PLMN search procedure is the cell search procedure. We saw in Section 14.4.1 that the PLMN search could request the location of a cell in a single PLMN or it could request the search for all available PLMNs in the coverage area.

**Table 14.6.** *Summary of transitions to PLMN SEARCH state*

| From substate | Comment |
| --- | --- |
| NO IMSI | SIM inserted and MM entity moves to perform a PLMN search. |
| Any except NO IMSI, NO CELL AVAILABLE, NORMAL SERVICE | User has requested a PLMN selection. |
| Any except NO IMSI, NO CELL AVAILABLE | Coverage lost from previously used cell, MM entity requests a PLMN search. |
| ATTEMPTING TO UPDATE | The LAU counter has exceeded its maximum number of attempts for this time period. The MM entity requests a PLMN search to find a different PLMN that can be used. |

The main part of the procedure for cell selection and reselection is performed by the AS and consequently this is covered in greater detail in Section 14.5. The NAS, however, does have some involvement in the cell selection and reselection process beyond the requesting of the AS selections and reselections. The main area of involvement concerns the maintenance of the list of forbidden areas. These are forbidden LAs, which are tracked by the NAS in the UE and normally received as part of the LR procedure. The list of forbidden areas is deleted when the UE is powered off. The AS recovers the NAS information from the SIB1 for a cell and passes it to the NAS. The NAS can then decode the LAI for the cell and decide if the cell is in a forbidden area.

### 14.4.3 LR procedure

Returning to the idle mode procedure summary diagram of Figure 14.5, we see that the third important MM idle mode function is the LR procedure. The order of events is: select a PLMN; find a suitable cell on that PLMN; and register in that cell on that PLMN. It is at this stage that the LR function becomes involved.

The LR procedure is concerned with the location updating of a specific UE. Figure 14.9 shows the LR state diagram, which illustrates the location update states defined previously in Section 14.4.1, and the transitions between these different location registration states.

In this section, we explore the states, the transitions between the states and how these relate to the other procedures and substates that we have considered so far. We focus on the states LR PENDING, UPDATED (U1), NOT UPDATED (U2) and ROAMING NOT ALLOWED (U3).

**LR PENDING**

In this state there is an on-going location update in progress. There are many possible outcomes, depending upon the success or otherwise of the location updating

## 14.4 NAS idle mode functions and interrelationship

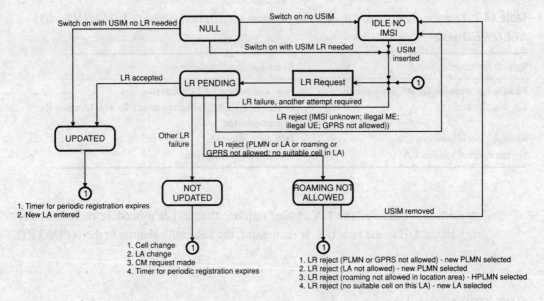

**Figure 14.9** LR state diagram.

progress. In Figure 14.9, five possible outcomes are defined. We now consider each in turn.

The outcome LR accepted occurs when the LR is successful. The LR entity moves to the UPDATED state (U1). If the outcome is an LR failure for some reasons that may be procedural, but do not constitute a registration rejection, then the entity moves to the NOT UPDATED state (U2). However, if the LR entity receives a rejection for reasons 'PLMN not allowed', 'location area not allowed', 'roaming not allowed in location area', 'GPRS not allowed', or 'no suitable cell in location area' then the entity moves to the ROAMING NOT ALLOWED STATE (U3). If the LR entity receives a rejection for reasons 'IMSI unknown in HLR', 'illegal ME', 'illegal MS', then the LR entity moves to the IDLE NO IMSI state and waits for a valid IMSI to be inserted. If the LR entity receives an LR failure response to the LR procedure, but one indicating that another attempt should be made, the LR entity moves back to the LR REQUEST state and tries another LR attempt.

## UPDATED (U1)

The UPDATED state is the state that the UE arrives in when the LR procedure is successful. From the UPDATED state, the LR entity needs to perform periodic registrations based on the expiry of a timer. In this case, it starts at the LR REQUEST state and follows the LR procedure as described previously. The use of the periodic location updates is described in more detail in Chapter 13. If successful, the LR entity returns to the UPDATED state.

The second state change whilst in the UPDATED state occurs as a consequence of a change in the LA of the new cell that the AS has selected as part of a cell

**Table 14.7.** *Triggers that cause LR entity to perform registration whilst in ROAMING NOT ALLOWED state*

| Reason for entering state | Event to trigger registration |
|---|---|
| PLMN not allowed, GPRS not allowed | New cell in new PLMN selected. |
| LA not allowed | Cell in new PLMN selected or cell in new location area selected. |
| Roaming not allowed in LA | HPLMN selected. |
| No suitable cell on this LA | Cell in new LA selected. |

reselection procedure. The LA change requires that an LR procedure be followed to register the UE in the new LA. If successful, the LR entity returns to the UPDATED state.

### NOT UPDATED (U2)

In the NOT UPDATED state, we assume that the LR attempt has failed due to some procedural failure (this might be due to a fault in the Node B, for instance). The LR entity attempts a new LR as soon as an appropriate event has occurred. The events that can trigger a new LR attempt are a cell change (cell reselection by the AS), an LA change, a CM request made (MM connection requested by CM sublayer), and expiry of periodic update timer.

Upon successful completion of the LR procedure, the LR entity moves to the UPDATED state. If the procedure is unsuccessful, it ends in the state described previously depending upon the nature of the failure.

### ROAMING NOT ALLOWED

In this state the LR entity has received an LR rejection from the network, and it depends on the reason for entering the state as to the events that need to occur for the LR entity to initiate a new registration attempt. Table 14.7 presents the reasons for entering the state, and the events that needs to occur for the LR entity to perform a new procedure. If the LR is successful, the LR entity moves to the UPDATED state. If the registration is not successful, the LR entity moves to the state defined above, based on the reasons for the unsuccessful attempt.

### Summary of functions and actions allowed in states U1, U2 and U3

Table 14.8 summarises the functions and actions that are applicable for a UE that is in one of the states shown in Figure 14.9. The ROAMING NOT ALLOWED state has been broken into four different cause types, which are outlined in Table 14.7.

**Table 14.8.** *Presenting registration actions and call functions based on LR states*

| State | Registration if changes to | | | | Norm calls | Paging resp |
|---|---|---|---|---|---|---|
| | Cell | Area | PLMN | Other | | |
| NULL | No | Yes | Yes | | No | No |
| IDLE NO IMSI | No | No | No | No | No | No |
| U1 – UPDATED | No | Yes | Yes | Periodic | Yes | Yes |
| U2 – NOT UPDATED | Yes | Yes | Yes | Periodic CM Req | Possible | Via IMSI |
| U3 – PLMN not allowed | No | No | Yes | No | No | If via IMSI |
| U3 – LA not allowed | No | Yes | Yes | No | No | If via IMSI |
| U3 – Roaming not allowed in LA | No | Yes | Yes | No | No | If via IMSI |
| U3 – No suitable cells in LA | No | Yes | Yes | No | No | If via IMSI |

## 14.5 AS idle mode functions and interrelationship

In this section we consider the AS idle mode functions. We saw earlier that the idle mode functions are divided into three main functions: PLMN selection and reselection, cell selection and reselection, and location registration.

As we saw in Table 14.2, the main AS idle mode functions are related to the cell selection and cell reselection. The support from the AS for PLMN selection and reselection is limited to functions such as PLMN identification and reporting to the NAS. Support for the functions related to the LR procedure is limited to the detection of NAS information (the information has no meaning to the AS) that is passed to the NAS for interpretation. The NAS is then able to detect a change in LAs.

In this section we focus on the AS issues of cell selection and reselection. We start by examining some of the key definitions and what makes a cell a 'suitable cell'.

### 14.5.1 Cell selection and reselection

In radio systems such as UMTS, cells are not hard defined quantities and so it is important to have a set of criteria that can be used to define whether a specific cell can be used by a UE or whether it should not be used. There are two aspects to this decision making process. The first part is the initial decision to be made about specific cells and which to select for a specific UE in a specific place. Then there are subsequent decisions that need to be made about whether that specific cell is better or worse than other potential cells.

The first procedure is referred to as cell selection and is typically done once, either at switch on or on returning to coverage from a region without coverage; the second

# 528  Idle mode functions

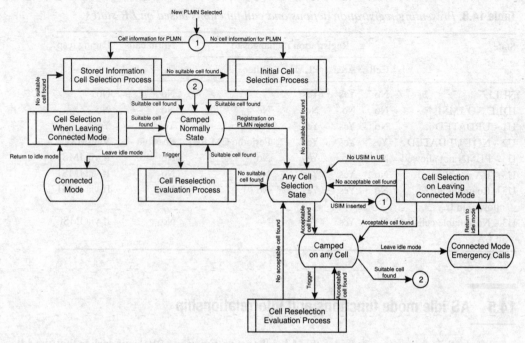

**Figure 14.10**  AS cell selection procedure.

is referred to as cell reselection, which is done continuously whilst a UE is in the idle mode and in certain connected mode states.

We will see in this section that the act of changing cells is the fundamental trigger to the NAS procedures that we reviewed in the earlier part of the chapter. In selecting a new cell, the AS could be selecting a new cell which is in a new LA, thus triggering the LR procedure in the NAS.

We start by considering the cell selection process. There are two basic types of cell selection: stored cell selection, in which the cells are known and details stored in the UE, and initial cell selection, in which the UE must scan a number of cells prior to a selection being made by the NAS.

We next consider the cell reselection procedure, which is the on-going process in the UE. There are two types of cell reselection procedure: one that uses hierarchical cell structures (HCSs) (i.e. macro-cells, micro-cells and pico-cells, all in the same area) and one that does not. The presence of all these cells has an impact on the cell reselection procedure. We consider the cell reselection procedure both when HCS is used and when it is not.

## 14.5.2  Cell selection overview

The cell selection procedure is illustrated in Figure 14.10. The procedure is invoked once the NAS has selected the PLMN, which the AS must attempt to locate. We start

## 14.5 AS idle mode functions and interrelationship

by considering the initial stages of the cell selection. There are two possible variations of the procedure for this first stage of cell selection. In the first variation the ME locates a 'suitable cell' based on a list of cells that the AS has stored in the USIM or the ME from the previous time that it was activated. In the second variation, the ME cannot locate a suitable cell from the stored list and needs to perform a cell search.

The objective in either of these two variations is to locate a suitable cell. The term 'suitable cell' refers to the characteristics that the cell must possess in order for the AS to select that cell for possible use by the NAS. We first consider what we mean by a suitable cell, and then proceed to examine the stored information cell selection procedure.

### Suitable cell and cell camping definition

Part of the cell selection procedure is the task of identifying a suitable cell. A suitable cell is defined as a cell with the following characteristics:

- The cell signal level selection criteria are acceptable for use.
- The cell is part of a PLMN regarded as acceptable by the NAS.
- The cell is not a barred cell as defined by system information broadcast within that cell.
- The cell is not part of a forbidden LA as defined by the NAS.

Apart from the first criterion, the elements in the list can be obtained from system information messages or signalling exchanges. The first element in the list requires that the UE performs measurements and makes decisions on these measurements. It is this specific aspect that we focus on here.

### Cell selection criteria

One of the problems with any radio system, whether it be GSM or UMTS, is that the quality of the signal between the user's handset and the radio network is constantly changing due to propagation phenomenona associated with the radio system. One of the many tools in the GSM/UMTS toolset to provide a quasi-static connection is referred to as 'cell camping'. Cell camping is where a mobile selects a cell and associates itself to that cell. The mobile remains associated to that cell until the quality of the cell is degraded below certain predefined thresholds and the mobile needs to find a new cell that can be regarded as a suitable cell.

The benefit to the network in performing these procedures is that the mobile keeps a constant point of contact with the radio network for long periods of time, even though the quality of the link may vary. A typical situation is illustrated in Figure 14.11, where a UE has started the process of looking for a suitable cell.

### 14.5.3 Stored information cell selection criteria

Returning to the cell selection procedure presented in Figure 14.10 we start by considering the stored information cell selection procedure. The UE starts with the cell

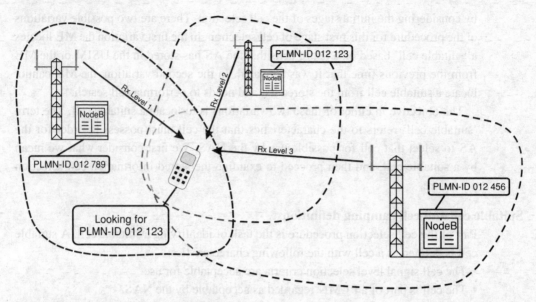

**Figure 14.11** UE looking for a suitable cell.

information that was stored in the ME or the USIM. As mentioned earlier, the UE has access to a list of previously used cells that are stored in the USIM in the elementary file $EF_{NETPAR}$ and potentially a list of cells that are stored in the ME. These lists can contain cell specific information on GSM, FDD mode cells and TDD mode cells.

The UE proceeds by checking the cells for which it has stored information, looking for a suitable cell. For UMTS, a suitable cell is defined in terms of a received signal level measurement (FDD mode and TDD mode) and a received signal quality measurement (FDD mode only).

FDD mode suitable cell

$Srxlev > 0$

$Squal > 0$

TDD mode suitable cell

$Srxlev > 0$

### Srxlev definition and calculation

The quantity *Srxlev* is a cell selection criteria that is based on the received signal level for the cell. Srxlev is defined by:

$$Srxlev = Qrxlev_{meas} - Qrxlevmin - Pcompensation$$

where $Qrxlev_{meas}$ is the RSCP measured using the CPICH for the FDD mode and the PCCPCH for the TDD mode.

The RSCP is a measurement made by the UE of the power that is assigned to the specific coded physical channel that the UE is attempting to receive. The measurement is an absolute measurement in dBm and is based on a number of measurements averaged over a specific time period. The period over which the measurements are made is related to the DRX process, and is defined in [23]. The rationale for this is that if a UE is using DRX, it should only make measurements at a rate that is compatible with the DRX sleep periods.

*Qrxlevmin* is the minimum received signal level defined for the cell. It is defined in broadcast information messages, which the UE needs to read as part of the initial cell selection procedure. The values for *Qrxlevmin* range from $-115$ dBm to $-25$ dBm for both the FDD and TDD modes of operation.

$$Pcompensation = \text{Max}(UE\_TXPWR\_MAX\_RACH - P\_MAX, 0) \quad (14.1)$$

i.e. it is the maximum value of two quantities (UE_TXPWR_MAX_RACH-P_MAX) or 0, where UE_TXPWR_MAX_RACH is the broadcast value of the maximum RACH power in the range from $-50$ dBm to $+33$ dBm, and P_MAX is the UE maximum Tx power.

**Example *Srxlev* calculation**

From (14.1), we see that *Pcompensation* is 0 until the RACH Tx power parameter is greater than the maximum Tx power of the UE. In this situation, this means that the received signal level needs to be greater than the minimum required by the difference between the UE_TXPWR_MAX_RACH parameter and the P_MAX value. The reason for this is illustrated in Figure 14.12. In this example, we assume that a cell has been configured for operation with UEs that have the capability to have a Tx power of $+33$ dBm. The UE that is trying to select the cell only has a Tx power of $+21$ dBm.

The problem with this scenario is that the UE can receive the downlink channels, but may not have sufficient Tx power on the uplink as illustrated. The solution (and hence the reason for *Pcompensation*) is for the mobile to only use the cell if the parameter *Srxlev* is greater than 0. The derived parameter *Pcompensation* can be used to force the UE to only select the cell if it has sufficient Rx power and consequently enough Tx power. In this example, the UE will have enough Tx power if the Rx power is greater than $-103$ dBm as indicated by the inner area, where *Srxlev* is greater than zero. The benefit to the system design is that cells can be defined for higher power mobiles, but can prevent lower power mobiles selecting them as suitable cells unless they are sufficiently close.

**Squal definition and calculation**

The definition for *Squal* is

$$Squal = Qqual_{meas} - Qqualmin$$

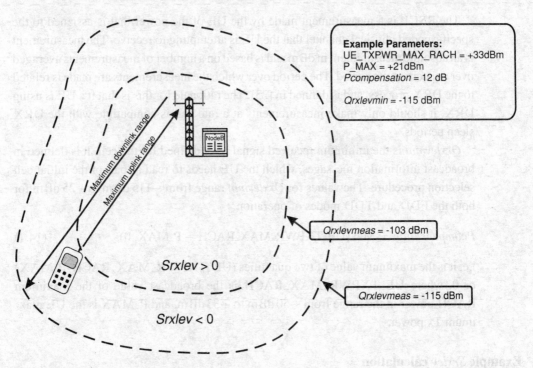

**Figure 14.12** Cell selection considerations using *Srxlev*.

*Qqual*<sub>meas</sub>

*Qqual*$_{meas}$

This is the measured received quality defined as the time averaged measurement of the CPICH $E_c/N_o$ estimate (strictly this should be $E_c/I_o$ as defined in [23], but referred to as $E_c/N_o$ in [24]). $E_c/I_o$ is a measure defined as the ratio of the received energy per chip and the total receiver interference plus noise power spectral density. As in the $Qrxlev_{meas}$ measurement, the averaging period for the measurement is related to the DRX duty cycle and defined in more detail in [23]. This measurement is only valid for the FDD mode.

*Qqualmin*

This is the minimum received signal quality defined for the cell. This quantity is contained in broadcast information messages [24] that the UE needs to read as part of the initial cell selection procedure. The values for *Qqualmin* range from $-20$ dB to $0$ dB, and it is valid for the FDD mode of operation only.

The objective for the quality measurement is to ensure that the cell quality is adequate for communications even in the presence of interference from other cells and possibly even the same cell (due to multipath propagation).

**Example *Srxlev* and *Squal* calculation**

Let us consider an example of how the UE can locate a suitable cell. In Figure 14.13, the UE has found an FDD mode UMTS cell and has been able to decode the broadcast

## 14.5 AS idle mode functions and interrelationship

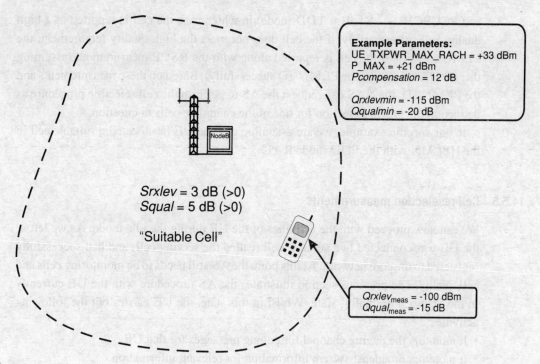

**Figure 14.13** Suitable cell selection summary.

information from that cell. The broadcast parameters define the maximum RACH Tx power for the cell as well as *Qrxlevmin* and *Qrxqualmin* as shown in the diagram. The UE maximum transmit power is +21 dBm and the UE has measured $Qrxlev_{meas}$ as −100 dBm and $Qqual_{meas}$ as −15 dB. From the measurements, the broadcast values and knowledge of the maximum Tx power, the UE can estimate the two quantities *Srxlev* and *Squal* to define whether the cell is a suitable cell. In this example, *Srxlev* and *Squal* are both positive, and therefore, the cell is a suitable cell, and the UE can now camp on this cell.

### 14.5.4 Initial cell selection procedure

The second variation of the cell selection procedure is the initial cell selection process, the procedure used when no preexisting information about the cells exists. With this procedure, the UE is requested to find either a specific PLMN with a specific RAT, or it is asked to find cells from all available PLMNs of a specified RAT. In either case, the AS needs to scan all RF channels to locate a suitable cell.

Let us consider an example where the UE is searching for a UMTS cell with an adequate signal level and for which the UE can decode the PLMN identity. There are two alternatives for the UE selecting and reporting a specific cell to the NAS. The first is if the cell has a sufficient quality (defined as CPICH RSCP > −95 dBm FDD mode and PCCPCH > −84 dBm TDD mode) in which case the cell is reported as a high quality cell. Alternatively, if the cell does not meet the high quality requirement, the PLMN identity of the cell is reported along with the RSCP measurement (assuming that the UE can decide the PLMN-ID successfully). Based on these measurements and the PLMN-ID, the NAS can request the AS to perform the cell selection procedure as defined in the previous section for one of the candidate cells in question.

In our specific example, we are assuming that the UE has located a suitable cell in the HPLMN with the FDD mode RAT.

### 14.5.5 Cell reselection measurements

We can now proceed with the activities of the UE during the idle mode. As we left it, the UE was connected to a suitable cell (called the serving cell) and had successfully registered to the core network. At this point the AS still needs to be monitoring cells and cell quality. The previous section illustrates the AS procedure with the UE currently in the 'camped normally' state. Whilst in this state, the UE carries out the following activities:

- It monitors the paging channel for paging messages for that UE.
- It monitors broadcast system information for relevant information.
- It makes appropriate measurements for the cell reselection evaluation procedure.
- It follows the cell reselection evaluation procedure when UE internal triggers are activated as defined in [23] for the FDD mode or [39] for the TDD mode.

The triggers for whether a cell reselection should occur are defined in [23] and [40] for the FDD mode. The UE is required to evaluate the cell selection criteria for the serving cell. There are two stages to the cell reselection procedure: a serving cell measurement stage and a second neighbour cell measurement stage triggered by measurements from the serving cell. In the first stage [23] the rate at which measurements should be made on the serving cell is defined. The measurement rate is related to the DRX duty cycle, with at least one measurement on the serving cell per DRX cycle required. For the FDD mode, the UE must measure the CPICH RSCP and the CPICH $E_c/I_o$ and filter the measurements.

For the second stage there are a number of triggers that can force the UE to make measurements on the neighbouring cells. First, the UE must compute the cell selection criterion ($s$), which is defined as *Squal* for the FDD mode and *Srxlev* for the TDD mode:

$s = Squal$ (FDD mode)

$s = Srxlev$ (TDD mode)

Next, the UE must evaluate a number of conditions to see if neighbour cell measurements are triggered: these conditions are outlined below.

- If serving cell selection criterion *s* fails for *Nserv* DRX cycles, the UE initiates neighbour cell measurements on all cells. *Nserv* is a parameter that is defined in [23] and relates to the DRX cycle length.
- If $s \leq Sintrasearch$, the UE performs intrafrequency measurements based on the cell list that the UE received from the UTRAN. *Sintrasearch* is a parameter defined in the SIB 3/4 broadcast messages.
- If $s \leq Sintersearch$, the UE performs interfrequency cell measurements based on the cell list that the UE received from the UTRAN. *Sintersearch* is a parameter defined in the SIB 3/4 broadcast messages.
- If $s \leq SsearchRATm$, the UE performs measurements on the list of cells of RAT type m that the UE received from the UTRAN. The parameter *Ssearch*RAT*m* is a parameter defined in the SIB 3/4 broadcast messages.

### 14.5.6 Cell reselection decision – HCS not used

Having made the measurements for the cell reselection, the UE now makes a decision on which cells to reselect based on these measurements. To do this, the UE needs to rank the cells in order, using criteria defined in [40].

For the FDD mode, the measurements could be either CPICH RSCP or CPICH $E_c/N_o$ (the quantity to use is defined by a parameter in the broadcast message). For the TDD mode the measurement is based on the PCCPCH RSCP. Initially in this process for the FDD mode, the measurements and parameters are based on the CPICH RSCP measurement irrespective of the measurement type requested.

The ranking criteria are:

$$Rs = Q_{meas,s} + Qhysts$$
$$Rn = Q_{meas,n} - Qoffsets, n - TOn(1 - Ln)$$

*Rs* is the serving cell ranking and *Rn* is the neighbour cell ranking. $Q_{meas,s}$ is the measurement on the serving cell, $Q_{meas,n}$ is the measurement on the neighbour cell, the types of measurements are the same as for the serving cell as defined above, *Qhysts* is a hysteresis value that is used to reduce 'ping-pong' effects between serving and neighbour cells and *Qoffsets,n* is an offset parameter between the serving and the neighbour cells. *TOn* is a temporary offset that is set to zero for the non-HCS case and *Ln* is a parameter that relates to the HCS process. The UE estimates the cell rank for all the relevant cells and RAT types, and reselects the new best ranked cell (except for additional FDD mode criteria defined below) if the cell is ranked better than the serving cell for a period defined by the parameter *Treselections* (this is defined in the SIB 3/4 messages) and if the UE has been camped on the serving cell for more than 1 second.

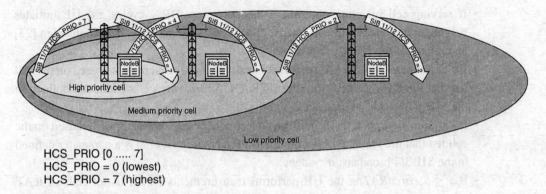

**Figure 14.14** Example of HCS.

If it is an FDD mode cell that is ranked as the best cell and the measurement type parameter indicates that the measurements are to be based on the CPICH $E_c/N_o$ measurements, the UE performs a second ranking for the FDD mode cells, but this time using the CPICH $E_c/N_o$ measurements and alternative values for $Qhyst_s$ and $Qoffset_{s,n}$. After this second process, the UE reselects the best-ranked FDD mode cell, provided that the additional time conditions defined above are met.

### 14.5.7 HCSs

An example of an HCS is shown in Figure 14.14. HCSs are used in systems such as UMTS to manage the traffic distributions between cells. In areas where there is a high user density, it is advantageous to have a number of small cells, with each small cell being able to accommodate the traffic within that cell. The problem with small cells, however, is that they are not suitable for users that are travelling at high speed and which consequently need a larger cell area.

Large cells cannot support high user densities, but they are suitable for high speed users; small cells are capable of supporting high user densities, but are not suitable for high speed users. The solution to this problem is to define a number of overlapping cells, of different sizes – a hierarchical cell structure. The objective is to place the user in the cell that best meets the needs of the user (i.e. slow moving users in small cells and high speed users in large cells).

To define which cell layer of the hierarchical cell structure a UE should be in, a HCS priority is defined (HCS_PRIO). The priority information is either transmitted as part of the cell broadcast information (SIB11 and SIB12 for neighbour cells and SIB3 and SIB4 for serving cells), or in the case of the neighbour cells, as part of an RRC measurement control message (MEASUREMENT CONTROL).

In Figure 14.14, the smallest cell (pico-cell) is allocated the highest priority, the next size cell (micro-cell) the second level priority and the largest cell (macro-cell) the lowest priority. A UE uses the cell priority and the UE's velocity to define which cell layer it should be in, assuming that HCS is in operation in that part of the system.

### 14.5.8 Cell reselection with HCS

In the case where hierarchical cell selections are to be included, the cell reselection procedure is complicated by the need to rank cell priority and also to include the effects related to the velocity of the user and hence the UE. Unless high speed UE rules are activated, while in the HCS mode and with the cell selection criteria below certain thresholds (defined in [40]) the UE should make measurements on all cells (intrafrequency and interfrequency). If the cell selection criteria are above these thresholds, the UE should only measure the higher priority cells. The high speed UE rule is activated if the number of cell reselections exceeds a quantity (NCR defined in SIB3/4 messages) measured over a period of TCRmax (defined in SIB3/4 messages). In this situation, measurements are made on cells (intrafrequency and interfrequency) that have the same or lower priority.

The reselection priority is given to the cells with the lower priority over those with the same priority. The rationale for this procedure is that the lower priority cells will have a larger area, and are therefore better equipped to deal with the high speed UEs.

For the HCS case, the cell ranking procedure is similar to the non-HCS case except for an additional quality threshold that relates to the cell hierarchy, and also the inclusion of a temporary offset that is present to bias the measurements away from neighbour cells until the expiry of a timer. This offset procedure helps to reduce reselections to those where the neighbour cells are much better, or to those where the neighbour cell is better after some defined time period (PENALTY_TIME defined in the SIB11/12 messages).

## 14.6 Example of idle mode procedures

To finish this chapter we examine an example of the operation of the idle mode procedures. We start with a review of the USIM EFs that we will use, then consider the PLMN selection and finally the cell selection procedure.

### EFs in USIM used for PLMN selection

In Section 14.1, the USIM consisted of a number of EFs. These EFs are a little like small data files that could be stored on a conventional PC, but in the USIM they are stored in the non-volatile memory on the UICC. There are a large number of these EFs defined for the USIM application, but here we will focus only on those that have some direct involvement in the PLMN selection process.

# 538 Idle mode functions

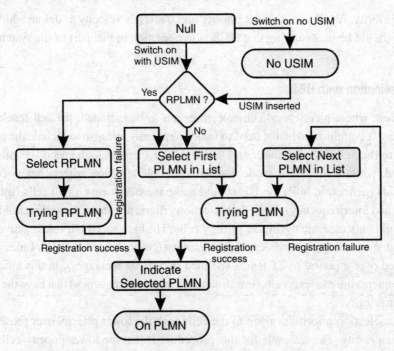

**Figure 14.15** NAS PLMN selection procedure.

The contents of the EFs are read by the ME as part of the USIM initialisation process. The size of each of the EFs used in the PLMN selection process varies according to the specific information that is stored in it. Changes in the EFs can occur due to updates performed by the network. Having defined the EFs used in the PLMN selection procedure, we can now proceed to explore the procedure itself.

## Example of automatic PLMN selection

To start, we are going to examine the NAS part of the PLMN selection process. The NAS is the part of the protocol layer that is activated soon after all of the various self checks and USIM initialisation phases have occurred.

As we saw in Section 14.4.1, there are two main modes for the selection procedure: automatic selection and manual selection. Here we focus on the automatic selection procedure. We start by exploring the first few stages of the power up sequence. The complete PLMN selection process is defined in [37]. The key stages for the PLMN selection procedure are illustrated in Figure 14.15. The first stage is a check on the RPLMN for the ME followed by the HPLMN. The RPLMN is the PLMN that the UE was last registered with prior to switching off. The RPLMN defines a PLMN and also the RAT. The HPLMN is the home PLMN; this is the PLMN that the UE is registered with. If neither of these is available, the UE needs to perform a cell search procedure.

## Registered PLMN check

In the first stage (we are assuming that the USIM is present in the ME), the ME checks to see whether there is a registered PLMN available. To do this, the elementary files $EF_{LOCI}$ and $EF_{RPLMNAcT}$ can be used. $EF_{LOCI}$ defines the LA and the location update status (U1, U2, U3 defined in Section 14.3.1) of the LA that the ME was previously registered on.

If the location information is defined as being 'updated' according to the USIM (update status U1), the ME can use the PLMN defined by the mobile country codes (MCCs) and MNCs contained in the LAI.

Having identified the RPLMN, the ME attempts to register with the RPLMN. We considered the registration procedure in greater detail in Chapter 13. If the location information is anything other than 'updated' then the ME proceeds with the rest of the procedure in which it needs to define a list of PLMNs in priority order.

In our specific example, we are assuming that the location information is not updated and therefore a RPLMN is not available and the UE must pass to the next stage of the proceedings.

## Registered PLMN not available

If, on checking, the RPLMN is found not to be available the NAS part of the ME needs to select a PLMN that can be passed to the AS part of the UE to determine whether that PLMN is available within the immediate coverage area. To facilitate this, the NAS defines a list in priority order for the PLMNs that the NAS part of the ME can use. The ME creates the prioritised list with information obtained from the USIM. The priority order for the PLMNs is shown in Table 14.4.

When defining the lists, each PLMN can be defined with access technology with the same priority, or alternatively each access technology can be individually defined with a relative position in the list and hence a relative priority order. The specification [37] defines a number of conditions for the selection of the PLMN based on the types of services and the capability of the ME.

In the case in which a defined PLMN has multiple access technologies with equal priority, it is an implementation issue as to which access technology is used first. If, on the other hand, the PLMN is the HPLMN, and the USIM does not contain the field 'HPLMN selector with access technology', then the UE searches for all technologies starting with GSM as the highest priority access technology. In our example, we assume that the ME has selected the HPLMN and the access technology is UMTS (defined as 'UTRAN' in the access technology field) of the information stored in the EF in the USIM.

In the next stage of the procedure, the NAS part of the PLMN selection process needs to request the AS to locate a suitable cell with the correct PLMN identity.

## 540  Idle mode functions

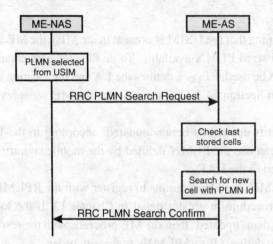

Figure 14.16  NAS–AS PLMN selection message flows.

### 14.6.1  AS part of cell selection

Here, we look at the AS part of the PLMN selection procedure. We assume that the previous tasks were completed successfully, and that the NAS has identified the PLMN that it wishes to register with. The NAS will, therefore, send an RRC PLMN search request message to the AS. The PLMN search request message contains the details of the PLMN that the ME has to locate, as well as the type of access technology (also referred to as RAT in the specifications).

In our specific example, the PLMN identity is the HPLMN, and the RAT is the UMTS technology (defined as UTRAN). These quantities are passed from the NAS to the AS via the PLMN search request message and this triggers the AS to look for the HPLMN with the specified RAT. Figure 14.16 illustrates the basic relationship between the NAS and the AS for the initial PLMN selection procedure.

Once the AS receives the search request, it can use the previously stored list of cells to search for the required PLMN. If this proves unsuccessful, the ME can search for a PLMN with the required identity. The ME has access to a list of previously used cells that are stored in the USIM in the elementary file $EF_{NETPAR}$. This list of information can contain GSM, FDD and TDD information. For the FDD mode the information can be for the same frequency (intrafrequency) or possibly for different frequencies (interfrequency). In addition, the FDD mode also includes possible scrambling codes for the cells.

Based on the list of cells previously used, the ME may attempt to locate the previous cells. Again, in this example, we are concentrating on the UMTS FDD mode of operation, and the procedures that the ME follows when locating a cell for the FDD mode of operation.

When the ME has located a suitable cell with the correct PLMN identity, the NAS is notified. The next stage of the procedure is the registration of the UE on the HPLMN and the establishment of services. The registration procedure and service request were considered in Chapter 13.

## 14.7 Summary

In this chapter we have reviewed the ideas, states and procedures surrounding the idle mode. The idle mode procedures are important to the operation of the UE because they define the activities that the UE takes under its own control without requiring intervention from the network except for the supply of parameters via broadcast channels.

The idle mode procedures are separated into NAS and AS elements. The NAS is responsible for the high level decision making, selecting and rejecting cells based on, for instance, current or previous signalling exchanges. The AS is responsible for making measurements and identifying new cells and PLMNs and reporting these to the NAS.

Together the cell selection and reselection and PLMN selection and reselection ensure that the UE is always attached to the most appropriate cell in the most appropriate network using the most appropriate technology.

# Appendix

The following is a list of numbers and titles for the 3GPP R99 specifications. TS is a technical standard and TR is a technical report. The specifications relevant to this book are highlighted in bold.

## 21 Series

| | | |
|---|---|---|
| TS | 21.101 | Technical Specifications and Technical Reports for a UTRAN-based 3GPP System |
| TS | 21.111 | USIM and IC Card Requirements |
| TS | 21.133 | 3G Security; Security Threats and Requirements |
| TR | 21.810 | Report on Multi-Mode UE Issues; Ongoing Work and Identified Additional Work |
| TR | 21.900 | Technical Specification Group Working Methods |
| TR | 21.904 | User Equipment (UE) Capability Requirements |
| TR | 21.905 | Vocabulary for 3GPP Specifications |
| TR | 21.910 | Multi-Mode UE Issues; Categories, Principles and Procedures |
| TR | 21.978 | Feasibility Technical Report; CAMEL Control of VoIP Services |

## 22 Series

| | | |
|---|---|---|
| TS | 22.001 | Principles of Circuit Telecommunication Services Supported by a Public Land Mobile Network |
| TS | 22.002 | Circuit Bearer Services (BS) Supported by a Public Land Mobile Network (PLMN) |
| TS | 22.003 | Circuit Teleservices Supported by a Public Land Mobile Network (PLMN) |
| TS | 22.004 | General on Supplementary Services |
| TS | 22.011 | Service Accessibility |
| TS | 22.016 | International Mobile Equipment Identities (IMEI) |
| TS | 22.022 | Personalisation of Mobile Equipment (ME); Mobile Functionality Specification |
| TS | 22.024 | Description of Charge Advice Information (CAI) |
| TS | 22.030 | Man–Machine Interface (MMI) of the User Equipment (UE) |
| TS | 22.031 | Fraud Information Gathering System (FIGS); Service Description; Stage 1 |
| TS | 22.032 | Immediate Service Termination (IST); Service Description; Stage 1 |
| TS | 22.034 | High Speed Circuit Switched Data (HSCSD); Stage 1 |
| TS | 22.038 | USIM/SIM Application Toolkit (USAT/SAT); Service Description; Stage 1 |
| TS | 22.041 | Operator Determined Call Barring |
| TS | 22.042 | Network Identity and Time Zone (NITZ) Service Description; Stage 1 |
| TS | 22.057 | Mobile Execution Environment (MExE) Service Description; Stage 1 |
| TS | 22.060 | General Packet Radio Service (GPRS); Service Description; Stage 1 |

## Appendix

| | | |
|---|---|---|
| TS | 22.066 | Support of Mobile Number Portability (MNP); Stage 1 |
| TS | 22.067 | Enhanced Multi-Level Precedence and Pre-emption Service (eMLPP); Stage 1 |
| TS | 22.071 | Location Services (LCS); Stage 1 |
| TS | 22.072 | Call Deflection (CD); Stage 1 |
| TS | 22.078 | Customized Applications for Mobile network Enhanced Logic (CAMEL); Service Description; Stage 1 |
| TS | 22.079 | Support of Optimal Routing; Stage 1 |
| TS | 22.081 | Line Identification Supplementary Services; Stage 1 |
| TS | 22.082 | Call Forwarding (CF) Supplementary Services; Stage 1 |
| TS | 22.083 | Call Waiting (CW) and Call Hold (HOLD) Supplementary Services; Stage 1 |
| TS | 22.084 | MultiParty (MPTY) Supplementary Service; Stage 1 |
| TS | 22.085 | Closed User Group (CUG) Supplementary Services; Stage 1 |
| TS | 22.086 | Advice of Charge (AoC) Supplementary Services; Stage 1 |
| TS | 22.087 | User-to-User Signalling (UUS); Stage 1 |
| TS | 22.088 | Call Barring (CB) Supplementary Services; Stage 1 |
| TS | 22.090 | Unstructured Supplementary Service Data (USSD); Stage 1 |
| TS | 22.091 | Explicit Call Transfer (ECT) Supplementary Service; Stage 1 |
| TS | 22.093 | Completion of Calls to Busy Subscriber (CCBS); Service Description; Stage 1 |
| TS | 22.094 | Follow Me Service Description; Stage 1 |
| TS | 22.096 | Name Identification Supplementary Services; Stage 1 |
| TS | 22.097 | Multiple Subscriber Profile (MSP) Phase 1; Service Description; Stage 1 |
| TS | 22.100 | UMTS Phase 1 |
| TS | 22.101 | Service Aspects; Service Principles |
| TS | 22.105 | Services and Service Capabilities |
| TS | 22.115 | Service Aspects Charging and Billing |
| TR | 22.121 | Service Aspects; The Virtual Home Environment; Stage 1 |
| TS | 22.129 | Handover Requirements Between UTRAN and GERAN or Other Radio Systems |
| TS | 22.135 | Multicall; Service Description; Stage 1 |
| TS | 22.140 | Multimedia Messaging Service (MMS); Stage 1 |
| TR | 22.945 | Study of Provision of Fax Service in GSM and UMTS |
| TR | 22.971 | Automatic Establishment of Roaming Relationships |
| TR | 22.975 | Advanced Addressing |

## 23 Series

| | | |
|---|---|---|
| TS | **23.002** | **Network Architecture** |
| TS | **23.003** | **Numbering, Addressing and Identification** |
| TS | 23.007 | Restoration Procedures |
| TS | 23.008 | Organisation of Subscriber Data |
| TS | 23.009 | Handover Procedures |
| TS | 23.011 | Technical Realization of Supplementary Services |
| TS | **23.012** | **Location Management Procedures** |
| TS | 23.014 | Support of Dual Tone Multi Frequency (DTMF) Signalling |
| TS | 23.015 | Technical Realization of Operator Determined Barring (ODB) |
| TS | 23.016 | Subscriber Data Management; Stage 2 |
| TS | **23.018** | **Basic Call Handling; Technical Realization** |
| TS | 23.031 | Fraud Information Gathering System (FIGS); Service Description; Stage 2 |
| TS | 23.032 | Universal Geographical Area Description (GAD) |
| TS | 23.034 | High Speed Circuit Switched Data (HSCSD); Stage 2 |
| TS | 23.035 | Immediate Service Termination (IST); Stage 2 |
| TS | 23.038 | Alphabets and Language-Specific Information |
| TR | 23.039 | Interface Protocols for the Connection of Short Message Service Centers (SMSCs) to Short Message Entities (SMEs) |
| TS | **23.040** | **Technical realization of Short Message Service (SMS)** |
| TS | **23.041** | **Technical realization of Cell Broadcast Service (CBS)** |

| | | |
|---|---|---|
| TS | 23.042 | Compression Algorithm for SMS |
| TS | 23.057 | Mobile Execution Environment (MExE); Functional Description; Stage 2 |
| **TS** | **23.060** | **General Packet Radio Service (GPRS) Service Description; Stage 2** |
| TS | 23.066 | Support of GSM Mobile Number Portability (MNP) Stage 2 |
| TS | 23.067 | Enhanced Multi-Level Precedence and Pre-emption Service (eMLPP); Stage 2 |
| TS | 23.072 | Call Deflection Supplementary Service; Stage 2 |
| TS | 23.078 | Customized Applications for Mobile network Enhanced Logic (CAMEL); Stage 2 |
| TS | 23.079 | Support of Optimal Routing (SOR); Technical Realization; Stage 2 |
| TS | 23.081 | Line Identification Supplementary Services; Stage 2 |
| TS | 23.082 | Call Forwarding (CF) Supplementary Services; Stage 2 |
| TS | 23.083 | Call Waiting (CW) and Call Hold (HOLD) Supplementary Service; Stage 2 |
| TS | 23.084 | MultiParty (MPTY) Supplementary Service; Stage 2 |
| TS | 23.085 | Closed User Group (CUG) Supplementary Service; Stage 2 |
| TS | 23.086 | Advice of Charge (AoC) Supplementary Service; Stage 2 |
| TS | 23.087 | User-to-User Signalling (UUS) Supplementary Service; Stage 2 |
| TS | 23.088 | Call Barring (CB) Supplementary Service; Stage 2 |
| TS | 23.090 | Unstructured Supplementary Service Data (USSD); Stage 2 |
| TS | 23.091 | Explicit Call Transfer (ECT) Supplementary Service; Stage 2 |
| TS | 23.093 | Technical Realization of Completion of Calls to Busy Subscriber (CCBS); Stage 2 |
| TS | 23.094 | Follow Me; Stage 2 |
| TS | 23.096 | Name Identification Supplementary Service; Stage 2 |
| TS | 23.097 | Multiple Subscriber Profile (MSP) Phase 1; Stage 2 |
| **TS** | **23.101** | **General UMTS Architecture** |
| **TS** | **23.107** | **Quality of Service (QoS) concept and architecture** |
| TS | 23.108 | Mobile Radio Interface Layer 3 Specification Core Network Protocols; Stage 2 (structured procedures) |
| **TS** | **23.110** | **UMTS Access Stratum Services and Functions** |
| TS | 23.116 | Super-Charger Technical Realization; Stage 2 |
| TS | 23.119 | Gateway Location Register (GLR); Stage 2 |
| **TS** | **23.121** | **Architectural Requirements for Release 1999** |
| **TS** | **23.122** | **Non-Access-Stratum Functions Related to Mobile Station (MS) in Idle Mode** |
| TS | 23.127 | Virtual Home Environment (VHE) / Open Service Access (OSA); Stage 2 |
| **TS** | **23.135** | **Multicall Supplementary Service; Stage 2** |
| TS | 23.140 | Multimedia Messaging Service (MMS); Functional Description; Stage 2 |
| TS | 23.171 | Location Services (LCS); Functional Description; Stage 2 (UMTS) |
| TR | 23.814 | Separating RR and MM Specific Parts of the MS Classmark |
| TR | 23.908 | Technical Report on Pre-Paging |
| TR | 23.909 | Technical Report on the Gateway Location Register |
| TR | 23.910 | Circuit Switched Data Bearer Services |
| TR | 23.911 | Technical Report on Out-of-Band Transcoder Control |
| TR | 23.912 | Technical Report on Super-Charger |
| TR | 23.923 | Combined GSM and Mobile IP Mobility Handling in UMTS IP CN |
| TR | 23.930 | Iu Principles |
| TR | 23.972 | Circuit Switched Multimedia Telephony |

## 24 Series

| | | |
|---|---|---|
| TS | 24.002 | GSM-UMTS Public Land Mobile Network (PLMN) Access Reference Configuration |
| **TS** | **24.007** | **Mobile Radio Interface Signalling Layer 3; General Aspects** |
| **TS** | **24.008** | **Mobile Radio Interface Layer 3 Specification; Core Network Protocols; Stage 3** |
| TS | 24.010 | Mobile Radio Interface Layer 3 – Supplementary Services Specification – General Aspects |

| TS | 24.011 | Point-to-Point (PP) Short Message Service (SMS) Support on Mobile Radio Interface |
|---|---|---|
| TS | 24.022 | Radio Link Protocol (RLP) for Circuit Switched Bearer and Teleservices |
| TS | 24.030 | Location Services (LCS); Supplementary Service Operations; Stage 3 |
| TS | 24.067 | Enhanced Multi-Level Precedence and Pre-Emption Service (eMLPP); Stage 3 |
| TS | 24.072 | Call Deflection Supplementary Service; Stage 3 |
| TS | 24.080 | Mobile Radio Layer 3 Supplementary Service Specification; Formats and Coding |
| TS | 24.081 | Line Identification Supplementary Service; Stage 3 |
| TS | 24.082 | Call Forwarding Supplementary Service; Stage 3 |
| TS | 24.083 | Call Waiting (CW) and Call Hold (HOLD) Supplementary Service; Stage 3 |
| TS | 24.084 | MultiParty (MPTY) Supplementary Service; Stage 3 |
| TS | 24.085 | Closed User Group (CUG) Supplementary Service; Stage 3 |
| TS | 24.086 | Advice of Charge (AoC) Supplementary Service; Stage 3 |
| TS | 24.087 | User-to-User Signalling (UUS); Stage 3 |
| TS | 24.088 | Call Barring (CB) Supplementary Service; Stage 3 |
| TS | 24.090 | Unstructured Supplementary Service Data (USSD); Stage 3 |
| TS | 24.091 | Explicit Call Transfer (ECT) Supplementary Service; Stage 3 |
| TS | 24.093 | Call Completion to Busy Subscriber (CCBS); Stage 3 |
| TS | 24.096 | Name Identification Supplementary Service; Stage 3 |
| **TS** | **24.135** | **Multicall Supplementary Service; Stage 3** |

## 25 Series

| TS | 25.101 | **User Equipment (UE) Radio Transmission and Reception (FDD)** |
|---|---|---|
| TS | 25.102 | User Equipment (UE) Radio Transmission and Reception (TDD) |
| TS | 25.104 | **Base Station (BS) Radio Transmission and Reception (FDD)** |
| TS | 25.105 | UTRA (BS) TDD: Radio Transmission and Reception |
| TS | 25.113 | Base Station and Repeater Electromagnetic Compatibility (EMC) |
| TS | 25.123 | **Requirements for Support of Radio Resource Management (TDD)** |
| TS | 25.133 | **Requirements for Support of Radio Resource Management (FDD)** |
| TS | 25.141 | Base Station (BS) Conformance Testing (FDD) |
| TS | 25.142 | Base Station (BS) Conformance Testing (TDD) |
| TS | 25.201 | **Physical Layer – General Description** |
| TS | 25.211 | **Physical Channels and Mapping of Transport Channels onto Physical Channels (FDD)** |
| TS | 25.212 | **Multiplexing and Channel Coding (FDD)** |
| TS | 25.213 | **Spreading and Modulation (FDD)** |
| TS | 25.214 | **Physical Layer Procedures (FDD)** |
| TS | 25.215 | **Physical Layer; Measurements (FDD)** |
| TS | 25.221 | Physical Channels and Mapping of Transport Channels onto Physical Channels (TDD) |
| TS | 25.222 | Multiplexing and Channel Coding (TDD) |
| TS | 25.223 | Spreading and Modulation (TDD) |
| TS | 25.224 | Physical Layer Procedures (TDD) |
| TS | 25.225 | Physical Layer; Measurements (TDD) |
| **TS** | **25.301** | **Radio Interface Protocol Architecture** |
| **TS** | **25.302** | **Services Provided by the Physical Layer** |
| **TS** | **25.303** | **Interlayer Procedures in Connected Mode** |
| **TS** | **25.304** | **User Equipment (UE) Procedures in Idle Mode and Procedures for Cell Reselection in Connected Mode** |
| **TS** | **25.305** | **User Equipment (UE) Positioning in Universal Terrestrial Radio Access Network (UTRAN); Stage 2** |
| **TS** | **25.306** | **UE Radio Access Capabilities Definition** |
| **TS** | **25.307** | **Requirements on UEs supporting a Release-Independent Frequency Band** |
| **TS** | **25.321** | **Medium Access Control (MAC) Protocol Specification** |

| | | |
|---|---|---|
| TS | 25.322 | Radio Link Control (RLC) Protocol Specification |
| TS | 25.323 | Packet Data Convergence Protocol (PDCP) Specification |
| TS | 25.324 | Broadcast/Multicast Control (BMC) |
| TS | 25.331 | Radio Resource Control (RRC) Protocol Specification |
| TS | 25.401 | UTRAN Overall Description |
| TS | 25.402 | Synchronisation in UTRAN Stage 2 |
| TS | 25.410 | UTRAN Iu Interface: General Aspects and Principles |
| TS | 25.411 | UTRAN Iu Interface Layer 1 |
| TS | 25.412 | UTRAN Iu Interface Signalling Transport |
| TS | 25.413 | UTRAN Iu Interface Radio Access Network Application Part (RANAP) Signalling |
| TS | 25.414 | UTRAN Iu Interface Data Transport & Transport Signalling |
| TS | 25.415 | UTRAN Iu Interface User Plane Protocols |
| TS | 25.419 | UTRAN Iu-BC Interface: Service Area Broadcast Protocol (SABP) |
| TS | 25.420 | UTRAN Iur Interface: General Aspects and Principles |
| TS | 25.421 | UTRAN Iur Interface Layer 1 |
| TS | 25.422 | UTRAN Iur Interface Signalling Transport |
| TS | 25.423 | UTRAN Iur Interface Radio Network Subsystem Application Part (RNSAP) Signalling |
| TS | 25.424 | UTRAN Iur Interface Data Transport & Transport Signalling for CCH Data Streams |
| TS | 25.425 | UTRAN Iur Interface User Plane Protocols for CCH Data Streams |
| TS | 25.426 | UTRAN Iur and Iub Interface Data Transport & Transport Signalling for DCH Data Streams |
| TS | 25.427 | UTRAN Iur and Iub Interface User Plane Protocols for DCH Data Streams |
| TS | 25.430 | UTRAN Iub Interface: General Aspects and Principles |
| TS | 25.431 | UTRAN Iub Interface Layer 1 |
| TS | 25.432 | UTRAN Iub Interface: Signalling Transport |
| TS | 25.433 | UTRAN Iub Interface NBAP Signalling |
| TS | 25.434 | UTRAN Iub Interface Data Transport & Transport Signalling for CCH Data Streams |
| TS | 25.435 | UTRAN Iub Interface User Plane Protocols for CCH Data Streams |
| TS | 25.442 | UTRAN Implementation-Specific O&M Transport |
| TR | 25.832 | Manifestations of Handover and SRNS Relocation |
| TR | 25.833 | Physical Layer Items Not for Inclusion in Release 99 |
| TR | 25.853 | Delay Budget within the Access Stratum |
| TR | 25.921 | Guidelines and Principles for Protocol Description and Error Handling |
| TR | 25.922 | Radio Resource Management Strategies |
| TR | 25.925 | Radio Interface for Broadcast/Multicast Services |
| TR | 25.931 | UTRAN Functions, Examples on Signalling Procedures |
| TR | 25.941 | Document Structure |
| TR | 25.942 | RF System Scenarios |
| TR | 25.944 | Channel Coding and Multiplexing Examples |
| TR | 25.993 | Typical Examples of Radio Access Bearers (RABs) and Radio Bearers (RBs) Supported by Universal Terrestrial Radio Access (UTRA) |

## 26 Series

| | | |
|---|---|---|
| TS | 26.071 | AMR Speech Codec; General Description |
| TS | 26.073 | AMR Speech Codec; C-Source Code |
| TS | 26.074 | AMR Speech Codec; Test Sequences |
| TS | 26.090 | AMR Speech Codec; Transcoding Functions |
| TS | 26.091 | AMR Speech Codec; Error Concealment of Lost Frames |
| TS | 26.092 | AMR Speech Codec; Comfort Noise for AMR Speech Traffic Channels |

| | | |
|---|---|---|
| TS | 26.093 | AMR Speech Codec; Source Controlled Rate Operation |
| TS | 26.094 | AMR Speech Codec; Voice Activity Detector for AMR Speech Traffic Channels |
| TS | 26.101 | Mandatory Speech Codec Speech Processing Functions; Adaptive Multi-Rate (AMR) Speech Codec Frame Structure |
| TS | 26.102 | Adaptive Multi-Rate (AMR) Speech Codec; Interface to Iu and Uu |
| TS | 26.103 | Speech Codec List for GSM and UMTS |
| TS | 26.104 | ANSI-C Code for the Floating-Point Adaptive Multi-Rate (AMR) Speech Codec |
| TS | 26.110 | Codec for Circuit Switched Multimedia Telephony Service; General Description |
| TS | 26.111 | Codec for Circuit Switched Multimedia Telephony Service; Modifications to H.324 |
| TS | 26.131 | Terminal Acoustic Characteristics for Telephony; Requirements |
| TS | 26.132 | Narrow Band (3,1 kHz) Speech and Video Telephony Terminal Acoustic Test Specification |
| TR | 26.911 | Codec for Circuit Switched Multimedia Telephony Service; Terminal Implementer's Guide |
| TR | 26.912 | Codec for Circuit Switched Multimedia Telephony Service; Quantitative Performance Evaluation of H.324 Annex C over 3G |
| TR | 26.915 | Echo Control For Speech and Multi-Media Services |
| TR | 26.975 | Performance Characterization of the Adaptive Multi-Rate (AMR) Speech Codec |

## 27 Series

| | | |
|---|---|---|
| TS | 27.001 | General on Terminal Adaptation Functions (TAF) for Mobile Stations (MS) |
| TS | 27.002 | Terminal Adaptation Functions (TAF) for Services using Asynchronous Bearer Capabilities |
| TS | 27.003 | Terminal Adaptation Functions (TAF) for Services Using Synchronous Bearer Capabilities |
| TS | 27.005 | Use of Data Terminal Equipment – Data Circuit Terminating Equipment (DTE-DCE) Interface for Short Message Service (SMS) and Cell Broadcast Service (CBS) |
| TS | 27.007 | AT Command Set for 3G User Equipment (UE) |
| TS | 27.010 | Terminal Equipment to User Equipment (TE-UE) Multiplexer Protocol |
| TS | 27.060 | Packet Domain; Mobile Station (MS) Supporting Packet Switched Services |
| TS | 27.103 | Wide Area Network Synchronization |
| TR | 27.901 | Report on Terminal Interfaces – An Overview |
| TR | 27.903 | Discussion of Synchronization Standards |

## 29 Series

| | | |
|---|---|---|
| TS | 29.002 | Mobile Application Part (MAP) Specification |
| TS | 29.007 | General Requirements on Interworking Between the Public Land Mobile Network (PLMN) and the Integrated Services Digital Network (ISDN) or Public Switched Telephone Network |
| TS | 29.010 | Information Element Mapping between Mobile Station – Base Station System (MS – BSS) and Base Station System – Mobile-Services Switching Centre (BSS – MCS) Signalling |
| TS | 29.011 | Signalling Interworking for Supplementary Services |
| TS | 29.013 | Signalling Interworking between ISDN Supplementary Services Application Service Element (ASE) and Mobile Application Part (MAP) Protocols |
| TS | 29.016 | Serving GPRS Support Node SGSN – Visitors Location Register (VLR); Gs Interface Network Service Specification |

| | | |
|---|---|---|
| TS | 29.018 | General Packet Radio Service (GPRS); Serving GPRS Support Node (SGSN) – Visitors Location Register (VLR); Gs Interface Layer 3 Specification |
| TS | 29.060 | General Packet Radio Service (GPRS); GPRS Tunnelling Protocol (GTP) across the Gn and Gp Interface |
| TS | 29.061 | Interworking between the Public Land Mobile Network (PLMN) supporting Packet Based Services and Packet Data Networks (PDN) |
| TS | 29.078 | Customized Applications for Mobile network Enhanced Logic (CAMEL); CAMEL Application Part (CAP) Specification |
| TS | 29.108 | Application of the Radio Access Network Application Part (RANAP) on the E-Interface |
| TS | 29.119 | GPRS Tunnelling Protocol (GTP) Specification for Gateway Location Register (GLR) |
| TS | 29.120 | Mobile Application Part (MAP) Specification for Gateway Location Register (GLR); Stage 3 |
| TS | 29.198 | Open Service Architecture (OSI) Application Programming Interface (API) – Part 1 |
| TR | 29.994 | Recommended Infrastructure Measures to Overcome Specific Mobile Station (MS) and User Equipment (UE) Faults |
| TR | 29.998 | Open Services Architecture API Part 2 |

## 31 Series

| | | |
|---|---|---|
| TS | 31.101 | **UICC-Terminal Interface; Physical and Logical Characteristics** |
| TS | 31.102 | **Characteristics of the USIM Application** |
| TS | 31.110 | Numbering System for Telecommunication IC Card Applications |
| TS | 31.111 | Universal Subscriber Identity Module Application Toolkit (USAT) |
| TS | 31.120 | **UICC-Terminal Interface; Physical, Electrical and Logical Test Specification** |
| TS | 31.121 | UICC-Terminal Interface; Universal Subscriber Identity Module (USIM) Application Test Specification |
| TS | 31.122 | Universal Subscriber Identity Module (USIM) Conformance Test Specification |

## 32 Series

| | | |
|---|---|---|
| TS | 32.005 | Telecommunications Management; Charging and Billing; 3G Call and Event Data for the Circuit Switched (CS) Domain |
| TS | 32.015 | Telecommunications Management; Charging and Billing; 3G Call and Event Data for the Packet Switched (PS) Domain |
| TS | 32.101 | Telecommunication Management; Principles and High Level Requirements |
| TS | 32.102 | Telecommunication Management; Architecture |
| TS | 32.104 | 3G Performance Management |
| TS | 32.106–1 | Telecommunication Management; Configuration Management (CM); Part 1: Concept and Requirements |
| TS | 32.106–2 | Telecommunication Management; Configuration Management (CM); Part 2: Notification Integration Reference Point (IRP): Information Service |
| TS | 32.106–3 | Telecommunication Management; Configuration Management (CM); Part 3: Notification Integration Reference Point (IRP); Common Object Request Broker Architecture (CORBA) |
| TS | 32.106–4 | Telecommunication Management; Configuration Management (CM); Part 4: Notification Integration Reference Point (IRP); Common Management Information Protocol (CMIP) Solution |

| | | |
|---|---|---|
| TS | 32.106–5 | Telecommunication Management; Configuration Management (CM); Part 5: Basic Configuration Management Integration Reference Point (IRP): Information Model |
| TS | 32.106–6 | Telecommunication Management; Configuration Management (CM); Part 6: Basic Configuration Management Integration Reference Point (IRP): Common Object Request Broker |
| TS | 32.106–7 | Telecommunication Management; Configuration Management (CM); Part 7: Basic Configuration Management Integration Reference Point (IRP): Common Management Information |
| TS | 32.106–8 | Telecommunication Management; Configuration Management (CM); Part 8: Name Convention for Managed Objects |
| TS | 32.111–1 | Telecommunication Management; Fault Management; Part 1: 3G Fault Management Requirements |
| TS | 32.111–2 | Telecommunication Management; Fault Management; Part 2: Alarm Integration Reference Point (IRP): Information Service |
| TS | 32.111–3 | Telecommunication Management; Fault Management; Part 3: Alarm Integration Reference Point (IRP): Common Object Request Broker Architecture (CORBA) Solution Set |
| TS | 32.111–4 | Telecommunication Management; Fault Management; Part 4: Alarm Integration Reference Point (IRP): Common Management Information Protocol (CMIP) Solution Set |

## 33 Series

| | | |
|---|---|---|
| TS | **33.102** | **3G Security; Security Architecture** |
| TS | **33.103** | **3G Security; Integration Guidelines** |
| TS | **33.105** | **Cryptographic Algorithm Requirements** |
| TS | 33.106 | Lawful Interception Requirements |
| TS | 33.107 | 3G Security; Lawful Interception Architecture and Functions |
| TS | 33.120 | Security Objectives and Principles |
| TR | 33.901 | Criteria for Cryptographic Algorithm Design Process |
| TR | 33.902 | Formal Analysis of the 3G Authentication Protocol |
| TR | 33.908 | 3G Security; General Report on the Design, Specification and Evaluation of 3GPP Standard Confidentiality and Integrity Algorithms |

## 34 Series

| | | |
|---|---|---|
| TS | **34.108** | **Common Test Environments for User Equipment (UE) Conformance Testing** |
| TS | 34.109 | Terminal Logical Test Interface; Special Conformance Testing Functions |
| TS | 34.121 | Terminal Conformance Specification, Radio Transmission and Reception (FDD) |
| TS | 34.122 | Terminal Conformance Specification, Radio Transmission and Reception (TDD) |
| TS | 34.123–1 | User Equipment (UE) Conformance Specification; Part 1: Protocol Conformance Specification |
| TS | 34.123–2 | User Equipment (UE) Conformance Specification; Part 2: Implementation Conformance Statement (ICS) Specification |
| TS | 34.123–3 | User Equipment (UE) Conformance Specification; Part 3: Abstract Test Suites (ATSs) |
| TS | 34.124 | Electromagnetic Compatibility (EMC) Requirements for Mobile Terminals and Ancillary Equipment |
| TR | 34.901 | Test Time Optimisation Based on Statistical Approaches; Statistical Theory Applied and Evaluation of Statistical Significance |

TR 34.907 Report on Electrical Safety Requirements and Regulations
TR 34.925 Specific Absorption Rate (SAR) Requirements and Regulations in Different Regions

## 35 Series

TS 35.201 **Specification of the 3GPP Confidentiality and Integrity Algorithms; Document 1: f8 and f9 Specifications**
TS 35.202 **Specification of the 3GPP Confidentiality and Integrity Algorithms; Document 2: Kasumi Algorithm Specification**
TS 35.203 Specification of the 3GPP Confidentiality and Integrity Algorithms; Document 3: Implementers' Test Data
TS 35.204 Specification of the 3GPP Confidentiality and Integrity Algorithms; Document 4: Design Conformance Test Data

# References

[1] D. McDysan and D. Spohn. *ATM Theory and Applications*. McGraw-Hill, 1999.
[2] 3GPP Technical Specification 25.410: UTRAN Iu Interface: General Aspects and Principles.
[3] P. Calhoun et al. Diameter Base Protocol. RFC 3588, September 2003.
[4] 3GPP Technical Specification 33.105: Cryptographic Algorithm Requirements.
[5] J. Proakis. *Digital Communications*. McGraw-Hill, 1989.
[6] R. Ziemer and R. Peterson. *Introduction to Digital Communication*. Prentice-Hall, 2001.
[7] 3GPP Technical Specification 25.213: Spreading and Modulation (FDD).
[8] 3GPP Technical Specification 25.101: UE Radio Transmission and Reception (FDD).
[9] F. Halsall. *Data Communications, Computer Networks and Open Systems*. Addison-Wesley, 1996.
[10] 3GPP Technical Specification 25.104: UTRA (BS) FDD; Radio Transmission and Reception.
[11] U. Rohde, J. Whitaker and T. Bucher. *Communications Receivers*. McGraw-Hill, 1997.
[12] B. Razavi. Design considerations for direct-conversion receivers. *IEEE Trans. on Circuits and Systems*, **44**: 428–435, 1997.
[13] C. Iversen. A UTRA/FDD Receiver Architecture and LNA in CMOS Technology. Ph.D. thesis, Aalborg University, Denmark, November 2001.
[14] 3GPP Technical Specification 25.306: UE Radio Access Capabilities Definition.
[15] 3GPP Technical Specification 25.212: Multiplexing and Channel Coding (FDD).
[16] P. Elias. Coding for noisy channels. *IRE Conv. Rec., Part 4*, 37–47, 1955.
[17] A. Viterbi. Error bounds for convolutional codes and an asymptotically optimum decoding algorithm. *IEEE Trans. Inf. Theory*, **IT-13**: 260–269, 1967.
[18] C. Berrou, A. Glavieux and P. Thitimajshima. Near Shannon limit error-correcting coding and decoding: Turbo codes. *ICC'93 Geneva, Switzerland*, 1993.
[19] C. Berrou and A. Glavieux. Reflections on the Prize Paper: 'Near optimum error-correcting coding and decoding: turbo codes'. http://www.ieeeits.org/publications/nltr/98_jun/reflections.html
[20] Texas Instruments, Implementation of a WCDMA rake receiver on a TMS320C62x DSP Device. Application Report SPRA680, July 2000.
[21] B. Vejlgaard, P. Mogensen and J. Knudsen. Performance analysis for UMTS downlink receiver with practical aspects. *IEEE VT Conf. Amsterdam* pp. 998–1002, 1999.
[22] D. Parsons. *The Mobile Radio Propagation Channel*. Wiley, 1994.
[23] 3GPP Technical Specification 25.133: Requirements for Support of Radio Resource Management (FDD).
[24] 3GPP Technical Specification 25.331: Radio Resource Control (RRC) Protocol Specification.
[25] J. Iinatti and K. Hooli. Effect of signal quantisation on WCDMA code acquisition. *Electron. Lett.* **36**(2): 187–189, 2000.

[26] Siemens and Texas Instruments. Generalised hierarchical Golay sequence for PSC with low complexity correlation using pruned efficient Golay correlators. 3GPP TSG-RAN W1 Tdoc 99–554, June 1999.

[27] R. Yarlagadda and J. Hershey. *Hadamard Matrix Analysis and Synthesis with Applications to Communications and Signal/Image Processing*. Kluwer Academic Publishers, 1997.

[28] 3GPP Technical Specification 25.322: Radio Link Control (RLC) Protocol Specification.

[29] 3GPP Technical Report 25.853: Delay Budget within the Access Stratum.

[30] M. Degermark *et al.* IP header compression. *RFC2507*, February 1999.

[31] C. Bormann *et al.* RObust Header Compression (ROHC): Framework and four profiles: RTP, UDP, ESP, and uncompressed. *RFC3095*, July 2001.

[32] 3GPP Technical Specification 24.008: Mobile Radio Interface Layer 3 Specification; Core Network Protocols; Stage 3.

[33] 3GPP Technical Specification 24.007: Mobile Radio Interface Signalling Layer 3; General Aspects.

[34] 3GPP Technical Specification 24.011: Point-to-Point (PP) Short Message Service (SMS) Support on Mobile Radio Interface.

[35] 3GPP Technical Specification 24.010: Mobile Radio Interface Layer 3 – Supplementary Services Specification – General Aspects.

[36] 3GPP Technical Specification 23.060: General Packet Radio Service (GPRS) Service Description; Stage 2.

[37] 3GPP Technical Specification 23.122: Non-Access-Stratum Functions Related to Mobile Station (MS) in Idle Mode.

[38] 3GPP Technical Specification 31.120: UICC-Terminal Interface; Physical, Electrical and Logical Test Specification.

[39] 3GPP Technical Specification 25.123: Requirements for Support of Radio Resource Management (TDD).

[40] 3GPP Technical Specification 25.304: UE Procedures in Idle Mode and Procedures for Cell Reselection in Connected Mode.

[41] 3GPP Technical Specification 21.905: Vocabulary for 3GPP Specifications.

[42] J. Hagenauer, E. Offer, and L. Papke. Iterative decoding of binary block and convolutional codes. *IEEE Trans. on Inf. Theory*, **42**(2): 429–445, 1996.

# Index

2 Mb/s, introduction to 8
3G
   evolution from 2G 4
   IMT2000 1–3
   origins and terminology 1–3
   other technologies 1

access point name (APN) *see* session management protocol
access preambles *see* physical random access channel and physical common packet channel
access service class selection *see* MAC
access stratum, introduction to 9–10, 15
acknowledged mode
   configuration 322
   DATA PDU (AMD) 317
   entity structure 315–17
   logical channel usage 315
   piggyback STATUS PDU 318
   RESET and RESET ACK PDU 320–1
   state diagram 321
   state variables 321
   STATUS PDU 317
acknowledged mode (AM RLC) 314–35
acknowledged mode operation 329–35
   interactions with MAC 330–2
   use of
      Timer_Poll 332
      Timer_Poll_Prohibit 332
      VT(A) 332
      VT(DAT) 332
      VT(MS) 332
      VT(S) 332
      VT(SDU) 332
acknowledged mode polling 322–6
   every Poll_PDU PDU 326
   every Poll_SDU SDU 326
   last PDU in buffer 323
   last PDU in retransmission buffer 323
   poll timer 326
   timer based polling 326
   window based polling 326

acknowledged mode RESET 335
acknowledged mode timers 326–9
   Timer_Discard 328
   Timer_EPC 327
   Timer_MRW 328
   Timer_Poll 327
   Timer_Poll_Periodic 328
   Timer_Poll_Prohibit 327
   Timer_RST 328
   Timer_Status_Periodic 328
   Timer_Status_Prohibit 328
acquisition indication channel (AICH)
   acquisition indicators 126
   acquisition signatures 132
addressing
   cell radio network temporary identifier (c-RNTI) 47, 261
   IMSI 46
   introduction to 46–7
   mobile country code (MCC) 46, 452
   mobile network code (MNC) 46, 452
   mobile subscriber identity number (MSIN) 46
   P-TMSI 47, 51
   serving radio network temporary identifier (s-RNTI) 47
   TMSI 46, 51, 447
   UTRAN radio network temporary identifier (u-RNTI) 47, 261
adjacent channel leakage ratio (ACLR) *see* transmitter
AICH transmission timing (ATT) parameter *see* random access
analogue to digital converter (ADC) 184–7
   dynamic range 186–7
   sampling rate 184
anonymity key (AK) *see* security
asynchronous cells 88, 93
asynchronous transfer mode 13, 14
attach flag (ATT) 450, 453–5
audio service, typical characteristics of 7
authentication *see* security
authentication and key management field (AMF) *see* security

553

# Index

authentication token (AUTN) *see* security
authentication vector *see* security

bandwidth expansion from spreading *see* codes
Bayes' theorem *see* turbo decoder
bit error rate, definition of 6
broadcast and multicast control 27
broadcast channel (BCH) 123, 251
   as part of CCTrCH 219
broadcast control channel (BCCH) 250
broadcast control domain, introduction to 10
broadcast multicast control 344–8
   CBS message PDU 347
   network architecture 344
   protocol architecture 344–5
   schedule message PDU 347

C/T multiplexing *see* MAC
calculated transport format combination (CTFC) 292, 391–2
call control 440
   multi-call 440
   radio access bearer identity (RAB-Id) 440
   stream identifier (SI) 440
   transaction identifier (TI) 440
call control protocol 456–67
   connection management sublayer 456
   introduction to 13
   mobile network layer 456
   mobile originated call establishment 456–60, 461–7
   mobile terminated call establishment 460–1
   primitives 456
   state machine 456
call session control function 18–19
carrier to interference ratio (C/I) *see* signal to interference ratio
CC messages
   ALERTING 459, 461, 467
   CALL CONFIRMED 460
   CALL PROCEEDING 459, 465
   CONNECT 460, 461, 467
   CONNECT ACKNOWLEDGE 460, 461, 467
   PROGRESS 459
   SETUP 445, 460, 462, 465
CDMA, introduction to 68–9
Cdma2000 3
cell broadcast short message service *see* services
cell change order *see* handover
cell phase reference 124
cell radio network temporary identifier (c-RNTI) *see* addressing
cell timing 134–5
CELL_DCH state *see* RRC mode
CELL_FACH state *see* RRC mode
CELL_INFO_LIST *see* measurement variables
CELL_PCH state *see* RRC mode
channel coding, transport channels 224

channel estimation 204, 206
   architecture for 205
channelisation codes *see* codes
chip rate, definition of 70–1
chips
   definition of 71
   relationship to modulation *see* codes
cipher key (CK) *see* security
ciphering *see* security
circuit switched, introduction to R99 service domain 10–13, 15
code block segmentation 224
code groups 130, 133–4, 209
code sets 133–4
coded composite transport channel (CCTrCH) 218–21, 228
   downlink aspects 219–21
   uplink aspects 218–19
codes
   allocation of 70
   bandwidth expansion 72, 74
   channelisation code 34, 85, 109
   code length 74–6
   complex scrambling and modulation 105–7
   complex scrambling code generator 107–8
   complex spreading 105
   creation of OVSF code tree 82, 83
   crest factor *see* codes, peak-to-mean
   data rate control through 69
   definition of spreading 72
   downlink channelisation code use 130–1, 132
   downlink interference control via scrambling codes 90–1
   downlink scrambling code use 90–1, 130–1, 132
   effects of gain factors 86
   Fibonacci configuration scrambling code generator 107, 108
   Galois configuration scrambling code generator 108
   Gold codes 95, 96–7, 107, 108, 112, 132
   Hadamard 82, 130, 209
   introduction to channelisation 69–70, 71–87
   introduction to scrambling 69–70
   introduction to scrambling code 34
   Kasami codes 95–6, 114
   loss of orthogonality 86
   M-sequences 95
   multiuser despreading 78–80
   multiuser spreading 76–8
   need for scrambling code 88
   orthogonal variable spreading factor (OVSF) 82–5, 86, 87, 91, 111, 128
   orthogonality 80–2
   OVSF code tree 82–5
   PCPCH message scramble code 130
   peak-to-mean 87
   PRACH message scramble code 130, 268
   pseudo-noise (PN) sequence 93, 94–5

# Index

quadrature modulation 85–6
relationship to modulation 76
scrambling code
  autocorrelation properties 93, 94–5
  cross-correlation properties 94–5
  downlink interference control 88
  implementation 95–7, 113
  introduction 87–97
  properties 93
spreading factor 71, 74–6, 85, 204, 269
spreading modulation 76
synchronization requirements 74
uplink capacity 92
uplink channelisation codes 86–7, 111, 128–30
uplink interference control 69
uplink scrambling code use 76, 91–2, 111, 113, 128–30
user separation 69
variable data rate 83–5
Walsh 82, 83
coding gain *see* receiver
collision detection *see* physical common packet channel
collision detection / channel assignment indicator channel (CD/CA ICH) *see* physical common packet channel
common control channel (CCCH) 249, 362
common packet channel (CPCH) 118, 251
  as part of CCTrCH 219
  busy table 280
  data transmission 280–2
  operation 277–82 *see also* physical common packet channel
common traffic channels (CTCH) 250
compressed mode
  introduction to 37–8
  methods of achieving 38
  slot structures 38
  use of
    puncturing in 38
    spreading factor reduction in 38
connection frame number (CFN) 363
connection management protocol, introduction to 13
connection management sublayer 439–40
  call control (CC) entity 440 *see also* call control
  GPRS short message service (GSMS) entity 440 *see also* short message service
  service access points (SAPs) *see* service access point
  session management entity 439–40 *see also* session management
  supplementary service entity 440
control plane, basic definition of 10
convolutional codes 229–35
  introduction 229
  semi-static transport format information 289
  code rate 229
  decoder 233–5

encoder 229–32
  maximum likelihood sequence estimation (MLSE) 229, 233
  soft decision decoder 234–5
  state diagram 230
  tail bits 231–2
  trellis diagram 231
  Viterbi algorithm 229, 233–4
  WCDMA 1/2 rate 181
  WCDMA 1/3 rate 232–3
  WCDMA encoder 232–3
core network, release 99 overview 9
correlator *see* rake receiver
COUNT-C *see* security
COUNT-I *see* security
CPCH control channel (CCC) *see* physical common packet channel
cyclic redundancy check
  decoder
  defined sizes 222
  encoder
  generator polynomial 223
  semi static transport format information 288
cyclic redundancy check (CRC) bits 221–3

data stream, definition of 7
DECT 3
dedicated channel (DCH) 127, 251
dedicated control channel (DCCH) 250
dedicated physical channel (DPCH)
  downlink 85, 102, 127–8
  downlink multicode transmission 127
dedicated physical control channel (DPCCH) 109, 110–11
  uplink 122
  uplink FBI bits 121
  uplink pilot bits 121
dedicated physical data channel (DPDCH) 34, 109, 110–11
  codes for 69–70
  uplink 120–2
  uplink data rate 76, 113, 121
  uplink multicode transmission 121–2, 129
  uplink spreading code, channelisation code and scrambling code *see* codes
dedicated traffic channels (DTCH) 251
direct transfer procedures *see* RRC
discontinuous reception (DRX) 44–5
  DRX cycle 360
  page indicator 358
  paging occasion 360
discontinuous transmission (DTX) *see* rate matching
downlink, data rate 76
downlink shared channel (DSCH) 127–8, 251
  as part of CCTrCH 219
DRX cycle length coefficient 368
dynamic resource allocation control (DRAC) 392–4

$E_b/N_o$ see receiver
$E_c/N_o$ see receiver
EDGE 8
encryption see security
error detection 221–3
error vector magnitude see transmitter
expected message authentication code (XMAC) see security
expected response (XRES) see security
extrinsic information see turbo decoder

fast Hadamard transformer (FHT) see system acquisition
FDD mode, introduction to 3
FDMA, introduction to 66–7
Fibonacci configuration see codes
fingers see rake receiver
finite impulse response (FIR) filter see root raised cosine filter
fixed positions for transport channels see rate matching
flexible positions for transport channels see rate matching
forward access channel (FACH) 123, 251
  as part of CCTrCH 219
frame error rate, definition of 6
FRESH see security

Galois configuration see codes
GERAN 3
GGSN, introduction to 12
GMM messages
  ATTACH ACCEPT 476, 478
  ATTACH REJECT 476
  ATTACH REQUEST 476, 477, 478–9
  AUTHENTICATION AND CIPHERING REQUEST 479
  AUTHENTICATION AND CIPHERING RESPONSE 479
  DETACH ACCEPT 479
  DETACH REQUEST 479
  ROUTING AREA UPDATE ACCEPT 481
  ROUTING AREA UPDATE REJECT 481
  ROUTING AREA UPDATE REQUEST 481
Golay sequence 123, 133, 210
Gold codes see codes
GPRS mobility management (GMM) protocol 467–81
  authentication 478–9
  combined procedures (GPRS and non-GPRS) 474–5
  detach 478–9
  GPRS attach 476–8
  introduction to 14, 28
  links between GMM states and PMM states 472
  PMM CONNECTED

PMM IDLE
procedures 472–6
routing area updating 45, 475, 480–1
states 469–72
update status 468–9
GSM to UMTS handover see handover
GU1, GU2, GU3, update status see GPRS mobility management protocol

Hadamard see codes
handover 380–91
  cell change order 391
  GSM to UMTS 388–90
  hard 385
  inter-RAT cell change order 391
  soft 382
  UMTS to GSM 385–8
header compression see packet data convergence protocol
home location register 12, 19
home subscriber server 19
hybrid phase shift keying (HPSK) see modulation
hyper frame number 57, 513

idle mode 514–15
  access stratum functions 527–37
  cell camping 529
  cell selection and reselection 523–4, 527, 540
  cell selection criteria 529
  example of idle mode procedures 537–41
  forbidden PLMNs and location areas 517
  initial cell selection 533
  introduction to 43
  location registration 524
  PLMN selection and reselection 520–4, 534–7, 538
  $Qqual_{meas}$ and $Qqualmin$ 532
  $Squal$ 531
  $Srxlev$ 530–1
  stored information cell selection 529–33
  substate machine 515–19
  suitable cell 529
  update status U1, U2, U3 516
idle mode functions 508–41
IMSI see addressing
IMT2000 frequency band 137
integrity key (IK) see security
integrity protection see security
interleaving
  first 225
  second 229
IP multimedia subsystem 16, 17–19
Iu interface
  link to BC domain 11
  link to CS domain 11
  link to PS domain 11, 14
  overview of

Kasami codes *see* codes

layer 2 protocols
    BMC *see* broadcast multicast control
    definition of 26–7
    MAC *see* medium access control
    PDCP *see* packet data convergence protocol
    RLC *see* radio link control protocol
Layer 3 protocols
    NAS *see* nonaccess stratum protocol
    RRC *see* radio resource control protocol
location area
    introduction to 45–6
    updating *see* mobility management protocol
location area code 452
location area identifier (LAI) 448, 451
location update status 452
log likelihood ratio *see* turbo decoder
log MAP algorithm *see* turbo decoder
logical channel mapping *see* MAC
logical channels 248–51
    common control 249–50
    common traffic 250
    dedicated control 250
    dedicated traffic 251
    introduction to 31–3
logical to transport channel mapping *see* MAC
lossless SRNS relocation *see* SRNS relocation

M-sequences *see* codes
MAC
    access service class selection 254, 260, 273 *see also* random access
    architecture 251–7
    backoff parameters ($N_{BO1}$) 274
    broadcast (-b) 256–7
    ciphering 260 *see also* security
    common and shared (-c/sh) 253
    control / traffic multiplexing (C/T MUX) 256, 261
    dedicated (-d) 254–6
    distributed in UTRAN
    dynamic persistence level 274
    functions and services 257–61
    logical channel mapping
    logical to transport channel mapping 258–9
    measurements 260, 264
    overview
    PDUs 261–3
    primitives 261–3, 294
    priority handling 259
    protocol
    random access procedure 264–76 *see also* random access
    SDU (RLC PDU) 261
    target channel type field (TCTF) 261
    transport channel switching 260–1

transport format selection (TFS) 253, 256, 259, 273
    *see also* transport format combination
    UE identification management 259
MAP algorithm *see* turbo decoder
maximum likelihood sequence estimation *see* convolutional codes
max-log MAP algorithm *see* turbo decoder
measurement control 400–4
    FACH measurement occasion 400
    MEASURMENT CONTROL message 402–4
    SIB11
    SIB12
measurement reporting 414–25
    additional measured results 415
    cell individual offset 417
    event triggered periodic reporting 415
    forbid cell to affect reporting 417
    GSM measurements 431–3
    hysteresis 416
    interfrequency reporting 421–3
    intrafrequency reporting (FDD mode) 417–20
    intrafrequency reporting (TDD mode) 421
    measurement filtering 414
    periodic reporting 417
    quality measurement reporting 425
    report criteria 414
    reporting range 415
    reports 414
    terminology 414–17
    time-to-trigger 416
    traffic volume reporting 424
    UE internal measurement reporting 425
measurement sets 399
    active set 399
    detected set 399
    monitoring set 399
    virtual active set 399
measurement terminology
measurement types 396–9
    quality measurements 398
    signal measurements 397–8
    traffic volume measurements (TVM) 398
    UE internal measurements 398–9
    UE positioning measurements 399
measurement variables 404
    CELL_INFO_LIST 404
    MEASUREMENT_IDENTITY 405
MEASUREMENT_IDENTITY *see* measurement variables
measurements 395–435
    BCCH RSSI measurement 429, 431
    BSIC reconfirmation 427, 430
    cell ID location measurements 433
    GSM measurements using compressed mode 429
    initial BSIC identification 427, 430
    interfrequency cell measurements 408, 409

measurements (*cont.*)
  inter-RAT GSM cell measurements 408, 409, 425–33
  intrafrequency cell measurements 408
  location services 433–5
  network assisted GPS positioning 435
  observed time difference of arrival (OTDOA) location measurements 434–5
media gateway 16, 19
media gateway control function 19
media resource function 19
MM messages
  AUTHENTICATION REQUEST 464
  AUTHENTICATION RESPONSE 464
  CM SERVICE ACCEPT 454
  CM SERVICE REQUEST 444, 453–4, 462, 464
  IDENTITY REQUEST 449
  IDENTITY RESPONSE 449
  IMSI DETACH INDICATION 450
  LOCATION UPDATING ACCEPT 452
  LOCATION UPDATING REQUEST 452
  TMSI REALLOCATION COMMAND 448
  TMSI REALLOCATION COMPLETE 448
mobile country code *see* addressing
mobile equipment (ME) 509
mobile network code *see* addressing
mobile subscriber identity number *see* addressing
mobile switching center, introduction to 12–13
mobile termination (MT) 509
mobility management (MM) protocol 442–5
  authentication 448–9
  common procedures 447–50
  connection information transfer 455
  connection release 455
  connection transaction identifier 455
  identification 449
  IMSI attach 443–4, 453
  IMSI detach 450
  introduction to 13, 28
  location area updating 451, 475
  LOCATION UPDATE 443–4
  MM connection management procedures 447, 453–5
  MM entity states 442–5
  MM IDLE state 443–5
  mobile originated (MO) connection establishment 453–4
  mobile terminated (MT) connection establishment 454–5
  periodic updating 453
  primitives 445–6
  procedures 447–55
  protocol discriminator 444–5
  specific procedures 447, 450–3
  states for IMSI detach 445
  states for location updating 443–4
  states for mobile originated (MO) MM connections 444–5
  states for mobile terminated (MT) MM connections 445
  TMSI reallocation 447–8
  transaction identifier 444–5
mobility management sublayer 438–9
  functions 438
  GPRS mobility management (GMM) entity 438
  mobility management (MM) entity 438
  service access points (SAPs) *see* service access point
mode CS/PS *see* MS classes
mode I *see* network modes
mode II *see* network modes
mode PS *see* MS classes
modulation
  combined spreading and scrambling 104–7
  complex scrambling *see* codes
  downlink 102–8
  hybrid phase shift keying (HPSK) 111–13
  in-phase and quadrature representation 101–2
  orthogonal complex QPSK (OCQPSK) 111
  phase constellation 104, 106, 107
  phasor diagram representation 101
  QPSK 97, 129
  simple BPSK 101–2
  spreading 102–8
  uplink 108–14
MS classes 441–2
  and network modes 441–2
  mode CS/PS 442
  mode PS 442
MSC server 16, 17
multicode transmission in downlink *see* dedicated physical channel
multicode transmission in uplink *see* dedicated physical data channel
multimedia service, definition of 6–7
multiple access techniques 66–8
multiuser de-spreading *see* codes
multiuser spreading *see* codes

NAS procedures
  authentication *see* security
  call establishment 61
  registration 61
network architecture
  introduction to 8–19
  main network elements 9
  release 4 16–17, 19
  release 5 16–19
  release 99 8–16
network modes 442
  mode I 442, 475
  mode II 442
network services access point identifier (NSAPI) *see* session management protocol

Node B
    introduction to 16
    receiver adjacent channel filter requirements 157
    receiver adjacent channel selectivity 157–9, 160
    receiver blocking requirements 160–1
    receiver characteristics 154–65
    receiver dynamic range 161–2
    receiver dynamic tests 166–8
    receiver intermodulation specification 162–5
    receiver performance 165–8
    receiver phase noise 158, 159–60
    receiver sensitivity 154–7
    receiver spurious signal requirements 160
    receiver static tests 166
non-access stratum (NAS) 437–507
    architecture 437–41
    connection management sublayer 437 *see also* connection management sublayer
    introduction to 9–10
    mobility management sublayer 437 *see also* mobility management sublayer
    radio access bearer manager (RABM) entity 437
non access stratum protocol
    call control *see* call control protocol
    GPRS mobility management *see* GPRS mobility management protocol
    mobility management *see* mobility management protocol
    overview of 28
    session management *see* session management protocol
    short message service *see* short message service protocol
non-real time service, definition of 5
Nyquist filter 98, 99

orthogonal complex QPSK (OCQPSK) *see* modulation
orthogonal variable spreading factor (OVSF) *see* codes
orthogonality *see* codes

packet data convergence protocol 337–44
    architecture 337
    header compression 27, 339
    layer to layer protocol 339–42
    lossless SRNS relocation 342–4
    overview of 27
    PDUs 337
    SAPs 337
    sequence number management 342
    sequence number synchronization 343
packet data protocol (PDP) context *see* session management protocol
    creation 61
    introduction to 51

packet switched communications
    introduction to 4
    introduction to R99 service domain 10–15
paging 358–62
    paging message 362
paging channel (PCH) 123, 251
paging channel, as part of CCTrCH 219
paging control channel (PCCH) 249
paging indication channel (PICH) 125–6, 358–61
    paging indicators 126
    paging message 126
Paging indicators *see* paging indication channel
parallel concatenated convolutional code *see* turbo codes
peak code domain error *see* transmitter
phasor diagram *see* modulation
physical channel mapping 229
physical channel segmentation 228
physical channels
    introduction to 29–31
    mapping to/from transport channels 115
    physical common packet channel (PCPCH) 118–20
    physical random access channel (PRACH) 115 *see also* physical random access channel
    uplink 115–22
physical common packet channel (PCPCH) 118–20
    access preamble signatures 119, 126–7, 130
    channel assignment active 118, 119–20, 278, 279
    channel assignment not active 118, 119–20, 248
    collision detection 119
    collision detection/channel assignment indicator channel (CD/CA ICH) 120, 126–7, 278–9
    collision detection preamble signature 119, 130, 279
    CPCH control channel (CCC) 120
    CPCH status indication channel (CSICH) 119, 250
    message scramble code *see* codes
    PCPCH set 119
    physical layer transmissions 119
    resource availability (PRA) 278
    timing 136
    versatile channel assignment mode (VCAM) 278, 279
physical downlink shared channel 127
physical layer 115–36
    compressed mode *see* compressed mode
    definition of 25–6
    physical channels *see* physical channels
    power control *see* power control
    random access procedure *see* random access
    soft-handover *see* soft-handover
    spreading and scrambling *see* spreading and scrambling
    transport channel combining *see* transport channel combining

physical random access channel (PRACH) 115, 362
  access frame 116
  access preamble signatures 116, 130, 268
  access slots 116, 117
  message scramble code *see* codes
  message structure 117–18
  physical control information 118
  timing 136
  transport format combination indication (TFCI) 118
  use of 276
  use of AICH 126
power control
  downlink 42
  introduction to 39–42
  SIR target 39
  uplink 39–42, 93
  uplink algorithm 1 40–1
  uplink algorithm 2 41
  uplink closed loop 40–1
  uplink outer loop control 42
  uplink power control commands (TPC) 40
preferred pair 96, 97–114
primary common control physical channel (PCCPCH) 122–3, 131
primary common pilot channel (PCPICH), uses of 122, 204
primary synchronization channel (P-SCH) 122, 123–4, 131, 133
  use in system acquisition 208, 213
processing gain *see* receiver
protocol architecture
  control plane 24
  division between access and non-access stratum 9–10
  layered via of 24–8
  user plane 24
pseudo-noise (PN) sequence *see* codes
P-TMSI *see* addressing
puncturing 118

quality of service (QoS) 484, 487, 491, 498
  definition of 4–6
  introduction to 4–7
  radio access bearer manager 441

radio access bearer (RAB) establishment 377–80
  CS connection 378
  PS connection 380
radio access bearer manager (RABM) 441
  PDP contexts 441
  QoS 441
radio access bearers
  creation 61, 490
  overview to 28, 50
radio bearer (RB) establishment 377–80
radio bearers, introduction to 48–9
radio frame equalisation 225

radio frame segmentation 225
radio link
  introduction to 37, 48
  set of 37
radio link control PDU *see* transparent mode PDU or unacknowledged mode PDU or acknowledged mode PDU
radio link control protocol (RLC) 300–36
  acknowledged mode (RLC AM) *see* acknowledged mode
  automatic repeat request 27
  flow control 27
  overview of 27
  transparent mode (RLC TM) *see* transparent mode
  unacknowledged mode (RLC UM) *see* unacknowledged mode
radio network controller, introduction to 15, 16
radio network subsystem, introduction to 16
radio resource control (RRC) protocol 349
  architecture 349
  overview of 28
  system information broadcasting 352–8, *see also* system information
raised cosine filter
  alpha roll-off factor 98
  frequency response 98
  impulse response 98
  introduction 98
rake fingers
  algorithms
  architecture
  chip rate combining
  operation
  symbol rate combining
rake receiver 189–204
  channel estimation *see* channel estimation
  code correlation 95
  correlator 192, 194, 196, 200
  descrambling 202
  despreading 202
  diversity 198–200
  equal gain combining diversity 200
  fingers 201 *see also* rake fingers
  maximal ratio combining diversity 200
  multipath channel 189–90
  optimum receiver 190–2
  overview of operation 192–200
  receiver signals 201–4
  selection diversity 198
  structure 200
  use in soft-handover 36
  use in WCDMA 201–4
random access 264–76
  access service class 269
  AICH transmission timing (ATT) parameter 269, 270
  basic principles 264–5

cell update 267
initial random access 266–7
layer 1 information required 267–73 *see also*
    physical random access channel
MAC configuration 273–5
MAC procedure 275–6
message length (TTI) 268
message spreading factor 269
physical layer procedure 276
power setting 271, 272–3
preamble retransmission 271
preamble signatures 268 *see also* physical random
    access channel
scrambling codes used 268, 273 *see also* codes
subchannels 269
transfer of data 267
transport formats 273
uses of 266–7
random access channel (RACH) 115, 251, 362
    as part of CCTrCH 219
random number (RAND) *see* security
rate matching 225–8
    discontinuous transmission (DTX) on downlink
        227–8
    downlink fixed positions for transport channels
        227–8
    downlink flexible positions for transport channels
        227–8
    introduction to 35
    on downlink 227–8
    on uplink 226–7
    puncturing 226
    repetition 226
real-time service, definition of 5
receiver
    baseband filtering 99, 187–8, 216
    basics de-spreading 78–80
    binary phase shift keying 147
    coding gain 150–2
    convolutional code performance 150
    $E_b/N_0$ 148
    $E_c/N_0$ 148
    effects of channel coding 150–2
    energy per bit 147–8
    energy per chip 148
    integrator or summer 78
    Node B 12.2 kb/s test channel 152–4
    noise power spectral density $N_0$ 147
    processing gain 148–9
    received signals 146
    simple bit error rate performance 148
    turbo code performance 150
receiver blocking *see* Node B *or* UE receiver
receiver dynamic tests *see* Node B *or*
    UE receiver
receiver phase noise *see* Node B *or* UE receiver
receiver sensitivity *see* Node B *or* UE receiver
receiver static tests *see* Node B *or* UE receiver

recursive systematic convolutional code *see* turbo
    codes
Reed Muller code 118
registration, introduction 61
root raised cosine filter 99, 100, 102, 110, 188
    FIR implementation 100–1, 188
    impulse response 100
    use in uplink and downlink 129, 131
routing area
    introduction to 45–6
    updating *see* GPRS mobility management
        protocol
RR messages, PAGING RESPONSE 445, 454
RRC connection establishment 362–74
    activation time
    capability update requirement 368
    completion 373–4
    default DPCH offset (DOFF) 367
    DRX cycle length coefficient 368
    initial UE identity 363, 366
    introduction 60
    layer 1 setup 372–3
    layer 2 setup 372
    message flows 363
    RRC state indicator 368
    selection of SCCPCH 364
    setup 363–73
    signaling radio bearer setup 368–9
    transaction identifier 367
    u-RNTI, c-RNTI allocation 367
RRC connections, overview to 28, 48
RRC direct transfer 374–6
RRC messages
    ACTIVE SET UPDATE 381, 384–5
    ACTIVE SET UPDATE COMPLETE 384–5
    CELL CHANGE ORDER FROM UTRAN 391
    DOWNLINK DIRECT TRANSFER 376, 464,
        465
    HANDOVER FROM UTRAN COMMAND
        387
    HANDOVER TO UTRAN 389
    HANDOVER TO UTRAN COMPLETE 389
    INITIAL DIRECT TRANSFER 375–6, 464
    MEASURMENT CONTROL 402–4
    PHYSICAL CHANNEL
        RECONFIGURATION 385
    PHYSICAL CHANNEL
        RECONFIGURATION COMPLETE 385
    RADIO BEARER SETUP 283, 287, 292, 377,
        465
    RADIO BEARER SETUP COMPLETE 465
    RRC CONNECTION REQUEST 362–3
    RRC CONNECTION SETUP 287, 363–73
    RRC CONNECTION SETUP COMPLETE
        373–4
    SECURITY MODE COMMAND 57, 465
    SECURITY MODE COMPLETE 465
    SYSTEM INFORMATION 353, 354, 355–6

RRC messages (*cont.*)
   TRANSPORT CHANNEL
     RECONFIGURATION 292
   UPLINK DIRECT TRANSFER 465, 498
   URA UPDATE 343
   URA UPDATE CONFIRM 343
   UTRAN MOBILITY INFORMATION
     CONFIRM 344
RRC modes
   CELL_DCH, transport channel switching 261
   CELL_DCH state 43–4
   CELL_FACH, transport channel switching 261
   CELL_FACH state 44, 119
   CELL_PCH state 44
   connected mode 43
   idle mode *see* idle mode
   introduction to 43–5
   URA_PCH state 44–5

scrambling code *see* codes
searcher 206–8
   architecture 206–8
   finger allocation and validation 207–8
   finger cancellation 208
   role of 206
secondary common control physical channel
   (SCCPCH) 123
   use for paging 126
secondary synchronization channel (S-SCH) 122,
   123–4, 131, 133
   use in system acquisition 208, 213
secret key (K) *see* security
security
   algorithms 52
   anonymity key (AK) 54
   authentication and key management field (AMF) 54
   authentication basics 51
   authentication procedures 52–4, 61
   authentication token 54
   authentication vector 52–4
   authentication vector generation 52–4
   cipher key (CK) 54
   ciphering 52, 260
   COUNT-C 58
   COUNT-I 57
   expected message authentication code (XMAC) 55
   expected response (XRES) 54
   FRESH 57
   hyper frame number *see* hyper frame number
   integrity key (IK) 54
   integrity protection 51–2
   message authentication code 54
   message integrity protection 56–8
   random number (RAND) 52
   secret key (K) 53
   sequence number (SQN) 54
   START, $START_{CS}$ and $START_{PS}$ 513

   THRESHOLD 59
   user identity confidentiality 51
SECURITY MODE COMMAND *see* RRC
   messages
sequence number (SQN) *see* security
service access point
   dedicated control (DC) 439
   definition of 24–5
   GPRS mobility management access stratum
     (GMMAS) 438
   GPRS mobility management session management
     (GMMSM) 497
   high priority signalling radio bearer 3 (SRB3) 439
   introduction to 29–33
   logical channels *see* logical channels
   low priority signalling radio bearer 4 (SRB4) 439
   mobile network call control (MNCC) 456
   mobility management call control (MMCC) 445
   mobility management location services (MMLCS)
     445
   mobility management short message service
     (MMSMS) 445
   mobility management supplementary service
     (MMSS) 445
   physical channels *see* physical channels
   radio access bearer session manager (RABSM)
     495
   radio resource (RR) 438
   session management registration (SMREG) 485
   transport channels *see* transport channels
service domains, introduction to 10–15
services
   bearer service 21–3
   cell broadcast short message service 20–1, 27, 250
     *see also* broadcast multicast control protocol
   short message service 20
   supplementary services 23
   teleservices 19–21
serving radio network temporary identifier (s-RNTI)
   *see* addressing
session description protocol 18, 19
session initiation protocol 18, 19
session management (SM) 439–40
   network service access point identifier (NSAPI)
     439
   PDP contexts 439–40
   radio access bearer identity (RAB-Id) 440
   radio access bearer manager (RABM) 439
   transaction identifier (TI) 439
session management (SM) protocol 483–94, 498
   access point name (APN) 484–5, 490, 493, 498
   ACTIVE PENDING state 487–8
   ACTIVE state 488–9
   GPRS mobility management session management
     primitives 497
   INACTIVE PENDING state 489
   INACTIVE state 486–7

introduction to 14, 28
MODIFY PENDING state 489–90
network services access point identifier (NSAPI) 484–5, 487
packet data protocol (PDP) contexts
packet filters 494
PDP address 484, 487, 493, 498
PDP context activation example 497–8
PDP context transaction identifier (TI) 493
primary PDP context establishment 490–2
primitives 485–6, 495
radio access bearer manager primitives 495–7
secondary PDP context establishment 493–4
states 486–90
traffic flow template (TFT) 487, 493, 494
SGSN, introduction to 12
shared channel control channel (SHCCH) 250
short message service (SMS) 440
  PS and CS domain entity 440
Short message service (SMS) protocol 498–507
  architecture 498–500
  CP primitives 500
  example mobile originated SMS via SMC-CS 505–7
  example mobile terminated SMS via SMC-GP 505, 507
  introduction to 13, 28
  mobility management entity (SMSMM) 498
  relay protocol 499
  RL primitives 500–1
  short message control circuit switched entity (SMC-CS) 498
  short message control GPRS protocol entity (SMC-GP) 498
  SMC-CS states 501–2
  SMC-GP states 502–3
  SMR states 503–4
signal to interference ratio (SIR), effects of codes on 80
Signalling, basic definition of 9–10
signalling radio bearers 250, 256
signatures *see* PRACH, PCPCH, AICH
site selection diversity transmission (SSDT) 121
SM messages
  ACTIVATE PDP CONTEXT ACCEPT 488, 491, 492, 498
  ACTIVATE PDP CONTEXT REJECT 487
  ACTIVATE PDP CONTEXT REQUEST 487, 490, 492, 498
  ACTIVATE SECONDARY PDP CONTEXT ACCEPT 488
  ACTIVATE SECONDARY PDP CONTEXT REJECT 487
  ACTIVATE SECONDARY PDP CONTEXT REQUEST 494
  DEACTIVATE PDP CONTEXT ACCEPT 489
  DEACTIVATE PDP CONTEXT REQUEST 489
  MODIFY PDP CONTEXT ACCEPT 489
  MODIFY PDP CONTEXT REJECT 489
  MODIFY PDP CONTEXT REQUEST 488–9
SMS messages
  CP-ACK 500, 504, 506, 507
  CP-DATA 500, 504, 505, 507
  RP-ACK 500, 504, 506, 507
  RP-DATA 500, 504, 505, 507
  RP-ERROR 505
  RP-SMMA 504
soft-handover
  introduction to 36–7 *see also* handover
  softer-handover, variation of 37
space time transmit diversity (STTD) 124
spectrum emission mask *see* transmitter
spreading and scrambling
  basic principles of 34–5
  channelisation code *see* channelisation code
  scrambling code *see* scrambling code
spreading codes *see* codes
spreading factor *see* codes
spreading modulation *see* codes
SRNS relocation
  after soft handover 37
  lossless 27
START *see* security
symbol rate processing, transmission path 217–21
system acquisition 208
  stage 1, efficient Golay correlator approach 213, 216
  stage 1, hierarchical matched filter approach 213
  stage 1, matched filter approach 210–12
  stage 1 P-SCH detection 209, 210–13
  stage 2, fast Hadamard transformer 216
  stage 2 code group detection 209, 213–15
  stage 3 code set detection 209
  three-stage process 209–10
system frame number
  uses for paging 125
  uses for random access 269
  uses for system information broadcasting 356
system information
  BCH transport block 354, 356
  block type 1–block type 18 358
  broadcast channel (BCH) 353
  master information block (MIB) 353
  message 353
  primary common control physical channel (PCCPCH) 352
  scheduling 355–6
  scheduling blocks 354
  segmentation and concatenation 354
  SFNPrime 356
  structure of broadcast messages 352–3
  structure of SYSTEM INFORMATION message 356–8

target channel type field *see* MAC
TDD mode, introduction to
TDMA, introduction to 68
terminal adaptation function (TAF) 509
terminal equipment (TE) 509
timer
   T3212 453
   T3250 448
TMSI *see* addressing
transmission time interval
   for random access 268
   use of 224, 225
transmitter
   adjacent channel leakage ratio in Node B 144
   adjacent channel leakage ratio in UE 142–3
   baseband filtering 99
   error vector magnitude in Node B 144
   error vector magnitude in UE 140–2
   Node B specifications 143–5
   peak code domain error in Node B 144
   peak code domain error in UE 142
   spectrum emission mask in Node B 144–5
   spectrum emission mask in UE 143
   UE specifications 140–3
transparent mode (TM RLC) 300–5
   buffering and discard 304
   entity structure 300–1
   operation 303–5
   PDU 301
   segmentation 303
   state diagram 302
transport block *see* MAC PDU
transport block concatenation 224
transport channel combining, introduction to 35
transport channel multiplexing 228
transport channel switching *see* MAC
transport channels 251
   introduction to 31
   mapping to/from physical channels 115
transport format combination (TFC) 282, 292
   blocked state 294–6
   configuration 282–3
   dynamic information 289
   example operation 283
   excess power state 294–6
   indicators (TFCI) 122
   introduction 282
   number of transport blocks 289
   physical layer procedures 297
   selection 121, 294
   semi-static information 288–9
   set (TFCS) 292–3
   transport block size 289
   transport format set (TFS) 289
transport format selection (TFS) 259 *see also*
   transport format combination
tunnelling 14, 15

turbo codes 235
   decoder 236–9
   interleaver 236
   maximum a posteriori probability (MAP) 235
   origins 235
   parallel concatenated convolutional code (PCCC) 235
   principles 235–6
   recursive systematic convolutional code (RSC) 235
   semi-static transport format information 289
turbo decoder
   Bayes' theorem 238–9
   bit probability 242
   data 236
   log likelihood ratio 238, 239
   log MAP 244
   MAP algorithm 240–3
   MAP state probabilities 240–1
   MAP transition probabilities 240, 242
   max-log MAP 244
   parity information 236
   performance 247
   reduced complexity algorithms 243
   SISO extrinsic information 238–9
   SISO soft information 238
   soft-in soft-out (SISO) decoder 236, 237

U1, U2, U3, update status *see* idle mode
UE
   application architecture 510
   receiver adjacent channel selectivity 170
   receiver analogue to digital converter 179
   receiver architecture 177–80
   receiver balun 178
   receiver bandpass filter 178
   receiver baseband filter 179–80
   receiver blocking 170–2
   receiver characteristics 169–73
   receiver dynamic tests 174
   receiver intermodulation specification 172–3
   receiver low noise amplifier 178
   receiver performance 174–5
   receiver quadrature mixer 179
   receiver sensitivity 169
   receiver static tests 174
   transceiver architecture study 176–83
   transmitter architecture 181
   transmitter automatic gain control (AGC) 182
   transmitter digital to analogue converter
   transmitter filtering
   transmitter I/Q modulator 181
   transmitter isolator and coupler 182
   transmitter low pass filter 181
   transmitter power amplifier
   transmitter RF mixer

UICC 509–10
    application dedicated files (ADF) 510
    dedicated files (DF) 510
    elementary file (EF) 510
    USIM application dedicated file 510 *see also* USIM
UMTS
    definition of 2–3
    frequency bands 137
    to GSM handover *see* handover
unacknowledged mode (UM RLC) 306–14
    buffer status 311
    buffering and discarding 313–14
    configuration primitives 307
    entity creation 310
    entity structure 306
    operation 309–14
        transmitter actions 310
    PDU 306–7
    segmentation 311–13
    state diagram 307–9
    state variables 309
uplink
    near–far problem 93
    scrambling codes *see* codes
    spreading codes *see* codes
uplink physical channels 115–22
uplink shared channel 251
URA_PCH state *see* RRC mode

US PCS frequency bands 137
user data, basic definition of 9–10
user plane, basic definition of 10
USIM
    $EF_{DIR}$ 513
    $EF_{LI}$ 513
    $EF_{LOCI}$ 452
    $EF_{PL}$ 513
    $EF_{PLMN}$ 452
    HPLMN 513
    ME activation procedure 511–13
    RPLMN 513
USIM and UICC, architecture and operation 509–13
UTRA absolute radio frequency channel number (UARFCN) 137–9
UTRAN, release 99 overview 9, 15–16
UTRAN radio network temporary identifier (u-RNTI) *see* addressing
UTRAN registration area, introduction to 45
UWC136 3

variable data rate *see* codes
video service
    introduction to 6
    typical characteristics of 7
Viterbi algorithm *see* convolutional codes

Walsh *see* codes